SWEETENERS AND ENHANCERS

SWEETENERS AND ENHANCERS

Nicholas D. Pintauro

NOYES DATA CORPORATION
Park Ridge, New Jersey, U.S.A.
1977

Library of Congress Catalog Card Number: 77-71661
ISBN: 0-8155-0652-X
Printed in the United States

Published in the United States of America by
Noyes Data Corporation
Noyes Building, Park Ridge, New Jersey 07656

FOREWORD

The detailed, descriptive information in this book is based on U.S. patents that deal with sweeteners and sweetness enhancers and their commercial technology. To round out the complete technological picture, 11 British patents have been included.

This book serves a double purpose in that it supplies detailed technical information and can be used as a guide to the U.S. patent literature in this field. By indicating all the information that is significant, and eliminating legal jargon and juristic phraseology, this book presents an advanced, technically oriented review of sweeteners and sweetness enhancers.

The U.S. patent literature is the largest and most comprehensive collection of technical information in the world. There is more practical, commercial, timely process information assembled here than is available from any other source. The technical information obtained from a patent is extremely reliable and comprehensive; sufficient information must be included to avoid rejection for "insufficient disclosure." These patents include practically all of those issued on the subject in the United States during the period under review; there has been no bias in the selection of patents for inclusion.

The patent literature covers a substantial amount of information not available in the journal literature. The patent literature is a prime source of basic commercially useful information. This information is overlooked by those who rely primarily on the periodical journal literature. It is realized that there is a lag between a patent application on a new process development and the granting of a patent, but it is felt that this may roughly parallel or even anticipate the lag in putting that development into commercial practice.

Many of these patents are being utilized commercially. Whether used or not, they offer opportunities for technological transfer. Also, a major purpose of this book is to describe the number of technical possibilities available, which may open up profitable areas of research and development. The information contained in this book will allow you to establish a sound background before launching into research in this field.

Advanced composition and production methods developed by Noyes Data are employed to bring our new durably bound books to you in a minimum of time. Special techniques are used to close the gap between "manuscript" and "completed book." Industrial technology is progressing so rapidly that time-honored, conventional typesetting, binding and shipping methods are no longer suitable. We have bypassed the delays in the conventional book publishing cycle and provide the user with an effective and convenient means of reviewing up-to-date information in depth.

The Table of Contents is organized in such a way as to serve as a subject index. Other indexes by company, inventor and patent number help in providing easy access to the information contained in this book.

15 Reasons Why the U.S. Patent Office Literature Is Important to You —

1. The U.S. patent literature is the largest and most comprehensive collection of technical information in the world. There is more practical commercial process information assembled here than is available from any other source.

2. The technical information obtained from the patent literature is extremely comprehensive; sufficient information must be included to avoid rejection for "insufficient disclosure."

3. The patent literature is a prime source of basic commercially utilizable information. This information is overlooked by those who rely primarily on the periodical journal literature.

4. An important feature of the patent literature is that it can serve to avoid duplication of research and development.

5. Patents, unlike periodical literature, are bound by definition to contain new information, data and ideas.

6. It can serve as a source of new ideas in a different but related field, and may be outside the patent protection offered the original invention.

7. Since claims are narrowly defined, much valuable information is included that may be outside the legal protection afforded by the claims.

8. Patents discuss the difficulties associated with previous research, development or production techniques, and offer a specific method of overcoming problems. This gives clues to current process information that has not been published in periodicals or books.

9. Can aid in process design by providing a selection of alternate techniques. A powerful research and engineering tool.

10. Obtain licenses — many U.S. chemical patents have not been developed commercially.

11. Patents provide an excellent starting point for the next investigator.

12. Frequently, innovations derived from research are first disclosed in the patent literature, prior to coverage in the periodical literature.

13. Patents offer a most valuable method of keeping abreast of latest technologies, serving an individual's own "current awareness" program.

14. Copies of U.S. patents are easily obtained from the U.S. Patent Office at 50¢ a copy.

15. It is a creative source of ideas for those with imagination.

CONTENTS AND SUBJECT INDEX

INTRODUCTION 1

MIRACULIN, GLYCYRRHIZIN AND ARTICHOKE SWEETENERS 3
Miraculin 3
Stable Solid Powder 3
Improved Extraction Process 8
Enzyme Elimination 11
Solubilizing Agent 12
Acid Activator 15
Bicarbonate-Effervescent Activator 16
Monellin 17
Basic Commercial Process 17
UV Radiation Process 19
Glycyrrhizin 23
Sucrose Combination 23
Citrus Juice Sweetener 24
5′-Nucleotide Combination 25
Jerusalem Artichoke Sweeteners 27
Basic Extraction Method 27
Artichoke Bread Flour 29
Chlorogenic Acid Inducer 32

DIPEPTIDES 35
Aspartic Acid Alkyl Esters 35
L-Aspartyl-L-Tyrosine Methyl Ester 35
L-Aspartyl-L-Phenylalanine Esters 37
L-Aspartyl-L-(β-Cyclohexyl)Alanine 39
α-L-Aspartyl Derivatives 40
Isobutyl and Isopropyl Esters 40
Hexahydrophenylalanine Esters 41
Aminosuccinamic Acid Esters 43
Dipeptide Combinations 44
Alum and Naringin 44

Glucono-Delta-Lactone/Buffer 45
Special Buffer Salts 47
Dipeptide Solubilization 48
Dipeptide Codried with Polymaltose and Dextrins 48
Dipeptide Codried with Polyglucose 49
Dipeptide Coground with Acid 51
Citric Acid Melt 52
Bulk Density and Stability 54
Encapsulation Process 54
Codried Starch Hydrolysate 56
Special Cereal Applications 58
Foam Method 58
Coating Method 59

CHALCONES AND MALTOLS 61
Flavanone Glycoside Dihydrochalcones 61
Hesperetin Dihydrochalcone 63
Synthesis and General Use 63
Organic and Aqueous-Organic Solvents 65
Glycerol Solvent 68
Flavor Oil Solvents 69
Sugar-Dihydrochalcone Solvent 71
Derivatized Dihydrochalcones 72
Trihydroxyl-4'-Alkoxy Alkyl Sulfonate 72
4-Alkoxy Substituted 73
Trihydroxy-4'-Alkoxy-4-Carboxyl Substituted 74
Special Neohesperidoside 75
Dihydrochalcone Galactoside 76
Dihydrochalcone Xyloside 78
Saccharin and Dihydrochalcone 79
Maltol Flavor and Sweetener Enhancers 80
Basic Gamma-Pyrone Process 80
Maltol Synthesis Intermediates 82
Isomaltol Derivatives 85
Maltol and Flavoring Oils 87
Maltol-Like Sweetener 88

OTHER SYNTHETIC SWEETENERS 91
Diacetone Glucose 91
Saccharin Amine Salts 94
Tryptophane Derivatives 95
6-(Trifluoromethyl)Tryptophane 95
6-Chlorotryptophane 97
Tryptamine Compounds 99
Substituted Tetrazoles 105
5-(3-Hydroxyphenoxy)-1H-Tetrazole 105
5-Carbocyclicaminotetrazoles 108
8,9-Epoxyperillartine Sweeteners 112
5-Imino-4,4-Dimethyl-2-Imidazolidinone 115
Kynurenine Derivatives 116
Heliotropyl Nitrile 119
2-(3-Bromopropoxy)-5-Nitroaniline 121
3-Amino-4-n-Propoxybenzyl Alcohol 122

p-Methoxycinnamaldehyde (PMCA) 123
Special Oximes 125
Unsaturated Aldoximes 125
1,4-Cyclohexadiene-1-Carboxyaldehyde syn-Oxime 126
Stevioside Extraction 128
Maltitol Sweetener 130
Maltitol and Maltotriitol Mixture 132
Lactitol Sweetener 137
Soluble Maltose Polymers 139

SACCHARIN COMBINATIONS AND SPECIAL FORMULATIONS 143
Saccharin and Dipeptide 143
Simple Combinations 143
Synergistic Combinations 147
Saccharin and Glucono-Delta-Lactone 150
Blend with Buffer 150
Potassium Bitartrate Buffer 151
Sodium Gluconate Buffer 153
Bicarbonate Buffer 155
Gluconate and Fructose 156
Saccharin and Calcium Gluconate 156
Saccharin and Citrate Buffer 158
Saccharin and Lactose 159
Critical Lactose Level 159
Cream of Tartar Addition 160
Saccharin and Galactose 161
Saccharin and Ribonucleotides 163
Sodium Chloride Insoluble Fraction 165
Calcium Chloride-Starch Hydrolysate Combinations 167
Saccharin and Adipic Acid 168
Saccharin and Mannitol-Gum 168
Saccharin and Pectin 171
Saccharin and Maltol 173
Saccharin and Piperazine 175
Saccharin and Tryptophane 176
Saccharin and Oxazolidinones 177

METHODS FOR INCREASING BULK OF MIXES 179
Pillsbury Low Calorie Drink Mix 179
Basic Process 179
Mix with Fumaric Acid 185
Partially Synthetic Sweeteners 186
Foam System with Sucrose 186
Foam System with Lactose 189
Maltodextrin Extender 191
Agglomerated Spray-Dried Process 192
Pulverized Sugar Combination 193
Agglomerated Sugar plus Sweetener 195
Drum Drying with Starch 197
Codried Dipeptide-Organic Acid 202
Low Bulk Density Processes 203
Anticaking Lactose Combination 203
Peebles Process 206

Foremost-McKesson Process 210
Other Bulking Processes 219
Spray Drying Foamed Material 219
Granular Powders 222
Blaw-Knox Process 226
Spray-Dried with Whipping Agent 232
Spray-Dried with Dextrin 233
Drying with Malto-Dextrin 235
Granular Clusters 238

SUGAR SUBSTITUTES AND SPECIALTY INGREDIENTS 242
Sugar Substitutes and Extenders 242
Methyl Glucoside Sweetener 242
Bulking Agent for Colored Mixes 248
Cellulose Bulking Agent 250
Polyose Bulk Extender 260
Arabinogalactan Bulking Agent 262
Starch and Arabinogalactan 265
Coated Sugars 268
Saccharin-Sugar Product 269
Starch Hydrolysates 272
Maltol Sweetness Potentiator 274
Aerosol Sweetener 277
Stabilized Saccharin Dry Emulsion 280
Saccharin Complex 282
Amino Acid Intensifier 284

DRINKS, JELLIES, FRUITS AND CHEWING GUM 288
Drink Products 288
Dry Cola Beverage Mix 288
Basic Effervescent Process 289
Modified Bicarbonate Method 294
Gum Coated Bicarbonate 297
Hydrophilic Gum Additive 300
Glycerol Sweetener 305
Artificially Sweetened Jellies 306
Carrageenan-Pectin Base 306
Polyose Base 309
Fruit Treatments 312
Frozen Fruit—Low Density Syrup 312
Conditioning Prior to Freezing 315
Fruit Sections—Canned Product 318
Freeze Dried Fruit 320
Impregnation with Sweetener 320
Flavor Enhancement 325
Rapid Sweetening Method 329
Honey-Malt Flavoring 332
Sugarless Chewing Gum 335
Xylitol Formulation 335
Miraculin Coating 336
Dipeptide Sweetener 337
Gum Acacia for Improved Texture 340
Releasable Phosphate 343

Sugarless Candy. 345
Xylitol Dentifrice Formulation 348

DESSERTS AND BAKED GOODS. 350
Frozen Desserts. 350
Polyose as Freezing Point Depressant 350
Low Calorie Dessert 353
Dry Ice Cream Mix 356
Organic Acid Salt Additive for Milk Base Products 359
Fluffy Frosting 360
Dietary Dry Cake Mix 366
Dietetic Flour 370
Insoluble Protein Approach. 370
Low Calorie Doughs 372
Dietetic Low Protein Bread. 378
Polyose for Baked Goods 381
Polyose Thickening Agent. 382

COMPANY INDEX. 386
INVENTOR INDEX 387
U.S. PATENT NUMBER INDEX 390

INTRODUCTION

Based on the premise that Americans are becoming less physically active and are therefore expending less energy, the need to reduce daily caloric intake becomes increasingly important. Ignoring caloric/energy balance inevitably leads to the inability to maintain a healthful weight and other such health related problems.

Many diet plans have been developed and widely publicized as easy pathways to weight loss. Most plans are based on reduction of caloric intake or reduction in total food consumption. Fats, starches and sugars are the foods which are always cited as the culprits in overweight. A food that is low in protein and high in fats and carbohydrates is labeled "high caloric," obviously because its contribution to the diet is more calorific than nutritious.

Sugar in the form of sucrose and corn sugars has also been implicated as a causative agent in a spectrum of disorders ranging from dental caries to hyperglycemia, the aggravation of diabetes and cardiovascular diseases. The per capita annual consumption of sucrose in the United States is estimated at 100 to 125 pounds which would correspond to some 20 to 25% of total caloric intake.

If corn sugars were to be included in this estimate the percentage would increase to 35 and if starches were also to be included, the percentage for the entire group of carbohydrates would increase to 45% of total calorie intake. Based on these calculations, fats would account for some 40% of calorie intake and proteins the remaining 15 or so percent. It therefore becomes apparent that significant reduction of caloric consumption can be realized by the reduction of sugars in the diet.

Sugars can be defined as simple carbohydrates which are water-soluble and are sweet to the taste. This group of simple carbohydrates includes sucrose, glucose, fructose, lactose and maltose. Many complex carbohydrates, such as starch and plant gums, yield simple sugars in the metabolic processes of human nutrition.

Sugars are used in food formulations not only for sweetness (and flavor) but also for such functional purposes as, for example, imparting body and texture, developing color and serving as a dispersant, fixative, bulking agent, etc. In addition, the many complexing reactions of sugars with proteins, fats, emulsifiers and other food ingredients can be designed to generate desirable sensory qualities, that is appearance, flavor and texture. Sugars therefore are important to foods because they contribute to their acceptability in many different ways. Sugar substitute formulations containing saccharin and other synthetics thus must replace the functional characteristics of sugar as well as the sweetness.

It is theorized that the sweet taste of simple carbohydrates is produced by the polyhydroxy nature of sugars. The strength or intensity of sweetness cannot be measured quantitatively. It must be evaluated subjectively using humans. Therefore, since no absolute rating for sweetness is possible, only relative sweetness can be reported. Usually sucrose is used as the standard for this purpose.

Prior to 1970, saccharin and cyclamates were used in foods and beverages as nonnutritive sweeteners mainly as combined mixtures. In October 1969, an order was issued by the Food and Drug Administration stating that cyclamates could no longer be "generally recognized as safe" and could not be used in foods. In November 1973, Abbott Laboratories, the principle manufacturer of cyclamates submitted its last petition to lift the ban, asserting that the artificial sweetener was safe as a food additive. In reply, the Food and Drug Administration on May 11, 1976 stated that after two years of intensive study, it still could not assure the public that cyclamate was safe for everyday use.

No processes which use cyclamates as the sweetening agent are given in this publication. However, in several instances processes and examples are given which were originally developed for saccharin-cyclamate combinations, and in such cases the technology and formulations were modified for use of saccharin as the sole sweetening ingredient.

Over the past ten years a great deal of research has been conducted to identify and develop other nonnutritive sweeteners to be used as supplements with saccharin or as replacements for saccharin. The most recently developed sweeteners are the synthetic dipeptides, particularly L-aspartyl-L-phenylalanine (trade name Aspartame) which is 150 to 250 times sweeter than sucrose, monellin which is 300 times sweeter than sucrose, and the dihydrochalcone derivatives which are some 300 to 500 times sweeter than sucrose. The most important sugar substitute is still, of course, saccharin and perhaps more than half of this book is devoted to the use of saccharin in food products and beverages as a replacement for natural sugar.

The method provides, in addition to a broad coverage of basic industrial approaches to synthetic sweeteners, examples in the important areas of ingredient technology, product formulation and processing operations necessary to meet ingredient and product specifications.

The methods of synthesizing sugar substitutes are generally not given except in some special cases where these methods are part of the description of product application. Where the synthesis is not covered, the reader is referred to the original patent copy for further information. The reader is further reminded to consult the official literature of the Food and Drug Administration on the status of saccharin combinations, on proper labeling procedures and on the use of other synthetic sweeteners in foods and beverages.

MIRACULIN, GLYCYRRHIZIN AND ARTICHOKE SWEETENERS

Since the cyclamate ban there has been considerable interest in natural sweeteners with high sweetness intensities to replace ordinary sugars to reduce calories. In this chapter the important classes of natural materials are reviewed. These materials are extracted from natural products, and if used without chemical modifications they are acceptable by the Food and Drug Administration for use in food products.

MIRACULIN

Stable Solid Powder

Synsepalum dulcificum Daniell, Sapotaceae is a plant indigenous to West Central Africa which bears a red ellipsoid fruit commonly known as "miracle fruit." The fruit has a palatable pulp and skin and contains a large seed. It is characterized by a pleasant taste and by the unique property, well-recognized for over 200 years, of modifying the sweet and sour tastes in an unusual manner. It has been found that a component in the fruit depresses the sour taste and accentuates the sweet taste of any normally sour food eaten within a short period after first contacting the tongue with the pulp of fresh miracle fruit, thus causing the normally sour food to taste pleasantly sweet.

By exposing the taste receptors on the tongue to miracle fruit, any sour tasting food can be made to taste sweet without the addition of sugar or artificial sweeteners. For example, fresh lemon can be made to taste pleasantly sweet by first eating a miracle fruit berry. The taste-modifying principle in the miracle fruit berry known as miraculin binds itself to the taste receptors thus altering the sensory perception of the sour taste in foods eaten after the miracle fruit.

It has been determined that miraculin is a glycoprotein having a molecular weight of about 44,000. A wide variety of approaches have been explored in attempts to isolate the active component in miracle fruit for subsequent use as a taste-modifying material. The product obtained by these methods is less effective than the natural fruit because it was found to be highly unstable at normal room temperatures under normal atmospheric conditions. This instability necessitated either very quick use after isolation or storage at very low temperatures.

The miraculin or miracle fruit principle (MFP) is present in the pulp and on the inner surface of the skin of the miracle fruit and in its natural environment is quickly deactivated especially when exposed to the air once the skin is broken at room temperatures. Furthermore, after the fruit has been picked, even prior to breaking the skin, the active material begins to degrade but at a slower rate than when the skin is broken. While the process by which degradation proceeds is not known exactly, it is now believed that certain enzymes and/or acids present in the fruit accelerate degradation in the presence of air at normal room temperatures, and apparently even at temperatures below the freezing point of water.

It has been found that when the pulp of miracle fruit is frozen and subsequently lyophilized to form a granular or powder material, the product had to be refrozen in order to maintain the activity of the material that remained. Even when the pulp had been lyophilized, its effectiveness was not nearly as great, either on a weight basis or on a quality basis, as the active principle in the fresh fruit.

J.R. Fennell and R.J. Harvey; U.S. Patent 3,676,149; July 11, 1972; assigned to Meditron, Inc. described a process for extraction of miraculin and preparation of a stable product in powder form. A stable miraculin-rich composition is obtained by comminuting depitted ripe miracle fruit containing miraculin and then separating the vaporous and liquid components including acids and enzymatic components of the ripe fruit that degrade miraculin from the miraculin-rich material. The liquid and vaporous components are separated by dehydration and the enzymatic components are separated by any means that effects separation on the basis of density.

To minimize miraculin loss after picking, the whole fruit can be frozen to very low temperatures to await processing or the pulp and skin can be processed immediately after picking to obtain the concentrated miraculin. Comminution of the fruit serves to fracture the cell walls and thereby expose substantially all of the miraculin and facilitate subsequent processing. Dehydration can be effected in any convenient manner wherein low temperatures can be obtained including lyophilization, foam separation, spray drying or similar dehydration processes and can proceed or follow the separation step based on density.

It is preferred to separate the high density miraculin from the low density enzyme-rich material following dehydration because of the increased efficiencies obtained. If the miraculin is not separated from the material containing the enzyme, the product is unstable and will be degraded quickly at normal room conditions so that it loses its taste-modifying effect.

The miraculin-rich material, substantially free of the degrading enzyme and/or acids, has a substantially higher density than the material containing the enzyme. Therefore, the separation of the miraculin, that may contain some cellulosic material, from material containing the enzyme is effected by processes that separate materials on a density basis. To facilitate this separation, the mixed pulp and miraculin is preliminarily comminuted and screened to obtain uniformly small particle size.

The process is based upon the discovery that degradation of miraculin in the fruit is initiated immediately after the ripe fruit is picked, and that degradation of the active principle in its natural environment is accelerated by increased temperature and by contact with air. Thus, it is preferred to process the ripe miracle fruit as quickly as possible, at as low a temperature as possible, and in a nonoxidizing atmosphere, to obtain a high yield of miraculin. Preferably, the picked fruit is washed in water and then depitted at about 1°C to 4°C.

The fruit can be stored in a frozen state to await processing or can be processed immediately to obtain the active principle. When stored, temperatures of about -40°C or less are employed to arrest degradation since it has been found that degradation of the active principle in the frozen fruit occurs even when stored at temperatures of about -15°C. Since it is difficult to remove the pit or seed from the frozen berry, it is preferred to depit the berry prior to frozen storage. The depitted berry, regardless of whether it has been stored previously or whether it is processed directly after having been picked, is comminuted in a frozen state either alone or together with Dry Ice or ice formed from pyrogen-free distilled water.

When the berry is processed immediately after having been picked, the pulp and skin obtained from the depitting step are directed into a container placed in a low temperature bath which itself may contain crushed Dry Ice. The pulp and skin are then comminuted at low temperatures such as by blending, grinding or ball-milling with ball-milling in a shell freezer being preferred.

The pulp is comminuted until the average particle size of the mixture is about 600 microns or less. It is preferred to comminute the pulp to a particle size on the order of 100 to 125 microns to insure breaking of substantially all of the cell walls. Preferably, the mixture then is lyophilized under vacuum at a temperature of about -40° or less to remove liquid and vaporous components to include certain organic acids such as formic acid. Freeze drying is continued until there is no significant weight change in the material over about a four hour period.

At this point in the process, the dried pulp contains less than about 5 weight percent moisture. To remove the remainder of the liquid and volatiles from the pulp by lyophilization would require an inordinately long period of time. Therefore, it is best to complete the dehydration in a desiccator at normal room temperatures. The small concentration of moisture in the pulp during desiccation will not cause significant degradation of miraculin during the final drying period. The miraculin then is separated from the enzyme-rich material on the basis of density.

The separation of the miraculin-rich material from the enzyme-rich material is based upon the fact that the active principle is considerably more dense than the enzyme-rich material in the order of about 10 times as dense. Thus, the mixture of inert and active material described above can be separated by any convenient density separation method including settling from a suspension of the mixture in liquid, the use of fluidizing bed technique or through the use of cyclone type centrifuge. Some separation of miraculin-rich material from enzyme-rich material can be effected prior to dehydration by placing the comminuted pulp in settling pans at a temperature of 1° to 4°C until the highest density material has settled in the bottom. The settling is complete in a short period of about 20 minutes with the miraculin-rich material forming the lowest layer which is then separated. The material then is frozen to below about -40°C and dehydrated. The preferred separation method is conducted after dehydration and utilizes a cyclone type centrifuge for dry powder following dehydration as described in the examples.

The miraculin-rich material has a density in excess of 1 g/cc while the enzyme-rich cellulosic material has a density of less than about 0.5 g/cc. Since the high density material is white and the low density material is brown the separate layers can be easily and quickly identified on the basis of color. To obtain the desired product stability, enzyme-rich material concentration should be as low as possible with removal in the order of about 95% having been found to be adequate.

The active principle obtained by this process is insoluble in water and only partially soluble in saliva. To be effective in suppressing sour taste and enhancing sweet and salt taste, the product must be applied to the sour taste receptors on the tongue. The miraculin-rich material can be applied conveniently as a powder or in a unit dosage form mixed with inert solids such as a tablet, capsule or gum, or coated on the unit dosage form or mixed with water or solvents for use as a liquid spray or the like. The active principle is retained on the tongue a sufficient period to contact essentially all of the sour taste receptors. When used in unit dosage form, as little as about 0.1 mg of miraculin with a product of small particle size is required to obtain the taste-modifying effect.

Example 1: Ripe miracle fruit berries are picked and washed at 1° to 4°C in a water-ice bath in an insulated container. The fruit then is depitted at about 1° to 4°C in a juicer comprising a perforated cylinder housing a rotating brush extending along the cylinder length the ends of which contact the inside cylinder wall. During rotation, the brushes tumble and press the berries against the perforated housing causing the juice and pulp material to pass through the holes leaving the pits in the cylinder. The juice and pulp flow into containers that contain Dry Ice and are immersed in an alcohol-Dry Ice bath. Crushed Dry Ice (solid CO_2) is added directly to the fruit pulp obtained from the depitting step, and the mixture is thoroughly ground in a ball mill to a particle size of less than 150 microns (100 sieve size) while being maintained at a temperature of about -40° to -50°C.

The mixture then is placed in a freeze drying flask, placed in a shell freezer, and allowed to come to thermal equilibrium at a temperature of about -55°C. The material is then connected to a freeze-dryer vacuum system with a refrigerated condenser for condensing liquids and condensable vapors, where it remains until there is no significant weight change in the material over a four hour period. The material is then removed from the flask and placed in a desiccator cabinet in trays at room temperature for further drying or storage until the moisture content is substantially zero.

After the powder is thoroughly dry, it is placed in a temperature controlled milling machine, where the average particle size is reduced preferably to less than about 150 microns. The material is periodically screened and that retained by the 100 sieve size is returned to the milling machine until it can pass the 100 sieve size.

The fine powder is introduced into a pneumatic cyclone-type separator, whereby the dense miraculin is concentrated near the inside wall and the cellulosic material is concentrated closer to the center of the cyclone. The mixture to be separated is introduced into the top of the cyclone and caused to move in a circular path down the inside wall. The miraculin-rich material is separated from the lower density material by a baffle located at the interface of the miraculin and lower density material. The lower density material is recycled until substantially all the miraculin is separated. The concentrated miraculin can be recycled if necessary, to achieve any degree of separation from the lower density material. The miraculin powder obtained from the cyclone separator is room-temperature-stable even when stored in the open atmosphere for at least about 8 months and can then be used to produce unit dose forms including tablets or aqueous sprays.

Example 2: This example illustrates a typical miraculin formulation, a method for preparing chewable tablets and the results of tests on subjects ingesting the tablets. The formulation used to make the tablets is set forth in the following table.

Identification	Amount (mg)
Lactose, direct tableting grade	248.3
Sorbitol, direct tableting grade	80.0
Flavoring	7.0
Coloring	0.7
Magnesium stearate	13.0
Per tablet	349.0

The following procedure was carried out at a temperature of 68° to 75°F with relative humidity of less than 50% to prepare the tablets. The ingredients set forth in the table were mixed and blended with miraculin prepared as described in Example 1 at a concentration of 50 mg miraculin per tablet. The result of the mixture was screened to pass through a 20 sieve size. The tablets were made by pressing the formulation in a Stokes Rotary Tablet Press using a standard $^{12}/_{32}$ inch concave punch. The tablets had a hardness (Monsanto) of 3.0 to 3.5 and weighed 399 mg.

The tablets were tested for their taste-modifying effects by a procedure that determined the apparent sweetness effected by a standard citric acid solution after ingesting the miracle fruit tablet and compared this sweetness to sugar solutions of varying concentration.

Each subject rinsed his mouth for one minute with distilled water. The miracle fruit tablet then was thoroughly chewed for one minute. Then the subject rinsed his mouth with distilled water for thirty seconds and waited two minutes. The subject then tasted a standard solution comprising citric acid (0.00926 M) which had a sourness equivalent of 0.01 M hydrochloric acid.

After rinsing with the citric acid solution, the subject rinsed with distilled water for thirty seconds. After experiencing the sweetness of the citric acid solution, the subject was then required to compare the sweetness experienced with one of eleven standard sugar solutions. The sugar solutions had varying concentrations as follows: 0.1000 M, 0.1175 M, 0.1379 M, 0.1620 M, 0.1993 M, 0.2236 M, 0.2626 M, 0.3083 M, 0.3622 M, 0.4256 M, and 0.5000 M. The subject was then asked to write down which, if any, standard solution most closely compared to the sweetness experienced with the citric acid after taking the miracle fruit tablet. The biological assay procedure indicates that the tablets are effective in substantially increasing the sweetness of the normally sour citric acid solution. Usually the citric acid is comparable in sweetness to either the 0.2626 M or 0.3083 M sugar (sucrose) solution.

Although there were 200 different subjects tested, several having used the tablet in their daily diet, approximately 95% found the tablet to be effective in causing the citric acid to taste sweet, in sweetening other sour tasting foods and enhancing the flavor of almost all types of fruits and vegetables tasted. After six months of storage in a glass bottle at room condition with no particular care being taken in handling and storing of the tablets to maintain them dry, the tablets were still stable and effectively modified taste.

Improved Extraction Process

More detail is given by *R.J. Harvey and J.R. Fennell; U.S. Patent 3,920,815; November 18, 1975; assigned to Mirlin Corporation* for the preparation of miraculin powders without the need for comminuting the pulp and skin at low temperatures. The active enzymatic components of the ripe fruit that normally degrade the active principle, are either separated from, or deactivated, in the final concentrate. The liquid and vaporous components are separated by spray-drying and the enzymatic components are separated by means that effects separation on the basis of density.

Referring to Figure 1.1, washed depitted ripe miracle fruit berries are directed through conduit **1** to comminuter **2** wherein the skin and fruit pulp are blended to an average particle size of about 50 to 300 microns. The comminuted berries then are directed to filter **4** wherein the particles having an average size less than 150 microns are separated from larger size particles. The larger size particles are recycled to the comminuter while the small size particles in slurry form pass into hopper **7**. The particles are then washed several times in a solvent which dissolves carbohydrates such as a mixture of ethanol (75%) and water (25%) which solvent can also remove various pigments, some fat, and other more soluble components.

Removal of these components increases the concentration of the active principle. The essentially colorless, tasteless concentrate which is very nearly nonhygroscopic, does not result in rancidity due to oxidation of the fat components. The solid residue is pumped to spray drier **10**. The fruit is spray dried in chamber **12** under conditions so that the particle temperature is carefully controlled to prevent excessive oxidization or heat degradation of the taste-modifying principle. A cold gas is introduced into plenum **16** so as to reduce the temperature in the spray dryer **10**. The dry particles then are directed to cyclone **20** wherein the particles are separated from the gas. The pressurized air carries the particles into a separator **25** wherein the particles are separated on the basis of density.

The particles are introduced into the separator **25** tangentially along the walls so that the particle mixture passes downwardly in a spiral path. At a vertically intermediate height, a secondary air stream is introduced into plenum **26** which then is passed through permeable wall **27** which extends around the circumference

FIGURE 1.1: IMPROVED EXTRACTION PROCESS FOR MIRACULIN

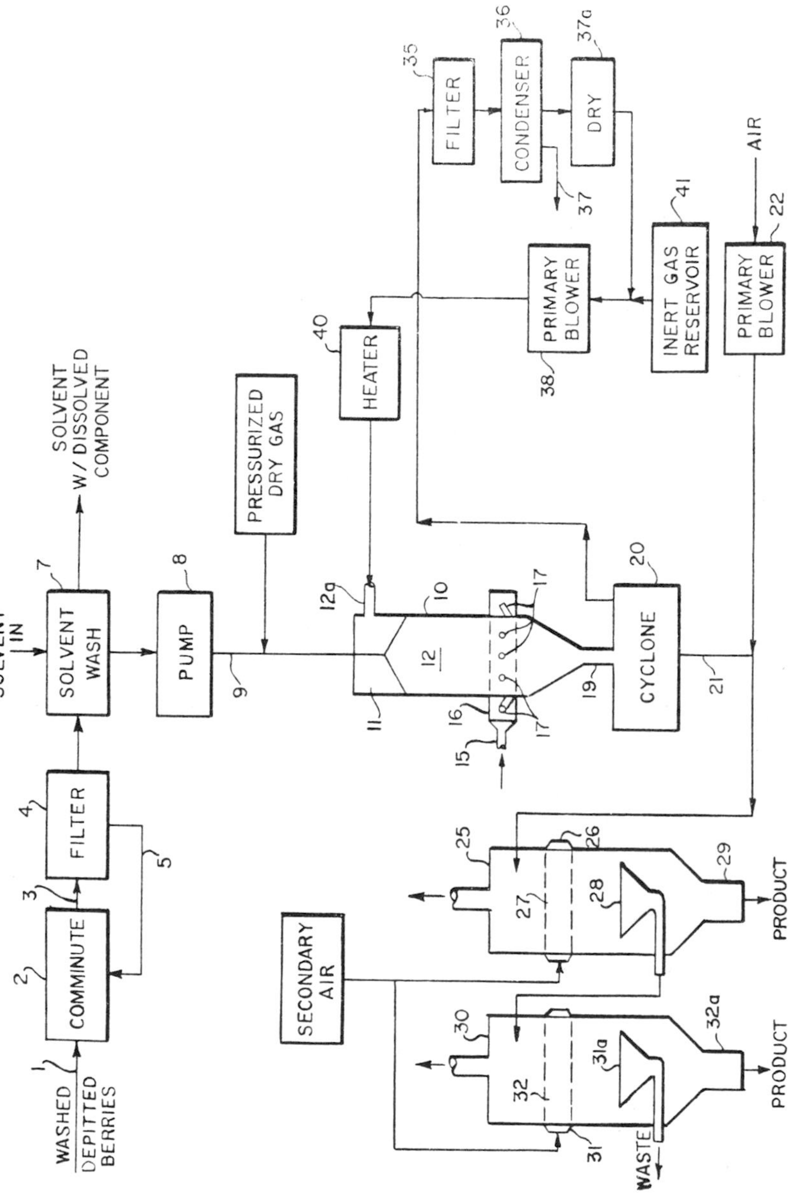

Source: U.S. Patent 3,920,815

of separator **25**. When the downwardly moving particles contact the secondary air stream which is introduced radially, the lower density particles are accelerated towards the center of the separator **25** and are confined and segregated from the higher density particles by means of hopper **28**. The higher density particles bypass hopper **28** and are collected in collecting means **29**. The lower density particles are directed to a second separator **30** and introduced tangentially so that they are passed downwardly in a spiral path. A secondary gas stream is introduced into plenum **31** and through permeable wall **32** extending around the circumference of the separator **30**. The low density particles are collected in hopper **31a** and the higher density particles are collected in collector **32a**.

The solvent extraction to remove carbohydrates can be conducted either prior to or subsequent to the spray-drying step. The solvent dissolves the carbohydrate without dissolving or degrading a substantial portion of the active principle. Representative suitable solvents include water, and ethanol-water mixtures. When employing water or solvents containing water, prior to the spray-drying step, the solvent extraction step should be conducted relatively quickly, i.e., for less than about 10 or 15 minutes since water tends to accelerate degradation of the active principle by the enzymes in the fruit. It is preferred to conduct the solvent extraction step prior to spray drying since residual solvent in the comminuted fruit can be removed in the spray-drying step.

Example 1: Ripe miracle fruit berries are picked and washed in cold water. The fruit then is depitted in a juicer comprising a perforated cylindrical housing with a rotating brush extending along the cylindrical length, the ends of which contact the inside cylinder wall. During rotation, the berries tumble and are pressed against the perforated housing causing the juice and pulp material to pass through the holes, leaving the pits in the cylinder. The juice and pulp are pumped to a mill to comminute the particles. The particles are screened to pass particles having an average size less than 150 microns (100 mesh) while the particles larger than 150 microns are recycled to the grinding step. Cooling is desirable during the grinding to prevent thermal degradation.

The particles are then washed twice in a cold solvent containing 75% ethanol and 25% distilled water. The solid material is separated from the liquid phase. The slurry (solid phase) is then pumped to a spray drier wherein the particles contact an incoming stream of dry air, at a temperature such that the particle temperature does not exceed more than about 80°C, usually from about 60°C to about 70°C. The average contact time of the particles and gas in the spray drier particles is less than about a second. The moisture content of the final spray dried powder is about 3 to 5 weight percent.

The fine dried powder obtained from the spray drier is introduced into a pneumatic cyclone-type separator, whereby the material rich in the active principle is concentrated near the inside wall and the other material is concentrated closer to the center of the cyclone. The mixture to be separated is introduced into the top of the cyclone and caused to move in a circular path down the inside wall. The product is separated from the lower density material by a baffle located at the interface between the high and low density material. The lower density material is recycled until substantially all the active principle is separated. The concentrated product can be recycled if necessary, to achieve any degree of separation from the lower density material.

The powder obtained by this process has a moisture content of 1 to 3 weight percent; it is room-temperature-stable even when stored in the open atmosphere for at least a year or more; and it can be used to produce unit dose forms including tablets or aerosol sprays. The dry powder obtained contains about 40% protein, about 25% carbohydrates, about 3% moisture, and the remainder fat and inerts.

Example 2: The procedure of Example 1 was followed with an additional step of adding sufficient sodium sulfite during the blending step so that the amount of sodium sulfite added represents 1% of the dry weight of the final product. The sodium sulfite was added to bleach the normally red or pink pulp and skin so that the final powdered product obtained is white. The sodium sulfite also functions as an antiseptic agent for microorganisms.

Enzyme Elimination

A process for purification of the active principle of miraculin is described by *R.I. Henkin and E.L. Giroux; U.S. Patent 3,925,547; December 9, 1975; assigned to the U.S. Secretary of Health, Education and Welfare.* Extraction is accomplished in a basic medium with polyvinyl pyrrolidone. The concentrated extract is acidified after addition of 6-aminocaproic acid with glacial acetic acid. The filtered solution containing the active principle is chromatographed on a carboxymethyl polyacrylamide gel adsorbent, employing liquid column chromatography techniques, following which the active principle is eluted with sodium phosphate.

The active principle is then placed on a column of QAE-Sephadex A-50, an ion exchange medium of diethyl-(2-hydroxypropyl) aminoethyl groups coupled to a modified dextran matrix. The purpose of this step is to purify further the active principle by placing it through a cellulose or polyacrylamide-based strongly basic anion exchanger. The active principle is eluted from the column and may then be subjected to a third liquid column chromatography procedure on a column of carboxymethyl polyacrylamide gel. The active principle is again eluted with sodium phosphate to provide the final purified product.

Berries of the plant, including pulp and skin, were allowed to soak for several hours at 0° to 4°C with stirring in a suspension of ⅕ part insoluble polyvinyl pyrrolidone (PVP) in 10 parts 0.1 M Na_2CO_3 buffer, pH 10.5. The pH of the initial suspension is in the range of about 10 to 11. Complete solubilization of the crude extract was found to occur only above the isoelectric point of the active principle, i.e., above pH 9. The homogenate was filtered in the cold (4°C). A typical yield was 40 mg of trichloroacetic acid (TCA)-precipitable nitrogen per 100 g wet weight of berries. The supernatant was then made 0.1 M with respect to 6-aminocaproic acid and additional insoluble PVP was added, the amount of additional PVP being equivalent to the amount employed in the initial suspension.

Over a half-hour period glacial acetic acid was added dropwise to the stirred suspension at 0° to 4°C to lower the pH to 6.0 to 6.5. Other equivalent acids may be used in this step, other than glacial acetic acid. After filtration, a solution containing the active principle, determined by bioassay, was obtained, a typical yield being 20 mg of TCA-precipitable nitrogen per 100 g wet weight of berries. The active principle was then adsorbed onto a short column of BioGel-

CM-30 resin, a trade name for a support material containing sodium carboxymethyl groups covalently linked to a polyacrylamide matrix, equilibrated with 0.1 M NaH_2PO_4 buffer, pH 6.0, at 0° to 4°C. The column was next washed with 0.1 M buffer, pH 6.5, then the active principle was eluted stepwise with 0.1 M Na_2HPO_4. A typical yield was 5 mg of TCA-precipitable nitrogen per 100 gm wet berry weight. The active principle was adjusted to pH 10.5 by addition of 0.1 N NaOH in 2% sodium carbonate. QAE Sephadex A-50 resin sufficient to form a 150 ml column was equilibrated with 0.1 M sodium carbonate buffer, pH 10.5. Half of the resin was added to the solution of active principle and the slurry was poured onto a 2.5 cm diameter column formed from the other portion of resin.

After packing, the column was washed with one column volume of 0.05 M NaCl in 0.1 M sodium carbonate buffer, pH 10.5. Elution was effected by a linear gradient of 0.05 to 0.65 M NaCl in 0.1 M sodium carbonate buffer, pH 10.5. The active principle was concentrated and solvent was exchanged for 0.1 M sodium phosphate buffer, pH 6.0. The concentrate was pumped onto a 2.5 x 20 cm column of carboxymethyl polyacrylamide gel. A shallow pH gradient in Na_2HPO_4 buffer eluted the active principle over the pH range 7.0 to 7.3. A typical overall yield was 20 to 25 mg of protein per 100 g wet weight of berries. An overall recovery of ⅕ of the taste changing activity present in the crude extract was observed. This represents a three-fold increase in specific activity.

In the use of the active principle of the fruit of *Synsepalum dulcificum* to control obesity, it has been found that the active principle can be administered orally with good results. For the human adult an oral dosage of about 10 to 400 μg has been found to be quite effective. In particular, to provide the desired effect over a time period of 10 to 30 minutes, the length of time required to consume a short meal, about 10 to 100 μg of the purified principle may be administered to a human adult. To produce the desired effect over longer periods of time, say 1 to 3 hours, about 100 to 400 μg should be used. These dosages may be administered as frequently as 5 times a day without adverse effect. Upon being administered orally, the active principle should be held in the mouth for at least two minutes and then either expectorated or swallowed. A sweetening effect is observed for the time intervals as stated above.

Solubilizing Agent

In the process of *J.N. Brouwer, G.J. Henning and H. van der Wel; U.S. Patent 3,682,880; August 8, 1972; assigned to Lever Brothers Company* a solubilizing agent is used to treat the flesh, pulp, or seeded fresh or frozen fruit. Suitable solubilizing agents fall into two groups: (1) compounds which dissolve the active factor in a form having a molecular weight as determined by gel filtration) of above about 50,000, generally above 200,000, and apparently representing a bound form of the active factor; and (2) compounds which dissolve the active factor in a form having a molecular weight of below about 50,000 and from which the pure active factor is preferably prepared.

The first group includes: (a) tannin-binding substances, for example, polymers such as polyvinylpyrrolidone, sorbitan mono-oleate dipolyethyleneglycol ether, and polyethyleneglycols with mean molecular weights above about 200, proteins such as gelatin, casein, and albumins, peptones, caffeine, and salts such as aluminum sulfate; (b) hydroxycarboxylic acids such as ascorbic acid and tartaric acid, and acylated neuraminic acids such as N-acetylneuraminic acid and N-glycolylneuraminic acid. Solubilizing agents of the second group include: (a) protamines, such as salmine; (b) polypeptides prepared from basic amino acids, such as polyarginine; (c) polyamines. When solubilizing agents of the second group are used, the pH of the homogenate should be 3 or above, with protamines 6 to 10, and with polyamines 6 to 8.

The constituents remaining insoluble after homogenization are removed by centrifuging at 10,000 to 30,000 *g* and/or by filtration through glass. (Since the active factor is absorbed on paper, filtration through paper filters must be avoided.) If the more or less clear solution so obtained contains no physiologically unacceptable substances, it can be used as such. Other dissolved substances that the extract contains, including the solubilizing agent, are removed wholly or partially by methods known as such for the separation of protein materials, which gives a further concentration. Such methods are precipitation with acetone, ethanol, or ammonium sulfate, and techniques such as gel filtration and ion exchange.

These concentrates can be used in the products when any physiologically unacceptable solubilizing agents have been completely removed. If desired, the concentrates can be concentrated further or evaporated to dryness at a temperature of not greater than 30°C, room temperature or lower: for example, concentration or evaporation to dryness is carried out in a thin-film evaporator or by freeze drying the frozen solution in high vacuum.

The concentrates obtained by dissolving the active factor with the aid of solubilizing agents of the first group contain the active factor in a form having an isoelectric point between about 4 and 7; the form of the active factor obtained with the aid of solubilizing agents of the second group has a lower molecular weight and an isoelectric point of about 9.

From a solution of the active factor obtained by means of solubilizing agents of the second group, a preparation has been obtained which exhibits only one band in the analytical ultracentrifuge. This preparation, which has no taste of its own but still shows the sweetening activity, is therefore the almost pure active factor, which will be called miraculin. The processes give a yield of up to 100 mg of miraculin from 1 kg of berries.

The molecular weight determined by means of the analytical ultracentrifuge is about 42,000 (± 3,000). On hydrolysis, amino acids and saccharides are formed. Saccharides have been detected in the band of miraculin after electrophoresis on polyacrylamide gel by oxidation with periodic acid and by the Schiff method. This also confirms the finding of Inglett et al that miraculin is a glycoprotein.

The physiologically acceptable concentrates and miraculin itself can be added to various foodstuffs and beverages such as yoghurt, buttermilk, Junket, mayonnaise, fruit juices, jam or marmalade. The amount of concentrate or active factor to be incorporated depends on the miraculin content and on the foodstuff or other product. It can be determined rapidly by a simple test. The addition of almost pure miraculin to yoghurt, for example, in an amount of 2.5 mg/l gives a sweet taste. With smaller amounts, the sour taste is reduced, but the sweet taste is not yet present. In berry juice, about ten times as much is needed: 20 ml/l gives the sweet taste. The stability of the substance in milk products is surprisingly good: after four days at room temperature or 14 days at 4°C the same activity was found. The stability in berry juice was less good, however, which is ascribed to the lower pH of the juice (pH 2.8). Preferably, the products should have a pH value higher than 3.

Example 1: 1.1 to 2.2 g of fruit flesh (wet weight) from 2 to 4 g of berries of *Synsepalum dulcificum* was homogenized with a 1% solution of polyethyleneglycol (mean molecular weight 20,000) in water. The pH of the homogenate was brought to 7 with a saturated sodium carbonate solution. Finally it was centrifuged at 10,000 *g* for 30 minutes. The liquid obtained in this way was treated at 20°C with an equal volume of acetone with stirring. The precipitated active factor was centrifuged off for 10 minutes at 1,200 *g*. The precipitate was washed with two volumes of acetone-water (2:1) and taken up in 6 ml of 0.1 M potassium phosphate buffer, pH 7. Insoluble material was centrifuged off at 10,000 *g* for 30 minutes.

The resulting solution, when added to 1 liter of yoghurt gave a pleasant sweet taste to this product. The same result was obtained by extraction with a 5% solution of polyethyleneglycol (mean molecular weight 400) in water, 1% of caffeine in 0.1 M potassium phosphate, pH 7, 0.1% of sorbitan mono-oleate dipolyethylene glycol ether in 0.1 M potassium phosphate, pH 7, and with 3% of peptone in water.

Example 2: 4.4 to 8.8 g of fruit flesh (wet weight) from 8 to 16 g of berries was homogenized with 20 ml of a 1% gelatin solution in water or in 0.1 M potassium phosphate buffer, pH 7. The pH of the homogenate was brought to 7 with a dilute KOH solution, the mixture was centrifuged for 30 minutes at 10,000 *g*. The solution was purified further by gel filtration through Sephadex G-25, a modified dextran [swollen with water and filled into a 3 x 11 cm column (volume of the bed 78 ml)]. The active factor (25 ml) was percolated through the column with water at 150 ml/hr. The eluate with volumes 25 to 50 ml inclusive contained the active factor; it was used for sweetening purposes.

Example 3: 2.2 g of the fruit flesh (wet weight) was homogenized with 5 ml of a 1% solution of salmine sulfate in water. After homogenization, the pH of the mixture was 3 to 3.5. The resulting mixture was centrifuged at 10,000 *g* (30 minutes).

Similar results were obtained when the pH of the mixture after homogenization but before centrifuging was brought to values of up to 10. The pH value required for the further treatment of the extract can therefore be adjusted before or after the insoluble constituents are centrifuged off. The extract obtained in this way contained a large amount of salmine, which was removed either by ion exchange or by gel filtration.

The removal of the salmine by ion exchange proceeded as follows. The pH of the solution was adjusted to about 10, and the solution was percolated through a column of CM-Sephadex C-50, a carboxymethyl derivative of modified dextran which possesses cation-exchanging properties. The CM-Sephadex C-50 was washed with 0.02 M sodium glycinate buffer (pH 10.5) and filled into a column (1 x 34 cm, volume of the bed 27 ml). The solution of the active factor was added to the column and was percolated with the same buffer at a rate of flow of 25 ml/hr. The active factor was present in the eluate with elution volumes of 16 to 35 ml, inclusive. The solution was free from salmine (according to starch gel electrophoresis at pH 4.3).

The removal of the salmine by gel filtration proceeded as follows. The solution of the active factor was brought to pH 7, after which sufficient solid NaCl was added to it to make it 0.1 M in NaCl. 1 ml of this mixture was subjected to gel filtration through Sephadex G-50, being eluted with a 0.1 M solution of NaCl in 0.02 M potassium phosphate buffer, pH 7, from a 1.5 x 21 cm column (volumn of the bed is 36 ml; rate of elution is 20 ml/hr). The eluate with elution volumes of 13 to 19 ml, inclusive, contained the active factor and that with elution volumes of 20 to 31 ml, inclusive, contained the salmine. The solution of the active factor obtained in this way was freed from salts by filtration through a water-washed column of Sephadex G-25. It could be used as such.

Acid Activator

R.J Harvey; U.S. Patent 3,849,555; November 19, 1974; assigned to Mirlin Corp. details the mechanism to obtain the sweet taste of miraculin by activating the taste receptors with an acid, sour or bitter food containing the acid activator. In the case of bitter tasting foods, the sour acid is rendered sweet tasting thereby masking the bitter taste. In the case of sour tasting foods already containing sour acids, the addition of acidic compositions dramatically improves the sweetness of the food ingested.

Suitable nontoxic acids include naturally occurring carboxylic acids such as citric, malic, ascorbic, formic, acetic, tartaric or inorganic acids in low concentrations or mixtures. The acids can be added as pure acids or in their natural form. Thus, the acids can be added to the foods as lime, lemon or apple juice. The acid is added to the food in amounts to obtain an effective acid molarity in the food of between 0.005 M to 0.03 M. By "effective molarity" it is meant the amount of acid in the food which when dissolved produces the same sweetening effect as an aqueous acid solution having the molarity set forth above.

When the acid is added in equivalent amounts below 0.001 M or above 0.1 M, it is ineffective to produce the desired result in conjunction with miraculin and renders the food to which it is added sour-tasting. The effective molarity of acid is obtained by mixing aqueous acid concentrates of about 0.1 M to 0.3 M with the food. Acid concentrates in excess of about 0.3 M are undesirable because the risk of denaturing the food is greatly increased. Generally, the desired effective molarity can be attained by mixing the food with from 0.1 to 0.7 weight percent acid solution. The particular amount of acid and desired effective molarity will depend upon the particular food eaten, the degree of sweetness desired and the particular acid used.

Example: Unsweetened tea drunk both before and after having contacted the tongue with miraculin has substantially the same sweetness. However, after having ingested miraculin a cup of tea mixed with between 0.15 and 0.50 g of a composition comprising 10% citric acid, 25% malic acid and 65% sucrose the sweetness of the tea is the same as when it is mixed with from 1 to 2 teaspoons of sugar. The same sweetening effect can be attained when the tea is mixed with lemon juice or the acid composition. Furthermore, the same sweet effect is noted when the acid composition or acidic fruit juice is added to such food items as cookies, jellies or fruits.

Bicarbonate-Effervescent Activator

In another approach the acid activating mechanism is derived from a bicarbonate-effervescent mixture as described by *J.R. Fennel and R.J. Harvey; U.S. Patent 3,898,323; August 5, 1975; assigned to Mirlin Corporation.* These compositions are effervesced by the contact in an aqueous solution of a nontoxic acid, and an alkaline material that spontaneously evolves carbon dioxide when dissolved in an acidic aqueous solution. The acid can be either incorporated in the effervescent composition or form part of the liquid being drunk.

For example, carbonated liquids contain carbonic acid which will react with sodium bicarbonate and water to form carbon dioxide. Thus, in this example, it would not be necessary to incorporate an acid when using sodium bicarbonate in the miraculin-containing composition. On the other hand, when the liquid being drunk contains little or no acid, the acidic component can be incorporated into the miraculin-containing composition.

Example 1: A powdered mixture of sodium bicarbonate and stable miraculin is prepared by the following procedure: The fruit is depitted at about 1 to 4°C in a juicer comprising a perforated cylinder housing a rotating brush extending along the cylinder length the ends of which contact the inside cylinder wall. During rotation, the brushes tumble and press the berries against the perforated housing causing the juice and pulp material to pass through the holes, leaving the pits in the cylinder. The juice and pulp flow into containers immersed in an alcohol-dry ice bath; a small amount of crushed dry ice (solid CO_2) is added directly to the fruit pulp obtained from the depitting step, and the mixture is thoroughly ground in a ball mill to a particle size of less than 150 microns (No. 100 sieve size) while being maintained at a temperature of about -40° to -50°C.

The powdered frozen pulp and juice of the berries are added to an aqueous solution of sodium bicarbonate, pH 7.8. The mixture is then placed in a freeze-drying flask, placed in a shell freezer, and allowed to come to thermal equilibrium at a temperature of about -55°C. The material is then connected to a freeze-dryer vacuum system with a refrigerated condenser for condensing liquids and condensable vapors, where it remains until there is no significant weight change in the material over a four-hour period. The material is then removed from the flask and placed in a desiccator cabinet in trays at room temperature for further drying or storage until the moisture content is between 0.5 and 1.0%. The powder and sodium bicarbonate comprise a homogeneous mixture.

After the powder is thoroughly dry, it is placed in a temperature controlled milling machine, where the average particle size is reduced preferably to about 150 microns. The material is periodically screened and that retained by the No. 100 sieve is returned to the milling machine until it can pass the No. 100 sieve size.

The fine powder is introduced into a pneumatic cyclone-type separator, whereby the dense miraculin and associated sodium bicarbonate is concentrated near the inside wall and the cellulosic material and associated sodium bicarbonate is concentrated closer to the center of the cyclone. The mixture to be separated is introduced into the top of the cyclone and caused to move in a circular path down the inside wall. The miraculin-rich material is separated from the lower density material by a baffle located at the interface of the miraculin and lower density material. The lower density material is recycled until substantially all the miraculin is separated. The concentrated miraculin-sodium bicarbonate can be recycled if necessary, to achieve any degree of separation from the lower density material. The miraculin-sodium bicarbonate obtained from the cyclone separator is room-temperature-stable even when stored in the open atmosphere for at least about eight months, and can then be used to produce unit dose forms, including tablets or aqueous sprays.

Example 2: An effervescent miraculin composition was prepared by first forming a slurry of ascorbic acid and sodium bicarbonate in the following proportion:

Composition A

Ascorbic acid	176.12 g (1 mol)
Sodium bicarbonate	168.04 g (2 mols)

Composition A was dissolved in dry ethyl alcohol. The resultant slurry was mixed with 646 g stable miraculin and 58 g of a binder comprising Maltrin-10 which is essentially a starch. After the components were thoroughly mixed, the alcohol was evaporated by heating the resultant mixture to a temperature to about 50°C in a dry atmosphere. Unit dosage forms of the resultant composition then were individually sealed in a moisture-proof wrapping.

Example 3: An effervescent coating for use with an acid-containing liquid is prepared by mixing an aqueous solution of 15 weight percent sodium bicarbonate with 65 weight percent stable miraculin and 20 weight percent of a binder comprising Maltrin-10. After being thoroughly mixed, the water was evaporated by heating the resultant mixture in a dry-air atmosphere to a temperature of about 50° to 60°C. Unit dosage forms of the resultant dry mixture were then individually sealed in a moisture-proof wrapping.

MONELLIN

Basic Commercial Process

In the process of *R. Dobry; U.S. Patent 3,878,184; April 15, 1975; assigned to Beech-Nut, Inc.*, monellin, the sweet tasting principle of Serendipity berries is recovered by the following steps:

(1) Clusters of berries; stems, skins, seeds and all are macerated.

(2) The macerated berry clusters are extracted with water at about ambient temperature using from about 3 to 8 weight units of water per unit weight of berry clusters. The water is acidified to a pH of from about 3 to about 5.

(3) Insoluble matter is separated by centrifugation.

(4) The clarified extract is contacted with carboxymethylcellulose free acid.

The carboxymethylcellulose (CMC) free acid with which the extract is contacted is fine (-80 mesh). This material is a white granular powder derived from the sodium salt of CMC. Since it is an acid and not the sodium salt, it is completely insoluble in water. This material is obtainable as Hercules CMC Free Acid.

Contacting the clarified extract with the CMC free acid serves to adsorb the monellin on the CMC free acid. The CMC free acid may be either a fine or coarse particle size, either in its "native" state as the free acid or partly neutralized with a base. A fine particle size is one that will pass through an 80-mesh screen, whereas a coarse particle size is one that will be retained on a 50-mesh screen. The CMC free acid containing the adsorbed monellin may be used as such as a sweetening agent, especially if it is fine enough and therefore not gritty in the mouth, or as an intermediate for further purification of monellin.

The contacting between extract and adsorbent may take place by passing the clarified extract through a column containing neutralized CMC free acid or by agitating a mixture of the clarified extract and the CMC free acid. After having adsorbed monellin from the clarified extract, the CMC free acid is washed with water having a pH of from about 3 to about 5 in order to displace any unadsorbed colloidal matter, color bodies and bitter-tasting ingredients. The CMC free acid containing adsorbed monellin is then dried to yield a material having an intensely sweet taste.

A liquid sweetening preparation may be prepared by contacting the CMC free acid containing adsorbed monellin with any buffer or basic solution which will adjust the pH from about 7 to about 10. By such treatment the monellin is desorbed from the CMC free acid. Filtration of the resulting slurry produces a liquid concentrate free of adsorbent which resembles molasses in color, sweetness and viscosity. The liquid concentrate may be used as such as a sweetening agent. When the monellin has been desorbed from the CMC free acid by means of buffer solutions rather than alkali or ammonia, the liquid concentrate may be treated to remove mineral salts, for example, by dialysis and either dried to yield a more purified extremely sweet product or the dialyzed concentrate may be further concentrated by ultrafiltration to yield a highly concentrated liquid sweetening agent.

Example 1: Forty pounds of frozen whole berry clusters freshly rinsed with tap water and allowed to drain are macerated in a motorized meat grinder at a fine setting. The ground berries are added to 28 gallons of water and agitated for twenty minutes. The pH of the resulting slurry is adjusted to 4.2 with concentrated HCl and insoluble matter is separated by a two-stage centrifugation: Sharples Super-D-Canter continuous centrifuge (screw conveyor type) followed by Sharples supercentrifuge (tubular bowl type). The clarified extract is slightly hazy and pinkish in color being both bitter and sweet to the taste.

Fine (-80 mesh) CMC free acid is slurried in water and neutralized to pH 6.0 by dropwise addition of 2.7 ml of 1 N NaOH per gram of dry CMC. A glass column, measuring 30 mm outside diameter and 400 mm in length is wet-filled with the neutralized CMC adsorbent to a height of 55 mm and about 1,400 ml of the clarified extract are passed downward through the column. By the time all of the primary extract is fed to the column, the adsorbent is visibly loaded to a depth of 20 mm, as indicated by a change in color, and remains in that condition after a rinse with water acidified to pH 4.2. The column is then drained and its contents extruded by a moderate back-pressure of compressed air. The loaded segment of the adsorbent is carefully sliced off with a sharp knife, dried in vacuo, and pulverized by a mortar and pestle. The final material is a light fluffy solid, reddish-brown in color and intensely sweet when placed in the mouth.

Example 2: About 2,000 ml of clarified extract obtained according to the first paragraph of Example 1 are passed through a glass column having the same dimensions as in Example 1, which is loaded with coarse CMC free acid (having a particle size such that substantially all is retained on a number 50 mesh screen) to a height of 240 mm. The adsorbent is prepared by neutralizing CMC free acid with aqua ammonia to pH 7.0, and storing the neutralized product moist under refrigeration until use. Following adsorption the column is rinsed with water having a pH of 4.2. The column is then eluted with 0.1 M phosphate buffer at pH 9.5 and the monellin collected as a liquid concentrate.

Example 3: About 18 grams of coarse (+50 mesh) dry CMC free acid, not neutralized with base, are slurried in water and allowed to stand for several hours until fully swollen. The CMC free acid is then drained, added to 420 ml of clarified extract having a pH of 4.0 and agitated for 30 minutes. The CMC free acid is separated from the spent extract by settling and the supernate decanted with care. This procedure is repeated six times, adding 420 ml of clarified extract to the CMC free acid each time until the adsorbent is dark in color and essentially no longer able to pick up sweetness from the extract. The adsorbent is then rinsed with several volumes of water which has been acidified with citric acid to pH 4.0. The rinsed adsorbent is titrated with 1 M sodium carbonate until the pH reaches 9.0. Separation of the liquid from the solid yields a liquid concentrate similar to that of Example 2. Aliquots of this liquid are reacidified to pH 5.0, adsorbed on fine (-80 mesh) CMC, washed with an acetone mixture (4:1), vacuum dried and pulverized. The dried product is lighter in color than the final product of Example 1.

UV Radiation Process

O.A. Essiet; U.S. Patents 3,687,693; August 29, 1972; and 3,826,795; July 30, 1974 describes a process for extraction and recovery of a natural sweetening agent on a commercial basis. The berries of the African plant *Dioscoreophyllum cumminsii,* also called Nigerian berries, contain substantial amounts of water-soluble sweetening material which, in its purified form, has a sweetening effect on a dry weight basis about 1,500 times as great as sucrose.

The chemical composition of the sweetening agent is not definitely known. It is known, however, to be a carbohydrate having a molecular weight on the order of 10,000. It occurs as part of a carbohydrate-protein complex having a molecular weight of the order of 100,000. This complex has a sweetening effect of the order of 150 times that of sucrose on a dry weight basis.

So far as known, the sweetening material, either by itself or in combination with the protein of the complex, has no detrimental physiological effect when ingested by humans, particularly in the amounts in which it must be used for producing an acceptable sweetening effect. This sweetening material in either form has no unpleasant aftertaste.

Figure 1.2 is a flow sheet describing the extraction process. The seeds of the Nigerian berries have a thin shell covered with very short spines surrounding a kernel having an appearance and consistency similar to that of a very small garden pea. These seeds not only interfere with treatment of the berries to separate the aqueous sweetening material from the solids of the berries, but the kernels contain an extremely bitter material which is liberated if the shell of the seed is cut or broken. This bitter material is difficult to remove from the recovered sweetening material and it is important to remove and discard the seeds at an early stage in the process.

FIGURE 1.2: PROCESS FOR EXTRACTING A SWEETENING AGENT FROM BERRIES

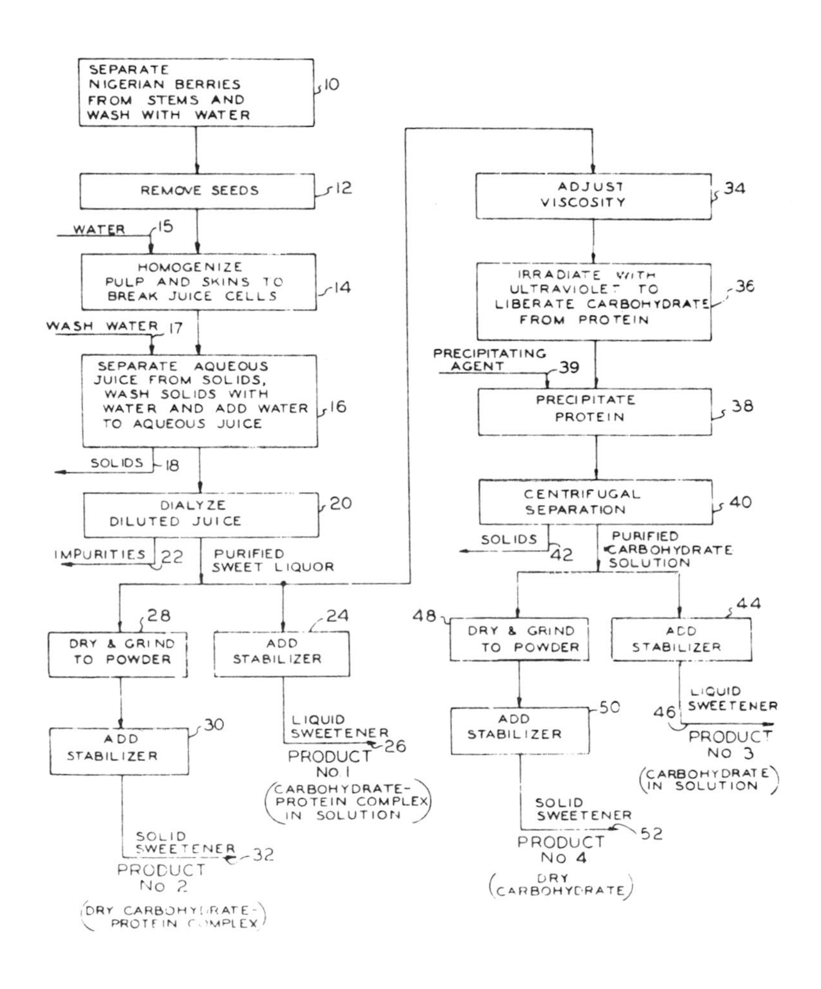

Source: U.S. Patent 3,687,693

A seed removal step is indicated at **12** using a machine analogous to a cherry pitting machine in which the seed is pushed or punched out of the fruit leaving the pulp and skin. It is almost impossible to remove all of the seeds from the berries as almost inevitably a few of the berries will be missed, and care must be taken not to cut or break the seeds in further steps of the process.

The remaining material, including any berries from which the seeds have not been removed, are further treated as indicated at **14** to break down the cells of the pulp to liberate the water-soluble carbohydrate-protein complex so that it can be removed in water solution. An amount of water approximately equal to twice the weight of the original berries being processed is added as indicated at **15** and the mixture subjected to a homogenizing operation. This homogenization can be continued for a sufficient length of time to produce a semiliquid material in which substantially all of the natural cells of the pulp of the fruit have been ruptured and the homogenized material is primarily a suspension of finely divided solids in an aqueous solution of water-soluble material including the carbohydrate-protein complex.

An alternate procedure is to stop the homogenizing operation as soon as the residual seeds have been liberated from the berries and the pulp and skins become a fluid mass. The seeds can then be screened from this fluid mass, after which the separated material can be subjected to more intense homogenization in the same or similar apparatus to more thoroughly break down the cells of the pulp.

As indicated at **16**, the aqueous material or partly diluted juice is separated from the solids, including any residual seeds which may be present, by a filtering operation followed by a high-speed centrifugal clarification operation. In general, such an operation will not remove all of the very finely divided suspended solids from the separated aqueous liquid. These solids can, however, be removed in high-speed centrifugal separators now available commercially to clarify the aqueous liquid.

The solid residue from either a filtering operation or a centrifugal separation is washed with water, as also indicated at **16**, and this wash water is separated from the solid residue and added to the separated juice to provide a diluted juice. The washing is repeated several times with decreasing amounts of water. The total amount of water indicated at **17** used for washing will, in general, be approximately equal to the original weight of the berries being processed, so that the total weight of water added to the diluted juice will be approximately three times that of the original berries being processed. The separated solids are discharged from the process as indicated at **18**.

The diluted juice is then subjected to a dialyzing operation as indicated at **20**. Water-soluble impurities, most of which have molecular weight below 500, will diffuse through the diaphragm into the water and can be discarded from the process, as indicated at **22**. Best results are obtained if the temperature of the liquids in contact with the diffusion membranes is maintained between approximately 4° and 10°C. With the amount of water discussed above as being added in the prior homogenizing and washing operations, the purified sweet liquor from the dialyzing operation will have a viscosity similar to that of a light corn syrup commercially sold for table use. A purified sweet liquor having the viscosity discussed above contained 42 milligrams of the complex per 100 grams of water, and 1 gram of such purified sweet liquor has the sweetening effect of approximately 6 grams of dry cane or beet sugar. This purified sweet liquor is chemically stable at ambient temperature below about 65°C and can be stabilized against fungicidal action by the addition of a stabilizer as indicated at **24** to produce a stabilized Product No. 1. This stabilized product can be used in liquid form as a liquid sweetener either in the concentration at which it is recovered from the dialyzer or in more dilute or more concentrated form.

Alternatively, the purified sweet liquor from the dialyzing operation **20** can be subjected to a drying operation and then ground to a powder, as indicated at **28**. The powder is essentially the carbohydrate-protein complex in purified form. Sodium chloride in powdered form is an effective stabilizer and can be added in small amounts; for example, 1 percent on weight basis to the dried and powdered material to produce Product No. 2, which can also be used as a solid sweetener as indicated at **32**.

The carbohydrate portion of the carbohydrate-protein complex may, however, be separated from the protein. A method of liberating the carbohydrate from the protein is to irradiate a shallow body or layer of an aqueous solution of the complex with ultraviolet light. The irradiation is carried out when the solution has viscosity of between approximately 1.5 to 11 centipoises.

The intensities and principal wavelengths of the ultraviolet radiation from different types of ultraviolet sources vary over a substantial range and the intensity of the radiation even for the same type of source varies from unit to unit. The best test of the optimum time for irradiation treatment is, therefore, the extent of denaturing of the carbohydrate-protein complex (breaking of the chemical bond between the carbohydrate and protein) and the amount of recovery of the carbohydrate after precipitation of the protein in the irradiated solution and removal of the precipitated material, as described below. Recoveries of 80 to 90 percent can be obtained using a quartz mercury vapor lamp and radiation time for ten minutes to one hour or more, depending upon the intensity of the radiation.

The liberated protein can be precipitated, as indicated at **38**, by one of a number of protein precipitating agents, indicated as being added at **39**, a preferred agent being polyvinyl alcohol of sufficiently low molecular weight to be a thin liquid, for example, in an amount equal to approximately 20 percent by volume of the resulting solution containing the protein and liberated carbohydrate. These precipitation agents also precipitate any of the residual carbohydrate-protein complex which has not been broken down.

The precipitated material is centrifugally separated from the resulting carbohydrate solution since it can be substantially completely removed, as indicated at **40**, by continuous centrifugal separation in a commercial type of high speed continuous centrifugal separator. Thus, the precipitated solids have sufficient flow characteristics to be continuously discharged from the bowl of the separator and from the process, as indicated at **42**. The separated solids are washed with small amounts of water and the washing water added to the liquid material separated from the solids.

The resulting liquid material is a purified carbohydrate solution. This solution can be adjusted in viscosity, if desired, and may have a stabilizer, such as methylparaben, added as indicated at **44** to produce a liquid sweetener, indicated at **46** as Product No. 3. This is essentially a carbohydrate solution in which the carbohydrate has a sweetening effect upon a dry weight basis about 1,500 times that of sucrose.

GLYCYRRHIZIN

Sucrose Combination

R.E. Muller and R.J. Morris, Jr.; U.S. Patent 3,282,706; November 1, 1966; assigned to MacAndrews & Forbes Company provide a means for potentiating the sweetness of sugar by adding small amounts of glycyrrhizin. Licorice root contains from about 6 to 14 percent of a triterpenoid glycoside called glycyrrhizin. This compound is present in the root as the mixed calcium and potassium salt of glycyrrhizic acid.

Glycyrrhizin has a sweetness value about 50 times greater than that of sucrose and is perhaps the sweetest chemical processed commercially that is found in nature. Glycyrrhizic acid is obtained in 90 percent or more purity by grinding the root, extracting the ground material with hot water, and treating the extract to recover the acid-insoluble fraction containing the glycyrrhizic acid.

Glycyrrhizic acid can be ammoniated, to provide ammoniated glycyrrhizin, by replacing one or more of the three acid hydrogen atoms with ammonium. Ammoniated glycyrrhizin, therefore, ranges from a monoammoniated product to an essentially fully (tri)ammoniated product and mixtures thereof. Ammoniated glycyrrhizin is well known and widely used, and also has a sweetness value about 50 times that of sucrose.

Ammoniated glycyrrhizin, of course, has the characteristic licorice flavor and it is primarily for this that this material has found widespread use as a flavoring agent in, for example, confections. Because of the licorice flavor, this material has not been used alone as a sweetening agent except in some licorice flavored confections, since the amount required for sweetening also imparts the characteristic licorice flavor. This material has also been used in very minute quantities as a foaming agent in beverages.

It has been found, however, that ammoniated glycyrrhizin potentiates the sweetness of sucrose in sucrose-containing foods, confections and beverages at levels which do not impart appreciably the licorice flavor. By "potentiate" is meant that the sweetness value of the combination of sucrose and ammoniated glycyrrhizin is over and above the sum or mere additive effects of the known sweetness values and is, therefore, the result of synergism between the sucrose and ammoniated glycyrrhizin in certain relative proportions of one to the other.

Example 1: A root beer flavored soft drink beverage, which normally contains 12 percent by weight of sucrose, is prepared using: (a) 8 percent of sucrose and 0.04 percent of ammoniated glycyrrhizin; and (b) 5 percent of sucrose and 0.08 percent of ammoniated glycyrrhizin. Both beverages (a) and (b) had a sweetness equivalent to the normal 12 percent sucrose beverage. However, using known sweetness values for sucrose and ammoniated glycyrrhizin, the sweetening value computed from mere additive effects would be equivalent to 10 percent sucrose in (a) and 9 percent sucrose in (b).

Example 2: A cake whose recipe calls for 2 cups of sucrose (450 g) is prepared using 1 cup of sucrose (225 g) and ½ teaspoon of ammoniated glycyrrhizin (1 g). The cake had a sweetness equivalent to the same cake prepared from 2 cups of sucrose. However, using known sweetness values, the sum of the sweetening values, if merely additive, should be equivalent to 1½ cups of sucrose.

Example 3: Five parts by weight of ammoniated glycyrrhizin are thoroughly mixed with 500 parts of sucrose to form a sweetening agent. When used as a sweetening agent in beverages, confections, and foods, this sweetener is equivalent to 1,000 parts of sucrose. On the basis of known sweetening values, this sweetener should be equivalent to 750 parts of sucrose. As a sweetening agent, this material may be mixed with a diluent or extender, like water, invert sugar, corn syrup, and the like, which does not alter its sweetening properties.

For example, a mixture of 333 parts by weight of sucrose, 665 parts of corn syrup and 2 parts of ammoniated glycyrrhizin is equivalent in sweetness to 1,000 parts of sucrose. However, on the basis of the known sweetening values (the corn syrup having a sweetness value of 40 percent of that of sucrose), this mixture should be equivalent to 767 parts of sucrose.

Example 4: This example illustrates the use of the sweetening agent described in the preceding example. A caramel confection is prepared from:

Corn syrup	1,470 g
Sucrose	1,000 g
Ammoniated glycyrrhizin	2 g
Butter	230 g
Evaporated milk	2.7 l
Salt	3 g
Vanillin	0.5 g

The regular recipe without the ammoniated glycyrrhizin calls for 1,350 grams of corn syrup and 1,350 grams of sucrose.

Flavored carbonated and noncarbonated beverages (grape, strawberry, orange, and root beer flavored) were prepared using varying proportions of sucrose and ammoniated glycyrrhizin to determine the range of proportions of ammoniated glycyrrhizin to sucrose to provide the synergistic potentiation of the sweetness of sucrose without imparting significant licorice flavor. These preparations were then subjected to sensory testing by a taste panel using the "rank order test" and the "scalar difference from control test." It was determined that the critical proportions are not less than 0.2 and not more than 2.2 parts by weight of ammoniated glycyrrhizin per 100 parts by weight of sucrose.

Citrus Juice Sweetener

Another method to prepare an extract of glycyrrhizin is described by *D.H. Koski and J.B. Koski; U.S. Patent 3,585,044; June 15, 1971.* The extract is used to prepare a citrus juice beverage. Any suitable source of licorice may be used. However, it is preferred to utilize the natural licorice root as the licorice source. The licorice root is reduced to a pelleted form prior to mixing with water. However, licorice in powder form may also be used. Twenty to thirty parts by weight of licorice are mixed with fifty to seventy parts by weight water and the mixture is heated to its boiling point. Next, sugar (sucrose) in an amount of between about eight and thirty-five parts by weight is added to the licorice-water mixture at essentially the first sign of ebullition. Sucrose is preferred; however, sugar-substitutes may be used if desired, for example, saccharin and the like.

The mixture comprising licorice, water and sugar is boiled in a substantially closed system for a period of time. The boiling time period may vary depending upon the exact amounts of the various materials that are employed. A suitable time period includes between about one and three minutes. Following the boiling period, the heating of the mixture is discontinued and the mixture is allowed to steep with the system still substantially closed for a relatively short period of time, for example, between about one and three minutes. Preferably, the boiling time and the steeping time are approximately equal in length. Finally, the licorice-containing mixture is combined with citrus juice, for example, a mixture of orange and lemon juices.

The citrus juice mixture consists of about 120 parts by weight of orange juice and about 40 parts by weight of lemon juice. Either fresh or concentrated juices may be used. However, it is best that the juices be strained through a suitable filtering means, for example, steel strainer, to remove a substantial portion of solids prior to mixing the citrus juice with the licorice-containing mixture.

The resulting beverage may be dehydrated so as to form a concentrate for packaging and preservation purposes and later reconstituted with water. The resulting mixture may also be frozen and packaged in the form of a sherbet. The beverage product may be used as a juice drink or may be utilized, for example, as a mix for alcoholic beverages.

5'-Nucleotide Combination

Ammoniated glycyrrhizin has been used to potentiate the sweetness of sucrose in sucrose-containing foods, confections and beverages at levels which do not impart appreciably the licorice flavor as in U.S. Patent 3,282,706. Nevertheless, obtaining the desired degree of sweetness still requires the use of considerable, even though reduced, amounts of sucrose.

M.K. Cook; U.S. Patent 3,851,073; November 26, 1974; assigned to MacAndrews & Forbes Company reports that it is possible to use a 5'-nucleotide to repress the licorice flavor of ammoniated glycyrrhizin. The quantity required to obtain this result will vary to some degree with the particular 5'-nucleotide selected, some being more effective than others. Generally on the order of about 1%, by weight, of the 5'-nucleotide, based on the weight of ammoniated glycyrrhizin will be sufficient to repress the licorice flavor. Greater amounts of 5'-nucleotide can be utilized, but once the amount used is sufficient to repress the licorice flavor, little is to be gained by using substantially larger amounts.

Since ammoniated glycyrrhizin and 5'-nucleotides are solids, they can be prepared in finely divided form and mixed together to form a powdered product. The ammoniated glycyrrhizin salt-5'-nucleotide sweetening agent may be added to products at levels sufficient to provide the desired sweetening effect. The amount of sweetening agent required in a given product depends on a variety of factors, including the relative bitterness or tartness of the product, the presence of any undesirable flavors which must be masked, and the level of sweetness generally found acceptable in the product. An acceptable sweetening effect can be achieved with as little as 0.2% or less, or as much as 1% or more by weight, based on the weight of the sweetened product of the ammoniated glycyrrhizin salt-5'-nucleotide sweetener.

Example 1: A chewing gum was prepared in the normal manner by formulating a mixture of 22% chewing gum rubber base, 50% sorbitol solution, 25% mannitol solution, 1.8% calcium carbonate, 1% spearmint oil and 0.2% ammoniated glycyrrhizin-5'-nucleotide sweetener; all percentages being by weight, based on the weight of the chewing gum formulation. The sweetener contained 1% of the 5'-nucleotide based on the weight of ammoniated glycyrrhizin. The chewing gum was evaluated by a flavor test panel of 20 people and found to be completely free of any licorice nuance and equivalent in sweetness to a gum formulation prepared in the same manner, containing 0.1% saccharin in place of the ammoniated glycyrrhizin-5'-nucleotide sweetening agent.

Example 2: A sugar-free candy gum-drop was prepared by formulating in the usual manner a mixture of 54% gum arabic, 30.7% water, 2% USP glycerine, 12% sorbitol solution, 0.1% citric acid, 1% distilled spearmint oil, 0.2% ammoniated glycyrrhizin-5'-nucleotide sweetener and certified color additions, all percentages being by weight, based on the weight of the candy gum-drop formulation. The sweetener contained 1% of the 5'-nucleotide based on the weight of ammoniated glycyrrhizin. The gum-drop was free of any licorice nuance and equivalent in sweetness to a gum-drop prepared in the same manner containing 0.1% saccharin in place of the ammoniated glycyrrhizin-5'-nucleotide sweetening agent.

Example 3: A sugar-free dentifrice cream was prepared in the usual manner by formulating a mixture of 50% dicalcium phosphate, dental grade, 30% USP glycerine, 1% gum tragacanth, 1% sodium lauryl sulfate, 0.05% methyl parahydroxybenzoate, 1% peppermint oil, 16.7% water and 0.2% ammoniated glycyrrhizin-5'-nucleotide sweetener, all percentages being by weight, based on the dentrifrice composition. The sweetener contained 1% of the 5'-nucleotide, based on the weight of ammoniated glycyrrhizin. The dentifrice cream was free of any licorice nuance and equivalent in sweetness to a dentifrice cream formulated in the same manner containing 0.1% saccharin in place of the ammoniated glycyrrhizin-5'-nucleotide sweetening agent.

Example 4: A sugar-free, low-calorie chocolate pudding was prepared by initially forming a mixture of 15.2% Margel, 13.3% calcium gluconate USP, 7.9% tetrasodium pyrophosphate (food grade), 0.8% vanillin, 61.5% cocoa (10 to 12% fat, dutched), 0.4% anhydrous malt and 0.2% ammoniated glycyrrhizin-5'-nucleotide sweetener, all percentages being by weight, based on the weight of the pudding formulation.

The sweetener contained 1% of the 5'-nucleotide based on the weight of ammoniated glycyrrhizin. Twenty-six grams of the above mixture was well-blended with 16 ounces of cold skim milk until a smooth, lump-free pudding formed. The pudding was free of any licorice nuance and equivalent in sweetness to a pudding formulated in the same manner containing 0.1% saccharin in place of the ammoniated glycyrrhizin-5'-nucleotide sweetening agent.

JERUSALEM ARTICHOKE SWEETENERS

Basic Extraction Method

A different approach to the utilization of naturally occurring polysaccharides is given by *M. Rubin; U.S. Patent 2,782,123; February 19, 1957.* The naturally occurring polysaccharides which are proposed as the source of the sweetening agent are those obtainable from the tubers of the Jerusalem artichoke plant. One method of separating or extracting the polysaccharides is by means of extraction with hot water. The tubers of the Jerusalem artichoke, when freshly harvested, contain up to about 20% (on the wet basis) of carbohydrate materials.

The freshly harvested root exhibits poor keeping qualities as such because of enzymes naturally present, and must, therefore, be promptly utilized for extraction, or else dried. On the other hand, the tuber when dehydrated may be kept for long periods without deterioration or substantial loss of carbohydrate content.

The natural polysaccharides required for the preparation of the new sweetening agent may be separated from the tubers by heating three pounds of freshly harvested roots with 3 liters of water at close to boiling point. The extraction process is continued for three hours, when the supernatant solution is mechanically separated. The root residue is further extracted with one liter of water, and the first and second extracts are combined.

Alternatively, there may be utilized for this stage of the sweetening agent preparation process, one pound of ground dehydrated artichoke root. This eliminates the necessity for processing of the freshly harvested root within one week following its removal from the ground to avoid spoilage. Such dehydrated artichoke flour may be stored even as long as two years after its preparation, particularly when it is stored in sealed cans, at room temperature. It had been customary to add milk of lime or calcium carbonate as a buffering agent, followed by addition of oxalic acid to remove excess calcium.

The aqueous extract of artichoke root containing the polysaccharide materials, obtained as described above, serves as a starting point in the new process for the preparation of the natural sweetening agent. By suitable treatment of the extract with acids, transformations are brought about in the polysaccharide materials derived from the natural plant, which result in their conversion into potential sweetening agents of great intensity and efficiency.

The fact that the products of acid treatment, whether in the form of solution, syrup, or dried powder, are not of themselves sweet, suggests that they are not, to any significant extent, composed of monosaccharides. Their inherent sweetening power is potential and must be developed further when they are utilized as sweeteners.

It is believed that the acid treatment of the polysaccharide materials derived from the plant tubers results in the formation of fragments or partial degradation products of intermediate molecular weight, which have negligible sweetness in themselves, but which are readily convertible by heating and other means into an agent of extraordinary sweetness.

Example 1: The aqueous extract of polysaccharide materials obtained from the Jerusalem artichoke tubers and having an optical rotation to the sodium D line in a 0.5 dm polariscope tube at room temperature of +0.5°, which may be either neutral, faintly acid or faintly alkaline, is strongly acidified as by gradual addition (with stirring) of 20 ml of concentrated hydrochloric acid, to the total extract volume of 4 liters. The acidified solution is then warmed to 90°C for a period of one-half hour. During this time the optical rotation of aliquot samples of the solution changes from +0.5° to -0.6°. When this point is reached, the solution is cooled to room temperature and the excess of mineral acid is neutralized by addition of solid calcium carbonate. The neutralized solution is clarified by filtration or centrifugation and is ready for use as a sweetening agent.

The degree of transformation of the starting extract corresponded to about 30% of that which would have been possible had the reaction been permitted to go to the final completion. This was determined by separately allowing an aliquot portion of the acidified solution to continue its conversion until substantial constancy of the optical rotation reading indicated that the reaction had ceased. For the particular quantities of material utilized above, the point of completed conversion was attained at an optical rotation of -2.5°. Thus, arresting the reaction at an optical rotation of -0.6° indicated roughly that the reaction had proceeded about one-third of the way to completion. Obviously, by stopping the reaction at other stages, various degrees of conversion may be achieved.

For the purpose of utilization of the solution as a sweetening agent, the limits of utility of the conversion reaction have been established as between 25% and 85% of complete conversion. The rapidity of the conversion reaction is directly related to the acidity (pH) of the solution and to temperature, and increases as pH is lowered and temperature increased. While the most rapid and optimum results were obtained when the pH was lowered to from 1 to 2, the process can also be advantageously carried out at higher pH values ranging up to 5. At these higher values, however, the reaction proceeds slowly. Thus, at a pH of 4 brought about by the addition of 85% phosphoric acid, the conversion reaction for the solution used in the foregoing example required about two hours.

It is preferable to use as acidifying agents those acids whose anions can be removed after hydrolysis by formation of insoluble compounds upon neutralization, as for example, calcium compounds. The use of phosphoric acid is advantageous in providing a buffering action, as well as desirable phosphate ions, which may be useful in food fortification.

As a neutralizing agent and for purposes of arresting the reaction, there may be used in place of the calcium carbonate other alkalis and alkaline salts, such as lime, dilute sodium hydroxide or potassium hydroxide solutions. By the use of stoichiometric quantities of base as calculated from the amount of acid used initially, it is possible to eliminate final filtration of the hydrolyzed solution.

In the use of soluble alkalis for neutralization, attention must be given to the possibility of added salt content attributable to the neutralization. In some instances, the enhanced sweetening effect of the solutions prepared may be up to as much as 25 times the sweetness of sucrose in actual use. It may be noted that the sweetening agent thus produced is not particularly sweet in itself but only exhibits its sweetening effect on heating or equivalent processing in connection with food processing.

Concentration by evaporation by conventional methods may be used to obtain a syrup especially suitable for use in the preparation of jams and jellies. Alternatively, the concentration process may be carried on during the conversion operation by raising the temperature of the mixture to the boiling point, or by lowering the pressure in order to maintain the temperature of the conversion mixture at a predetermined temperature.

The converted product may be obtained in the form of a powder by application of spray drying techniques. In this form, the sweetening agent has particular utility as an ingredient in dry mix baking preparations. The preparation of preformed cake and biscuit mixes by this means has proved especially useful and important. The application of the sweetening agent in the development of sweetness in situ in processed foods and baked goods is illustrated by the following example.

Example 2: A solution of 30% converted polysaccharide, equivalent to one-half pound of dehydrated Jerusalem artichoke root or to three pounds of freshly harvested root was added to the following baking mixture: 60 lb high gluten flour, 75 lb whole eggs, 40 lb cottonseed oil and 20 lb water. The mixture was baked in the usual manner. The resulting product was a pleasantly sweet, even flavored, well-textured biscuit. To obtain a comparable degree of sweetness in a mixture using the same quantities listed above required (in a separate experiment) the addition of 25 lb of sugar in place of the extract of ½ lb of dried root powder, above described. Alternatively, a comparable sweetening effect using a synthetic sweetener would require, for example, the addition of 1¼ oz of saccharin to the mix.

A matter of considerable significance is the negligible change in caloric value brought about by the addition of the sweetening agent of this process. By total combustion it was determined that 100 grams of product baked as described had a caloric value of 586. The same baking mixture without the addition of the sweetening solution had a caloric value of 578 per 100 grams. The difference due to the addition of the sweetening agent is thus minimal.

The products may be used for the sweetening of any processed food providing a substantial amount of heat is used in the processing. As for example, the sweetening products may be used in the manufacture of baked goods, candy and confectionery; in the preservation of fruits, and for cooking purposes. They are of great value in the sweetening of heat processed foods for diabetics. Depending upon the nature of the end use of the sweetening agent, varying amounts of converted extract, and extracts representing varying degrees of conversion, may be used. For materials that may be subjected to only a moderate degree of heating, such as fruit prepared for canning in diabetic foods, it has been found best to utilize a more highly converted preparation of the order of 80% of total conversion.

Artichoke Bread Flour

Artichoke flour is used to prepare diet bread in the process of *H.J. Schaefer and J.W. Tintera; U.S. Patent 3,497,360; February 24, 1970.* One of the important features which makes bread palatable and desirable as a food is the leavening which is usually provided by yeast. The fermentation of the yeast renders the gluten of the flour more elastic and changes some of the insoluble proteins of the gluten into soluble forms. The flavor, color and texture of the bread are all controlled, in large measure, by the action of the yeast. The most important basic substances which are consumed in the yeast fermentation are the simple or monosaccharide sugars. In order to provide this constituent in the bread dough for the growth of the yeast, sugar must be added to the dough mix, or reliance must be placed upon conversion of wheat starch to sugar by a constituent of the starch called diastase. This conversion of the starch to sugar can be enhanced by the addition of malt to the dough mix, since the malt contains a high concentration of diastase. It is common to add sugar or malt, or both, to bread mixes to satisfy this requirement.

Unfortunately, the monosaccharide sugars which must be present, or produced, in the bread dough for consumption by the yeast in the leavening process are the very sugars which are rapidly absorbed by the body of the consumer and which, therefore, provide rapid changes in blood sugar.

The main constituents of white wheat flour (patent flour) are gluten and wheat starch. The gluten is primarily a protein material which has coherent and adherent properties which tend to hold together the bread dough and the finished loaf. The wheat starch is primarily carbohydrate material, which provides the carbohydrate nutrients. These wheat flour constituents are both important in providing leavened bread which has the appearance and flavor appealing to consumers.

Many of the high-protein and low-carbohydrate food substances which are recognized as valuable for those who have carbohydrate metabolism difficulties, and which are available in the form of flours which could be incorporated into bread, are not particularly palatable or desirable in flavor when incorporated in bread.

One of the most important features of the process is that a yeast leavened bread can be produced successfully from a dough mix in which the wheat starch comprises as little as 8% of the total dry ingredients of the dough by weight. When the wheat starch is present in this low proportion, it is very substantially consumed by the yeast in the leavening process so that there is very little wheat starch remaining in the leavened loaf of bread after baking. The wheat starch may be present initially in a proportion up to approximately 14% by weight. However, it is preferably kept to an absolute minimum, which appears to be approximately 8%. In order to make the bread dough rise sufficiently with so little wheat starch available to "feed" the yeast, a high ratio of yeast to wheat starch is used. This ratio is preferably from one-third to two-thirds. This refers to commercially available compressed yeast.

In bread-making processes, it has been common not only to add sugar as a yeast food, but also to add various other materials, such as malts, which are effective in converting wheat starch to sugar so that it may be directly effective as a food for the yeast. In accordance with this process, it is particularly desirable to avoid any conditions or any constituents which may cause the presence of residual sugar in the finished bread product.

Malts and other so-called yeast "foods" which promote the conversion of wheat starch to sugar are likely to do just this. These materials enhance the action of the yeast simply by accelerating the conversion of wheat starch to sugar so that there is plenty of sugar available as a direct food for the yeast.

The use of malts and other yeast food materials which enhance the conversion of wheat starch to sugar is avoided and, instead, a high proportion of yeast is used in order to obtain the desired leavening action by consuming virtually all of the sugar which is available from the normal conversion of wheat starch to sugar. Such conversion is due to the normal content of a substance called diastase in the wheat flour, which is present in the mix without the addition of malt.

Stated another way, when malts and other materials of a similar nature are added to the composition, the objective is generally to obtain sufficient leavening of the bread with a smaller amount of initial yeast added to the mix. In those instances, with the malt present, the conversion of wheat starch to sugar is likely to outrun the consumption of that sugar by the yeast in the leavening process so as to result in a higher proportion of residual sugar in the finished product. No sugars are added to the composition to enhance the fermentation action of the yeast. Basically, the only carbohydrate present which supports the growth of the yeast in any substantial way is the wheat starch.

In order to obtain a satisfactorily light loaf with so little carbohydrate available to sustain the leavening action of the yeast, the leavening must be carried out quickly and the bread promptly baked after the fast-leavening process before the gases have a chance to escape and the bread has a chance to fall. Generally, the proofing time is only about thirty minutes, or until the bread mix has approximately doubled in volume.

Example 1:

Ingredients	Percent of Dry Ingredients
4 lb wheat flour (gluten)	17.1
2 lb wheat flour (vital gluten)	8.5
12 lb oat flour	51.5
2 lb soy flour	8.5
1½ lb Jerusalem artichoke flour	6.4
9 oz salt	2.4
¾ oz sodium propionate	0.3
1 lb 4 oz yeast	5.3
9 qt water	

All of the ingredients except the water are mixed together in a dry blend. Immediately after dry blending, the water is added at lukewarm temperature and mixed in for about seven or eight minutes. A standard four-speed bread-mixing machine is used. It is run on low speed for about six minutes and then at second speed for about two minutes. After mixing, the dough is scaled into 1 lb 2 oz loaf sizes and then proofed for about thirty minutes in steam so that the volume of the scaled dough approximately doubles to fill the pan.

The bread is then baked at 360°F for a maximum of about 55 minutes. The baking is carried out without steam. With the quantities given above, 36 loaves of bread are obtained. They are very palatable and appetizing in appearance and can be tolerated in surprising quantities by individuals having serious carbohydrate metabolism difficulties.

Example 2: A prepackaged dry mix for the production of yeast bread in accordance with the process is packaged as follows: A first inner package is provided containing the following ingredients:

	Parts
Wheat flour (gluten)	34
Wheat flour (vital gluten)	17
Oat flour	103
Soy flour	17
Jerusalem artichoke flour	13
Salt	5

A second inner package is provided containing 10.7 parts of compressed yeast. A complete package of a typical size has a total weight of approximately two pounds. The instructions for use of the mix provide for adding one and one-half pints of lukewarm water, mixing into a dough, and scaling into three loaves in separate pans. The individual loaves are then proofed for a period sufficient to allow the dough to rise to fill the pan, and then the breads are baked in a dry oven for about fifty minutes at 360°F.

The prepackaged dry mix of Example 2 is convenient for sale as a "mix" product from grocery shelves and diet food stores for the preparation of conveniently small family-sized batches of bread in accordance with the process. Thus, fresh bread can easily be available to every household of a diabetic or hypoglycemic, whether or not such bread is available from local bakeries.

The Jerusalem artichoke flour referred to in the above-mentioned examples is a dehydrated product which consists of pulverized tuberous roots of the Jerusalem artichoke plant. This material is commercially available under the name "American Jerusalem Artichoke Flour." It is well-known for its high content of a non-nutritive sweetener called inulin.

Jerusalem artichoke flour is incorporated for the purpose of imparting a sweet and palatable flavor to the bread without using sugars, which are readily assimilated by the body. It is believed that the Jerusalem artichoke flour also contains various beneficial nutrient substances, although the caloric value of the conventional proteins, fats, and carbohydrates which are available for assimilation by the body are virtually nil.

Chlorogenic Acid Inducer

Chlorogenic acid and cynarine, components of artichoke extract, are claimed to be sweetness inducers as described by *C.-H. Lee, R.J. Scarpellino and M.M. Murtagh; U.S. Patent 3,916,028; October 28, 1975; assigned to General Foods Corp.*

It is known that many acids, for example, citric acid, sulfuric acid and hydrochloric acid, exhibit after tasting a sweetness-inducing quality after the initial sourness has subsided. However, where such acids undergo structural modification, they are known to lose their sweetness-inducing properties. The drawbacks to using chemically unmodified acids at the levels required for the sweetness-inducing properties to become evident are obvious. Firstly, the initial sour taste with which one must contend is undesirable in many food systems and, in most cases, is difficult to disguise. More important, however, is the fact that the required levels of such acids as hydrochloric and sulfuric are physiologically intolerable and, therefore, cannot be considered for use in edible systems. In contradistinction, chlorogenic acid and cynarine do not induce sweetness of themselves as is common with other acids. However, when these particular acids are structurally modified, in this case as a salt, they do exhibit the sweetness-inducing property.

The compounds are extremely stable as well as being very soluble in both aqueous and nonaqueous media such as glycerol, propylene glycol and 1,3-butylene glycol. This makes the salts of these compounds ideal for incorporation into systems such as chewing gum, where a longer-lasting and pleasant sweetness is desired. One may also conceive of such compounds being incorporated in freeze-dried form into a tablet or lozenge which would not only cause a sweet sensation to be experienced after having allowed the medicinal lozenge to dissolve in the mouth, but would effectively mask any unpleasant taste of and impart a sweet taste to nonsweet foodstuffs, beverages and pharmaceuticals.

Essentially, the compositions are readily prepared by suspending the unsubstituted acid in water, after which a sufficient amount of alkali substance or amine is added to solubilize the acid, the pH being subsequently adjusted to a range of between four and eight in order to abolish the sour taste of the acid. As little as 0.3 to 0.5% of the acid itself is needed in some instances for the induction of sweetness to be observed. However, this is in the case of more sensitive tasters, the average individual having a detection range between 1 to 3%.

When the salts of these acids are to be utilized in dry form for incorporation into solid systems, the solution of the acid salt is preferably freeze-dried. The compositions are readily oxidized by air. Consequently, spray-drying and drum-drying tend to lessen the activity of the compositions due to the slight oxidation that takes place during these processes. The freeze-dried forms of the compositions, however, maintain their activity and are stable for an indefinite length of time at room temperature.

It has been determined that the longer one keeps a compound in the mouth, the more intense the sweetness induced. This finds significant utility in foodstuffs such as chewing gum, where the problem of rapidly decreasing sweetness is not only alleviated but a product having a longer-lasting and possibly increasing sweetness intensity is derived.

Example 1: Chlorogenic acid is extracted from artichokes as follows: 250 g (9 oz) of frozen artichoke hearts are boiled in 60 ml of water for ten minutes. To the entire mixture is added 100 ml of water, and it is transferred to a Waring blender. After blending it at high speed for five minutes, the suspension is transferred into a beaker and 150 ml of ethanol is added.

After one hour of stirring, the suspension is centrifuged at 10^4 rpm for 30 minutes. The volume of the supernatant is then reduced in vacuo at 50°C until all ethanol evaporates. The solution is freeze-dried to give 11 grams of dry powder, which possesses sweetness-inducing properties.

Example 2: Two grams of cynarine was suspended in 50 ml of water. Approximately 0.5 grams of potassium bicarbonate was added to solubilize the acid, after which the pH was adjusted to 6.0. Water was added to the solution in order to bring the final volume to 100 ml.

A small amount of the salt solution was placed in the mouth for approximately one minute. After expelling this solution from the mouth, milk was subsequently sampled and was found to taste intensely sweet.

Example 3: Three grams of chlorogenic acid was suspended in 70 milliliters of water and enough potassium hydroxide added to solubilize the acid and to adjust the pH to 5 to destroy the sour taste of the acid. Enough water was subsequently added to bring the final volume to 100 milliliters. A small amount of this solution was placed in the mouth for a period of one-half minute. After expelling this solution, water was taken into the mouth and found to taste intensely sweet.

DIPEPTIDES

ASPARTIC ACID ALKYL ESTERS

The sweeteners discussed in this chapter are extracted from plant materials, but additional chemical modifications are necessary to improve the quality and intensity of the sweetening characteristics. In many instances, chemical modification is necessary to obtain pure sweetness in place of bitter or disagreeable off-taste.

L-Aspartyl-L-Tyrosine Methyl Ester

R.H. Mazur, A.H. Goldkamp and J.M. Schlatter; U.S. Patent 3,475,403; Oct. 28, 1969; assigned to G.D. Searle & Co. give a coupling reaction of an aspartic acid derivative with an appropriate amino acid ester to form a dipeptide. The dipeptide derivatives possess a sweet taste. Aspartic acid containing dipeptide lower alkyl esters are represented by the following structural formula:

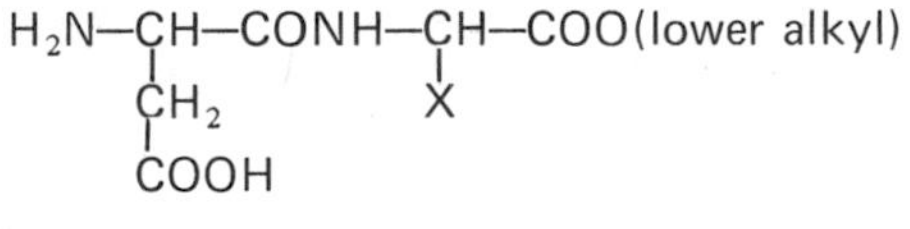

$$H_2N{-}\underset{\underset{\displaystyle COOH}{|}}{\underset{|}{CH_2}}\!\!\!\!\!\!\!\!\overset{}{CH}{-}CONH{-}\underset{\underset{\displaystyle X}{|}}{CH}{-}COO(\text{lower alkyl})$$

where X is one of the following radicals

$CH_2{-}C_6H_4{-}OR$ (para-substituted benzene ring) R = H or lower alkyl

$\underset{\underset{\displaystyle OH}{|}}{CH}{-}CH_3$

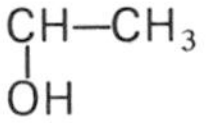

$(CH_2)_nS(O)_m(\text{lower alkyl})$ n = 0, 1, 2; m = 0, 2

The compounds are produced by first contacting a suitable aspartic acid derivative, where protecting groups have been attached to the amino and β-carboxy groups and the α-carboxy group has been converted to a reactive ester function, with the appropriate amino acid ester. The protecting groups are then removed by suitable means from the aspartic acid portion of the resulting intermediate product. When the amino group is protected by a benzyloxycarbonyl function and the β-carboxy group by a benzyl ester moiety, catalytic hydrogenation is a preferred method for effecting removal.

These processes are specifically illustrated by the reaction of N-benzyloxycarbonyl-L-aspartic acid α-p-nitrophenyl, β-benzyl diester, with L-tyrosine methyl ester to yield β-benzyl-N-benzyloxycarbonyl-L-aspartyl-L-tyrosine methyl ester, which is hydrogenolyzed in aqueous acetic acid, utilizing palladium black catalyst, to afford L-aspartyl-L-tyrosine methyl ester.

The dipeptides characterized by a sulfone group are obtained by reaction of the protected aspartic acid derivative with the appropriate amino acid ester containing the sulfone group. The latter substances are conveniently produced by esterification of the corresponding amino acid sulfones. For example, L-methionine sulfone is converted to its methyl ester hydrochloride by reaction with a solution of thionyl chloride in methanol and the free ester is allowed to react with N-benzyloxycarbonyl-L-aspartic acid α-p-nitrophenyl, β-benzyl diester to yield β-benzyl-N-benzyloxycarbonyl-L-aspartyl-L-methionine methyl ester sulfone, which is hydrogenolyzed with palladium to L-aspartyl-L-methionine methyl ester sulfone.

The dipeptide sweetening agents are water-soluble, stable substances which can be utilized in a variety of physical forms, e.g., as powders, tablets, syrups, etc. Liquid or solid carriers such as water, glycerol, starch, sorbitol, salt, citric acid and other suitable nontoxic substances can also be used.

It has been determined that the property of sweetness is affected by the stereochemistry of the individual amino acid units comprising the dipeptide structure. The L-L isomers, for example, L-aspartyl-L-tyrosine methyl ester, are especially sweet. It is thus apparent that mixtures containing the L-L isomers, i.e., DL-DL, L-DL or DL-L share that property also.

The sweetening agents are particularly useful to diabetics as substitutes for sugar. They are, moreover, lacking in the unpleasant aftertaste exhibited by such synthetic sweeteners as saccharin. The absence of toxic properties results from their derivation from natural sources, i.e., the naturally occurring amino acids utilized by the animal body in the manufacture of essential proteins.

An example of a typical sweetened orange soda is shown by the following preparation. A stock supply of bottler's syrup is prepared by mixing 5.5 ml of a 50% aqueous citric acid solution with 150 ml of water, dissolving 2 grams of L-aspartyl-L-tyrosine methyl ester in that solution, adding successively 7.02 ml of the orange flavor base and 2.7 grams of sodium benzoate and diluting that mixture to 200 milliliters with water. One ounce samples of the bottler's syrup are transferred to 6 ounce bottles and 110 ml of cold tap water is added to each bottle. To each bottle 42 ml of cold charged bottling water (5 volumes carbon dioxide) is then added to achieve carbonation. Each bottle is capped and the contents mixed. Comparison of the latter samples with orange soda containing a quantity of sucrose 50 times that of the named dipeptide derivative reveals no detectable difference in sweetness.

Example 1: A mixture containing 10.05 parts of N-benzyloxycarbonyl-L-aspartic acid α-p-nitrophenyl, β-benzyl diester, L-tyrosine methyl ester and 45 parts of ethyl acetate is stored at about 65°C for approximately 24 hours, then is cooled and washed successively with 50% aqueous potassium carbonate, water, dilute hydrochloric acid and water. Drying of that solution over magnesium sulfate followed by distillation of the solvent under reduced pressure affords a gummy residue, which is purified by crystallization from ether-ethyl acetate to afford β-benzyl-N-benzyloxycarbonyl-L-aspartyl-L-tyrosine methyl ester, melting at 125° to 127.5°C and exhibiting an optical rotation, in methanol, of -5°.

Example 2: To a solution of 20 parts of β-benzyl-N-benzyloxycarbonyl-L-aspartyl-L-tyrosine methyl ester in 250 parts of 75% aqueous acetic acid are added 2 parts of palladium black and the resulting mixture is shaken with hydrogen at atmospheric pressure and room temperature until the uptake of hydrogen ceases. The catalyst is removed by filtration and the filtrate is concentrated to dryness under reduced pressure. The resulting residue is crystallized from aqueous ethanol to afford L-aspartyl-L-tyrosine methyl ester, which exhibits a double melting point at about 180° to 185°C and 230° to 235°C with decomposition. This compound displays an optical rotation, in water, of +4° and is represented by the following structural formula:

$$H_2NCH{-}CONHCHCOOCH_3$$

(first CH bears CH_2–COOH; second CH bears CH_2–C$_6$H$_4$–OH, para-hydroxyphenyl)

L-Aspartyl-L-Phenylalanine Esters

Another series of peptides was developed by *J.M. Schlatter; U.S. Patent 3,492,131; January 27, 1970; assigned to G.D. Searle & Co.* The sweetening property of the dipeptide substances is dependent also upon the stereochemistry of the individual amino acids, i.e., aspartic acid and phenylalanine, from which the dipeptides are derived. Each of the amino acids can exist in either the D or L form, but it has been determined that the L-aspartyl-L-phenylalanine esters are sweet while the corresponding D-D, D-L and L-D isomers are not. Combinations of isomers which contain the L-L dipeptide, i.e., DL-aspartyl-L-phenylalanine, L-aspartyl-DL-phenylalanine and DL-aspartyl-DL-phenylalanine are sweet also.

The L-aspartyl-L-phenylalanine methyl ester is 100 to 200 times as sweet as sucrose. The corresponding ethyl ester is 25 to 50 times as potent as sucrose. These dipeptide esters furthermore do not result in the unpleasant aftertaste characteristic of synthetic sweeteners such as saccharin and cyclamate. In consequence of their derivation from natural sources, i.e., naturally occurring amino acids, these sweeteners are devoid of toxic properties.

The dipeptide esters are conveniently manufactured by methods suitable for the coupling of amino acids. The starting material is the aspartic acid derivative wherein the amino function is protected by a benzyloxycarbonyl group and the β-carboxy function by a benzyl ester group, and the α-carboxy group is converted to a p-nitrophenyl ester function. When that substance is allowed to react with a phenylalanine ester, displacement of the more reactive p-nitrophenyl ester group occurs to afford the protected dipeptide of the following formula:

```
CONH—CHCOOR
|      |
|      CH2—C6H5
|
CH—NHCOOCH2—C6H5
|
CH2COOCH2—C6H5
```

where R is a lower alkyl radical. Removal of the N-benzyloxycarbonyl and O-benzyl protecting groups is conveniently effected by hydrogenolysis at atmospheric pressure and room temperature, utilizing palladium as the catalyst. Those processes are specifically illustrated by the reaction of N-benzyloxycarbonyl-L-aspartic acid α-p-nitrophenyl, β-benzyl diester with L-phenylalanine methyl ester to afford β-benzyl-N-benzyl-oxycarbonyl-L-aspartyl-L-phenylalanine methyl ester and hydrogenolysis of that intermediate in aqueous acetic acid with palladium metal catalyst to produce L-aspartyl-L-phenylalanine methyl ester.

Example: A solution of 88.5 parts of L-phenylalanine methyl ester hydrochloride in 100 parts of water is neutralized by the addition of dilute aqueous potassium bicarbonate, then is extracted with approximately 900 parts of ethyl acetate. The resulting organic solution is washed with water and dried over anhydrous magnesium sulfate. To that solution is then added 200 parts of N-benzyloxycarbonyl-L-aspartic acid α-p-nitrophenyl, β-benzyl diester, and that reaction mixture is kept at room temperature for about 24 hours.

The reaction mixture is cooled to room temperature, diluted with approximately 390 parts of cyclohexane, then cooled to approximately -18°C in order to complete crystallization. The resulting crystalline product is isolated by filtration and dried to form β-benzyl-N-benzyloxycarbonyl-L-aspartyl-L-phenylalanine methyl ester, melting at about 118.5° to 119.5°C.

The use of these esters is described by *J.M. Schlatter; U.S. Patent 3,642,491; February 15, 1972; assigned to G.D. Searle & Co.*

Example 1: Carbonated Orange Soda – A stock supply of bottler's syrup is prepared by mixing 5.5 ml of a 60% aqueous citric acid solution with 150 ml of water, dissolving 2 grams of L-aspartyl-L-phenylalanine methyl ester in that solution, adding successively 7.02 ml of orange flavor base and 2.7 grams of sodium benzoate, then diluting that mixture to 200 ml with water. One ounce samples of the bottler's syrup are transferred to 6 ounce bottles and 110 ml of cold tap water is added to each bottle. To each bottle, 42 ml of cold charged bottling water (5 volumes carbon dioxide) is then added to achieve carbonation. Each bottle is capped and the contents mixed.

Example 2: Gelatin Dessert – A sample is prepared from 2.07 grams of plain gelatin, 0.34 gram of imitation raspberry flavoring, 0.34 gram of citric acid, 14.41 grams of lactose and 0.05 gram of L-aspartyl-L-phenylalanine methyl ester. The ingredients are combined and dissolved in 82.79 ml of boiling spring water. Thereafter, the solution is poured into dishes and chilled to set.

Example 3: Heat-Treated Peach Pack – Fresh peaches, after washing, removal of the pits and slicing into pieces of desired size (fruit content equals 40 to 80% by weight of the resulting puree), are added to a syrup prepared from 0.834 gram of L-aspartyl-L-phenylalanine methyl ester per 10 ounce can of puree. The mixture is then acidified by the addition of citric acid such that the ratio between the acid and the sweetness content is 1:30. The prepared puree is then packed cold in 10 ounce cans. Prior to sealing the container, the product undergoes a steam exhaustion for 5 minutes. After sealing, the cans are treated for 15 minutes at 100°C.

Combinations of the dipeptide sweetening agents with sugar or synthetic sweeteners such as saccharin likewise can be incorporated into the consumable materials.

L-Aspartyl-L-(β-Cyclohexyl)Alanine

The dipeptide is reported by *H. Gregory; British Patent 1,233,216; May 26, 1971; assigned to Imperial Chemical Industries Limited, England* to be 360 times sweeter than sucrose. It has the formula

$$
\begin{array}{ccc}
CH_2COOH & & CH_2\text{—}C_6H_{11} \\
| & & | \\
H_2N\text{—}CH\text{—}CO\text{—}NH\text{—} & & CH\text{—}COOR
\end{array}
$$

where R stands for an alkyl radical of 1 to 3 carbons, or a hydroxyalkyl radical of up to 3 carbons, or a salt of these, together with a nontoxic, orally-acceptable diluent or carrier. Since the optical isomers of lower alkyl esters of aspartyl-β-cyclohexylalanine other than the L-L isomers are not bitter, these may be present as part of the nontoxic, orally-acceptable diluent or carrier.

A suitable value for R is, for example, a methyl radical, and suitable salts are, for example, acid addition salts, for example hydrochlorides, or alkali metal salts, for example sodium salts. The pharmaceutical compositions may be in the form of tablets or capsules, or, more especially, in the form of solutions or suspensions intended for oral administration. The foodstuff composition may be in the form of a solid or a liquid food, and a preferred form of foodstuff composition is a so-called soft drink.

L-aspartyl-L-(β-cyclohexyl)alanine methyl ester was compared with sucrose by determining the concentration of an aqueous solution whose sweetness was the same as a standard aqueous solution of sucrose. Using this test this ester was found to possess about 360 times the sweetening power of sucrose. In the presence of the acidulant component of a soft drink, essentially citric acid, the above compound was found to be 500 times as sweet as sucrose.

Example: Aqueous solutions of L-aspartyl-L-(β-cyclohexyl)alanine methyl ester were prepared containing respectively 0.05 and 0.01% by weight per volume of the dipeptide ester. A panel of tasters could all detect the 0.05% solution as sweet, and half of the tasters could detect the 0.01% solution as sweet. The same tasters could all detect a 5% aqueous sucrose solution as sweet, but most of the tasters could not detect sweetness in a 1% aqueous sucrose solution. On this basis, L-aspartyl-L-(β-cyclohexyl)alanine methyl ester possesses between 100 and 500 times the sweetening power of sucrose.

α-L-ASPARTYL DERIVATIVES

Isobutyl and Isopropyl Esters

Y. Ariyoshi, N. Yasuda and T. Yamatani; U.S. Patent 3,920,626; November 18, 1975; assigned to Ajinomoto Co., Inc., Japan claim that the aqueous solution of α-L-aspartyl-O-isobutyryl-L-serine methyl ester is about 50 times as sweet as that of sucrose. The corresponding O-propionyl derivative is about 40 times as potent as sucrose. α-L-aspartyl-β-alanine-i-propyl ester in aqueous solution in comparison is about 6 to 8 times as sweet as sucrose.

The dipeptide esters do not have the unpleasant aftertaste characteristic of known synthetic sweeteners such as saccharin. When these esters and saccharin are used together, the unpleasant taste due to the latter is much improved, and furthermore, the sweetening potency of the latter is increased geometrically rather than arithmetically. The dipeptide esters are water-soluble substances which can be prepared in a variety of forms suitable for utilization as sweetening agents. Typical forms are tablets, powders, suspensions, solutions and syrups.

The dipeptide esters are produced by known methods of preparing aspartyl peptides. For example, the esters are produced by condensing an aspartic acid derivative where the amino group is masked by a protecting group such as carbobenzoxy, p-methoxycarbobenzoxy, t-butyloxycarbonyl, and formyl, and the β-carboxy group is masked by esterification with a suitable alcohol such as benzyl and t-butyl alcohol, with an alkyl ester of amino acid or a derivative thereof, or with a hydrohalide salt of the other amino acid ester in the presence of an equivalent amount of a base.

The condensation can be carried out by: (1) converting the α-carboxy group of the aspartic acid derivative to an activated ester with p-nitrophenol, pentachlorophenol, N-hydroxysuccinimide, chloroacetonitrile or the like; (2) using as condensing agents carbodiimides like N,N'-dicyclohexylcarbodiimide, carbonyldiimidazoles and isonitriles like i-propylisonitrile; or (3) using an agent capable of preparing mixed anhydrides such as ethyl chloroformate and i-butyl chloroformate. The esters are also prepared by contacting an N-protected aspartic anhydride with an alkyl ester of the other amino acid. The N-protecting groups are easily removed from the produced intermediates by hydrogenation in the presence of palladium-carbon catalyst or mineral acid treatment. When an N-protected aspartic anhydride, is used the β-isomer which has no sweet taste is produced simultaneously with the desired α-isomer. However, pure α-isomer may be readily obtained from the crude product by recrystallization because the amount of the α-isomer produced is much larger than that of the β-isomer. Furthermore, the dipeptide esters may be produced by reacting a strong acid salt of aspartic anhydride such as the benzenesulfonate, hydrochloride and hydrobromide, with an alkyl ester of the other amino acid or derivative.

The serine unit in the dipeptide ester can be acylated before or after the formation of the peptide bond. Acylation may be achieved by reacting acid chlorides and acid anhydrides with the dipeptide ester, or by reacting a serine ester derivative where the amino group is protected by carbobenzoxy, t-butyloxycarbonyl or formyl, with an acylating agent and then using the O-acyl serine ester for the formation of the peptide bond after removing the N-protecting group.

Example: Coffee – Enough α-L-aspartyl-O-isobutyryl-L-serine methyl ester was added to hot black coffee so that the amount of the dipeptide ester was 0.05% of the solution by weight. The coffee had a sweet taste similar to that produced by sucrose. Substitution of α-L-aspartyl-β-alanine isopropyl ester (1% by weight) in the above procedure resulted in the coffee having a pleasant aftertaste.

Gelatin Dessert – After 14 grams of plain gelatin powder were added to 360 ml water and dissolved by heating for 5 minutes, 2.5 grams of α-L-aspartyl-DL-β-aminobutyric acid methyl ester were added to the solution together with small amounts of imitation vanilla flavoring and coloring agent. The solution was poured into dishes and chilled to set. The obtained gelatin dessert was savory.

Imitation Ice Cream – Ice cream (butter fat content 4%, total fat content 8%) was prepared in a conventional manner from the following substances:

Skim milk powder, g	129
Unsalted butter, g	73
Shortening, g	60
Dextrose, g	75
Glucose syrup solids, g	30
Emulsifying agent, g	4
Stabilizing agent, g	4
Water, g	945
Imitation vanilla flavor, ml	4
α-L-aspartyl-O-isobutyryl-L-serine methyl ester, % by weight	0.5

A palatable and flavory ice cream was obtained.

Hexahydrophenylalanine Esters

Illustrations of the preparation of sweetened products using α-L-aspartyl-L-hexahydrophenylalanine to replace sucrose are given by *J.M. Schlatter; U.S. Patent 3,800,046; March 26, 1974; assigned to G.D. Searle & Co.*

Coffee: To rehydrated hot brewed coffee was added a sample of α-L-aspartyl-L-hexahydrophenylalanine methyl ester until the content of the dipeptide in the solution reached 0.033%. Upon comparison with similarly compared solutions of coffee sweetened with sucrose it was found that to achieve the sustained degree of sweetness a 4% solution of sucrose was required. Thus, in black coffee the dipeptide exhibited a sweetness potency of 150 times that of sucrose.

Powdered Beverage Concentrate: The powder was prepared by mixing 0.05 part of citric acid, 0.04 part of imitation strawberry flavoring, 0.090 part of α-L-aspartyl-L-hexahydrophenylalanine methyl ester, and 0.0609 part of lactose. The powder was then dissolved in 100 parts of spring water, and the resulting beverage was evaluated at room temperature. The sample was compared with a similar sample, prepared as indicated above except that 9 parts sucrose and 0.87 parts dextrose were substituted for the dipeptide ingredient. Upon tasting it was determined that the two samples achieved the same degree of sweetness and hence it was concluded that in powdered concentrates of this type, the dipeptide exhibited 125 times the sweetness potency of sucrose.

Carbonated Orange Soda: A stock supply of bottler's syrup is prepared by mixing 5.5 ml of a 50% aqueous citric acid solution with 150 ml of water, dissolving 2 grams of α-L-aspartyl-L-hexahydrotyrosine methyl ester in that solution, adding successively 7.02 ml of orange flavor base and 2.7 grams of sodium benzoate and diluting that mixture to 200 ml with water. One ounce samples of that bottler's syrup are transferred to six ounce bottles and 100 ml of cold tap water is added to each bottle. To each bottle 42 ml of cold charged bottling water (5 volumes carbon dioxide) is then added to achieve carbonation. Each bottle is capped and the contents mixed. Comparison of the latter samples with orange soda containing the quantity of sucrose 50 times that of the named dipeptide derivatives reveals no detectable difference in sweetness.

Sweetening Solution Formulation: 1.0 gallon of distilled or deionized water is warmed to 160° to 180°F and 0.35 ounce of benzoic acid and 0.175 ounce of methyl p-hydroxybenzoate are then added. After these preservatives are dissolved, 1.0 gallon more of distilled or deionized water is added. The solution is then brought to room temperature. 0.3 pound of α-L-aspartyl-L-hexahydrophenylalanine methyl ester is then added. Distilled or deionized water is added to bring the volume to 2.5 gallons. Each teaspoon of the sweetening solution is equivalent to about 1.6 teaspoons of sugar.

Milk Pudding: 1.14 ounces of the ingredients below was added to 2 cups of cold skimmed milk in a bowl. The mixture was slowly stirred with an egg beater to disperse the powder, then rapidly mixed until a smooth texture was obtained. It was then allowed to set.

Alginate	6 pounds
Cocoa	30 pounds
Tetrasodium pyrophosphate, anhydrous	3 pounds
Salt	1 pound 8 ounces
α-L-aspartyl-L-hexahydrophenylalanine methyl ester	1 pound 2 ounces
Vanillin	4.8 ounces

Preserves (100 Pound Batch):

Fruit	55 pounds
α-L-aspartyl-L-hexahydrophenylalanine methyl ester	1.5 ounces
Pectin	1 pound
Potassium sorbate	1 ounce
Water	5 gallons 1 pint

Dietetic Syrup:

	Percent
α-L-aspartyl-L-hexahydrophenylalanine methyl ester	0.30
Carboxymethylcellulose	0.50
Pectin	1.60
Flavor	8.50
Citric acid, anhydrous	2.00
Color	0.30
Sodium benzoate	0.10
Water	86.70

Dry Gelatin Mix: A sample is prepared from 2.07 grams of plain gelatin, 0.34 gram of imitation raspberry flavoring, 0.34 gram of citric acid, 14.41 grams of lactose and 0.10 gram of α-L-aspartyl-L-hexahydrophenylalanine methyl ester. Those materials are dry blended and packaged to form a dry gelatin mix package. At the time the gelatin dessert is prepared, the dry content of the package is combined with about 83 ml of boiling spring water. Thereafter the solution is poured into dishes and chilled to set.

It has been determined that the sweetening property of the dipeptide substances is dependent upon the stereochemistry of the individual amino acids, e.g., aspartic acid, phenylalanine, tyrosine and tyrosine O-alkyl esters from which the peptides are derived. Each of the amino acids, can exist in either the D or L form, but it has been determined that the L-L isomers, e.g., α-L-aspartyl-L-hexahydrophenylalanine ester derivatives, are especially sweet while the corresponding D-D, D-L, and L-D isomers are not. Moreover, mixtures containing the L-L isomers, i.e., DL-DL, L-DL or DL-L share that property of sweetness also.

Aminosuccinamic Acid Esters

Although no valid theory exists for predicting sweetness of chemical compounds, R.W. Moncrieff, *The Chemical Senses,* L. Hill Ltd., London, 1951 has proposed some generalizations for sweeteners as follows: (1) polyhydroxy and polyhalogenated aliphatic compounds are generally sweet; (2) α-amino acids are usually sweet, but the β or δ amino acids are not, and the closer together the amino group and the carboxy group, the greater the sweetness; (3) on ascending a homologous series, taste changes from sweet to bitter, and taste and water solubility disappear simultaneously; (4) alkylation of an amino or amido group often gives a sweet tasting compound; (5) one nitro group in a molecule often gives a sweet taste; (6) some aldehydes are sweet, ketones are never sweet.

On the basis of this theory, *M. Lapidus and M.M. McGettigan; U.S. Patent 3,814,747; June 4, 1974; assigned to American Home Products Corporation* synthesized the following compounds with the L-aspartic acid configuration. The patent gives methods of synthesis of these compounds.

N-(L-α-carboxyphenethyl)-3-(2,2,2-trifluoroacetamido)-L-succinamic acid, N-methyl ester;
N-(L-α-carboxyphenethyl)-3-(2,2,2-trifluoroacetamido)-L-succinamic acid, N-methyl ester, dicyclohexylamine salt;
L-3-(2,2,2-trifluoroacetamido)succinanilic acid;
L-3-amino-4'-chlorosuccinanilic acid;
L-3-amino-4'-cyanosuccinanilic acid.

The compounds of this process are useful wherever a sweet taste without caloric value is indicated. This would include comestibles such as food or medicines intended for human or animal consumption. Such items as soda, ice cream, coffee, tea and chewable medications would particularly lend themselves for use with compounds of this process. Additionally, the compounds of this invention may be combined with other synthetic sweetening agents or with sugars to provide sweetening compositions.

DIPEPTIDE COMBINATIONS

Alum and Naringin

H.R. Schade; U.S. Patent 3,934,047; January 20, 1976; assigned to General Foods Corporation found that when aluminum potassium sulfate, naringin or a mixture of these compounds is added to such sweeteners as for example, L-aspartyl-L-phenylalanine methyl ester, referred to as APM, in an amount effective to modify such aftertaste, or is used in foodstuffs including beverages having a pH or about 5 or higher and containing such sweeteners, the sweet aftertaste characteristic of these sweeteners is eliminated.

Aluminum potassium sulfate referred to as alum, and naringin, otherwise known as 4',5,7-trihydroxyflavanone-7-rhamnoglucoside have been found to exhibit unique flavor modifying properties and it is this use which serves to benefit the sweetener market. Since the flavor modifiers have a unique puckering effect on the mouth of the user when sampled alone, it is felt that this property in effect reduces the lingering sweetness by in fact altering the manner in which the sweetness of the dipeptide is physiologically perceived.

The astringent properties of both of these compounds prevent their use in bulking agent quantities but they can be used in minor amounts with both nutritive and nonnutritive sweeteners having a lingering sweet aftertaste and known bulking agents to provide for example, a bulked table sweetener devoid of lingering sweet aftertaste and low in calories. In addition, synergistic sweetening compositions devoid of persisting sweetness and which may be used in virtually any liquid or solid foodstuff may be formulated by combining alum, naringin or both with a dipeptide and other sweeteners known to enhance the sweetening potency of the dipeptide; for example, saccharin.

The components of the sweetening composition must be in combination in order to achieve the desired results. Hence, any suitable method of combining or complexing the components of the sweetening composition may be utilized.

Thus, while the sweetener and alum may be combined prior to addition to the foodstuff or added concurrently or intermittently, there will be some instances where it will be preferred that the sweetener either alone or in combination with alum be added to the foodstuff after the foodstuff has been either partially or completely processed. Thus, for example, where the particular product is to be subjected to increased temperatures, e.g., above 100°C, and where dipeptide sweeteners are to be used, it may be desirable to add the same to the product after such heat processing due to the sensitivity of these dipeptides to thermal degradation. The modified sweetener can be used in beverages, breakfast drinks, syrups, candies, cereals, desserts such as puddings, gelatin, and in virtually any dry, semimoist or moist foodstuff preferably having a pH of no lower than about 5 and in which a nonlingeringly sweet taste is desired.

Example: A puffed cereal product is presweetened with L-aspartyl-L-phenylalanine methyl ester (APM) in the following two-step method. A corn syrup system is prepared which contains:

	Percent
Corn syrup (42 DE)	56
Mor-Rex (10 DE)	18
Water	26

This syrup is homogenously mixed using a magnetic stirrer and thereafter sprayed onto the puffed cereal product at a ratio of one part syrup to one part cereal. An APM-oil system containing 81.810 grams safflower oil, 13.428 grams APM, and 4.762 grams aluminum potassium sulfate is mixed to form a homogenous oil system. This APM oil system is sprayed onto the precoated puffed cereal at a ratio of 97.7 grams cereal to 2.3 grams APM solution to give a final concentration of 60% cereal, 26.86% corn syrup, 10.84% Morrex, 1.88% safflower oil, 0.31% APM and 0.11% alum.

The resulting cereal product has an initial burst of sweetness when consumed but does not demonstrate any lingering sweet aftertaste in the mouth of the user.

Glucono-Delta-Lactone/Buffer

An artificial sweetener known as dipeptide sweetener, which is actually aspartyl phenylalanine methyl ester has recently been developed. Although this product has not as yet come into actual extensive use, it has already been determined that like the cyclamates and saccharin, the dipeptide sweetener although many times as sweet as sugar, actually about one hundred and fifty times as sweet as sugar, does not have the natural sweetness taste of ordinary sugar. The dipeptide sweetener exhibits a flat sweetness along with a slight bitter aftertaste. In addition, when used for the sweetening of beverages, for example, the sweetness is slightly delayed.

In other words, someone drinking a cup of coffee sweetened solely with dipeptide sweetener would at first not taste the sweetness. Shortly thereafter the sweet taste would appear, but this sweet taste would not be a natural sweet taste because of what may best be referred to as a flatness of taste, and in addition, a bitter aftertaste occurs in the mouth of the user.

M.E. Eisenstadt; U.S. Patent 3,875,311; April 1, 1975; assigned to Cumberland Packing Corporation claims an aspartyl-phenylalanine methyl ester (APM) combination that has the desired effect of sweetness approaching that of natural sugar without any flatness of taste and without any cloying sweetness or bitter aftertaste. The composition is prepared as follows: (a) the dipeptide sweetener is mixed with (b) cream of tartar and/or sodium bicarbonate and/or potassium bicarbonate, and also with (c) glucono-delta-lactone and/or sodium gluconate and/or potassium gluconate in a ratio of one part of (a) to $\frac{1}{25}$ to 6 parts of (b), most preferably $\frac{1}{4}$ to 2 parts of (b) and with 5 to 50 parts of (c), most preferably 10 to 20 parts of (c). It is noted that too great an amount of (c) can result in a composition which has a tendency towards sourness and even more important will actually cause milk to curdle so that the composition cannot be used for the sweetening of coffee, for example. It is therefore essential that the amount of (c) be within the limits set forth. Furthermore, if (c) were used alone, then even the maintaining of (c) within the limits set forth would not be sufficient to obtain a complete masking of the undesired unnatural sweetness of the dipeptide sweetener without the disadvantages of possible sourness and curdling of milk. For this reason, it is necessary to use (b) along with (c). However, if (b) were used alone, that is without (c) there would be no masking of the undesired taste characteristics of the dipeptide sweetener.

The masking of the undesired taste characteristics cannot be accomplished by (b) alone, but (b) in combination with (c) effects a masking of all undesired taste characteristics while preventing the undesired effects of the use of (b) alone. It is necessary to use (b) and (c) in the proportions indicated along with the dipeptide sweetener in order to obtain a masking of the slight sweetness and bitter aftertaste of the dipeptide sweetener and in order to achieve the development of the rapid natural sweetness. As indicated above, up to 50% of the dipeptide sweetener can be substituted by saccharin without any adverse effects. In percentages by weight, the composition consists essentially of about 5 to 7% of (a), about 2 to 8% of (b), and about 82 to 92% of (c).

Example 1: 53.9 pounds of glucono-delta-lactone, 36.3 pounds of sodium gluconate, 5.5 pounds of dipeptide sweetener, 1.07 pounds of cream of tartar and 3.23 pounds of sodium bicarbonate are thoroughly mixed to provide a uniform mixture. The resulting mixture is approximately ten times as sweet as sugar and can be used in the place of sugar to give a sweetening effect with no carbohydrate or caloric intake whatsoever. This composition can be used to sweeten beverages or in cooking, in all quantities, even to highly sweetened beverages, without causing any bitter aftertaste and without adversely affecting the taste of the food or beverage to which it is applied. Furthermore, the composition does not curdle milk.

Example 2: A sweetening composition is prepared as in Example 1, however using 4.30 pounds of sodium bicarbonate and no cream of tartar.

Example 3: A sweetening composition is prepared as in Example 1, however using potassium bicarbonate in place of the sodium bicarbonate.

Example 4: 6.2 pounds of dipeptide sweetener, 85.8 pounds of glucono-delta-lactone and 8.0 pounds of potassium bicarbonate are thoroughly mixed to provide a uniform mixture. The resulting mixture can be used in the same manner as the composition of Example 1.

Example 5: A sweetening composition is prepared as in Example 4, however substituting 6 pounds of sodium bicarbonate and 2 pounds of cream of tartar for the potassium bicarbonate.

Example 6: A sweetening composition is prepared as in Example 4, however using 4.2 pounds of dipeptide sweetener and 2.0 pounds of sodium saccharin.

Special Buffer Salts

In another process by *M.E. Eisenstadt; U.S. Patent 3,875,312; April 1, 1975; assigned to Cumberland Packing Corporation* the following composition is prepared to improve the sweetness characteristics of aspartyl-phenylalanine methyl ester (APM): (a) The dipeptide sweetener is mixed with (b) cream of tartar and/or sodium bicarbonate or potassium bicarbonate, and also with (c) lactose and/or dextrose in a ratio of one part of (a) dipeptide sweetener to 0.1 to 10 parts of (b), most preferably 0.5 to 5 parts of (b) and with 5 to 30 parts of (c), most preferably 10 to 20 parts of (c).

It is noted that too great an amount of lactose can result in the masking of the sweetness of the dipeptide sweetener as well as the masking of the taste of the food or beverage to which it is applied, so that it is essential that the amount of lactose be within the limits set forth. Furthermore, if lactose alone were used, in order to obtain a sufficient amount of lactose with the dipeptide sweetener to mask the undesired unnatural sweetness, the amount of lactose would be so great that it would mask the taste of the food or beverage to which the composition is applied. For this reason, it is necessary to use the cream of tartar or sodium bicarbonate or potassium bicarbonate along with the lactose. However, if cream of tartar or the bicarbonate were used without the lactose (or dextrose as discussed below) with the dipeptide sweetener, it would be impossible to mask the undesired taste characteristics of the dipeptide sweetener.

On the other hand, dextrose is actually a sweetening agent and dextrose alone does not have the effect of sufficiently masking the undesired sweetening effect of the dipeptide sweetener. It is only by using the dextrose in combination with the cream of tartar or sodium bicarbonate or potassium bicarbonate that it is possible to achieve the natural sweetness of the composition.

Thus, it is necessary to use cream of tartar powder (potassium bitartrate) and/or sodium bicarbonate and/or potassium bicarbonate with lactose and/or dextrose in the proportions indicated above along with the dipeptide sweetener in order to obtain a masking of the flat sweetness and bitter aftertaste and the development of a rapid natural sweetness. Up to 50% of the dipeptide sweetener can be substituted by saccharin without any adverse effects.

In percentages by weight, the composition consists essentially of about 80 to 98% of lactose and/or dextrose (preferably 90 to 97%), about 1 to 4% of the dipeptide sweetener (up to 50% of which may be substituted by saccharin), preferably about 2 to 3%, and about 0.1 to 6%, preferably about 1 to 2% of any one or a combination of cream of tartar, sodium bicarbonate and potassium bicarbonate.

Example: 37.5 pounds of lactose (powdered), 3 pounds of dipeptide sweetener (aspartyl-phenylalanine methyl ester) and 0.5 pound of cream of tartar are thoroughly mixed to provide a uniform mixture. The resulting mixture is approximately ten times as sweet as sugar and can be used in the place of sugar to give a sweetening effect with low caloric intake. This composition can be used to sweeten beverages or in cooking, in all quantities, even to highly sweeten beverages, without causing any bitter aftertaste and without adversely affecting the taste of the food or beverage to which it is applied.

DIPEPTIDE SOLUBILIZATION

Dipeptide Codried with Polymaltose and Dextrins

It has been found that polymaltose and polymerized dextrins which are nontoxic, bland, low-calorie synthetic polysaccharides prepared by acid catalyzed polymerization under vacuum at elevated temperatures when codried with a dipeptide serves multifunctional purposes. The resulting dry sweetening composition is low in calories, low in hygroscopicity, devoid of unpleasant aftertaste and at least two to six times faster dissolving than APM (L-aspartyl-L-phenylalanine methyl ester) alone.

The details of this process are given by *I. Furda and J.F. Trumbetas; U.S. Patent 3,934,048; January 20, 1976; assigned to General Foods Corporation.* Combining dipeptides with other carbohydrates such as fructose, invert sugar and dextrin by codrying techniques serves to increase the sweetness of the resulting composition by reason of contributing its own sweetness to that of the system.

Since, polymaltose and the polymerized dextrins are hygroscopic, and the dipeptides have extremely poor solubility, the multifunctional sweetening formulation derived results from the retention of only the desirable qualities of each starting material. The ability of APM to eliminate the tendency of these synthetic polymers to deliquesce under moist conditions can be explained by the hydrophobic character of the benzene ring on the dipeptide molecule which functions to reduce the hygroscopicity of the polymer by reason of contributing its own hydrophobicity to the system.

Thus, the higher the ratio of dipeptide to polymer, the less hygroscopic the system becomes. This does not involve any chemical interaction between the two components, however, and therefore, no relinquishment of sweetness by the dipeptide is experienced. Alternatively, the rate of solubility of the dipeptide, specifically APM, is increased by relying upon the innately high rate of solubility of these polymers. Thus the higher the ratio of polymer to dipeptide, the more rapidly soluble the sweetening composition becomes. Since the interests are conflicting in terms of acquiring the most desirable degree of solubility and hygroscopicity, the dipeptide and polymer should be in sufficient proportion to produce a sweetening composition which is both fast dissolving and low in hygroscopicity.

The dextrins utilized are those having a dextrose equivalent (DE) greater than 20. These dextrins are particularly preferred since polymerization occurs more readily with the greater amount of reducing groups present. In addition, the high DE polymerized dextrins are less susceptible to enzyme hydrolysis due to their highly branched structure.

The solution may be prepared by simply homogenously comixing the artificial sweetening agent with the polysaccharide and combining into one homogenous solution. Concentrations of these compounds are usually in the range of about 1:3 to about 1:4 by weight of the dipeptide to polysaccharide.

As the sweetness level of the multifunctional formulation may be adjusted according to the specific requirements of the foodstuff by changing the ratio of sweetener to polyglucose, so may the bulk density of the end product be adjusted by selecting the proper drying procedure. Freeze-drying produces the lowest bulk density product and eliminates the possibility of thermal degradation but is the most expensive means of codrying the composition. However, since this method does produce a final product with the fastest rate of solubility in terms of other drying methods, it is preferred where rate of solubility is a prime consideration.

In terms of deriving a matrix appropriate for use as a table sweetener, it is important that the drying conditions be carried out by a method which effectively bulks the matrix formed as by achieving a distinct blistering effect by spray-drying the solution under conditions which result in a bulk compatible with like sweetening effects of an equal volume of sucrose. Such a product is produced by spray-drying. However, it may also be produced by drum drying, either atmospherically or under a vacuum. The spheres produced by spray-drying are less dusty, more glossy in appearance, and more suggestive of a crystalline table sugar product than the drum dried product which has a relatively extreme degree of dustiness.

In all instances, the better quality product especially from a hygroscopicity and deliquescence point of view can be prepared when the polysaccharide is purified as by solvent precipitation or ultrafiltration. The ultrafiltrated polysaccharide is preferred not only because of the undesirability of using large amounts of flammable solvents, but mainly due to the fact that the molecular size of the product can more easily be controlled making it possible to eliminate virtually all of the low molecular weight reaction products. These products contain appreciable amounts of reducing groups which, when present, could react with the free amino group of the dipeptide sweetener to cause a reduction in sweetness. Polymaltose and polymerized dextrins are not currently available but are prepared by the polymerization of maltose or dextrins respectively by using preferably phosphoric acid as a catalyst at elevated temperatures and reduced pressure.

These polysaccharides may also be prepared by controlled polymerization of maltose or a dextrin with sorbitol and polycarboxylic acid at specified ratios which may result in two unpurified forms, an acid fused form and a bleached neutralized form. The latter demonstrates a faster rate of solubility when codried with APM than the acid form and is, therefore, preferred.

Generally, the low molecular weight fragments present in the unpurified polysaccharides account for the high degree of hygroscopicity. Consequently, when these synthetic polymers are purified as for example by ultrafiltration where such undesirable low molecular weight fragments are removed, the hygroscopicity of these polymers is reduced making it possible for higher ratios of polymer to dipeptide to be utilized without an accompanying increase in hygroscopicity.

Dipeptide Codried with Polyglucose

In another process *I. Furda and J.F. Trumbetas; U.S. Patent 3,971,857; July 27, 1976; assigned to General Foods Corporation* use polyglucose, a low-calorie synthetic glucan, in combination with APM to obtain a low-calorie sweetener, low in hygroscopicity.

Example 1: 2.5 grams of L-aspartyl-L-phenylalanine methyl ester (APM) and 7.5 grams nonpurified polyglucose N (neutral form) prepared by polycarboxylic acid catalyzed polymerization are dissolved in 200 ml water at room temperature and then freeze dried. The APM/polyglucose freeze-dried sample was compared to a polyglucose control in terms of hygroscopicity. Both samples were exposed to the atmosphere at a relative humidity of 70% at 90°F for 7 days and measured for moisture uptake. Moisture absorption in the APM/polyglucose sample was 15% lower than the control and occurred during the initial 2 to 3 day period. The control deliquesced during the initial 24 hours. At the conclusion of the 7 day period, the polyglucose was in a liquid state while the APM/polyglucose sample, while not devoid of moisture, was still a partially dry flowable mass. The water uptake by both samples ceased after one weeks' storage and the appearance of the samples did not change during additional three weeks' storage.

Example 2: A sample of APM/polyglucose as prepared in Example 1 was tested for its rate of solubility as against an APM control. The test was carried out at 1% concentration based on APM in room temperature water.

APM	10+ minutes
APM/polyglucose	1 minute 40 seconds

Example 3: 2.5 grams APM and 7.5 grams polycarboxylic acid-catalyzed nonpurified polyglucose A (acid form) are dissolved in 200 ml water at room temperature and the solution is freeze-dried. The polyglucose control and APM/polyglucose test samples were stored for 7 days at 90°F and a relative humidity of 70%. The test sample containing the acid form polyglucose had a moisture uptake which was 25% lower than the control which occurred in the first 3 days of the storage period. As in Example 1, the control sample deliquesced after 24 hours. At the conclusion of the 7 day period, the control was completely liquified and the test sample, although less moist than the control had some degree of caking and could not be considered an absolutely free-flowing mass.

Example 4: A sample of the APM polycarboxylic-acid-catalyzed polyglucose prepared in Example 3 was evaluated for its rate of solubility as against an APM control. The test was conducted in room temperature water at a concentration of 1% based on APM. The APM/polyglucose sample was only about 2 times faster to dissolve than APM alone indicating that the acid form of polyglucose is not preferred.

Example 5: Polyglucose N solution prepared by polycarboxylic acid polymerization is ultrafiltrated through an Amicon membrane UM 2 (1,000 molecular weight cut-off point) in order to remove the low molecular weight fragments comprising mostly unreacted material and by-products of polymerization and neutralization. The solution is then dried by spray-drying, freeze-drying or drum-drying. 7.5 grams of this dried material is dissolved with 2.5 grams APM in 200 ml of room temperature water. The solution was freeze-dried and stored with an ultrafiltrated polyglucose N control for a period of 8 days at 90°F and a relative humidity of 70%.

The APM/ultrafiltrated polyglucose test sample remained free-flowing after 8 days while the ultrafiltrated polyglucose control deliquesced after 48 hours. After 8 days of storage, water uptake by both samples ceased, the moisture uptake by the test sample having been about 80% lower than the control. Despite the minor amount of moisture absorption that did take place, the APM/polyglucose sample remained free-flowing. Freeze-dried samples of APM and APM/ultrafiltrated polyglucose were tested for their rate of solubility in room temperature water and at 1% concentrations. Results are shown below

APM	10 minutes
APM/polyglucose	1 minute 50 seconds

Example 6: 100 grams of glucose is dispersed in 150 ml of water. The solution is heated to slightly above room temperature in order that the glucose dissolve completely. Five drops of concentrated phosphoric acid is added to the mixture which results in a 2.0 pH solution. The entire solution is then evaporated on a rotary evaporator to form a thick syrup. The syrup is heated in a vacuum oven at 125°C and 10^{-2} to 10^{-3} mm Hg for about 10 hours. The bright yellow glossy material is then dissolved in water, adjusted to pH 6.5 by adding several drops of sodium hydroxide solution.

An aqueous solution containing 30% APM and 70% polyglucose prepared by this manner is formulated and thereafter freeze dried. A dry APM/polyglucose sample placed in a closed jar and stored at room temperature for 2 years remained stable and sweet. An APM control and the APM/polyglucose test sample were evaluated for their rate of solubility and the following results derived:

APM	10+ minutes
APM/polyglucose	1 minute

The test sample did not deliquesce when exposed to an atmosphere of 85% relative humidity and 90°F for 12 days although the water uptake was 20% as compared to a polyglucose control which did deliquesce under the same test conditions.

Dipeptide Coground with Acid

J.H. Berg and J. Trumbetas; U.S. Patent 3,868,472; February 25, 1975; assigned to General Foods Corporation reports that L-aspartyl-L-phenylalanine methyl ester (APM) exhibits improved solubility when coground with an acid in the presence of an organic solvent. When the solvent is evaporated, a coating of acid remains on the surface of the low-solubility dipeptide.

Due to the fact that this method involves a type of commingling of the solid acid and dipeptide, there is no chemical interaction, i.e., there is no protonation of the dipeptide due to the absence of protons in the dry system and ostensibly no salt formation. Consequently, the dispersibility of the dipeptide is increased while the peptide still remains structurally independent. Essentially, therefore, the rate of solubility of the dipeptide is increased by relying upon the innate solubility of the acid, the acid and dipeptide being in sufficient proportion to one another to permit the acid to coat the dipeptide thereby serving to partially or completely enclose the same in a soluble transport system.

The organic solvent functions both to reduce the electrostatic forces between the container vessel and the ground material and in some cases, such as in the case of citric acid which has a slightly soluble character in organic solvents, to partially solubilize the acid thus effecting a more uniform coating of the dipeptide. In all instances, the effect of cogrinding in the presence of the organic solvent allows the acid and dipeptide molecules to come into close contact which is further improved by subsequent evaporation of the solvent. The organic solvent functions in most cases to improve the efficiency of the grinding process and is believed to be the most effective method of achieving an intimate association of the acid and dipeptide. The sweetener/acid blends find application in such systems as cake mixes, pastries, gelatin desserts, Kool-Aid, and the like. An organic solvent-ground sweetener/acid preparation with a weight ratio of 1:2 respectively was found to dissolve at 0.07% sweetener concentration in cold carbonated water in 25 seconds. Consequently, such blends find utility in carbonated systems as well, e.g., carbonated beverages, and carbonated candies.

Example 1: Malic acid and L-aspartyl-L-phenylalanine methyl ester (1:4 weight ratio of sweetener to acid) were ground together in ethanol using a mortar and pestle for approximately 1 hour. The ethanol was subsequently evaporated to afford a completely dry mixture. Approximately 4.5 grams of the mixture was placed in a typical beverage formulation, whereby it comprised approximately 90% of the solid formulation of which 1.0 gram or 0.5% of the dry mixture was the dipeptide. In comparison to the two minute solubility time of sucrose in the beverage system, the dipeptide commingled with the acid was found to dissolve completely in 25 seconds.

Example 2: Three samples of L-aspartyl-L-phenylalanine methyl ester were prepared and the following systems evaluated for their solubility, the overall concentration of L-aspartyl-L-phenylalanine methyl ester being in all three instances 1%. The solubility tests were carried out in room temperature water with the same amount of stirring being applied to each sample.

Sample Number		H_2O Solubility
1	L-aspartyl-L-phenylalanine methyl ester	4.5 minutes
2	L-aspartyl-L-phenylalanine methyl ester + malic acid (1:3 respectively) added separately but simultaneously	4.25 minutes
3	L-aspartyl-L-phenylalanine methyl ester coground in ethanol with malic acid (1:3 respectively)	35 seconds

Citric Acid Melt

Another method for improving the solubility of APM is given by *M.D. Shoaf and L.D. Pischke; U.S. Patents 3,956,507; May 11, 1976 and 3,928,633; Dec. 23, 1975; both assigned to General Foods Corporation.* The method involves creating a hot melt (below 370°F) capable of forming a relatively amorphous matrix within which APM crystals are dispersed discretely and then causing that hot melt to undergo cooling to permanently fix the dipeptide in a form so that the APM is a dispersed phase.

The matrix will be formed by subjecting the APM and the matrix material, i.e., citric acid powder, to a high shear in a blender such as a vertical cutting machine, then the blend is sprinkled onto the heated surface of a calender roll. The hot melt ingredients are melted on the heat exchange surface of the heated roll and accumulate into a plastic molten mass. Eventually the mass will be ground after cooling to the intended particle size. In lieu of a roll fusing operation other means to effect a molten condition and conveniently distribute the APM particles throughout the matrix may be used such as a Sandvik bed dryer where the blend of ingredients is uniformly distributed as a bed onto a heated metal belt; after fusion takes place, there is a cooling effected by a blast of chilling or cooling air directed at the belt, the fused mass being ultimately doctored and ground as in the case of the roll fusing operation. It is believed to be important and essential that the blend ingredients be controlled in the moisture content. Whereas a certain amount of plasticity is desired in order to promote handleability of the molten mass such for instance as by blending anhydrous citric acid with monohydrate citric acid which tends to reduce the temperature at which melting occurs, a control of the amount of moisture either generated or present in the mix should be observed. Roller fusing is one of the simplest and most convenient methods that might be used since the moisture content may be readily controlled.

Typically, a blend of 30 to 60 parts citric acid monohydrate to 80 to 50 parts of anhydrous citric acid will be used as diffusing medium, the mixed melting points of the blends having softening and thermoplastic properties well below the melting point of the APM. Although one of the more preferred fusing mediums will be an anhydrous citric acid from the stability standpoint, this acid should be altered in its melting characteristics by the addition of other agents such as sorbitol or another lower polyhydric alcohol capable of providing the plasticizing characteristics intended for assuring encapsulation.

When viewed under a microscope, the APM particles will be seen to be virtually substantially completely imbedded in the matrix solids such that there is no tendency towards impairing flowability and agglomeration is virtually avoided. By the same token, however, the APM retains its solubility through the discrete dispersion of the moieties in the matrix material. Roll fusion is most preferred because in calendering the APM particles are postively submerged in the matrix.

The matrix can be ground to a predetermined particle size distribution; the composition can be milled to a desired size and will fracture in milling along the lines of fracture for the molten matrix per se. The grind can be so fine as to provide a material that will pass a 140 mesh screen, the size of grinding being within the limitations of grinding equipment. The composition is quite dense, the grind having a density of 0.96 gram per cubic centimeter at a grind of –70+120 mesh; such densification minimizes any increase in total product volume enabling a sweetening composition to be used as part of a mix having a volume equivalent to that of an unsweetened mix.

A blend of equal weights of anhydrous citric acid and citric acid monohydrate is prepared. This blend is then added at an equal weight to the APM. The ingredients are mixed using a Hobart blender. The mixed ingredients are then sieved through a No. 8 screen, U.S. Standard Screen Series onto a double drum dryer heated by steam at a pressure of between 22 to 28 psi, the air temperature above the rolls ranging between 116° and 122°F. The distance between the drums is maintained throughout at less than eight thousandths of an inch, one foot diameter drums being operated at approximately one revolution per minute.

The material is fused substantially completely as the temperature is elevated to an optimal operating steam pressure of about 26 psi and temperature of 120°F above the roll with minimal browning being observed at the doctor or scraper and maximal fusion of the citric acid; the material is doctored from the drum dryer and collects at the scraping blade in the form of continuous chunks or pieces which are quite frangible while being in an amorphous state. This mass is then ground and sieved to collect fractions of use. The material can be ground to a size where it will pass a No. 40 U.S. Standard Screen. In a solubility test, particles could be added to water at 45°F and go into solution in 30 to 35 seconds (employing the −120+140 mesh fraction). The finely divided material can be blended with flavor and coarser granular citric acid to prepare a beverage mix which is cold-water-soluble and quite flavorful. The mix can be hermetically packaged in a conventional polyethylene coated foil wrap under controlled humidity conditions and will be stable and retain its stability for months.

A variety of materials for use as fusing agents, the suitability of which will be dependent on their intrinsic melting characteristics and their use in combination, may be selected from a plasticizing polyol or like material serving to provide flowability and handleability in forming a spreadable matrix within which the APM particles may be submerged and fixed. The class of fusing agents will range from monosaccharides and polysaccharides to proteins and other materials or sources as the following list will indicate:

- Juice crystals such as freeze-dried or drum-dried orange juice or grape juice;
- Coffee extract such as freeze-dried or spray-dried extract;
- Tea extract such as freeze-dried or spray-dried tea extract;
- Vegetable solids concentrate such as tomato purees which have been dried and are high in naturally occurring amorphous polysaccharides;
- Dairy by-products such as whey;
- Gelable extracts of alginic or pectinic acid respectively derived from citrus or apple peel or seaweeds such as kelp;
- Low dextrose equivalent starch polymers, that is, those having a dextrose equivalency of 5 to 20 in an intermixture with glycerol or a like polyol;
- Ascorbic acid and its isomers, i.e., erythorbic acids and their salts;
- Fusible gums such as propoxylated cellulose; and
- Protein extracts of animal hide, hoof, bone or feather in the form of gelatin and keratin.

In addition, other sweetening agents, the so-called artificial sweeteners, have the ability to fuse and serve as a matrix for the APM particles. Another group of fusing agents is the class of relatively hard emulsifiers and triglycerides such as mono- and diglycerides of hard fats such as stearic acid; also naturally occurring matrix materials such as chocolate liquor.

BULK DENSITY AND STABILITY

Encapsulation Process

In another approach to improve the bulk density and stability of L-aspartyl-L-phenylalanine methyl ester (APM) *L.D. Pischke and M.D. Shoaf; U.S. Patent 3,962,468; June 8, 1976; assigned to General Foods Corporation* prepare an encapsulated powder that is readily blendable and resists caking. Spray-drying is the preferred method to dry and encapsulate individual droplets. A typical example of the operation is described.

A 25% solids slurry is prepared for processing through a Fryma mill, the stones of the mill being adjusted to a spacing of about 75 microns. A solution is prepared consisting of 60% APM solids and 40% corn syrup solids (maltodextrin), 10 DE. The dispersion at 70°F enters the mill and issues at a temperature of approximately 85°F therefrom, the mill being jacketed with water maintained at 160°F. The finished relatively cool aerated slurry is collected in a stainless steel kettle at a density of 0.75 gram per cubic centimeter and is pumped by a positive displacement pump and fed under 150 psig spray pressure through a 40/27 nozzle and produces droplets which when dried had a size of 50 to 300 microns. The droplets are contacted by inlet drying air at approximately 330°F entering a 16 foot diameter vertical drying tower, the air being fed at 14,000 cubic feet per minute and the outlet air temperature being at 210°F. The droplets dry as spherical beads collected at a moisture content generally less than 5% and typically 2 to 3%. The dry particles have the following particle size distribution:

Particle Size (U.S. Standard Sieve)	Weight %
+50	2.57
-50+70	27.33
-70+120	45.25
-120+140	10.60
-140+200	11.38
-200+300	1.74
10 Pan	1.13

The particles have a uniform particle size distribution, are free-flowing, white, and dissolve in 60 seconds in 45°F water when in the presence of a beverage acid such as citric acid at pH 3.5. The droplets had a loose bulk density of 0.246 gram per cubic centimeter and packed density of 0.282 gram per cubic centimeter with percent packing after tapping of 12.8%.

The particles can be readily blended to a flowable condition with beverage mix ingredients such as dry citric acid or equivalent acidulent powder, flavors and colors and will be stable against caking when packaged in a water vapor transmission packaging barrier such as a polyethylene coated foil having a water vapor transmission of about 0.04 gram per 100 square inches in 24 hours at 100°F (95% RH).

It is an important processing advantage that by producing a cool slurry-distribution of the aspartyl acid sweetening compound one is able to produce extraordinarily high concentrations of the derivative by weight (i.e., above 30%) of the carrier of fixative solids without encountering equipment plugging or blockage of drying equipment such as spray-drying atomizing nozzles and assuring a continuous and uniform sweetness in the ultimately dried composition. Concentrations ranging from 20 to as high as 60% are optimal and ideal by reason of the ability to afford a much higher ratio of sweetness to calories contributed by the carrier. Thus, not only does the process provide a composition that is dense, stable, soluble and flowable, but also one which for a given sweetness intensity has a very consequential reduction in calories that are contributed by the dextrin solution.

In accordance with its most preferred aspects, a beverage mix composition having an extremely low caloric value stemming from an admixture of edible food acids, buffer salts and natural or artifical colors and flavors is combined with a fixative which contributes less than 0.5 calorie per sweetness level equivalent to a comparable sweetness of sucrose. The concentration of aspartic acid derivative will vary in large measure in accordance with the sweetness thereof; of course, the sweeter the derivative, the less of it is required; however, it is estimated that a minimum dextrin content of 10% by weight of the derivative will be necessary to fix the slurried derivative, this level being in part functional in accordance with the dextrose equivalency, the higher the DE of the dextrin source the higher the fixation capacity and the lower the amount required.

The spray-dried form of the product can be quite distinctive. When viewed under normal microscopy (700 diameters), many of the particles will be discrete randomly nested crystals of sweetening compound; many of the crystals will appear to be interlaced while being bound or enveloped by the relatively transparent dextrin matrix. Whereas some particles will evidence a slight occasional protrusion of needle-like particles from this matrix, the large majority, if not the entirety, will appear to be effectively bound by the dextrin material.

In some spray-drying operations, the particles may not be spherical but thus may be angularly or oblately shaped depending upon the characteristics of the drying tower and drying conditions. Under natural light, the particles will have a glass, glossy appearance and will retain this appearance under typical storage conditions.

The spray-dried form of fixed sweetener will have an equilibrium relative humidity at room temperature (68°F) such that the particles do not pick up more than 9% moisture in cases where the composition is dried to about 3% moisture; at between 40 and 80% equilibrium relative humidities, the fixed product will pick up less moisture than the ground counterpart that is unfixed; this indicates less of a tendency to absorb moisture under normal packaging conditions; on the other hand, the relative difference in moisture pick-up will generally not be reduced by more than 2% moisture gain at any given equilibrium relative humidity.

Of great importance, however, is the fact that the fixed APM particles will not cake when stored in a dry blend with edible food acids such as citric acid, whereas the subdivided sweetening compound is quite prone to caking when comixed in a simple blend when packaged in a polyfoil pouch or equivalent packaging material.

Codried Starch Hydrolysate

The powders described by *J.A. Cella and W.H. Schmitt; U.S. Patent 3,753,739; August 21, 1973; assigned to Alberto-Culver Company* are effective in overcoming the instability characteristics of dipeptide sweeteners. Only those starch hydrolysates which have a DE of 0 to 20 are satisfactory for use. It has been found that the DE of the starch hydrolysates is essentially directly proportional to the instability of the starch hydrolysate-dipeptide composition in dry form as well as in solution. If starch hydrolysates having a DE in excess of about 20 are sought to be used, the rate of destruction of the dipeptide sweetener is unduly rapid, in the process of preparing a dry pulverulent sweetener composition, as well as in the dry pulverulent compositions, on standing under normal storage or shelf conditions, and on incorporation of compositions into hot aqueous drinks such as coffee and tea.

Dipeptide-starch hydrolysate slurries are prepared for spray-drying at a 55 to 60 percent solids concentration and a temperature of 50°C. Between the time of preparing the mixture and the time that it is introduced or fed into the preheater or preheaters of the spray drier, where drying is effected by means of a spray drier, only as short a period of time as is reasonably possible or practicable should be permitted to elapse. At most, only a few minutes should be permitted to elapse and, more desirably, such elapsed time should not exceed about 1 minute. The spray-drying operation is initiated promptly and the spray drier is operated under conditions such that the moisture content of the finished, spray-dried sweetener composition does not exceed about 1%, by weight of finished composition.

In general, the dipeptide sweetener will usually fall within the range of 2 to 5% of the finished spray-dried sweetener composition. The viscosities of the aqueous compositions, as fed into the preheater or preheaters of the spray drier, are variable, falling into the range of 50 to 1,000 cp as measured by a Brookfield Viscosimeter. The viscosities depend, of course, not only on the concentration of solids in the aqueous composition, for a particular starch hydrolysate and dipeptide sweetener, but also on the temperature of such aqueous composition or solution. Thus, in an illustrative case of an aqueous composition or solution containing 55% total solids, in which, on the total solids basis, the starch hydrolysate constitutes 95% and the dipeptide sweetener constitutes 5% the viscosity is about 300 cp at 37°C, about 100 cp at 50°C and about 80 cp at 55°C. The following are examples of illustrative aqueous compositions prior to drying. The percentages stated are by weight.

Example		% by Weight
1	Milo starch hydrolysate (DE 0)	38
	Aspartyl-phenylalanine methyl ester	2
	Water	60
2	Cornstarch hydrolysate (DE 5)	47.5
	Aspartyl-phenylalanine methyl ester	2.5
	Water	50
3	Cornstarch hydrolysate (DE 12)	57
	Aspartyl-phenylalanine methyl ester	3
	Water	40
4	Cornstarch hydrolysate (DE 10)	52.25
	Aspartyl-phenylalanine methyl ester	1.65
	Sodium saccharin	1.1
	Water	45
5	Milo starch hydrolysate (DE 11.5)	52.195
	Aspartyl-phenylalanine methyl ester	1.65
	Calcium saccharin	1.1
	Potassium sorbate	0.055
	Water	45

It has been found that the dipeptide sweeteners tend to react with the starch hydrolysate to cause a slight discoloration of the finished dry pulverulent dipeptide sweetener composition, the extent of which varies with the temperatures to which the mixtures of the dipeptide sweetener and the starch hydrolysate are subjected and the length of time of contact. Generally, if optimum conditions under which the sweetener compositions are prepared are observed, the discoloration is quite minimal. The discoloration can be substantially reduced by adding small proportions of hydrogen peroxide to the mixture of the starch hydrolysate and the dipeptide sweetener, or by adding glucose oxidase to the mixture, prior to the step of drying the mixture. This should be done under conditions such as not to substantially affect the DE of the starch hydrolysate. Using a cornstarch hydrolysate having a DE of 10 to 12, there are obtained finished dried dipeptide sweetener compositions with a bulk density of the order of 0.05 to 0.1 g/cc, and with caloric contents of the order of 1 to 2 calories per level teaspoon.

SPECIAL CEREAL APPLICATIONS

Foam Method

A sprayable foam containing dipeptide sweetener is used by *P.M. Rousseau; U.S. Patent 3,947,600; March 30, 1976; assigned to General Foods Corporation* to coat ready-to-eat cereals for sweetening. In preparing the foamy composition, a relatively saturated aqueous solution of the APM is prepared which in any event is operative to fully wet the particles of APM. This mixture may then be converted into a foam suspension having a light, uniform condition of high overrun as by passage through an homogenizer or otherwise; thus, instead of homogenization, the foam can be generated by mixing the APM and water in a pressurized vessel and causing the solution to undergo such expansion as creates a light delicate reticular structure which can then be applied as such in the form of a spray by atomization means.

Generally, the foam will be applied as a very minor weight percent of the food product, typically less than 3%, the level of foam applied being dependent upon the intended sweetness and the amount of water that may be indeed tolerated by the food product, since an excessive amount of water can contribute towards instability. For most dry alimentary food products and dry comestibles, the overall composition should have a moisture content less than 10% as the result of the APM foam emulsion being applied; and applications to products such as ready-to-eat breakfast cereal products and the like will involve adding that amount of foam which does not cause an increase in moisture content above 5%. The sweetening power of the APM is 100 to 300 fold that of sucrose and so the level in the foam will be used at that level for the intended sweetness comparable to sucrose.

Should too high a level of water be added by the dipeptide foam, the coated product may be dried by heat or otherwise, care being exercised to avoid heating the APM above 160°F or that elevated temperature where degradation may ensue.

A mixture of 80% water and 20% APM is prepared and mixed for a short period of time by gradually adding the APM to the water with spoon mixing thereby creating a solution of APM which is highly saturated. The APM level added should not be in that excess which limits its ability to be sprayed upon being homogenized and foamed. This mixture is then fed into a hand homogenizer and is pumped to discharge a light uniform foam suspension. Four to seven grams of the foam in a glass container is weighed out depending upon the percentage of sweetener to be applied to a breakfast cereal product; in this example a charge of 400 grams of toasted corn flakes was coated with the stated quantity of foam.

The filled glass container is part of a common atomizing spray mechanism adapted to be used to dispense fine particles, the spray nozzle being equipped with a depending tubular stem which is located within the body of the foam and under the action of the atomizer causes the foam to be sucked through the stem into the atomizing nozzle from the container and be transmitted in the form of a fine particulate dispersion. The foam is sprayed in this form onto the cereal flakes as they are tumbled in a conventional revolving coating reel. The foam created had a sweetening power equivalent to 150 parts of a comparable weight of sucrose. After the application of the foam, it is not necessary to dry the aqueous application as it is in such a small amount.

Coating Method

Combination of dipeptide and starch hydrolysate as a coating solution is another method for sweetening ready-to-eat cereals as described by *P.A. Baggerly; U.S. Patent 3,955,000; May 4, 1976; assigned to General Foods Corporation.*

A coating of a dried solution of such an hydrolysate of cereal solids having a fine dispersion of an L-aspartic acid derivative has been found to smooth out the taste impact generated by any sweetening imbalance attributable to the incomplete solution of the APM or nonuniformity of its dispersion. Such a solution when applied to a cereal base such as corn flakes, puffed cereal products, baked goods such as pastry mixes and a variety of confectionary foodstuffs intended to be sweetened with the sensation of sucrose will provide a uniform distribution of sweetness such that when eaten the foodstuff has minimized localized physiological response identified as hot spots; the starch hydrolysate contributes significantly to smoothing out the sweetness sensation.

A dextrin solution, so-called because the oligosaccarides are not completely dissolved but practically speaking are substantially dissolved or colloidally dispersed so as to have the gross appearance of a solution, has the L-aspartic acid derivative compound uniformly dispersed throughout. It will be practical to increase the temperature of the aqueous medium serving as a solvent for the dextrinous material and facilitate mixing to a uniform degree preparatory to having the sweetening compound dispersed therein.

Homogenization or other means to finely disperse the derivative throughout the solution is preferably employed to assure a uniform dispersion and permit application of the coating solution by atomization or other spray techniques known to skilled art workers. In some applications it may be practical to wet mill the L-aspartic acid derivative in the dextrin solution to assure a substantially discrete form of finely suspended particles. Generally, the coating solution will be maintained at a temperature below 170°F during its preparation and application to the dry comestible.

The L-aspartic acid derivatives, when used at a sweetening power equivalent to that of a sucrose application for which it is substituted in the coating solution, is present in sufficient quantity to exceed the solubility of the sweetening compound; thus, the sweetening derivative is present in both the form of a solution solute and a very fine dispersion. Although the dextrinous saccharides are not as sweet per se and generally contribute little noticeable sweetness, they do appear to balance the foodstuff to which they are applied as a solution and on which they are dried as a coating. A typical application is described below.

9.65 grams of hydrolyzed cereal solids (Mor-Rex) having a dextrose equivalency of 10 to 13 and composed of the following assay of carbohydrates on a dry basis are used to prepare a solution by addition to 14 grams of water at 110°F and 1.08 grams of APM.

Mor-Rex Analysis

DE	10 to 13
pH	4.5 to 5.5
Carbohydrate, % db	
Dextrose	1
Di-saccharide	4
Tri-saccharide	5
Tetra-saccharide	4
Penta-saccharide	4
Hexa-saccharide and above	82

The dextrin solution is prepared by stirring the warm solution to eliminate any lumps and facilitate mixing and insure solution of the dextrin; the APM is added to the dextrin solution and uniformly mixed and homogenized in a bench-top homogenizer to create a uniform suspension of the APM particles which is allowed to cool to ambient room temperature, i.e., 72°F. The solution is ready for spray application.

The solution thus produced can be sprayed on 444.5 grams of corn flakes and then dried at air temperature of 180°F for 25 minutes until a moisture content of approximately 2.5% is obtained. Homogenizing the mixture in water produces a very discrete finite dispersion of the APM such as would permit application as a fine slurry onto the corn flakes by atomization, 24 grams of the coating being to uniformly coat all of the cereal flakes used resulting in a coating of sweetening of about 0.24% by weight.

The coated cereal system had a sweetness quite comparable to that of sucrose-coated corn flakes and did not have the overly frosted appearance that many consumers associate with an undesirable or excessive amount of sucrose; the product when tested, in packaging, will be found to be stable over a period of at least 3 months storage when tested under accelerated packaging conditions of high and low relative humidity.

The level of the use of the starch hydrolysate itself will be dictated more by the intended appearance of the coating on the flake or other comestible rather than its functionality as such; thus, for some breakfast cereal applications, it may be desirable to use a larger amount of the hydrolysate for purposes of achieving a gloss simulating the gloss of a sugar-sweetened cereal product having a low level of reducing saccharides and highly suggestive therefore of a noncrystallized sugar coating. On the other hand, other applications may call for the incorporation of substances such as fats, starches, and such which are operative to create a dull or crystalline appearance suggestive of other sweetened cereal applications ranging from a fondant frosting or topping appearance to a thin light crystallization synonymous with a surface sanding which is common to many current ready-to-eat breakfast cereal applications.

CHALCONES AND MALTOLS

FLAVANONE GLYCOSIDE DIHYDROCHALCONES

The extraordinary sweetness of the dihydrochalcones was identified by *R.M. Horowitz and B. Gentili; U.S. Patent 3,087,821; April 30, 1963; assigned to the U.S. Secretary of Agriculture.* The dihydrochalcones have the following structure:

(1)

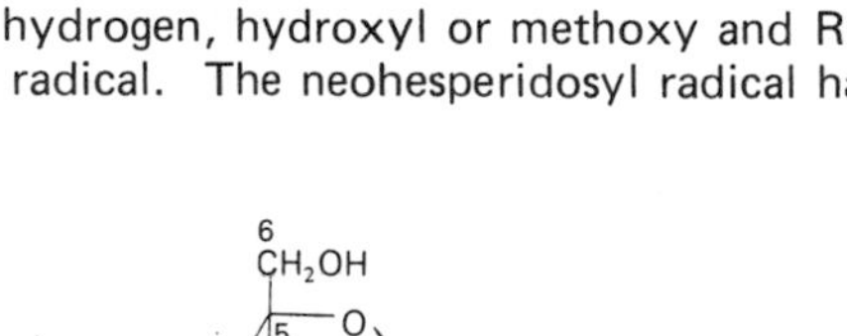

R is hydroxy; R' is hydrogen, hydroxyl or methoxy and R" is the glucosyl or the neohesperidosyl radical. The neohesperidosyl radical has the structure

(2)

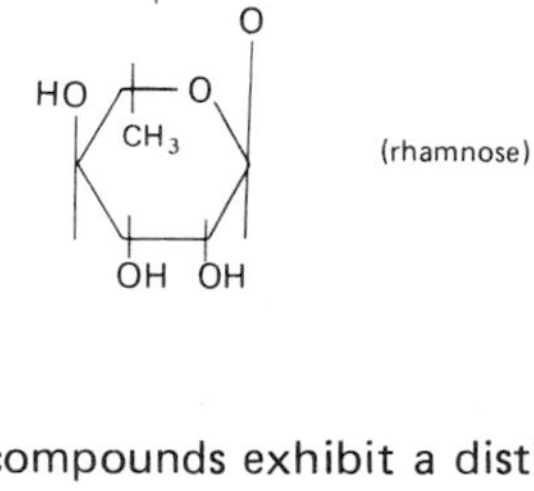

(glucose)

(rhamnose)

It was found that these compounds exhibit a distinguishing characteristic in that they are intensely sweet. The unusual nature of the compounds is demonstrated by the fact that sweetness is not a general characteristic of dihydrochalcones as a class. Indeed, many dihydrochalcones which differ from the compounds by the nature of the substituents on the basic dihydrochalcone nucleus are tasteless or even bitter. Included in this category are such compounds as hesperidin dihydrochalcone, phloridzin and poncirin dihydrochalcone. The dihydrochalcones are useful as sweetening agents for foods and other products of all kinds and they are 300 to 500 times sweeter than sucrose.

The dihydrochalcones can be prepared from the corresponding flavanone glycosides by known methods. The process involves two steps: conversion of the flavanone glycoside to the corresponding chalcone and reduction of this intermediate to the dihydrochalcone.

In the first step the flavanone glycoside is contacted with a relatively concentrated aqueous solution of an alkali, for example a 20 to 25% aqueous solution of sodium or potassium hydroxide at room temperature. This results in opening the heterocyclic ring between the 1 and 2 positions, producing the chalcone form of the flavanone glycoside. This intermediate is recovered from the reaction system in conventional manner, that is, by acidifying the system to precipitate the chalcone.

In the next step the aim is to hydrogenate the ethylenic double bond of the chalcone to produce the dihydrochalcone. This is readily accomplished by contacting the intermediate with hydrogen gas in the presence of a hydrogenation catalyst, typically finely divided platinum, palladium or Raney nickel.

The dihydrochalcone being soluble in water and stable, can be purified by recrystallization from water solutions in conventional manner. By applying this synthesis to the flavanone glycoside having the proper substituents, any of the dihydrochalcones of the process may be prepared. The reactions involved in the synthesis are as follows:

R"–O– ... OH O ... R, R' — flavanone glycoside

↓ alkali

R"–O– ... –OH ... –C(=O)–CH=CH– ... R, R' — chalcone

↓ H_2

R"–O– ... –OH ... –C(=O)–CH_2–CH_2– ... R, R' — dihydrochalcone

Typical examples of dihydrochalcones are listed on the following page. The structure of these compounds is as shown in Formula (1) where R, R' and R" have the stated values.

Naringin dihydrochalcone: R is hydroxyl at the 4' position, R' is hydrogen, R" is the neohesperidosyl radical.

Neohesperidin dihydrochalcone: R is hydroxyl at the 3' position, R' is methoxyl at the 4' position, R" is the neohesperidosyl radical.

Prunin dihydrochalcone: R is hydroxyl at the 4' position, R' is hydrogen, R" is the glucosyl radical.

Eriodictyol 7-neohesperidoside dihydrochalcone: R and R' are each hydroxyl at the 3' and 4' positions, R" is the neohesperidosyl radical.

Homoeriodictyol-7-neohesperidoside dihydrochalcone: R is hydroxyl at the 4' position, R' is methoxyl at the 3' position, R" is the neohesperidosyl radical.

There is substantial variation in sweetness among the individual compounds, depending on such factors as the nature of the glycosyl radical on ring A and the nature and the number of substituents on ring B. Of the compounds tested, neohesperidin dihydrochalcone is outstanding in its level of sweetness, being on a molar basis 15 to 20 times sweeter than saccharin.

Example: Dilute aqueous solutions of various concentrations of the dihydrochalcone were prepared. These solutions were tested by a taste panel in conjunction with a 2×10^{-4} molar solution of saccharin, as a control. The purpose of this test was to measure the sweetness of the compounds by determining the concentration of each solution required to provide a sweetness effect equal to that of the saccharin solution. From the resulting data, the sweetness of the compounds (in relation to saccharin) was determined by applying the formula:

$$\text{Sweetness} = \frac{2 \times 10^{-4}}{\text{Molar concentration of solution of dihydrochalcone having equal sweetness to saccharin solution}}$$

The results are tabulated below:

Compound	Molar Concentration*	Sweetness
Saccharin (control)	2×10^{-4}	1
Neohesperidin dihydrochalcone	1×10^{-5}	20
Naringin dihydrochalcone	2×10^{-4}	1
Prunin dihydrochalcone	5×10^{-4}	0.4

*Providing equal sweetness.

It is obvious from the above table that neohesperidin dihydrochalcone is by far the sweetest of the dihydrochalcones tested. Moreover, although the prunin derivative is less sweet than saccharin it still is intensely sweet.

HESPERETIN DIHYDROCHALCONE

Synthesis and General Use

In another synthesis by *R.M. Horowitz and B. Gentili; U.S. Patent 3,583,894; June 8, 1971; assigned to the U.S. Secretary of Agriculture* an enzyme step is included for converting of hesperidin to hesperetin dihydrochalcone. The following steps are applied.

(A) The starting compound, hesperidin, which is tasteless, is converted into hesperidin chalcone. This step is carried out in conventional manner as by contacting the starting compound with a solution of an alkali, for example, a 10 to 25% aqueous solution of NaOH or KOH at room temperature.

(B) The hesperidin chalcone is then converted into the corresponding dihydrochalcone, by conventional hydrogenation. Thus, for example, the chalcone is contacted with hydrogen gas in the presence of a hydrogenation catalyst such as finely divided platinum, palladium, or Raney nickel.

(C) The hesperidin dihydrochalcone is then converted into hesperetin dihydrochalcone glucoside by enzymatic hydrolysis. In this hydrolysis the sugar moiety of the hesperidin dihydrochalcone is attacked, resulting in splitting off the rhamnose portion. The remainder of the molecule remains the same. The hydrolysis of the β-rutinosyl radical is shown below.

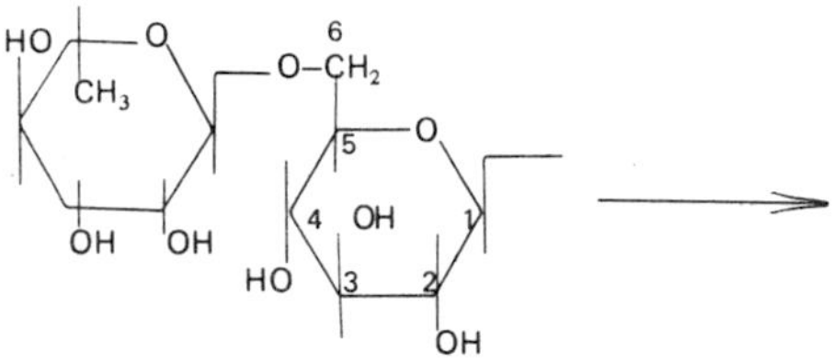

β-rutinosyl or 6-O-α-L-rhamnosyl-β-D-glucosyl radical

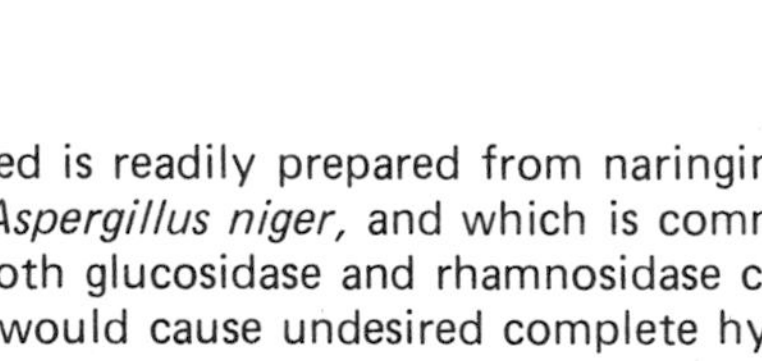

L-rhamnose

β-D-glucosyl radical

The special enzyme used is readily prepared from naringinase, an enzyme elaborated by the fungus *Aspergillus niger,* and which is commercially available. Naringinase contains both glucosidase and rhamnosidase components; therefore, if it were used as is it would cause undesired complete hydrolysis.

However, by a simple treatment, the glucosidase component can be inactivated without substantial loss of rhamnosidase activity. This is accomplished by heating an aqueous dispersion of naringinase (maintained at a pH of 6.4 to 6.8) to a temperature of 60° to 65°C and holding it at that temperature until the glucosidase component is inactivated.

In a typical application, the glucosidase-free naringinase is added to an aqueous solution of hesperidin dihydrochalcone. If the pH of the mixture is not at about neutrality (6.5 to 7) it is brought to this level by addition of a conventional buffer.

The reaction mixture is then held at a temperature of about 25° to 50°C until the desired reaction is essentially complete. After completion of the hydrolysis step, the product, hesperetin dihydrochalcone glucoside, can be readily separated from any completely hydrolyzed by-product (hesperetin dihydrochalcone) by applying conventional separation techniques capable of taking advantage of the fact that the product is soluble in water while the by-product is very insoluble in water.

The tasteless compound, hesperidin dihydrochalcone, is converted into the intensely sweet compound, hesperetin dihydrochalcone glucoside. This compound exhibits a sweetness about equal to that of saccharin, on a molar basis. Moreover, hesperetin dihydrochalcone glucoside imparts a more agreeable sweetness (i.e., less clinging and absence of bitter or other secondary taste effects) than some of the other dihydrochalcone sweeteners such as naringin dihydrochalcone or prunin dihydrochalcone.

Example: 50 g of hesperidin was dissolved in 250 ml of 10% aqueous potassium hydroxide. This solution, which contains hesperidin chalcone after brief standing, was used directly in the next step.

5 g of a hydrogenation catalyst (10% palladium-carbon) was added and the solution hydrogenated at about 30 psi at room temperature until the hydrogenation was complete (90 minutes). The product was filtered and brought to pH 6.7 by the addition of hydrochloric acid. An aliquot of this solution representing 6 g of hesperidin was diluted to 245 ml with water and used in the next step.

180 ml of the enzyme solution (prepared as described below) was added to the solution of hesperidin dihydrochalcone and the system held at 45°C for 13 hours. After 3 hours the mixture was seeded with hesperetin dihydrochalcone glucoside. The product was refrigerated overnight, then filtered through Celite (a diatomaceous earth filter aid). The filter cake was extracted with boiling methanol, the methanol extract decolorized with charcoal, filtered and taken to dryness. The residue crystallized from ethanol, giving 1.68 g of chromatographically pure hesperetin dihydrochalcone glucoside, MP 122°C.

An additional quantity was obtained from the aqueous filtrate by extracting first with ether (to remove traces of hesperetin dihydrochalcone) and then with ethyl acetate to extract the desired product. The combined yield of crystalline hesperetin dihydrochalcone glucoside was 2.15 g (47% based on hesperidin).

Preparation of the Special Enzyme: 10 g of naringinase A was added with shaking to a solution made up of 18.2 ml of 0.1 M potassium dihydrogen phosphate and 182 ml of 0.2 M disodium hydrogen phosphate. The mixture, after filtering through Celite, gave a clear solution having a pH 6.6. This was kept at 65°C for 2 hours to inactivate the β-glucosidase present. The resulting solution was used as described in the last step, above.

Organic and Aqueous-Organic Solvents

G.P. Rizzi; U.S. Patent 3,739,064; June 12, 1073; assigned to The Procter & Gamble Company developed an improvement in the use of hesperetin dihydrochalcone. Horowitz and Gentili teach that partial hydrolysis of the neohesperidosyl group, with the removal of the rhamnose moiety, results in the formation of glucosyl dihydrochalcone compounds having an intense sweetness comparable to that of saccharin. It is necessary that the dihydrochalcones have a sugar substituent in their structures to exhibit any useful degree of sweetness.

Complete removal of the sugar moiety from these compounds and replacement of the sugar group with hydrogen yields the aglycone, hesperetin dihydrochalcone, which is characterized as being only moderately sweet. Because of its low level of sweetness, hesperetin dihydrochalcone has not been employed as a sweetening agent. It has been discovered, however, that hesperetin dihydrochalcone, when solubilized in the presence of an organic solvent or aqueous-organic solvents, exhibits an enhanced and useful degree of sweetness 100 times that of sucrose. The aglyconic dihydrochalcones of the process have the following formula:

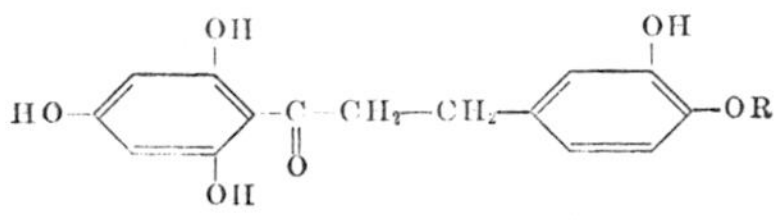

When R in the above formula is methyl, the compound is hesperetin dihydrochalcone. Hesperetin dihydrochalcone can be prepared from hesperidin by hydrolysis in aqueous mineral acid, followed by catalytic hydrogenation. Alternatively, hesperetin dihydrochalcone glucoside, prepared by the method of Horowitz, U.S. Patent 3,429,873, can be subjected to further hydrolysis with removal of the β-D-glucose moiety and recovery of the aglycone, hesperetin dihydrochalcone. Likewise, U.S. Patent 3,375,242 describes a process for condensing naringin with isovanillin to yield neohesperidin chalcone which, on hydrogenation and complete hydrolysis of the sugar, yields hesperetin dihydrochalcone.

When R in the above formula is ethyl, the compound is homohesperetin dihydrochalcone. When R is propyl, the compound is bis-homohesperetin dihydrochalcone. When R is any of the indicated groups other than methyl, the compounds are homologs of hesperetin dihydrochalcone and are referred to as hesperetin-like dihydrochalcones. These homologs can be prepared in much the same fashion as hesperetin dihydrochalcone.

Many of the ingestible polar, organic liquids which can be used to solubilize hesperetin dihydrochalcone and the hesperetin-like dihydrochalcones, especially the esters, have flavor properties of their own and are recognized as being major constituents in many natural flavor oils. Thus, when such solvents are used, even at a low concentration, they will impart to the food being sweetened some of their own flavor; this is sometimes desirable.

However, when it is desired to prepare artificial sweetening compositions containing hesperetin dihydrochalcone and hesperetin-like dihydrochalcones having essentially no flavor sensation other than that of sweetness, it is necessary to employ as the ingestible organic solvent for the dihydrochalcone a material having little or no flavor properties of its own.

Of course, the compounds must serve to solubilize and enhance the sweetness of the hesperetin dihydrochalcone and hesperetin-like dihydrochalcones. Two ingestible organic solvents which fulfill these requirements are ethyl alcohol and 1,2-dihydroxypropane.

Aqueous solutions of hesperetin dihydrochalcone and the hesperetin-like dihydrochalcones are almost imperceptibly sweet, but when properly solubilized with ingestible organic solvents, the sweetness of these compounds is enhanced and the solutions are suitable for use as artificial sweetening compositions.

This enhanced sweetness of solutions of hesperetin dihydrochalcone and the hesperetin-like dihydrochalcones, properly solubilized, was evaluated and a relative scale of sweetness established for these solutions when compared with other sweetening agents. These relative values were determined by a panel of 10 volunteer tasters who sampled solutions of various sweetening agents at various concentrations to establish a lower threshold concentration for the perception of sweetness. The data in the following table indicate the relative sweetness of some of the materials tested.

Substance	Taste Threshold Concentration (molar)	Approximate Relative Sweetness (molar basis)
Sucrose (standard)	4-6 x 10^{-2}	1
Sodium saccharin	8-10 x 10^{-5}	560
Naringin dihydrochalcone	6-10 x 10^{-4}	63
Neohesperidin dihydrochalcone	6-10 x 10^{-5}	630
Homoneohesperidone dihydrochalcone	2-6 x 10^{-5}	1,260
Bis-homoneohesperidin dihydrochalcone	1-4 x 10^{-4}	200
Hesperetin dihydrochalcone glucoside	1-4 x 10^{-4}	200
Hesperetin dihydrochalcone	*	**
Hesperetin dihydrochalcone plus 2.5% ethyl alcohol	5 x 10^{-4}	100

*Saturated.
**Almost imperceptible.

As can be seen from the table, hesperetin dihydrochalcone, properly dissolved in the presence of a small amount of ethyl alcohol, results in the formation of a solution which is about 100 times sweeter than an equivalent molar concentration of the standard sugar, sucrose. (The sweetness of sucrose is not enhanced by ethyl alcohol.)

This enhanced sweetness of hesperetin dihydrochalcone is also noted when the compound is dissolved in other organic solvents and solvent mixtures. For example, 5 parts of hesperetin dihydrochalcone can be dissolved in a mixture of 95 parts of water and 5 parts of ethyl acetate to yield a solution which is about 100 times as sweet as an equal concentration of sucrose. Likewise, a solution consisting of 1 part hesperetin dihydrochalcone, 2 parts of 1,2-dihydroxypropane and 98 parts of water is about 100 times as sweet as an equal concentration of sucrose.

A solution consisting of 2 parts of hesperetin dihydrochalcone dissolved in a solvent consisting of 1 part ethyl alcohol, 1 part benzaldehyde, and 97 parts water exhibits a sweetness about 100-fold greater than an equivalent amount of sucrose.

Example 1: A concentrated, nonaqueous sweetening composition having an intense sweetness is prepared in the following manner: 10 g of hesperetin dihydrochalcone is dissolved in 100 g of ethyl alcohol. The resulting solution is suitable, without further treatment, for use as a highly concentrated sweetening composition. The hesperetin dihydrochalcone is replaced by an equivalent amount of iso-homohesperetin dihydrochalcone and butoxy-homohesperetin dihydrochalcone, respectively, and sweetening compositions are secured.

Example 2: One-half gram of homohesperetin dihydrochalcone is dissolved in a mixture of 1000 g of water and 50 g of sorbitan monooleate polyoxyethylene with gentle warming. The resulting solution is suitable for use as a sweetening composition without further treatment.

Example 3: A vanilla-flavored sweetening composition suitable for simultaneously sweetening and flavoring foodstuffs is prepared as follows: one part hesperetin dihydrochalcone is dissolved in ten parts vanillin and 50 parts ethyl alcohol.

Glycerol Solvent

Sugar alcohols such as glycerol are used as the cosolubilizing agent for dihydrochalcone compounds as discussed by *G.P. Rizzi and J.S. Neely; U.S. Patent 3,743,716; July 3, 1973; assigned to The Procter & Gamble Company*. The sugar alcohols or polyols have a natural sweetness which is less than sucrose. When properly cosolubilized with dihydrochalcone compounds by means of ingestible polar, organic liquids, the natural sweetness of these polyol compounds is enhanced and the resulting solutions are suitable for use as artificial sweeteners. This enhanced sweetness of solutions of the polyols containing codissolved dihydrochalcones was evaluated by volunteer tasters who sampled solutions of various polyols codissolved with the dihydrochalcones.

In this way, it was found that a minimum concentration of about 5×10^{-5} molar hesperetin dihydrochalcone or homohesperetin dihydrochalcone is required to enhance the sweetness and improve the flavor perception of polyols. On a ratio basis, about 10^{-6} part dihydrochalcone codissolved with one part polyol, by weight, is sufficient to enhance the natural sweetness of the polyol.

Thus, a solution having the weight composition: 60% sorbitol, 1% ethyl alcohol, 6×10^{-5}% hesperetin dihydrochalcone, the remainder being water, is sweeter than a similar composition which does not contain the dihydrochalcone and is approximately equivalent in sweetness to a 60% aqueous solution of sucrose.

The polyols, i.e., glycerol and sugar alcohols, are soluble in water and ingestible polar, organic liquid solvents and solvent mixtures. The solid polyols can be dissolved in any of these solvents. Glycerol, due to its hygroscopicity and low melting point, is generally encountered as a 95% solution, rather than as a pure solid. As such, it need not be dissolved in a solvent. However, an ingestible polar, organic liquid must be used, to codissolve the dihydrochalcone and the 95% glycerol liquid. Alternatively, the solvents and solvent mixtures can be used to further dilute the liquefied glycerol if so desired.

Example 1: 0.2 g of hesperetin dihydrochalcone is dissolved in 100 g of ethyl alcohol and 1,000 g of water is admixed therewith; 80 g of sorbitol is then dissolved in the solution. The resulting solution is suitable for use as a sweetening composition without further treatment.

Example 2: ¾ g of hesperetin dihydrochalcone is dissolved in 20 g of 1,2-dihydroxypropane with gentle warming and 1,000 g of water containing 300 g dissolved xylitol is added. The resulting solution is suitable for use as a sweetening composition without further treatment.

Example 3: A concentrated, nonaqueous sweetening composition having an intense sweetness is prepared in the following manner: 0.1 g of hesperetin dihydrochalcone is dissolved in 100 g of ethyl alcohol and 1,000 g glycerol (95%) is added thereto. The resulting solution is suitable, without further treatment, for use as a highly concentrated sweetening composition.

Example 4: One-half gram of homohesperetin dihydrochalcone is dissolved in a mixture of 1,000 g of water and 50 g of sorbitan monooleate polyoxyethylene with gentle warming; 150 g of xylitol is dissolved in the solution. The resulting solution is suitable for use as a sweetening composition without further treatment.

Flavor Oil Solvents

Solvents for 3-(m-hydroxyphenyl)-phloropropiophenone (substituted hesperetin dihydrochalcone) include polar compounds in aqueous mixtures. Once dissolved the compounds exhibit their sweetening properties. *G.P. Rizzi; U.S. Patent 3,932,678; January 13, 1976; assigned to The Procter & Gamble Company* found that various naturally occurring and synthetically reconstituted flavor oils which are obtainable from plants are suitable to solubilize 3-(m-hydroxyphenyl)-phloropropiophenone. It is not possible to specify with certainty the compositions of these various oils other than that they are highly complex liquid mixtures containing polar organic compounds such as lactones, ketones, aldehydes, thiols, carboxylic acids and acid esters.

Some flavor oils contain nitriles, imides, organonitrates and the like. A long history of use by humans has shown that such flavor oils are physiologically acceptable and they are preferred for use as ingestible organic solvents. Often such flavor oils are employed with ethyl alcohol and 1,2-propylene glycol to provide various extracts, tinctures and concentrates containing the oils. These naturally occurring, ingestible organic solvent oils can also be used with water as a cosolvent.

Examples of flavor oils suitable for use as solubilizing agents for 3-(m-hydroxyphenyl)-phloropropiophenone include: oil of sweet birch, oil of spearmint, oil of wintergreen, oil of sassafras, cedar wood oil, anise oil, pine oil, dill oil, celery seed oil, various citrus oils including lemon, orange, lime, tangerine and grapefruit oils, clove oil, peppermint oil, cassia, carrot seed oil, cola concentrate, ginger oil, angelica oil, singly and in mixtures.

Any of the above polar, organic liquids can be used in conjunction with water to provide aqueous-organic solvent systems useful in the preparation of the artificial sweetener compositions. For example, 3-(m-hydroxyphenyl)-phloropropiophenone can be dissolved in ethyl alcohol and then diluted with water to yield a 0.002 molar solution of the phloropropiophenone containing 5% ethyl alcohol, which composition is suitable for sweetening foods and beverages.

The following examples illustrate the sweetener compositions. In each instance, the 3-(m-hydroxyphenyl)-phloropropiophenone is dissolved in the ingestible organic solvent by stirring at 30° to 50°C. When water is used in the composition as a cosolvent, the 3-(m-hydroxyphenyl)-phloropropiophenone is predissolved in the organic solvent and the resulting solution is mixed with the water.

Example 1: 5 g of 3-(m-hydroxyphenyl)-phloropropiophenone are dissolved in 100 g of ethyl alcohol and 1,000 g of water is admixed. The resulting solution is suitable for use as a sweetener composition without further treatment.

Example 2: ¾ g of 3-(m-hydroxyphenyl)-phloropropiophenone is dissolved in 20 g of 1,2-dihydroxypropane at 50°C. The resulting solution is suitable for use as a sweetener composition without further treatment.

Example 3: A concentrated, nonaqueous sweetener composition having an intense sweetness is prepared in the following manner: 10 g of 3-(m-hydroxyphenyl)-phloropropiophenone is dissolved in 100 g of ethyl alcohol. The resulting solution is suitable for use as a highly concentrated sweetener composition without further treatment.

Example 4: ½ g of 3-(m-hydroxyphenyl)-phloropropiophenone is dissolved in a mixture of 1,000 g of water and 50 g of sorbitan monooleate polyoxyethylene with gentle warming. The resulting solution is suitable for use as a sweetener composition without further treatment.

Example 5: ⅒ part of 3-(m-hydroxyphenyl)-phloropropiophenone is dissolved in 10 parts of isoamyl acetate and the resulting solution provides a banana-flavored sweetener composition.

Example 6: 0.2 g of 3-(m-hydroxyphenyl)-phloropropiophenone is dissolved in 100 g of ethyl alcohol and 1,000 g of water; 80 g of sucrose is then dissolved in the solution. The sweetness of the sucrose is substantially enhanced and the resulting solution is suitable for use as a sweetener composition without further treatment. The sucrose is replaced by an equivalent amount of fructose, glucose, lactose and cellobiose, respectively, and equivalent results are secured.

Example 7: 5×10^{-6} mol of 3-(m-hydroxyphenyl)-phloropropiophenone is dissolved in 15 ml of a synthetic pineapple oil (corresponding to winter fruit) consisting of 2.91 parts ethyl acetate, 0.61 part acetaldehyde, 0.45 part methyl n-valerate, 0.60 part methyl isovalerate, 1.40 parts methyl isocaproate and 0.75 part methyl caprylate and thence diluted to one liter with a solution of 50 parts mannose in 100 parts water. The natural sweetness of the mannose is enhanced and a pineapple-flavored sweetener composition is provided.

Example 8: One-half part of 3-(m-hydroxyphenyl)-phloropropiophenone is dissolved in 10 parts bitter almond oil and the resulting solution is added to a solution of 700 parts sorbitol in 1,000 parts water. The weak natural sweetness of the sorbitol is enhanced and an almond-flavored sweetener composition is provided.

The bitter almond oil is replaced by oil of sweet birch, oil of spearmint, oil of wintergreen, oil of sassafras, cedarwood oil, anise oil, pine oil, dill oil, celery seed oil, lemon oil, lime oil, orange oil, grapefruit oil, tangerine oil, peppermint oil, clove oil, cassia, carrot seed oil, cola concentrate, ginger oil and angelica oil, respectively, and sweetener compositions of the corresponding flavors are secured.

The sorbitol is replaced by an equivalent amount of xylitol, mannitol, glycerol and 1,2,3,4-tetrahydroxybutane, respectively, and equivalent results are secured.

Sugar-Dihydrochalcone Solvent

Simple mixture of sugars and the dihydrochalcones does not result in the desired sweetness-enhancing effect. Therefore, it is critical that a suitable cosolvent for the dihydrochalcone and the sugar be employed. The sugars are all very soluble in water and are also soluble in a variety of polar, organic liquids.

The dihydrochalcones are almost entirely insoluble in water but are soluble in various polar, organic liquids. The dihydrochalcones are also soluble in mixtures of polar, organic liquids and water. It is therefore possible to provide organic solvents and organic-aqueous solvents properly formulated to enhance the natural sweetness of the sugar and provide improved sweetening compositions.

G.P. Rizzi; U.S. Patent 3,751,270; August 7, 1973; assigned to The Procter & Gamble Company states that codissolution of the aglyconic dihydrochalcone and the sugar so as to enhance the natural sweetness of the sugar can be accomplished by simple mixture with any of the solvent compositions herein noted. Gentle warming can be used to speed the rate of dissolution but this is entirely optional as these compounds are found to be quite soluble in the solvents.

Preparation of aqueous-organic solutions of the dihydrochalcone and sugar can be accomplished by dissolving the dihydrochalcone in the ingestible polar, organic liquid and mixing the resulting solution with water containing the sugar, and also by simply adding the dihydrochalcone to a mixture of water, sugar and ingestible polar, organic liquid.

In the following test, four subjects were selected at random and tasted the test solutions, without knowledge of their composition. The results are as shown in the following table.

Sweetness Enhancement by Dihydrochalcones

Solution Composition	Degree of Sweetness
(A) 0.25% glucose in water	None
(B) 5 x 10^{-5} molar hesperetin dihydrochalcone (0.25% ethyl alcohol + 99.75% water solvent)	None
(C) 0.25% glucose + 5 x 10^{-5} molar hesperetin dihydrochalcone (codissolved in 0.25% ethyl alcohol + 99.75% water solvent)	Sweet

The enhanced sweetness of the combination of the dihydrochalcone and the sugar was apparent to all subjects. Because of the wide variations in the response to physiological stimuli, and the subjective nature of that response, it is inappropriate to attempt to quantify the results from the tests designed to demonstrate the enhanced sweetness properties of sugars when codissolved with the dihydrochalcones.

In general, it can be said that dissolution of the dihydrochalcones with the sugars, especially the preferred sugars, glucose, fructose and sucrose, in the ratios noted, causes a 1.5- to 2-fold enhancement of the natural sweetness of the sugars.

Example 1: 0.2 g of hesperetin dihydrochalcone is dissolved in 100 g of ethyl alcohol and 1,000 g of water is admixed therewith; 80 g of sucrose is then dissolved in the solution. The resulting solution is suitable for use as a sweetening composition without further treatment.

Example 2: ¾ g of hesperetin dihydrochalcone is dissolved in 20 g of 1,2-dihydroxypropane with gentle warming and 1,000 g of water containing 300 g dissolved glucose is added. The resulting solution is suitable for use as a sweetening composition without further treatment. The 1,2-dihydroxypropane is replaced by an equivalent amount of sorbitan monooleate polyoxyethylene and a sweetening composition is secured.

Example 3: ½ g of homohesperetin dihydrochalcone is dissolved in a mixture of 1,000 g of water and 50 g of ethyl alcohol with gentle warming; 150 g of fructose is dissolved in the solution. The resulting solution is suitable for use as a sweetening composition without further treatment.

Example 4: A vanilla-flavored sweetening composition suitable for simultaneously sweetening and flavoring foodstuffs is prepared as follows: one-tenth part hesperetin dihydrochalcone is dissolved in 10 parts vanillin and 500 parts water, and 50 parts cellobiose added.

DERIVATIZED DIHYDROCHALCONES

Trihydroxyl-4′-Alkoxy Alkyl Sulfonate

A group of highly soluble dihydrochalcones is described by *G.A. Crosby, G.E. Dubois and N.M. Weinshenker; U.S. Patent 3,974,299; August 10, 1976; assigned to Dynapol.* These materials have the chemical structure

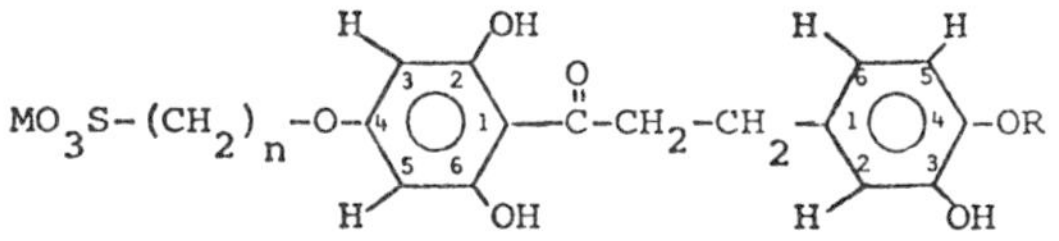

$$MO_3S-(CH_2)_n-O-C_6H_2(OH)_2-C(=O)-CH_2-CH_2-C_6H_3(OH)-OR$$

wherein M is a physiologically acceptable metal cation, n is 1, 2 or 3 and R is methyl, ethyl or propyl. These compounds have tastes characterized by sweetness and no appreciable aftertaste or companion taste. Compared to many other dihydrochalcones they are very water-soluble so that intense sweet flavors can be generated.

The materials of the above formula are conveniently formed, in a general sense, by the mechanism of alkylating the 7 hydroxyl group of the natural product hesperetin, shown on the following page, or a 4′ ethoxy or propoxy equivalent of hesperetin with an alkyl sulfonate group and thereafter opening and reducing the alkylated hesperetin's flavanone structure to the corresponding dihydrochalcone by contacting the alkylated hesperetin with base under reducing conditions, for example, hydrogen plus a catalyst.

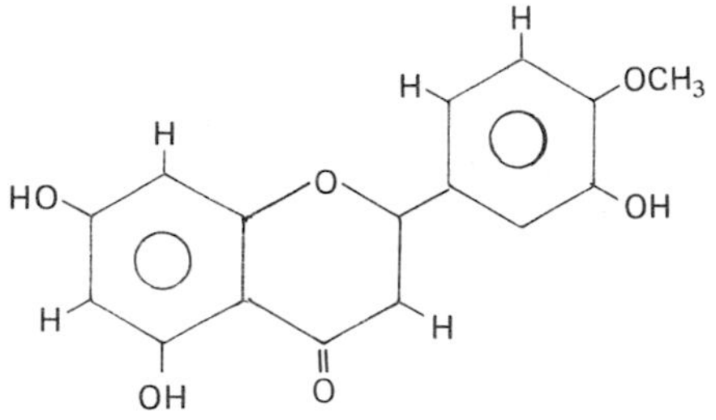

The ring opening and reduction may be carried out in two steps, the opening brought about by contacting the alkylation product with a relatively strong base such as an aqueous alkali metal hydroxide solution for 0.2 to 4 hours at 10° to 100°C, the hydrogenation being carried out with hydrogen gas and a catalyst such as a supported noble metal catalyst. Preferably the reduction and opening are carried out simultaneously with base, catalyst and hydrogen.

These dihydrochalcones find application as sweeteners of consumable materials. In this use they are admixed with edible materials such as foods, beverages, medicines, in amounts effective for affording the degree of sweetness desired.

Illustrations of the type of commercial products in which the sweetening agent or combinations with known sweetening agents can be used are granulated mixes which upon reconstitution with water provide noncarbonated drinks; instant pudding mixes; instant coffee and tea; pet foods; livestock feed; tobacco and consumable toiletries such as mouthwashes and toothpastes.

The amount of dihydrochalcone employed can vary widely, just as the amount of natural sugar sweetener employed varies from person to person and food application to food application. As a general rule, the weight of dihydrochalcone added will be about $^1/_{100}$ to $^1/_{1000}$ the weight of sucrose required to yield the same sweetness. Thus, additions of from about 0.001 to 0.50% by weight (basis edible substance) may be usefully employed. The materials offer the advantage that their solubility permits such addition to most food systems.

4-Alkoxy Substituted

Another derivative is given by *G.A. Crosby and G.E. DuBois; U.S. Patent 3,976,687; August 24, 1976; assigned to Dynapol* in which the alkoxy group at the 4 position contains a methoxy group to which is attached a carboxylic acid group (see U.S. Patent 3,974,299 above).

The amount of dihydrochalcone used can vary widely, just as the amount of natural sugar sweetener varies from person to person and food application to food application. As a general rule, the weight of derivatized dihydrochalcone added will be about $^1/_{100}$ to $^1/_{500}$ the weight of sucrose required to yield the same sweetness. Thus, additions of 0.0001 to 0.05% by weight (basis edible substance) may be used.

Trihydroxy-4'-Alkoxy-4-Carboxyl Substituted

The derivative described by *G.A. Crosby and G.E. DuBois; U.S. Patent 3,976,790; August 24, 1976; assigned to Dynapol* is a trihydroxyl with an alkoxyl at 4' position and a carboxyl group-substituted methoxy group at the 4 position (see U.S. Patent 3,974,299 above).

Aqueous solutions (0.009% by weight) of a variety of the products and intermediates are formed. They are tested by volunteers to determine their organoleptic properties, with the following results:

Compound	Taste Observed
HO, OH, OH, O, O-CH_3, OH	Moderately sweet*
HO, OH, OH, O, O-n-C_3H_7, OH	Tasteless
HOOC-CH_2O, OH, OH, O, O-CH_3, OH	Extremely sweet, without linger of aftertaste
NaOOC-CH_2O, OH, OH, O, O-CH_3, OH	Extremely sweet, clear, no aftertaste**
COOH, CH_2O, OH, O, O-CH_3, OH, O, CH_2COOH	Tasteless

*This compound had good sweetness but was so insoluble in water that a 0.009% solution could not be simply made up and thus its usefulness as a sweetener would be minimal.

**In a second, more precise comparison taste test, this material is found to be 480-500 times as sweet as sucrose on a pound-for-pound basis. In a blind study, it is found to be more sugar-like in taste than sucrose itself.

As can be seen from the above taste data, the compounds of the process have desirable sweetener properties.

SPECIAL NEOHESPERIDOSIDE

The compound 2',4',6',3-tetrahydroxy-4-n-propoxydihydrochalcone-4'-β-neohesperidoside was prepared from the corresponding flavanone glycoside by *L.O. Krbechek and G. Inglett; U.S. Patent 3,625,700; December 7, 1971; assigned to International Minerals & Chemical Corporation.* The compound has the structural formula

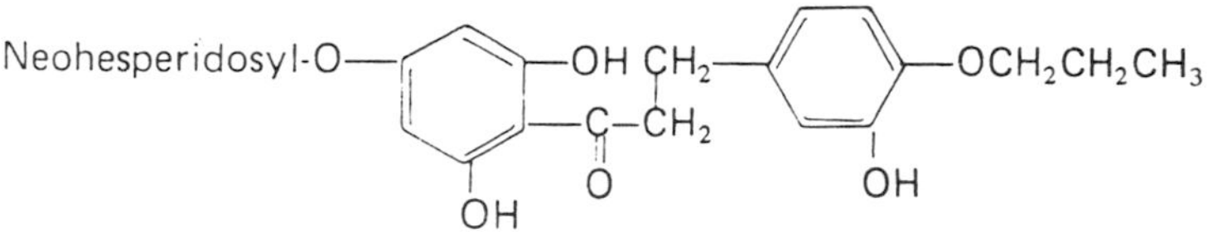

Neohesperidosyl-O—(ring: OH, OH)—C(=O)—CH_2—CH_2—(ring: OH)—$OCH_2CH_2CH_3$

This compound is a stable, water-soluble compound having intensely sweet characteristics. It has been found to be over 2,000 times sweeter than sucrose. In contradistinction to closely similar compounds, 2',4',6',3-tetrahydroxy-4-n-propoxydihydrochalcone-4'-β-neohesperidoside is not characterized by unattractive aftertaste qualities. When it is used at levels where aftertaste can be detected, the sensation observed by an individual is not one of bitterness or metallic taste, but a pleasing, cooling sensation.

Although the compound contains a sugar moiety in the molecule, its intensive sweetness characteristics allow it to be used as a sweetening agent for edible formulations in relatively small amounts (0.0001 to 1% by weight), whereby the incorporation of the compound contributes practically negligibly to the caloric value of the ultimate formulation.

The 2',4',6',3-tetrahydroxy-4-n-propoxydihydrochalcone-4'-β-neohesperidoside may be synthesized from readily available starting materials, i.e., 3-hydroxy-4-n-propoxy benzaldehyde and phloroacetophenone-4'-β-neohesperidoside, the latter being an alkaline degradation product of naringin.

The compound 2',4',6',3-tetrahydroxy-4-n-propoxydihydrochalcone-4'-β-neohesperidoside can be used as a sweetening agent in any material or formulation for human consumption in view of its intense sweetness, water solubility, stability and low caloric value. Representative examples of edible materials, i.e., materials intended for consumption or at least to come in contact with the mouth of the ultimate user include fruits, vegetables, meats, nuts, cereals, carbonated and noncarbonated soft drinks, alcoholic beverages, pastries, candies, frozen and nonfrozen desserts and dessert supplements, chewing gums, toothpastes, mouth rinses, tobacco products, and the like.

Example 1: Phloroacetophenone-4'-β-neohesperidoside in the amount of 7.5 g (0.016 mol) was added to a solution containing about 10.3 g of potassium hydroxide in 20 ml of water at 110° to 115°C. To the resulting hot, clear solution, 3.75 g (0.022 mol) of 3-hydroxy-4-n-propoxybenzaldehyde were added. The solution was maintained at 115° to 120°C for 5 minutes and was then poured onto 150 g of ice. The solution was acidified with cold, dilute hydrochloric acid to provide a solution having a pH of 6.5 and a temperature below 10°C. Ultraviolet examination of the cool, acidified solution showed an intense absorption at about 370 μ, which is characteristic of the chalcone structure.

Cyclization of the chalcone to the flavanone was effected by heating the solution to 80° to 90°C for one-half hour. The heated solution was then cooled to below 40°C (the boiling point of methylene chloride) and extracted with two 100-ml portions of methylene chloride, from which 1.5 g of unreacted 3-hydroxy-4-n-propoxybenzaldehyde were obtained. The precipitate was collected from the cold aqueous phase.

The crude product was recrystallized twice from 100 ml of water and filtered at 30°C, then triturated twice with 35 ml of water and filtered at 80°C to yield 1.3 g (13% yield based on the phloroacetophenone) of 3',5,7-trihydroxy-4'-n-propoxyflavanone-7-β-neohesperidoside, having a melting point of 222° to 224°C. Analysis—Calculated for $C_{30}H_{38}O_{15}$ (percent): 56.42 C; 5.99 H. Found (percent): 56.13 C; 6.40 H.

A solution of 1.2 g of purified 3',5,7-trihydroxy-4'-n-propoxyflavanone-7-β-neohesperidoside in 50 ml of 20% potassium hydroxide was hydrogenated on a Parr shaker in the presence of 0.5 g of 5% palladium on charcoal. The solution was acidified with cold, dilute hydrochloric acid to provide a solution having a pH of 5.5 and a temperature below 10°C.

The catalyst was removed from the solution and sodium chloride was added. The gum-like product was collected from the cold solution. The crude product was recrystallized four times with 50 ml of water to yield 0.6 g (50% yield) of 2',4',6',3-tetrahydroxy-4-n-propoxydihydrochalcone-4'-β-neohesperidoside, melting point of 142° to 144°C.

Example 2: In order to compare the sweetness of 2',4',6',3-tetrahydroxy-4-n-propoxydihydrochalcone-4'-β-neohesperidoside with that of sucrose, serially diluted aqueous solutions of 2',4',6',3-tetrahydroxy-4-n-propoxydihydrochalcone-4'-β-neohesperidoside were prepared and labeled for identification by a sample number.

The resultant samples were tasted by a panel of individuals with reference to a 5% aqueous solution of sucrose in order to determine the dilution of the sweetener of this process having an equivalent sweetness level to that of the sucrose control solution.

The averaged results of these tests revealed that a solution of 2',4',6',3-tetrahydroxy-4-n-propoxydihydrochalcone-4'-β-neohesperidoside having a concentration of about 0.00228% is equivalent in sweetness to the sucrose control solution, i.e., 2',4',6',3-tetrahydroxy-4-n-propoxydihydrochalcone-4'-β-neohesperidoside is characterized by a sweetness of about 2,200 times that of sucrose.

DIHYDROCHALCONE GALACTOSIDE

R.M. Horowitz and B. Gentili; U.S. Patent 3,876,777; April 8, 1975; assigned to the U.S. Secretary of Agriculture present the dihydrochalcone galactosides as high intensity sweetening agents. The dihydrochalcone galactosides have the structure shown on the following page, wherein X represents a radical selected from the category consisting of hydrogen, hydroxy, and lower alkoxy.

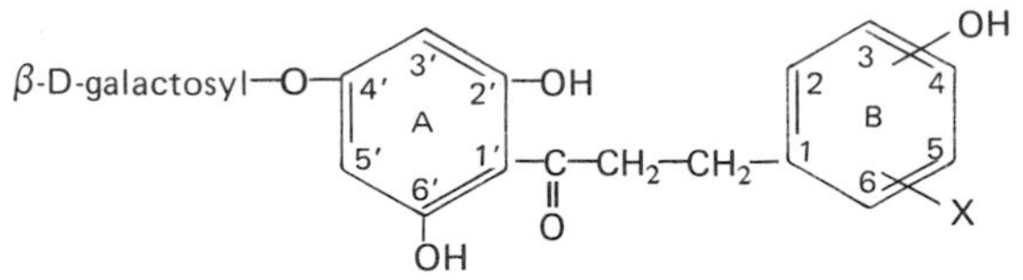

These compounds can be prepared from the corresponding flavanones by known methods, typically by applying the following steps: (1) Attachment of the galactosyl radical to the 7-hydroxy group of the flavanone, (2) conversion of the resulting flavanone galactoside to the corresponding chalcone, and (3) reduction of the latter to the dihydrochalcone.

Example: Hesperetin 7-β-D-Galactoside – A mixture of hesperetin (5.7 g), α-acetobromogalactose (15.5 g), anhydrous potassium carbonate (30 g), and acetone (200 ml) was shaken at room temperature for 20 hours. The mixture was filtered, the filtrate taken to dryness and the residue deacetylated by warming it in 3 N aqueous sodium hydroxide (40 ml) for 15 minutes. On cooling to room temperature, the crude potassium salt of the title compound crystallized out (2.3 g after washing with aqueous methanol and drying).

This was slurried in aqueous methanol, acidified with concentrated hydrochloric acid and kept overnight in a refrigerator. Hesperetin 7-β-O-galactoside was obtained as a tan-colored solid (1.7 g), MP about 205°C. Recrystallized from methanol it was obtained as light tan needles, MP 222° to 223°C. It gave the ultraviolet and proton magnetic resonance spectra expected for hesperetin substituted on the 7-hydroxy group by a β-D-galactosyl residue.

Hesperetin Dihydrochalcone 4'-β-D-Galactoside – A solution of hesperetin 7-β-D-galactoside (0.8 g) in 10% aqueous potassium hydroxide (10 ml) containing 10% palladium-carbon catalyst (0.3 g) was shaken with hydrogen at 30 psi for 3 hours. The catalyst was filtered out and the filtrate acidified with hydrochloric acid in an ice bath. The oily precipitate crystallized overnight.

It was recrystallized from water to give 0.47 g of hesperetin dihydrochalcone 4'-β-D-galactoside (or more precisely, 3,2',6'-trihydroxy-4-methoxy-4'-β-D-galactosyloxy dihydrochalcone) in the form of light tan needles, MP 129° to 131°C. It gave the ultraviolet and proton magnetic resonance spectra expected for hesperetin dihydrochalcone substituted on the 4'-hydroxy group by a β-D-galactosyl residue.

Taste Tests – Taste tests were conducted with hesperetin dihydrochalcone 4'-β-D-galactoside, known dihydrochalcone sweeteners, and saccharin. The results are tabulated below.

Compound	Relative Sweetness (molar basis)
Hesperetin dihydrochalcone 4'-β-D-galactoside	1.5-2.0
Hesperetin dihydrochalcone 4'-glucoside	1
Naringin dihydrochalcone	1
Neohesperidin dihydrochalcone	20
Saccharin (Na)	1

It was observed that although the compounds of the process are less sweet than neohesperidin dihydrochalcone, its sweetness is pleasant and less clinging than that of the latter compound. The process compounds are soluble in water and stable, even in aqueous solution. They are useful for sweetening all types of edible materials.

Typical illustrative examples of edible materials which may be sweetened with the compounds are fruits; vegetables; juices or other liquid preparations made from fruits or vegetables; meat products, particularly those conventionally treated with sweetened liquors, i.e., bacon and ham; milk products such as chocolate dairy drinks; egg products such as nogs, custards, angel food mixes; salad dressings; pickles and relishes; ice creams, sherbets, and ices; ice milk products; bakery products; icings; confections and confection toppings, syrups, and flavors; cake and pastry mixes; beverages such as carbonated soft drinks, fruit ades; wines; dietary-type foods; cough syrups and other medicinal preparations intended for oral administration; dental preparations such as pastes, powders, foams and denture-retaining adhesives; mouthwashes and similar oral antiseptic liquids; tobacco products; adhesives for gumming stamps, envelopes and labels, etc.

They are incorporated in the material to be sweetened in the amount required to attain the desired level of sweetness. Ordinarily, because of their intense sweetness, the compounds are employed in a very minor proportion, that is, in a concentration usually less than 0.5%. It is obvious, however, that there is nothing critical about the concentration of dihydrochalcone which is used; it is simply a matter of attaining a desired sweetness level appropriate to the material in question.

DIHYDROCHALCONE XYLOSIDE

In a similar approach *R.M. Horowitz and B. Gentili; U.S. Patent 3,826,856; July 30, 1974; assigned to the U.S. Secretary of Agriculture* prepared dihydrochalcone xylosides which have the following general formula:

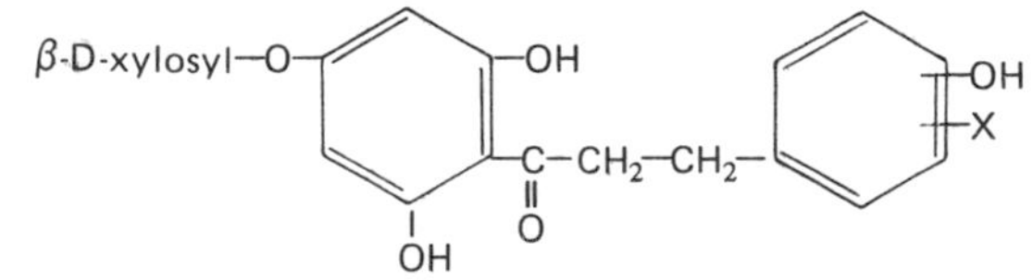

β-D-xylosyl—O— —OH —OH
—C—CH_2—CH_2— —X
‖
O
OH

Taste Test: Taste tests were conducted with hesperetin dihydrochalcone 4'-β-D-xyloside, known dihydrochalcone sweeteners, and saccharin. The results are tabulated below.

Compound	Relative Sweetness (molar basis)
Hesperetin dihydrochalcone 4'-β-D-xyloside	2
Hesperetin dihydrochalcone 4'-glucoside	1
Naringin dihydrochalcone	1
Neohesperidin dihydrochalcone	20
Saccharin (Na)	1

Moreover, it was observed that although the compound is less sweet than neohesperidin dihydrochalcone, its sweetness is pleasant and less clinging than that of the latter compound.

The dihydrochalcone xylosides are very sweet and they are useful for sweetening all types of materials which are intended for consumption or at least for contact with the mouth of the user, such materials being herein generically designated as edible materials. Typical illustrative examples of edible materials which may be sweetened with the compounds are fruits; vegetables; juices or other liquid preparations made from fruits or vegetables; meat products, particularly those conventionally treated with sweetened liquors.

The compounds are incorporated in the material to be sweetened in the amount required to attain the desired level of sweetness. Ordinarily, because of their intense sweetness, the compounds are employed in a very minor proportion, that is, in a concentration usually less than 0.5%. It is obvious, however, that there is nothing critical about the concentration of dihydrochalcone which is used; it is simply a matter of attaining a desired sweetness level appropriate to the material in question. Moreover, the technique of sweetening materials with the compounds offers no difficulty as the selected dihydrochalcone is simply incorporated with the material to be sweetened.

The dihydrochalcones may be added directly to the material or they may be first incorporated with a diluent to increase their bulk so that small amounts of the compounds may be metered into the material. As diluents one may use liquid or solid carriers such as water, glycerol, starch, sorbitol, salt, sugar, citric acid or other nontoxic substance compatible with the material to be sweetened.

SACCHARIN AND DIHYDROCHALCONE

K. Ishii, J. Toda, H. Aoki and H. Wakabayashi; U.S. Patent 3,653,923; April 4, 1972; assigned to Takeda Chemical Industries, Ltd., Japan report on the synergistic action between saccharin and neohesperidin dihydrochalcone to give a low calorie sweetening agent. The synergistic action was not observed in other combinations of known sweetening agents and chalcones and dihydrochalcones, e.g., between sucrose and neohesperidin dihydrochalcone, between saccharin and naringin dihydrochalcone, between saccharin and hesperetin 7-β-D-glucoside dihydrochalcone, between sucrose and naringin dihydrochalcone, etc.

The characteristic and synergistic action between a saccharin sweetener and neohesperidin dihydrochalcone in sweetening effect upon which the process is based is confirmed, for example, by the following test. The relative sweetness ranges of Samples A, B and C were analyzed by a panel consisting of 18 members each of whom has the ability for an extra judgement of sweetness. The results are summarized in the table below.

Sample				
	Solvent	Saccharin Sodium (% w/v)	Neohesperidin Dihydrochalcone (% w/v)	Main Effect*
A	Distilled water	-	0.005	-0.633
B	Distilled water	0.01	-	0.022
C	Distilled water	0.005	0.0025	0.611

*Significant at 1% level

As clearly seen from the table, Sample C shows a stronger sweetness than that of Sample A and that of Sample B, which means that a synergistic action in sweetening effect is exhibited between neohesperidin dihydrochalcone and saccharin sodium.

The sweetening composition may be prepared by mixing a saccharin sweetener with neohesperidin dihydrochalcone. The ratio of neohesperidin dihydrochalcone to the saccharin sweetener is from 0.1:1 to 3:1 by weight. When neohesperidin dihydrochalcone is mixed with a saccharin sweetener in a ratio outside the range above mentioned, there is not exhibited an eminent synergistic action but a tendency to a mere additive action. Most advantageous ratios of neohesperidin dihydrochalcone to the saccharin sweetener are from 0.3:1 to 3:1.

Incorporation of neohesperidin dihydrochalcone into a saccharin sweetener may be effected by simple and thorough mixing. The mixed materials may further be mixed with one or more additional ingredients. These are other sweetening agents such as sucrose or sorbitol; diluents such as carboxymethylcellulose, glucose, lactose or dextrin. The mixed materials can be in powdery or crystalline form, or in the liquid phase, as e.g., in solution in a suitable solvent (water, alcohol, etc.). The resultant powdery, crystalline or granular mixture may be put to use as such, or it may be further coated with a coating agent suitable for practical use. The sweeteners thus prepared are applicable to all kinds of foods to which conventional sweetening agents are applicable.

Ordinarily, because of their intense sweetness, the sweeteners are used in a relatively low concentration, as served, usually less than 0.03%. However, when concentrations of both the main ingredients of the sweetening compositions, i.e., neohesperidin dihydrochalcone and a saccharin sweetener are extremely low, the synergistic action in sweetening effect is not attained. It is best to incorporate the sweetening compositions in foods so as to make the concentration of neohesperidin dihydrochalcone and a saccharin sweetener at least 0.0002% and at least 0.0005%, respectively, relative to the total weight of the foods as served.

Example 1: 2.5 parts by weight of neohesperidin dihydrochalcone and 7.5 parts by weight of saccharin sodium are homogeneously mingled to give 10 parts by weight of powdery sweetening composition. This composition shows a delicious sweetness which is about 400 times as strong as that of sucrose.

Example 2: 25 parts by weight of neohesperidin dihydrochalcone, 25 parts by weight of saccharin sodium and 50 parts by weight of sodium cyclohexylsulfamate are homogeneously mingled. The mixture is kneaded with water, granulated, dried and sieved to give a granular sweetening composition. This composition has a high ability to deliciously sweeten, for example, orange drinks when added thereto in the range of from 0.04 to 0.06%.

MALTOL FLAVOR AND SWEETENER ENHANCERS

Basic Gamma-Pyrone Process

Maltol (3-hydroxy-2-methyl-gamma-pyrone) is one of the most valuable gamma-pyrones. Its utility derives from the fact that maltol enhances the flavor and aroma of a variety of food products, thereby making them even more acceptable to the consumer. Among the foods which are markedly improved in these respects by maltol may be mentioned baked products such as breads, cakes and pies; confections such as candies and ice creams and certain beverages such as coffee. In addition, maltol is used as an ingredient in perfumes and essences.

Maltol has been obtained in limited quantity from natural products by difficult and expensive processes. The commercial production of maltol has depended, for example, upon the destructive distillation of wood and, as is well known, these pyrolysis reactions generally provide low yields of the desired product. Furthermore, isolation processes are by their nature limited in capacity to the total supply of raw material readily available. In addition, there is a tendency for maltol obtained from such destructive distillation processes to contain certain impurities which adversely affect its use as an aroma enhancer.

B.E. Tate and R.L. Miller; U.S. Patent 3,130,204; April 21, 1964; assigned to Chas. Pfizer & Co., Inc. developed the chemical synthesis of maltol from kojic acid, 2-hydroxy-methyl-5-hydroxy-gamma-pyrone, a substance which is readily available in large supply from fermentation processes. The maltol prepared by the process has been found to be free of contamination by impurities ordinarily found in maltol prepared by destructive distillation processes.

The process consists of (1) treating kojic acid with oxygen in the presence of a noble metal catalyst at a pH of at least 10 to form comenic acid, (2) decarboxylating the comenic acid to form pyromeconic acid, (3) treating the pyromeconic acid with formaldehyde at a pH of at least about 5 to form hydroxymaltol and (4) reducing the hydroxymaltol under acidic to substantially neutral conditions to form maltol. This process results in a yield of maltol of about 50% based on kojic acid.

Example 1: In an 8-liter stainless steel vessel fitted with a stirrer and an air sparger is placed a suspension of 350 g of kojic acid in 3,500 ml of water. The pH is adjusted to 11.1 by addition of 256 ml of 50% aqueous sodium hydroxide and then 142 g (7.1 g as metal) of a 5% palladium on charcoal catalyst is added. Air is passed into the suspension at a rate of 2,100 ml per minute.

The reaction, which is slightly exothermic, is maintained at a temperature of 20° to 22°C by occasional application of external cooling. After 11 hours the reaction mixture is filtered to remove the catalyst and is treated with 600 ml of concentrated hydrochloric acid. The crystals of comenic acid which precipitate from the pH 0.5 mixture are removed by filtration, washed with a small amount of cold water and are air-dried. There is obtained 328 g of product. This is 85.3% of the theoretical yield. Titration data indicate the product to be 99.2% pure; therefore, there is obtained an 84.6% yield of comenic acid as corrected for purity.

Example 2: In a 5-gallon stainless steel reaction vessel fitted with a mechanical stirrer, thermometer and distillation receiver are placed 1,750 g of comenic acid prepared as described in Example 1 and 2,675 ml of dimethyltetraethylene glycol (Ansul Ether E-181). A vacuum of 130 to 160 mm of Hg is applied to the system and the reaction mixture is heated to 210° to 215°C during one hour and heating is continued for an additional eight hours. When the reaction temperature reaches 210°C, the vigorous evolution of carbon dioxide is observed and pyromeconic acid begins to distill from the vessel. At the end of the reaction, pyromeconic acid has ceased to distill. During the reaction the solvent which codistills with the product is returned periodically to the vessel. The pyromeconic acid is collected and is suspended in about 5 volumes of hexane and the product is recovered from the suspension by filtration. After drying there is obtained pyromeconic acid in an amount representing a 76% yield. The procedure is repeated reducing the amount of solvent to 2,175 ml and adding the comenic acid semicontinuously to the hot mixture during 2.7 hours. No vacuum is applied to the system. A yield of pyromeconic acid corresponding to 80% of theory is obtained.

Example 3: Pyromeconic acid, 1,000 g, 8.92 mols, prepared as described in Example 2 is dissolved in 4.9 liters of water in a stainless steel vessel. The pH is adjusted to 10.0 with 447 ml (8.47 mols) of 18.9 N sodium hydroxide solution. Then 669 ml, 8.92 mols, of 37% by weight aqueous formaldehyde is added. The temperature spontaneously rises to 45°C, then gradually decreases to 29°C during a two-hour stirring period.

The hydroxymaltol is precipitated from the clear solution by adjusting the pH to 5 with 705 ml of 12 N hydrochloric acid. The slurry is cooled to 5°C, stirred for 30 minutes and is filtered. The hydroxymaltol is washed with three liters of ice water and is vacuum dried. The first crop of hydroxymaltol, MP 154° to 155°C, weighs 949 g and represents 75% of the theoretical yield. Concentration of the 9.5 liters of mother liquor to 2 liters in a vacuum of 20 mm Hg yields a second crop of 152 g. The total yield of hydroxymaltol, therefore, is 87% of the theoretical.

Example 4: In a 300-ml Pyrex flask fitted with a magnetic stirrer, condenser and addition funnel is placed 10.0 (0.07 mol) of hydroxymaltol prepared as in Example 3, 10.6 g (0.16 g-atom) of zinc dust, 60 ml of water and 50 drops of a 1% aqueous solution of cupric sulfate. The reaction mixture is stirred and heated to reflux and 34 ml of concentrated hydrochloric acid is added dropwise during one-half hour. After two hours of refluxing, the reaction is filtered hot to remove unreacted zinc and the pH of the filtrate is adjusted to pH 10 by the addition of a 20% aqueous sodium hydroxide solution.

The precipitated zinc hydroxide is removed by filtration and is washed with 30 ml of water at 60°C. The combined filtrates are adjusted to pH 4 with 20% aqueous hydrochloric acid and, after cooling to 10°C, the crop of crystals of maltol are collected by filtration. There is obtained 3.97 g of maltol, MP 156° to 158°C. Extraction of the filtrate with 5 to 50 ml portions of chloroform and evaporation of the chloroform yields an additional 2.12 g of maltol. The combined weight of product is 6.09 g; this represents a 69% yield.

Maltol Synthesis Intermediates

In related work *B.E. Tate and R.P. Allingham; U.S. Patent 3,159,652; Dec. 1, 1964; assigned to Chas. Pfizer & Co., Inc.* give details for preparation of valuable intermediates and improved steps in the synthesis of maltol from kojic acid. These intermediates are the crystalline compounds 6-methylcomenic acid (MP 237.5° to 238.0°C), 6-methylkojic acid (MP 145° to 145.5°C), and 2,3-dihydro-6-methylcomenic acid (MP 156° to 157°C). In addition the process provides gamma-pyrones of the formula shown on the following page where P is selected from the group consisting of hydroxyl, di-lower alkylamino, the lower alkyl groups containing from 1 to 4 carbon atoms, piperidino, morpholino, chloro, bromo and iodo.

O
OH
HO_2C
CH_2P
O

Still another gamma-pyrone is di-n-butylaminomethylkojic acid. These intermediates and maltol may be readily obtained from kojic acid according to the routes outlined in the following sequence wherein X is chloro, bromo or iodo.

$HOCH_2$ … OH — Kojic acid — [O], Catalyst, pH≥10 (1) → HO_2C … OH — Comenic acid — HCHO, OH^- (2) → HO_2C … OH, CH_2OH — 6-hydroxymethylcomenic acid — halogenating agent (3) →

HO_2C … OH, CH_2X — 6-halomethylcomenic acid — [H] (4) → HO_2C … OH, CH_3 — 6-methylcomenic acid — Δ, $-CO_2$ (5) → OH, CH_3 — Maltol

Example 1: Comenic acid, 156 g, 1 mol prepared according to Example 1 in U.S. Patent 3,130,204 is mixed with 550 ml of water, and the pH is adjusted to 10 with a 50% aqueous sodium hydroxide solution. The mixture is treated with 83.1 g of 37% aqueous formaldehyde and is stirred at 25°C for 1.5 hours. The pH is then adjusted to 0.8 by the addition of concentrated hydrochloric acid, 300 ml of water is added and the suspension is cooled to 5°C and filtered. There is obtained 209 g of product; this represents a nearly quantitative yield. The product when analyzed indicates that 6-hydroxymethylcomenic acid is partially present in the form of its sodium salt.

The crude 6-hydroxymethylcomenic acid is converted completely to the free acid by dissolving 7.88 g in 175 ml of boiling acetic acid and treating the hot mixture with a solution of 1 g of concentrated sulfuric acid in 20 ml of acetic acid. The precipitate of inorganic salt which forms is removed by filtration and the filtrate is evaporated to one-half volume and the product is allowed to crystallize at 25°C.

The crystalline product is collected by filtration and weighs 2.5 g. Evaporation of the filtrate to one-half volume affords a second crop weighing 0.38 g. Recrystallization of the first crop from hot water yields a material with a melting point of 178° to 179°C and having a neutral equivalent of 187 and 94.1; values calculated for 6-hydroxymethylcomenic acid are 186 and 93. Analysis—Calculated for $C_7H_6O_6$: C, 45.16; H, 3.23. Found: C, 45.42, 45.16; H, 3.41, 3.32.

Example 2: Dry hydrogen bromide gas is passed rapidly into a stirred mixture of crude 6-hydroxymethylcomenic acid, 35 g, 0.18 mol, and glacial acetic acid, 210 cc. The temperature rises to 55° to 60°C within 15 minutes, then the mixture is heated to 90° to 110°C, is maintained at that temperature for 2 hours and then is cooled to 30°C.

The reaction mixture is filtered, the filtrate is concentrated in a vacuum corresponding to about 5 mm of Hg to one-sixth volume, and the crystals which precipitate are removed by filtration. There is obtained 39.1 g of 6-bromomethylcomenic acid corresponding to a 90% yield of theory. An additional small amount of product is obtained by concentration of the filtrate to one-half volume. Recrystallization of the crude product from ethyl acetate gives pure 6-bromomethylcomenic acid, MP 197° to 197.5°C. Analysis—Calculated for $C_7H_5O_5Br$: C, 33.75; H, 2.02. Found: C, 33.78; H, 1.94.

The same procedure is carried out substituting hydrogen chloride for hydrogen bromide and 20 g of 6-hydroxymethylcomenic acid is converted to 14.4 g of 6-chloromethylcomenic acid. The procedure is repeated substituting hydrogen iodide for the corresponding hydrogen bromide. 6-iodomethylcomenic acid is obtained.

Example 3: 6-bromomethylcomenic acid, 1.24 g, is mixed with 100 ml of glacial acetic acid, 0.5 g of a 5% palladium on carbon catalyst (50% in water) and 0.385 g of ammonium acetate and the mixture is shaken in an atmosphere of hydrogen at 25°C and at an initial pressure of 50 psi. The calculated amount of hydrogen is absorbed in 15 minutes. The reaction is stopped, the mixture is filtered and the filtrate evaporated in vacuo. There is obtained 1.18 g of crystalline residue. Recrystallization of the residue from a 1:1 acetone-water mixture affords 6-methylcomenic acid, MP 237° to 238°C (decomposition).

The procedure is repeated substituting 4.1 g of 6-chloromethylcomenic acid prepared as in the preceding example; 80% of the theoretical amount of hydrogen is absorbed in 75 minutes. Evaporation of the filtered reaction mixture yields an oily product; this is crystallized from acetone to yield 1.8 g of material melting at 228° to 232°C. The procedure is repeated substituting 6-iodomethylcomenic acid for the corresponding 6-bromomethylcomenic acid. Substantially the same results are obtained.

Example 4: 6-methylcomenic acid, 3.0 g, 0.012 mol, is suspended in 12 ml of dimethyl phthalate in a 25 ml 3-neck round-bottomed flask equipped with mechanical stirrer, thermometer and short distillation head connected in turn to a round-bottomed receiver. The mixture is stirred and heated and it is found that most of the methylcomenic acid has dissolved when the temperature reaches 150°C.

When the temperature reaches 215°C, vigorous evolution of carbon dioxide occurs and this continues for about 15 minutes. The external temperature is allowed to rise to 250°C, then the mixture is cooled to 100°C and a vacuum of about 20 mm of Hg is applied. The reaction mixture is distilled at an external temperature of 180° to 250°C and distillation is continued until very little material remains in the flask. Most of the distillation occurs at vapor temperature of 140° to 150°C. The distillate is cooled to 15°C and the crystalline material which precipitates is collected by suction filtration. The product is washed with 5 ml of ethyl acetate and is dried. There is obtained maltol in good yield, MP, 157° to 160°C.

The procedure is repeated using 10.0 g of 6-methylcomenic acid and 40 ml of dibutyl carbitol. The reaction mixture is heated in the range of 230° to 245°C for about 45 minutes, then is distilled as described in the preceding procedure. The distillate is cooled to 25°C and maltol is collected upon a filter, then is dissolved in 30 ml of hot water, the solution is filtered hot, is cooled to 10°C and maltol is allowed to crystallize. The maltol is collected upon a filter and air dried. Further concentration of the filtrate to one-third volume and filtration to remove the crystalline precipitate affords a small additional amount of product. The procedure is repeated substituting 25 ml of α-methylnaphthalene for the corresponding 40 ml of dibutyl carbitol. There is obtained maltol, MP, 160° to 161.5°C, in good yield.

Example 5: A mixture of morpholine, 46.8 g, 0.534 mol; formaldehyde, 37% aqueous, 42.8 g, 0.534 mol; and anhydrous ethanol, 534 ml, is stirred for 15 minutes. Comenic acid, 62.4 g, 0.4 mol, is added all at once, and the mixture is stirred at room temperature for one hour, cooled to 10°C, and filtered. The 6-morpholinomethylcomenic acid weighs 99 g. A recrystallized sample, MP 173° to 174°C, had the following analysis: C, 48.61; H, 5.51; N, 5.23. Calculated for $C_{11}H_{13}NO_6 \cdot H_2O$: C, 48.35; H, 5.53; N, 5.13. Calculated as the monohydrate, the crude yield was 91%.

The procedure is repeated substituting a molecular equivalent amount of piperidine for morpholine. 6-piperidinomethylcomenic acid is obtained. The procedure is repeated substituting a molecular equivalent of dimethylamine for the morpholine. 6-dimethylaminomethylcomenic acid is obtained. The procedure is repeated substituting a molecular equivalent of di-n-butylamine for the morpholine. 6-di-n-butylaminomethylcomenic acid is obtained. Further examples illustrate the preparation of other intermediates such as 6-methylkojic acid and 2,3-dihydro-6-methylcomenic acid.

Isomaltol Derivatives

Isomaltol was obtained in micro quantities by acidic extractive distillation of breads made of wheat flour. It possesses a caramel-like, somewhat fruity flavor very similar to that of maltol. However, it melts at 102°C rather than at 162°C and is considerably more volatile than maltol.

J.E. Hodge and E.C. Nelson; U.S. Patent 3,015,654; January 2, 1962; assigned to the U.S. Secretary of Agriculture developed a method to synthesize O-galactosylisomaltol which is an O-glycoside in which D-galactose is linked to the isomaltol aglycone through the strongly acidic enolic hydroxyl group of the latter. It is believed that O-galactosylisomaltol has the following structure:

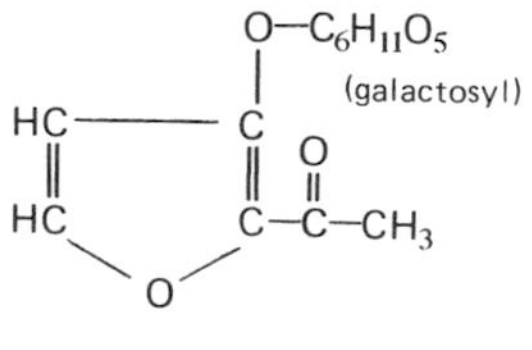

$O-C_6H_{11}O_5$ (galactosyl)

HC—C; HC=C—C(=O)—CH_3; O (furan ring)

This process for the preparation of isomaltol in favorable yields involves the discovery that it is possible to synthesize the intermediate compound O-galactosylisomaltol, which readily hydrolyzes in the presence of acid or alkali or under the influence of heat and moisture to liberate isomaltol and galactose. It is obvious that it would not be necessary to split the O-galactosylisomaltol and isolate the isomaltol for flavor use in pastries, baked goods, boiled candies, etc. (and especially those containing a hydrolysis-promoting acid such as citric or tartaric) since the galactoside is hydrolyzed to the flavorful isomaltol and the harmless galactose by the heat and steam present during the cooking or baking process.

The process involves the synthesis of the intermediate, O-galactosylisomaltol by reacting milk sugar, such as α-lactose hydrate, or other source of lactose, with the salt of a strongly basic secondary amine such as piperidine acetate, in an inert solvent medium such as methanol, and in the further presence of a nonreactant tertiary amine buffer such as trimethylamine or a sterically hindered nonreactant secondary amine such as diisopropylamine, at a reaction temperature in the range 60° to 110°C for from 10 to 24 hours. At the end of the reaction period O-galactosylisomaltol is crystallized and recovered from the reaction mixture by filtration.

Isolation of the O-galactosylisomaltol would not be necessary in a commercial process for preparing isomaltol. The reaction mixture can be acidified and then steam distilled or acidified, heated and solvent-extracted with benzene, chlorinated hydrocarbons, ethers, esters, ketones, hydrocarbons, or higher alcohols to obtain the isomaltol by evaporation of the solvent.

Example 1: In a 2-liter, 3-necked reaction flask, fitted with a motor-driven anchor-bladed stirrer, thermometer, and reflux condenser, 360 g (1 mol) of α-lactose hydrate, 85 g (1 mol) of piperidine, 60 g (1 mol) of glacial acetic acid, 100 ml of triethylamine, and 300 ml of absolute ethanol were heated and stirred at a constant temperature of 75°C. The last of the solids dissolved between 10 and 12 hours of heating.

After 15 hours of heating the reaction product was present and was isolated in 28% of the theoretical yield in a separate experiment. After 24 hours of heating at 75°C, 300 ml of absolute ethanol was added; then the dark brown reaction mixture was continually stirred for one hour while the flask was cooled in an ice-water bath to crystallize the product. The precipitate was filtered off with suction, washed several times with ethanol until nearly white and dried in a vacuum desiccator over anhydrous calcium chloride to a constant weight of 106 g (37% of theory); MP, 204° to 205°C.

Recrystallization from hot water or aqueous alcohol with the use of decolorizing charcoal gave pure white crystals of the same melting point, and with a specific optical rotation of –4.5° for a 2% solution in water with sodium light.

Acid hydrolysis or acid hydrogenolysis of the neutral compound, $C_{12}H_{16}O_8$ gave crystalline α-D-galactose, $C_6H_{12}O_6$, MP, 165° to 167°C, identified by its optical rotation in water and by its conversion by nitric acid to crystalline mucic acid, MP, 213° to 214°C. Methanolysis in anhydrous methanol-hydrogen chloride gave the known methyl β-D-galactopyranoside, $C_7H_{14}O_6$, MP 177° to 178°C, no optical rotation in water.

When the acid hydrolysate of the neutral compound, $C_{12}H_{16}O_8$, was extracted with ether and the ether extracts concentrated by evaporation, a colorless acidic compound was crystallized. Recrystallized and sublimed, the acidic compound melted at 100° to 101°C. By titration with standard base, the neutral equivalent was 124.

This same compound was sublimed and distilled from the neutral $C_{12}H_{16}O_8$ compound when it was heated to 205°C and caramelized. It was identified as isomaltol, by converting it to the O-methyl ether (MP 101° to 103°C), the O-benzoyl ester (MP 100° to 101°C), and the same green copper salt reported in the literature. Moreover, the stable violet color with ferric chloride, the acidity, the volatility, the solubilities and the reducing action toward Fehling solution conformed exactly to the literature reports.

Example 2: In the same apparatus described in Example 1, precooled to 1°C, the following were added in the order given: 500 ml absolute methanol, 48 g (1.06 mols) of anhydrous dimethylamine, 100 g trimethylamine, 360 g (1.00 mol) of α-lactose hydrate, and 60 g (1.00 mol) of glacial acetic acid was slowly dropped in with stirring.

The mixture was then continually stirred and heated under reflux at its boiling point for 24 hours. The temperature increased from 60°C (one hour) to 67°C (two hours) to 68°C (10 hours) to 71°C (20 hours) and to 72°C at 24 hours. The reaction flask was cooled to 2°C and held at this temperature for one hour until crystallization of the product was essentially complete. Isolated as described in Example 1, this first crop weighed 60.5 g.

The filtrate and washings were reheated at a constant reflux temperature of 77° ± 1°C for eight hours. After removing the solvents by distillation at atmospheric pressure over two additional hours, the dark solution was again cooled to 1°C and a second crop, isolated in the same way as the first, weighed 3 g. Both crops were identical, MP 204° to 205°C, representing the same compound, $C_{12}H_{16}O_8$, as was obtained in Example 1. The total yield in this experiment was 63.5 g, 22% of the theoretical amount.

Maltol and Flavoring Oils

A process by *J.M. Griffin; U.S. Patent 3,293,045; December 20, 1966; assigned to Chas. Pfizer & Co., Inc.* involves the use of maltol to amplify the intensity of flavoring oils such as anise, cinnamon and wintergreen.

Flavoring oils have been used at levels of up to 8,000 ppm based on the product to be ingested. According to this process, the content of the flavoring oil may be reduced to from 50 to 300 ppm based on the material to be flavored. In addition to providing for the ingestion of smaller quantities of the flavoring ingredient, the process also provides for a possible saving in the cost of some of the more expensive ingredients since less of the ingredient is required to give the desired flavor.

At the 50 to 300 ppm level of the flavoring oils, in the absence of maltol, relatively weak flavor strength is noted by those tasting the representative flavored confections, beverages and medicinal oils. It is found however, that if there is added from 15 to 100% of maltol by weight based on the normally ineffective amount of the flavoring oil, the flavor receives such a boost that the test subjects find it difficult to distinguish over-effects which are achieved only with amounts of the flavoring oil from 600 to 800 ppm and even more.

Example: A basic fondant plastic creme center formula for candy is prepared. In a vessel are placed 8 pounds of granulated sugar, 2 pounds of corn syrup, 1 pound of invert sugar and enough water to dissolve the sugar. All of the ingredients are then heated together; the batch is stirred occasionally until it boils. Any sugar grains which adhere to the vessel walls are washed down and the batch is heated until the temperature of the boiling mass reaches 240°F. The mixture is poured onto a clean, wet slab and is cooled to about 110°F. When it has reached this temperature it is creamed with a spatula and is stored until used.

Five batches of the fondant are mixed with methyl salicylate (synthetic wintergreen oil) until there are obtained, respectively, one batch each with 50, 150, 300, 500 and 800 ppm of methyl salicylate. Each of the batches containing methyl salicylate at 50, 150, 300 and 500 ppm is itself divided into five batches. One of the batches containing 50 ppm of methyl salicylate is left untreated and to the other four are added, respectively, enough maltol to provide batches containing 50, 75, 100 and 125 ppm of maltol. These five batches are tested by a panel and the flavors compared. It is found that the wintergreen flavor of all samples containing maltol is greater than the wintergreen flavor of the sample which does not contain maltol.

One of the batches containing 150 ppm of methyl salicylate is treated with enough maltol to provide 100 ppm; one of the batches containing 300 ppm of methyl salicylate is treated with enough maltol to provide 100 ppm. A taste panel evaluation of these three blends is made and compared with the taste of fondant samples flavored with 300 ppm, 500 ppm and 800 ppm of methyl salicylate and to which no maltol has been added.

The wintergreen taste with 150 ppm of methyl salicylate and 100 ppm of maltol is significantly greater than that of 150 ppm of methyl salicylate alone but less than that of 300 parts of methyl salicylate alone; the wintergreen flavor with 300 ppm of methyl salicylate and 125 ppm of maltol is greater than that of 500 ppm of methyl salicylate alone and less than that of 800 ppm of methyl salicylate alone; and the flavor of 500 ppm of methyl salicylate and 100 ppm of maltol is also greater than 500 ppm of methyl salicylate alone but less than that of 800 ppm of methyl salicylate alone.

Thus, it is clearly demonstrated that the flavor strength of 300 ppm of methyl salicylate plus 125 ppm of maltol (or 42% of maltol based on the methyl salicylate) is approximately equal to from at least 500 to 800 ppm of methyl salicylate alone. Furthermore, it is found that an appreciable boost in the wintergreen flavor is provided by maltol in amounts ranging from 20% (500 ppm of methyl salicylate with 100 ppm of maltol) to 100% and even higher (50 ppm of methyl salicylate with 50 ppm of maltol based on the methyl salicylate).

Maltol-Like Sweetener

H. Shimazaki, S. Tsukamoto, T. Saito, S. Eguchi and Y. Komata; British Patent 1,199,101; July 15, 1970; assigned to Ajinomoto Co., Inc., Japan describes a process for the production of a maltol-like substance, which comprises heating one or more amino acids selected from glutamic acid, aspartic acid, alanine, glycine and proline, or a salt, with a pentose selected from arabinose, ribose and xylose, or a precursor for such a pentose capable under the reaction conditions of producing pentose, so as to produce reaction products containing the desired maltol-like substance. This substance is prepared according to Example 1 and is referred to as substance A.

Suitable conditions for carrying out the Maillard reaction are as follows. One mol of the amino acid is heated with from 0.5 to 50 mols of the pentose in a solvent or suspending medium. This reaction should be carried out under relatively mild conditions. Usually, the temperature for the reaction is kept within the range from 85° to 100°C, and the solvent or suspending medium is generally water whose pH has been adjusted to a value of from 7 to 10.5.

The reaction is usually carried out for from 20 to 30 minutes until the solution becomes yellow or reddish yellow in color. When the pentose is xylose or a xylose-producing substance and the amino acid is glutamic acid, aspartic acid or a salt, the resulting reaction products in the solvent may be dried to powder, and the powder may be added directly to foodstuffs and medicines.

When using these particular amino acids and pentoses, the Maillard reaction may be carried out during a heating step in the production of the foodstuff if the amino acid and pentose are added to the raw materials of the foodstuff. The substance A is produced by the extraction and purification of the Maillard reaction products.

The substance A has a strong flavor and aroma, similar to that of maltol. When this substance A was diluted to a concentration of from 104 to 500 ppm, in water or pure ethanol, its flavor and aroma were more pleasant and sweeter than that of maltol.

Therefore, this substance A or the Maillard reaction products of glutamic acid, aspartic acid or a salt containing substance A, with xylose or a xylose-producing substance can be used for improving the flavor and aroma of foodstuffs such as breads, biscuits, candy, chocolate, meat and processed meat, milk products, processed egg products, smoked fish or meats, processed vegetables, powdered soups, dried fruits and nuts, canned fruits, soft drinks, alcoholic beverages such as liquors, wine or whisky, and instant coffee; and medicines for oral hygiene such as a gargle. Substance A or the reaction product containing substance A may also be used in dentifrices or tobacco products.

Example 1: 500 g of monosodium glutamate and 500 g of xylose were suspended in 500 ml of water and then steam was passed through the suspension for two hours in order to effect steam distillation. The resulting distillate was condensed and 2 liters of distillate were obtained. 5 g of an oily fraction were obtained by concentrating the distillate under reduced pressure at a temperature not in excess of 40°C. The oily fraction was then kept at 80°C under a reduced pressure of 10 mm Hg, and crystals appeared on the cold wall of the apparatus due to sublimation. The yield was 100 mg. This crystalline substance was the one referred to as substance A.

Example 2: A 1% solution of the substance A which was prepared in the same manner as that described in Example 1 was prepared, and this solution was added to a chocolate cake mixture in a concentration of 120 parts by weight of substance A per million parts by weight of chocolate. Chocolate cakes were produced from both the chocolate cake mixture containing the substance A and a mixture not containing any of substance A.

Both types of chocolate cakes were offered to a panel of 40 members for an organoleptic test. As a result of the test, 32 persons on the panel indicated that the flavor and aroma of the chocolate cake containing the substance A was preferable, and the remaining 8 persons indicated that there was no difference between the two types.

Example 3: A 1% solution of substance A was added to a marketed pineapple juice in a concentration of 25 ppm. The pineapple juice containing the substance A and a contrast which did not contain any substance A were offered to a panel of 25 members for an organoleptic test. Twenty-three persons on the panel indicated that the flavor and/or natural taste of the pineapple juice was improved by the addition of substance A.

OTHER SYNTHETIC SWEETENERS

This chapter reviews patents pertaining to chemicals which are derived almost completely by synthesis. In this sense, they are similar to saccharin as regards their synthetic nature.

DIACETONE GLUCOSE

Diacetone glucose can be used in pharmaceutical preparations as a substitute or replacement for glucose, sucrose or other sweetening agents. The material passes through the body substantially unmetabolized. The synthesis is given by *B. Osborne and H.E. Henneman; U.S. Patent 2,554,152; May 22, 1951; assigned to American Home Products Corporation.*

Preparation of Diacetone Glucose: Acetone (80 liters) is placed in an open, glass lined jacketed pan (Pfaudler, 30 gal capacity) and agitated efficiently (Lightening stirrer). Water is passed through the cooling jacket and the temperature of the acetone brought to 18°C. Concentrated sulfuric acid (3,500 cc, d = 1.84) is added gradually so that the temperature of the solution remains below 25°C. When the addition is complete, dextrose (4 kg) is added in one lot and the mixture stirred for 15 hours at 15°C. It is important that the temperature should remain low. Solution is slow but is ultimately complete.

The reaction mixture is neutralized with an aqueous solution of sodium hydroxide (5,300 grams of NaOH in the minimum quantity of water); addition is gradual and the temperature is carefully controlled so that it does not rise above 20° to 25°C. Addition of sodium hydroxide is stopped when a pH of 8.5 to 9 is attained. Sodium sulfate precipitates and is allowed to settle, the supernatant acetone is decanted or siphoned off and the sodium sulfate washed with acetone (12 liters) and the washings combined with the main bulk of acetone. The residual sodium sulfate is dissolved in water (20 gal) and the solution stirred with chloroform (6 to 12 liters); the aqueous layer is allowed to separate and is removed and discarded.

The combined acetone solutions are concentrated in a vacuum still (10 gal, Pfaudler, temperature of kettle ca 50°C). The concentrate is diluted with water (15 liters) and is then stirred with the chloroform which has been previously used to extract the sodium sulfate solution and the chloroform layer collected. The extraction of the aqueous phase is repeated thrice (3 x 6,000 cc $CHCl_3$) and the combined chloroform extracts dried (Na_2SO_4). The aqueous phase is discarded.

The chloroform extracts are filtered, concentrated under reduced pressure to a volume of about 6 liters and poured while hot into stirred hexane (10 liters) previously cooled to -5°C. The mixture sets to a crystalline semisolid mass. The product is collected, drained and washed with hexane (1½ liters). The product is a white crystalline powder. MP 105° to 107°C.

Preparation of 3-Methyl Diacetone Glucose: 300 grams of diacetone glucose are made into a paste with 150 cc of water and the reaction mixture heated and stirred at 70°C. During a period of 1 hour, 315 cc of 50% aqueous potassium hydroxide and 325 cc of dimethyl sulfate are added simultaneously through dropping funnels; this addition is gradual and is regulated so that the reaction mixture is maintained in an alkaline state.

After the addition is complete, the reaction mixture is stirred for 1 hour at 70° to 75°C, then cooled and extracted with ether (4 x 300 cc). The combined ether extracts are dried over potassium carbonate, filtered and the solvent removed. The residue is distilled under reduced pressure. Approximately 200 grams of colorless liquid are collected.

Preparation of 3-Methyl Glucose: Water (10 liters) and concentrated sulfuric acid (200 cc) are placed in a round bottomed 12 liter flask and agitated on a waterbath at 100°C. 3-methyl diacetone glucose (600 grams) is added and the mixture heated and stirred for 3 hours; complete solution is gradually attained.

Basic lead acetate (1,220 grams, 107 excess) is added to the reaction mixture which is then stirred for a few minutes and filtered while hot. The filtrate is cooled and hydrogen sulfide is passed through it until precipitation of lead sulfide is complete. The lead sulfide is removed by filtration, washed with water and the combined filtrate and washings concentrated under vacuum to a volume of about 2 liters. The hot solution is stirred with charcoal for 20 minutes and is then filtered and concentrated further under pressure. Finally successive small quantities of absolute ethanol are added and evaporated off under reduced pressure, in order to remove moisture. If pure, the product finally solidifies at this stage, but a thick syrup may be obtained.

The material is extracted by rubbing with absolute acetone and the acetone is decanted; after three such treatments the product is washed with a little cold absolute methanol and is finally rubbed with hot absolute methanol. A white crystalline solvent is finally obtained. The material is collected and washed with a little absolute methanol followed by dry ether. MP 163° to 167°C.

The product thus obtained is pure enough for all practical purposes but may smell of acetic acid. It can be further purified by boiling out with absolute methanol, filtering hot and reworking the filtrate. The thus improved and modified 3-methyl glucose may be taken alone or combined with other active ingredients and/or with a carrier so as to make the improved product more effective or especially attractive and desired by the diabetic, obese, edematous or other person.

Example 1: A gluten bread for diabetics may be made from the following ingredients.

Gluten flour, lb	6
Yeast, oz	1.7
3-methyl glucose, oz	1.4
Salt, oz	0.7
Water (boiling), pints	2

As variations, one may use some wheat flour for an equal amount of gluten flour; and milk may be substituted for the water. In addition, a shortening such as butter, lard or hydrogenated vegetable fat may be used to the extent of about 1 to 2 tablespoons.

Example 2: A lemon chiffon pie of normal sweetness may be made free of sugar by first preparing in the usual way an undercrust of the following ingredients and then baking at 500°F until done.

Flour, lb	1.5
Shortening, lb	0.5
Water (cold), oz	3.5
Salt, oz	0.25

The pie filling is prepared by mixing 23.5 oz of alpha-D-3-methyl glucose with 0.15 oz salt and 1 oz of lemon rind and 4 oz of lemon juice. To this mixture is added 8 oz of water and 7 oz of vegetable shortening and the mixture is heated to boiling. When the boiling point is reached, 12 oz of egg yolks are added and heating is continued long enough to thicken the mixture, heating being thereafter terminated. To the mixture is added 1 oz of gelatin previously soaked until soft in 5 oz of cold water. A meringue made with 10 oz of egg white and 10 oz of 3-methyl glucose is whipped into the mixture as prepared above, and after filling the baked pastry undercrust, more meringue may be added on top.

Example 3: To make a dehydrated gingerbread mix, 10 lb of molasses and 1.1 lb of shortening are mixed to obtain homogeneous dispersion. To the mixture, 10 lb of wheat flour or a mixture of 5 lb of wheat flour and 5 lb of gluten flour are gradually added. The dough is heat dried at about 160° to 170°F under reduced pressure and then ground and pulverized. To the pulverized mixture, one adds 1.5 lb of alpha-D-3-methyl glucose, 1 oz salt, 5 oz baking soda, 10 oz powdered whole egg, 1 oz powdered ginger and 0.5 oz of powdered cinnamon and thoroughly mixes to obtain a homogeneous dispersion.

Example 4: A gelatin dessert may be made by mixing 11 lb of alpha-D-3-methyl glucose, 1.5 lb of gelatin powder, 3.5 oz of citric acid powder, 0.6 oz of true pineapple flavor, 50 minims of butyl acetate, and 0.5 oz of certified food color such as Tartrazine color powder. Approximately 3 oz of this powder added to 1 pint of boiling water will make a pineapple flavored gelatin dessert.

Example 5: An artificial maple syrup may be prepared by mixing 3 lb of alpha-D-3-methyl glucose, 1 gal of water, 0.2 fluid oz of mapleine and sufficient caramel color to give the desired color to the product.

Example 6: A fondant cream useful as a center for candies may be made by heating 1 gal of water to which has been added 6 lb of alpha-D-3-methyl glucose, 12 lb of cane sugar, and 2 lb of Nulomoline syrup. The mixture is boiled rapidly to about 240° to 250°F and is thereafter cooled to 110° to 125°F and agitated to the desired consistency. Any flavoring agent may be added, if desired.

Example 7: A soft drink syrup may be prepared by combining 6.5 lb of alpha-D-3-methyl glucose, 10 fluid oz of distilled lime oil emulsion, 40 fluid oz of fresh lime juice, 0.5 oz of sodium benzoate, and 1 gal of water. To each fluid ounce of the syrup one adds 4 to 5 fluid oz of carbonated water.

SACCHARIN AMINE SALTS

Acceptable organic amine salts of saccharin are prepared by *L.C. Vacek; U.S. Patent 3,325,475; June 13, 1967; assigned to Maumee Chemical Company.* Such salts can, in general, be produced by preparing a slurry of saccharin (acid form), and adding the slurry to a solution containing an equivalent amount of the organic amine. Water is a suitable solvent system in which to conduct this ionic reaction. The pH of the salt can be determined experimentally, and the saccharin slurry can then be added to the organic amine solution until the indicated pH is achieved.

The desired saccharin salt can be recovered from the solution by evaporation of the water, preferably under vacuum to minimize the chance of hydrolysis. Examples of organic amines from which salts can be produced by reaction with saccharin include monoethanolamine, diethanolamine, amino sugars, for example, 1-amino-1-deoxy-D-glucitol, and 1-deoxy-1-(methylamino)glucitol, and alkylenediamines such as ethylenediamine, and including cyclics such as piperazine (diethylenediamine).

The saccharin salts may have a slightly higher caloric content than saccharin itself, particularly the amino deoxy sugar salts of saccharin. However, the slight caloric content is immaterial, because such a small quantity of the saccharin salts is required to perform a particular sweetening function that the effect is negligible. The results of taste tests indicate that saccharin salts are substantially free of any off taste which is characteristic of saccharin.

Example 1: The monoethanolamine salt of saccharin was prepared in a 250 cc beaker. The beaker was charged with 18.32 grams saccharin (acid form) slurried with 50 ml water. A 50 cc portion of a two normal water solution of monoethanolamine was then added to the beaker, with agitation. About ten minutes after the monoethanolamine solution addition was completed, a pale yellow solution was noted in the beaker. A 1 gram portion of C-190-N activated carbon was added to the light yellow solution in the beaker; the resulting mixture was stirred for 15 minutes; and the carbon was removed from the solution by filtration in a Buechner funnel, using filter aid and vacuum.

The resulting solution was water clear and colorless. The colorless solution was evaporated under vacuum at a maximum temperature of 60°C, and was then dried to constant weight at a temperature of about 60°C. The final drying required approximately 5 hours. A crude yield of 25.7 grams of the monoethanolamine salt of saccharin was obtained. The salt was a viscous, lightly yellow oil which solidified upon standing overnight into a light, pale yellow, crystalline solid mass.

Example 2: The procedure described in Example 1 has been used to produce various other organic amine salts of saccharin. The reactants which were used and the nature of the final product are set forth in the following table.

	Reaction			
Procedure	Amine Used	Amount of Amine, ml	Amount of Saccharin, g	Manner of Isolation
1	Diethanolamine	50*	18.32	Evaporation to constant weight.
2	Piperazine**	19.42***	36.64	By crystallization upon cooling to 20°C.
3	1-amino-1-deoxy-D-glucitol	100†	18.32	Evaporation to constant weight.

	Product		
	Identity	Yield, g	Description
1	Diethanolamine salt of saccharin	29.57	Viscous oily liquid which solidified on standing into a solid crystalline mass.
2	Disaccharin salt of piperazine	40.07	White crystalline solid.
3	Saccharin salt of 1-amino-1-deoxy-D-glucitol	37.37	Thick, light-amber colored oil which solidified on standing into a tacky solid having the appearance of glass.

*Added as 2 N water solution
**Reaction was conducted at a temperature of 70° to 80°C
***Grams
†Added as 1 N water solution

The procedures described above can also be used to produce saccharin salts of 1-deoxy-1-(methylamino)glucitol, ethylenediamine, pentamethylenediamine and tetramethylenediamine. In producing the diamine salts the procedure should be similar to No. 2, where two equivalents of saccharin are used per mol of the organic amine.

TRYPTOPHANE DERIVATIVES

6-(Trifluromethyl)Tryptophane

It is reported that the compound 6-(trifluoromethyl)tryptophane or its salt at a concentration of 0.075% by weight in foods, gives a sweetening effect comparable to that of a 10% concentration of sucrose. The compound can be used alone, or it can be combined with other known sweetening agents, most notably, sodium saccharin. The synthesis is described by *E.C. Kronfeld; U.S. Patent 3,535,336; October 20, 1970; assigned to Eli Lilly and Company.* The steps are given in the examples below. 6-(Trifluoromethyl)tryptophane has the formula

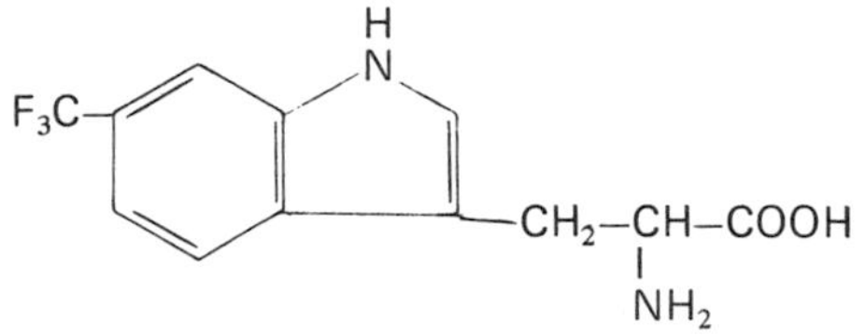

H
N
F_3C
CH_2–CH–COOH
NH_2

Example 1: 3-(Diethylaminomethyl)-6-(Trifluoromethyl)Indole – To a solution of 7.53 grams of diethylamine (0.15 mol) in 22.5 ml of cold 60% acetic acid was added 8.25 ml of 37% aqueous formaldehyde. 6-(Trifluoromethyl)indole (19.0 grams; 0.10 mol) was then added, and the resulting mixture warmed to 60°C. After two hours at this temperature, the solution was poured into 360 ml of 2 N sodium hydroxide, and the desired 3-(diethylaminomethyl)-6-(trifluoromethyl)indole product extracted with ether. The extract was dried, and solvent removed by distillation to separate the product, which was an oil.

Example 2: Diethyl ⟨[6-(Trifluoromethyl)-3-Indolyl] Methyl⟩ Formamidomalonate – 3-(diethylaminomethyl)-6-(trifluoromethyl)indole (12.5 grams; 0.046 mol), prepared as described in the foregoing example, and diethyl formamidomalonate (9.4 grams; 0.041 mol) were mixed in 31 ml of toluene.

Then, 0.76 gram of powdered potassium hydroxide was added. The resulting mixture was refluxed, while bubbling in nitrogen, for 1.5 hours. It was then cooled, resulting in precipitation of the diethyl ⟨[6-(trifluoromethyl)-3-indolyl]-methyl⟩ formamidomalonate product, which was separated by filtration and washed with water. A portion of the separated product was crystallized from ethanol, and the recrystallized portion found to melt at 182° to 185°C.

Example 3: 6-(Trifluoromethyl)Tryptophane – A solution of diethyl ⟨[6-(trifluoromethyl)-3-indolyl] methyl⟩ formamidomalonate (55.3 grams; 0.138 mol) in 330 ml of 1.5 N hydrochloric acid, 165 ml of glacial acetic acid, and 50 ml of ethanol was refluxed for 17 hours. The solution was then concentrated under vacuum, the residue neutralized with ammonium hydroxide, and the resulting 6-(trifluoromethyl)tryptophane product separated by filtration. The separated product was washed with water and ethanol, and recrystallized from acetic acid. The recrystallized portion melted at 267° to 270°C (dec).

6-(Trifluoromethyl)tryptophane has an asymmetric carbon atom; but resolution of the racemic mixture is not necessary, the mixture itself being satisfactory. However, only the d-enantiomorph is active as a nonnutritive sweetener. When for any reason it is desired to use one enantiomorph, resolution of the racemic mixture can be achieved by known procedures for the resolution of unsubstituted tryptophane.

Three such procedures are discussed and exemplified in detail in *Chemistry of the Amino Acids,* Greenstein et al, vol. 3, p. 2341 et seq., John Wiley and Sons, Inc., New York (1961). Particular attention is directed to the first two of these (illustrative procedure 39-5 and illustrative procedure 39-6).

6-Chlorotryptophane

Further details on 6-(trifluoromethyl)tryptophane type compounds are given by *T. Suarez, E.C. Kornfeld and J.M. Sheneman; U.S. Patent 3,899,592; Aug. 12, 1975; and British Patent 1,269,851; April 6, 1972; both assigned to Eli Lilly and Company.* Additional representative compounds to be used as active agent include the following: 6-chlorotryptophane, 6-bromotryptophane, 5-chloro-6-methyltryptophane, 6-methyltryptophane and other similar tryptophane derivatives. Mixtures of two or more compounds can also be used. Any given compound can be employed either as the d-enantiomorph or as a mixture of the d- and l-enantiomorphs.

The compounds can be used with any physiologically acceptable carrier, i.e., a substance or material which can safely be placed in the mouth, and, in appropriate cases, safely swallowed. Examples of such carriers are as follows: chewing gum, toothpaste, lip cosmetics, mouthwash, mouthspray, substances used in dentistry for cleaning of teeth, denture treating substances, chewing tobacco and other tobacco products.

Also included are pet toys, for example, rubber dog bones, as well as other mechanical devices temporarily retained within the mouth; glues and adhesives, as for use on stamps and envelopes; coffee, tea, fruitades such as lemonade, salad dressings, carbonated and noncarbonated beverages; baked goods such as bread, crackers, pretzels, pastries, or cake, cereal products; milk derived products, such as ice cream, ice milk, sherbets and custards; jelly and gelatin products.

For sucrose concentrations of from 6.5 to 45%, concentrations which occur in most foods and even many confectionary applications, the sweetening agent, when substituted for sucrose, gives approximately equivalent sweetness at concentrations of from 0.05 to 30%, as the racemic mixture. Where the d-enantiomorph is used alone, these rates can be reduced by a factor of about one-half. Higher or lower concentrations of the active agent can be used where the degree of sweetness, by a sucrose standard, is greater or lesser. However, where a high degree of sweetness is desired, it is generally preferred to use a combination of the active agent with one or more other known nonnutritive sweeteners.

When the active agent is used in conjunction with another nonnutritive sweetener, the exact ratio of the components is not critical and can vary considerably, depending upon the particular use.

In the combination of the active agent with 5-(3-hydroxyphenoxy)-1H-tetrazole, synergistic results are obtained at a ratio ranging from 1 part of active agent to 300 parts of the 5-(3-hydroxyphenoxy)-1H-tetrazole by weight to 1 part of the 5-(3-hydroxyphenoxy)-1H-tetrazole to 300 parts of the active agent; a particularly preferred ratio of combination is a ratio of from 1:50 to 50:1, by weight.

In synergism with the dihydrochalcone-type of sweetener, synergism is likewise noted throughout the ratio ranging from 1 part of the active agent to 30 parts of the dihydrochalcone-type sweetener, by weight, to a ratio of 1 part of the dihydrochalcone-type sweetener to 30 parts of the active agent by weight. A preferred ratio is from 1:5 to 5:1 by weight.

It is also possible to combine the active agent with sucrose or other nutritive sweeteners so as to obtain a sweetening substance of reduced caloric value. Such a mixture of nutritive sweetener and the active agent can be formulated as a foam, for a combination of sucrose and sodium saccharin. Such manner of formulating results in a composition having both volume and sweetening effect equivalent to sucrose, alone, but with reduced caloric value.

Example 1: An initial series of evaluations was carried out, each in essentially identical procedures. Each evaluation comprised the tasting of a small amount of the given compound, such amount being that which would adhere to a finger-tip. Compounds so evaluated were rated for degree of sweetness, and aftertaste, if any. The sweetness ratings were as follows:

Compound Evaluated	Sweetness Rating
dl-6-(trifluoromethyl)tryptophane	Very sweet
dl-6-bromotryptophane	Sweet
dl-6-methyltryptophane	Very sweet
dl-6-chlorotryptophane	Very sweet
dl-6-chlorotryptophane hydrochloride	Very sweet
dl-6-fluorotryptophane	Very sweet
dl-5,6-dimethyltryptophane	Sweet
dl-5,6-dichlorotryptophane	Very sweet
dl-6-isopropyltryptophane hydrochloride	Sweet
dl-6-methoxytryptophane	Sweet
dl-6-chlorotryptophane, sodium salt	Very sweet

No aftertaste was observed with any of the compounds evaluated.

Example 2: In further experiments, dl-6-chlorotryptophane was evaluated in a carbonated beverage; the compound was dissolved at different concentrations in a number of solutions, each having the following composition:

	Percent
Sodium citrate	0.025
Citric acid	0.22
Alcoholic lemon-lime extract	0.125
Sodium saccharin	0.01
Carbonated water, q.s.	

The various solutions were judged by a taste panel of five persons to determine which was comparable, in sweetness, to a standard solution. The standard solution contained sodium citrate, citric acid, and alcoholic lemon-lime extract in the same concentrations as above, but contained an amount of sodium saccharin to approximate the sweetness of a commercially available carbonated beverage.

This standard solution, therefore, was approximately equivalent to a commercially available carbonated beverage of this type sweetened with a known non-nutritive sweetener. The mixture containing the dl-6-chlorotryptophane in a concentration of 0.0125% was judged to be equivalent in sweetness to the standard. No off-flavor was noted in any of the solutions containing dl-6-chlorotryptophane.

Example 3: In other operations, dl-6-chlorotryptophane was evaluated in hot cocoa. In these operations, the compound was dissolved in a number of mixtures, at varying concentrations, each such mixture comprising 1 cup of milk, a tablespoon of commercially available cocoa powder, 0.02% of sodium saccharin, and the dl-6-chlorotryptophane in a given amount.

Each such mixture was heated to the boiling temperature and then permitted to cool somewhat. All mixtures were evaluated by a taste panel of five persons, to determine which of the mixtures afforded a degree of sweetness comparable to another mixture prepared as described but containing 1 cup of milk, 1 tablespoon of the same commercially prepared cocoa powder and 0.02% of sodium saccharin.

The consensus of the taste panel was that the mixture containing 0.0125% of the dl-6-chlorotryptophane afforded a sense of sweetness comparable to that of the standard control solution. No off-flavor was noted in any of the cocoa mixtures containing the dl-6-chlorotryptophane.

Example 4: In another evaluation, dl-6-chlorotryptophane was evaluated in a vanilla-flavored dessert sauce. In the evaluation, the compound was dissolved in various mixtures, each such mixture comprising the following ingredients:

1 cup milk
1 tablespoon of cornstarch
⅛ teaspoon salt
Sodium saccharin, in a concentration of 0.08%
dl-6-chlorotryptophane in a given amount

The concentration of the dl-6-chlorotryptophane was varied among the several mixtures. Each such mixture was then cooked until gelatinized, and thereafter cooled and 2 tablespoons of butter and ¼ teaspoon of vanilla added to each mixture.

A taste panel of five persons then evaluated the various dessert sauces; the consensus of the panel was that the sauce containing 0.08% of sodium saccharin and 0.08% of dl-6-chlorotryptophane afforded a sense of sweetness approximately comparable to that of the standard. Some off-flavor was noted by some members of the taste panel in the sauces containing the dl-6-chlorotryptophane and sodium saccharin; this off-flavor was believed to be due to the sodium saccharin.

Tryptamine Compounds

The process of *E.C. Kornfeld; U.S. Patents 3,615,700; October 26, 1971; and 3,737,436; June 5, 1973; both assigned to Eli Lilly and Company* is directed to compounds of the following formula:

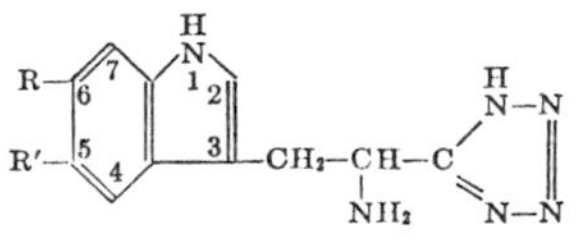

where R represents halo of an atomic weight of less than 85, lower alkyl, lower alkoxy, or trifluoromethyl; and R being trifluoromethyl, R' represents hydrogen, or R being halo as defined, lower alkyl, or lower alkoxy, R' represents hydrogen, halo as defined, lower alkyl, or lower alkoxy. The terms lower alkyl and lower alkoxy are employed to designate alkyl and alkoxy radicals of from 1 to 4, both inclusive, carbons.

The compounds are typically crystalline solid materials. The solubility of the compounds varies. Those which are salts are generally of moderate solubility in water and of lesser solubility in organic solvents, whereas those which are not salts are of only low to moderate solubility in water, but of higher solubility in organic solvents. As water solubility is desirable in the typical usage of a substance as a sweetening agent, the salts are often preferred.

The identity of the salt-forming moiety is not critical except that the salt be nontoxic and physiologically acceptable. Alkaline salt formation occurs at the site of the tetrazolyl proton; suitable alkaline salts are the ammonium, sodium, potassium, calcium, and magnesium salts. Acid salt formation occurs at the amino nitrogen atom, the ring nitrogen atoms being only weakly basic. It is necessary that the acid be a strong acid, that is, an acid having a pH of, numerically, below about 2.2 at a concentration of 0.1 N.

Tryptamine is the common name given to the compound of the following structural formula:

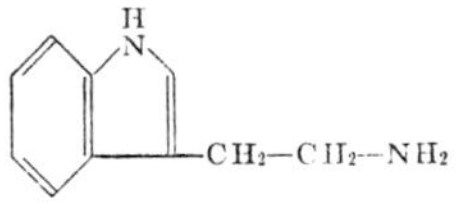

For convenience, all intermediates and products which share the essential moieties of tryptamine are named as derivatives of tryptamine. The free base compounds are prepared in accordance with the following reaction sequence:

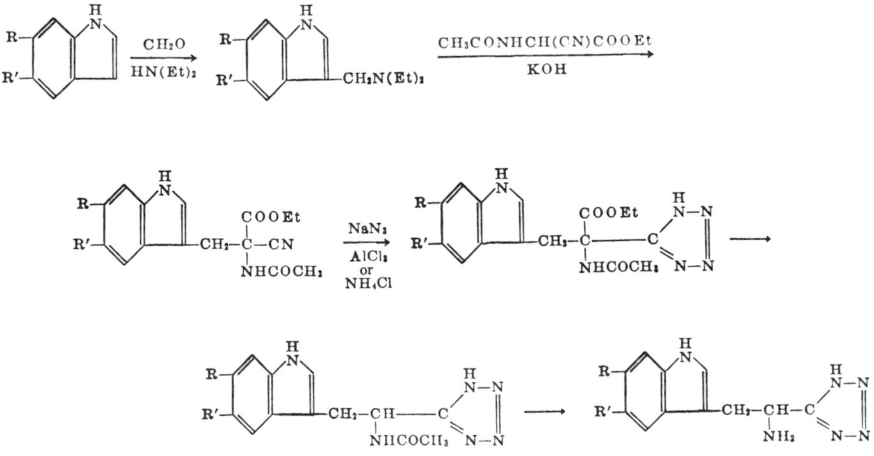

According to the foregoing reaction sequence, a 6-substituted indole is caused to undergo a typical Mannich reaction with formaldehyde and diethylamine and the resulting 6-substituted 3-(diethylaminomethyl)indole is then condensed with ethyl α-acetamido-α-cyanoacetate, to form the corresponding N-acetyl-α-cyano-α-carboethoxy-6-substituted tryptamine. Reaction with sodium azide and aluminum chloride, or preferably, where R or R' is lower alkyl or lower alkoxy, with sodium azide and ammonium chloride, followed by decarboxylation and hydrolysis of the resulting intermediate, produces the corresponding α-tetrazolyl-6-substituted tryptamine compound.

The compounds obtained in these methods can be reacted further to obtain corresponding salts. In such further reactions, the desired compound is reacted with a stoichiometric amount of a suitable acid or base to procure the corresponding salt.

Resolution of the compounds is not necessary, the racemic mixture itself being quite active as a nonnutritive sweetener. However, the d-form is the active moiety; therefore, resolution may be preferred to lessen the amount of substance needed for the desired sweetening effect. Moreover, due to the fact that typically only the l-form of amino acids is metabilized by the mammalian body, usage of the d-form, alone, may be preferred to preclude any opportunity for the mammalian body to incorporate the substance. When, for these or any other reasons, it is desired to use only the d-form, resolution of the racemic mixture can be achieved by procedures known for the resolution of unsubstituted tryptophane.

In the l-enantiomorphic form, the compounds are useful in that they can be racemized to obtain a mixture of d- and l-enantiomorphs. The racemization can be carried out by chemical or enzymatic means, and the resulting racemic mixture can then be employed as a starting material from which, by resolution, the d-enantiomorphic form useful as a nonnutritive sweetener is procured. Thus, both the d- and l-enantiomorphic forms are useful substances.

The amount of the nonnutritive sweetener to be used is not critical so long as an effective amount is used. Generally, an effective amount is that amount which provides a sense of sweetness comparable to that afforded by sucrose at a given usage rate.

Thus, for example, in confectionary products sucrose concentration may approach 100%, whereas in many common foods and liquids, the sucrose concentration may be as low as 1% or lower. Correspondingly, the amount of the active agent which will provide sweetness equivalent to that afforded by sucrose also varies widely. For sucrose concentrations of from 6.5 to 45%, concentrations which include most food and even many confectionary applications, the active agent, when substituted for sucrose, gives approximately equivalent sweetness at concentrations of from 0.05 to 30% as the racemic mixture.

Where the d-enantiomorph is used alone, these rates can be reduced by a factor of about one-half. Higher or lower concentrations of the active agent of this process can be used where the desired degree of sweetness, by a sucrose standard, is greater or lesser. However, usage of a high concentration of any nonnutritive sweetener generally increases the incidence of off-flavor and other undesirable side effects; for this reason, where a high degree of sweetness is desired, it is generally preferred to use a combination of the active agent with one or more other known nonnutritive sweeteners.

Conventional nonnutritive sweeteners with which the nonnutritive sweetener of this process can be suitably combined include saccharin and the substituted saccharin compounds which can be conveniently used in the form of salts; 5-(3-hydroxyphenoxy)-1H-tetrazole which can suitably be used in the form of the sodium, calcium, potassium, or ammonium salt; and the dihydrochalcone-type sweeteners which are described in U.S. Patent 3,087,821.

When the active agent is used in combination with one or more of previously known nonnutritive sweeteners, the exact ratio of the combination is not critical and can vary considerably. Synergism is sometimes noted, permitting a reduction of the amounts when used in combination.

Good results are generally obtained with combinations containing the active agent, on the one hand, and the known nonnutritive sweetener, on the other hand, in a ratio that ranges from 1:5 to 5:1, by weight.

It is known that the use of saccharin as a sweetening agent is accompanied by bitter aftertaste, experienced by a certain portion of the population. Since for many applications, the substance is ideally suited to usage as a sweetener, methods of diminishing the aftertaste have been studied. Therefore, in those unusual situations where administration of the active agent is accompanied by aftertaste, known methods of diminishing such aftertaste can be utilized. Furthermore, such methods can also be used where the active agent is combined with saccharin and/or other nonnutritive sweeteners.

It is also possible to combine the active agent with sucrose or other nutritive sweeteners, so as to obtain a sweetening substance of merely reduced caloric value. Such substance can be formulated as a foam, the purpose being to increase the bulk so that a given amount of the foam has a sweetness equivalent to the same amount of sugar, alone.

A preferred group of sweeteners are those of the formula:

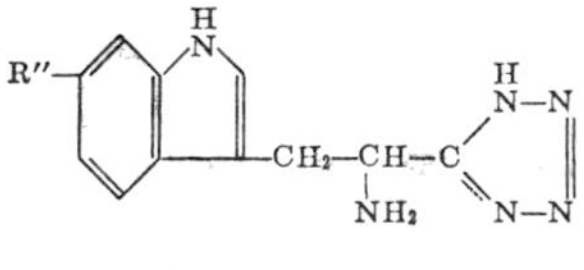

or their salts, wherein R" is chloro, fluoro, or methyl, and a particularly preferred sweetener is α-5-tetrazolyl-6-chlorotryptamine.

Example 1: dl-N-Acetyl-α-Cyano-α-Carboethoxy-6-Chlorotryptamine — 3-(Diethylaminomethyl)-6-chloroindole (8.0 grams; about 0.3 mol), ethyl α-acetamido-α-cyanoacetate (7.5 grams; about 0.044 mol), and powdered potassium hydroxide (6.5 grams) in 35 ml of toluene were refluxed for 1 hour under nitrogen. The reaction mixture was then cooled to room temperature, resulting in the precipitation of the desired ethyl dl-N-acetyl-α-cyano-α-carboethoxy-6-chlorotryptamine product. The product was separated by filtration and washed with toluene, water, ethanol, and ether, to yield 7.2 grams of material. A sample was recrystallized from ethanol, MP, 205° to 210°C.

Example 2: dl-N-Acetyl-α-Carboethoxy-α-5-Tetrazolyl-6-Chlorotryptamine – A solution of 2.8 grams of anhydrous aluminum chloride in 45 ml of tetrahydrofuran was added to a stirred suspension of 4.5 grams of sodium azide (about 0.07 mol) in 10.5 ml of the same solvent. The mixture was refluxed for 1 hour and then cooled to 25°C. dl-N-acetyl-α-carboethoxy 6-chlorotryptamine (6.94 grams, about 0.018 mol) was added, and the mixture was heated under reflux for 24 hours.

Water (32 ml) and concentrated hydrochloric acid (10.5 ml) were added with cooling. The aqueous layer was separated and extracted with 40 ml of 1:1 ether/tetrahydrofuran. The organic layers were combined, washed with water, and dried over magnesium sulfate, and the solvents were distilled to yield the desired dl-N-acetyl-α-carboethoxy-α-5-tetrazolyl-6-chlorotryptamine. The crude product was recrystallized from chloroform, yielding 7.3 grams, MP, 160°C.

Example 3: dl-N-Acetyl-α-5-Tetrazolyl-6-Chlorotryptamine – A solution of dl-N-acetyl-α-carboethoxy-α-5-tetrazolyl-6-chlorotryptamine (7.3 grams; 0.02 mol) and 3.2 grams NaOH in 30 ml of water was refluxed for 3 hours. The mixture was then cooled, and acidified with 10 ml concentrated hydrochloric acid, and then the crude acid (dl-N-acetyl-α-carboxy-α-5-tetrazolyl-6-chlorotryptamine) was separated by filtration and washed with water. It was then decarboxylated by heating in 200 ml of water for 2 hours. The solution was cooled, and the product was filtered, to yield 2.26 grams of the desired dl-N-acetyl-α-5-tetrazolyl-6-chlorotryptamine. A sample of this product was recrystallized from water, MP, 251° to 255°C.

Example 4: dl-α-5-Tetrazolyl-6-Chlorotryptamine – A solution of dl-N-acetyl-α-5-tetrazolyl-6-chlorotryptamine (2.0 grams; about 0.01 mol) in 25 ml of water containing 2 grams of sodium hydroxide was heated under reflux for 12 hours. The mixture was cooled and acidified to pH 5 with hydrochloric acid, to precipitate the desired dl-α-5-tetrazolyl-6-chlorotryptamine product. The product was separated by filtration, washed with water, and recrystallized from acetic acid, MP, 271° to 274°C (d).

Example 5: dl-N-Acetyl-α-Cyano-α-Carboethoxy-6-Fluorotryptamine – dl-N-acetyl-α-cyano-α-carboethoxy-6-fluorotryptamine, MP, 178° to 180°C, was prepared in accordance with the procedures of Example 1 from 3-(diethylaminomethyl)-6-fluoroindole.

Example 6: dl-N-Acetyl-α-Carboethoxy-α-5-Tetrazolyl-6-Fluorotryptamine – dl-N-acetyl-α-carboethoxy-α-5-tetrazolyl-6-fluorotryptamine, MP, 215° to 217°C, was prepared in accordance with the procedures of Example 2 from dl-N-acetyl-α-cyano-α-carboethoxy-6-fluorotryptamine.

Example 7: dl-α-5-Tetrazolyl-6-Fluorotryptamine – In accordance with the procedures of Example 3, dl-N-acetyl-α-carboethoxy-α-5-tetrazolyl-6-fluorotryptamine was reacted with sodium hydroxide and thereafter worked up in accordance with the procedures of Example 3 yielding the corresponding dl-N-acetyl-α-carboxy-α-5-tetrazolyl-6-fluorotryptamine. This product was then decarboxylated as in Example 3 and hydrolyzed as in Example 4, yielding dl-α-5-tetrazolyl-6-fluorotryptamine, MP, 267° to 270°C.

Example 8: dl-α-5-Tetrazolyl-6-Methoxytryptamine — dl-α-5-tetrazolyl-6-methoxytryptamine was prepared from 3-(diethylaminomethyl)-6-methoxyindole. The preparation was carried out in accordance with the procedures of Examples 1 through 4 except that in the reaction of the intermediate dl-N-acetyl-α-cyano-α-carboethoxy-6-methoxytryptamine to obtain the corresponding dl-N-acetyl-α-carboethoxy-α-5-tetrazolyl-6-methoxytryptamine intermediate, ammonium chloride was employed instead of aluminum chloride. The product so obtained melted at 264° to 267°C.

Example 9: An initial evaluation was carried out with various of the compounds serving as active agent in accordance with the process. The evaluation comprised the tasting of a small amount of the respective compound, the amount being that which would adhere to a finger tip. Each of the compounds was rated for degree of sweetness and aftertaste, if any. The sweetness ratings were as shown below. No aftertaste was observed with any of the compounds evaluated.

Compound Evaluated	Sweetness Rating
dl-α-5-tetrazolyl-6-chlorotryptamine	Very sweet
dl-α-5-tetrazolyl-6-fluorotryptamine	Very sweet
dl-α-5-tetrazolyl-6-methoxytryptamine	Sweet

Example 10: In another evaluation, dl-α-5-tetrazolyl-6-chlorotryptamine was tested for its sweetening effect in aqueous solutions. The evaluation enabled comparison with known sweetening agents.

In this evaluation, the compound was dissolved in each of several portions of water, thus obtaining several solutions containing the compound in varying concentrations. In addition, there was prepared an aqueous solution containing saccharin of a concentration equivalent to a 10% solution of sucrose. All solutions were of a pH of 6.7.

Thereafter, all solutions were taste tested. Comparison was made between the saccharin solution, as a standard and equivalent to a 10% solution of sucrose, and the various solutions of the dl-α-5-tetrazolyl-6-chlorotryptamine. The solution adjudged to be equivalent to the standard solution contained 0.1% of the dl-α-5-tetrazolyl-6-chlorotryptamine. No aftertaste was observed.

Example 11: Various of the compounds serving as the active agent were evaluated further, jointly with saccharin. In this evaluation, aqueous solutions were prepared containing saccharin and one of the selected compounds. The concentration of the saccharin in the solutions was uniformly 0.01%; the concentration of the selected compound was varied among the solutions.

In addition, as in Example 10, an aqueous solution containing sodium saccharin was prepared to serve as a standard equivalent to a 10% sucrose solution. All solutions were at a pH of 6.7. As in the evaluation reported in Example 10, the solutions containing the test compounds were taste tested. Taste was compared with the standard to determine which concentration of test compound and saccharin was equivalent to the standard. The concentrations of active agent in the solutions judged to be equivalent to the standard are listed in the following table. No aftertaste was noted.

Compound	% of Compound in Solution
dl-α-5-tetrazolyl-6-fluorotryptamine	0.02
dl-α-5-tetrazolyl-6-methoxytryptamine	0.05
dl-α-5-tetrazolyl-6-chlorotryptamine	0.015

SUBSTITUTED TETRAZOLES

5-(3-Hydroxyphenoxy)-1H-Tetrazole

It has been found by *W.L. Garbrecht; U.S. Patent 3,515,727; June 2, 1970 and U.S. Patent 3,597,234; August 3, 1971; both assigned to Eli Lilly and Company* that 5-(3-hydroxyphenoxy)-1H-tetrazole and its nontoxic, water-soluble salts alone or in combination with caloric materials (such as foodstuffs) or noncaloric materials (such as chemotherapeutic agents) for animal consumption, exhibit a pleasant sweet taste without concomitant bitter aftertaste.

The physiological mechanism whereby the compounds exhibit a sweet taste is not understood, and there appears to be no logical basis for determining the effect of structural variations among chemical compounds except by preparing the compounds and taste-testing them. In so doing, it has been found that simple structural changes completely destroy the sweet taste. The group consisting of 5-(3-hydroxyphenoxy)-1H-tetrazole has the following formula:

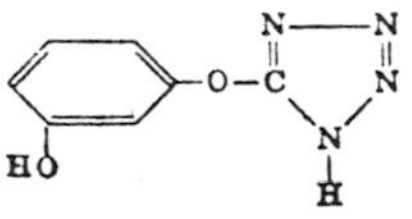

HO, O—C, N—N, N, N, H

Nontoxic, physiologically acceptable, water-soluble salts include the sodium, calcium, and ammonium salts. The compound, 5-(3-hydroxyphenoxy)-1H-tetrazole, is about two hundred times as sweet as sucrose. Both it and its nontoxic, water-soluble salts are useful as sugar substitutes or synthetic sweeteners.

The principal object of the process is to provide a 5-substituted tetrazole and a method of sweetening which will overcome the disadvantages found with the use of previously developed synthetic sweeteners.

The taste of several 5-aryloxy-1H-tetrazoles bearing different substituents is recorded in the following table:

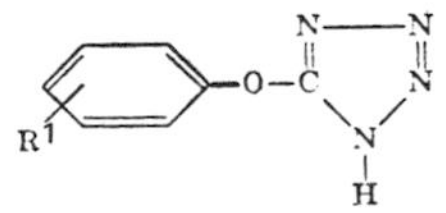

R^1, O—C, N—N, N, N, H

R^1	Taste	R^1	Taste
3-COOH	Tasteless	3-CH_3O	Tasteless
3-$COOCH_3$	Tasteless	4-CH_3O	Tasteless
3-$COOC_2H_5$	Tasteless	2-Cl	Tasteless

(continued)

R^1	Taste	R^1	Taste
4-$COOC_2H_5$	Tasteless	3-Cl	Tasteless
3-$CONH_2$	Bitter	4-Cl	Tasteless
3-$CONHCH_3$	Bitter	4-Br	Bitter
3-$C_6H_5CH_2O$	Tasteless	3-CH_3CONH	Bitter
4-$C_6H_5CH_2O$	Tasteless	4-CH_3CONH	Bitter
2-t-butyl	Tasteless	3,5-diCH_3	Tasteless
4-t-butyl	Tasteless	3,4-diCH_3	Bitter
3-t-butyl	Bitter	H	Very weakly sweet
2-CH_3	Tasteless		
3-CH_3	Tasteless	3-OH	Very sweet
4-CH_3	Tasteless	3-OH, Na salt	Very sweet
3-C_2H_5	Tasteless	3-OH, Ca salt	Very sweet
4-C_2H_5	Tasteless	3-OH, NH_4 salt	Very sweet

The preparation of 5-(3-hydroxyphenoxy)-1H-tetrazole is carried out in a manner similar to that described by E. Grigat et al, *Chem. Ber.,* 98, 3777 (1965). A mixture in water of a cyanogen halide (e.g., cyanogen bromide), sodium azide, and a resorcinol monoester in which one of the hydroxyl groups of resorcinol is protected by an acyl group is allowed to react. Suitable protective acyl groups include acetyl, benzoyl, 2,4-dinitrobenzoyl and 3,4-dichlorobenzoyl.

The resulting reaction product mixture is subjected to hydrolysis to remove the acyl group, conveniently by adjusting the pH to strongly basic and refluxing for about 1 to 3 hours. The hydrolysis mixture is acidified, the mixture exhaustively extracted with ether, and the combined ether extracts concentrated in vacuo to dryness. The residue is recrystallized from boiling water to yield 5-(3-hydroxyphenoxy)-1H-tetrazole.

Alternatively, the intermediate 5-(3-acyloxyphenoxy)-1H-tetrazole can be isolated by acidifying the reaction product mixture with an aqueous acid, for example, 6 N aqueous hydrochloric acid, refrigerated overnight, and the crystalline product, 5-(3-acyloxyphenoxy)-1H-tetrazole recovered by filtration. The acyloxy compound is hydrolyzed under basic conditions to remove the acyl group. Acidification and separation of the hydroxyphenoxytetrazole is done as before.

The term water-soluble salts includes those salts prepared from cationic materials sufficiently basic to react with 5-(3-hydroxyphenoxy)-1H-tetrazole to form salts. With respect to inorganic materials it has been found that the sodium, calcium, and ammonium cations are best suited for the formation of a salt. The salts of 5-(3-hydroxyphenoxy)-1H-tetrazole formed are water-soluble as indicated by their ability to form at room temperature, aqueous solutions containing concentrations of, respectively, up to 25% of the sodium or ammonium salt, and up to 50% of the calcium salt.

In addition, it must be appreciated that since the compounds of this process are acidic in character, they will react with those organic bases of sufficient basicity to form salts. Thus, for example, the compounds will also react with amines or alkaloids. As an example, consider the d-3-methyl-4-dimethylamino-1,2-diphenyl-2-propionoxybutane salt of 5-(3-hydroxyphenoxy)-1H-tetrazole (MP 167° to 168°C). This salt, unlike the propionoxybutane free base, is not bitter.

Example 1: 5-(3-Hydroxyphenoxy)-1H-Tetrazole – To a mixture of 75 ml of chloroform, 9 grams of resorcinol monoacetate, and 6.15 grams of cyanogen bromide, maintained at a temperature of about 0° to 5°C, were added 5.9 grams of triethylamine over a period of about 15 minutes. To the mixture thus obtained was added rapidly a solution of 3.8 grams of sodium azide in 50 ml of water without further cooling. The reaction mixture was then stirred for about 2 hours.

The reaction product mixture was acidified with excess 6 N aqueous hydrochloric acid, the chloroform layer separated, and the aqueous layer extracted twice with 50 ml portions of ether. The chloroform layer and ether extracts were combined, dried, and concentrated in vacuo. The solid residue which remained was recrystallized from aqueous ethanol to yield a crystalline product having a MP of about 100° to 102°C, which was identified by analysis as 5-(3-acetoxyphenoxy)-1H-tetrazole.

2 grams of 5-(3-acetoxyphenoxy)-1H-tetrazole were warmed with excess aqueous sodium hydroxide for about an hour. The hydrolysis mixture was cooled and acidified with concentrated aqueous hydrochloric acid, and the mixture extracted with several volumes of ether. The combined ether extracts were concentrated to dryness in vacuo. The residue was recrystallized from water to yield a crystalline product, 5-(3-hydroxyphenoxy)-1H-tetrazole, weighing 1.5 grams and having a MP of about 141° to 143°C.

Example 2: 5-(3-Hydroxyphenoxy)-1H-Tetrazole – To a mixture of 32.0 grams of bromine and 15 ml of water in a round bottom flask equipped with stirrer, condenser, thermometer, and dropping funnel, was added a solution of 10.4 grams of sodium cyanide in 50 ml of water, the addition being regulated to keep the temperature of the mixture at about 20° to 30°C.

A solution of 42.8 grams of resorcinol monobenzoate in 100 ml chloroform was added, the reaction mixture was cooled in an ice bath to 0° to 5°C, and 20.2 grams of triethylamine were added to the reaction mixture dropwise while maintaining the temperature thereof at 0° to 5°C.

A solution of 13.0 grams of sodium azide in 100 ml water was then added rapidly dropwise with stirring, allowing the temperature of the reaction mixture to rise to approximately 40°C, and stirring was continued for 30 minutes after the addition was complete. The mixture was made basic by the addition of sodium carbonate. A Dean-Stark collecting trap was attached to the reaction flask, and the reaction product mixture was refluxed and stirred for 2 to 3 hours to remove chloroform and to hydrolyze the benzoate ester.

The mixture was cooled, acidified with cold, concentrated aqueous hydrochloric acid, washed three times with 150 ml of benzene to remove benzoic acid, and the washings discarded. The aqueous layer was acidified and exhaustively extracted with ethyl ether until the copper acetate test was negative for the presence of tetrazole in the aqueous layer. The ether solution was evaporated and the residue recrystallized from boiling water (3 ml/g of residue), being decolorized with charcoal and stirred while the crystallization proceeded. The product, 5-(3-hydroxyphenoxy)-1H-tetrazole, was obtained as a solid weighing 33.7 grams (95% yield), and having a MP of about 141° to 143°C.

Example 3: 1 teaspoonful of a 0.015% aqueous solution of 5-(3-hydroxyphenoxy)-1H-tetrazole was added to a cup of coffee (180 ml). It was found that the aqueous solution mixed well with the coffee; and upon taste-testing, it was determined that the sweetness was about equivalent to that of coffee sweetened with 1 teaspoonful of sugar.

Example 4: To a glass of cold tea was added 1 teaspoonful of a 0.015% aqueous solution of 5-(3-hydroxyphenoxy)-1H-tetrazole. Again it was found as in Example 1 that the resulting product exhibited a sweet taste which was equivalent to that obtained with 1 teaspoonful of sugar (sucrose).

Example 5: Fresh sectioned grapefruit was sweetened to taste with 0.588% aqueous solution of 5-(3-hydroxyphenoxy)-1H-tetrazole, and the resulting product exhibited a sweet taste which closely resembled in taste that obtained with one teaspoonful of sugar.

5-Carbocyclicaminotetrazoles

R.M. Herbst; U.S. Patent 3,294,551; December 27, 1966; assigned to Eli Lilly and Company; and also British Patent 1,170,590 reports that certain 5-carbocyclicaminotetrazoles and nontoxic salts exhibit an intensely sweet taste without bitter aftertaste while other 5-substituted aminotetrazole compounds and the nontoxic salts exhibit no taste or are actually sour or bitter. The following compounds have been found to exhibit this unique sweet taste:

5-cyclohexylaminotetrazole;
5-o-tolylaminotetrazole;
5-m-tolylaminotetrazole;
5-m-methoxyphenylaminotetrazole;
5-m-chlorophenylaminotetrazole;
5-o-chlorophenylaminotetrazole;
5-phenylaminotetrazole.

In the preparation of 5-carbocyclicaminotetrazoles normally a 1-carbocyclic-5-aminotetrazole compound is prepared and then rearranged to form the product. The 5-carbocyclicaminotetrazoles also can be prepared directly.

Example 1: 1-Phenyl-5-Aminotetrazole – A sludge of 152 grams of N-phenylthiourea in 300 ml of absolute ethanol in a 1 liter round bottomed flask equipped with a reflux condenser was chilled in an ice bath. A total of 150 grams of methyl iodide was added to the sludge in several portions during half an hour taking care to mix the reactants thoroughly after each addition of methyl iodide.

The reaction mixture was allowed to warm to room temperature slowly as the ice in the cooling bath melted and was then kept at room temperature for 40 hours. Finally, after the mixture was boiled under reflux for one hour excess methyl iodide was removed while about 100 ml of ethanol was distilled from the reaction flask. The residual ethanolic solution of N-phenyl-S-methylthiouronium iodide was diluted with 200 ml of absolute ethanol.

The solution was cooled in an ice bath while 59 grams of 85% hydrazine hydrate solution was added slowly. An initial precipitate formed but redissolved as the reaction mixture was warmed slowly to reflux temperature. Methyl mercaptan was evolved rapidly as the solution was warmed on a steam bath and was absorbed in a suitable trap charged with aqueous sodium hydroxide.

The reaction mixture was boiled under reflux for two hours after the vigorous evolution of methyl mercaptan subsided and then stored at room temperature overnight. As much ethanol as possible was removed from the reaction mixture by distillation under reduced pressure from a warm water bath. The syrupy residue was dissolved in 300 ml of water and again subjected to distillation under reduced pressure until about 200 ml of distillate had collected.

The aqueous solution of N-phenyl-N'-aminoguanidine hydriodide was diluted with 500 ml of water, acidified with 10 ml of concentrated nitric acid and treated with a solution of 170 grams of silver nitrate in 250 ml of water, the latter added dropwise with continuous stirring during half an hour. Excess silver ion was removed by addition of 10 ml of concentrated hydrochloric acid after which the suspension was stirred at room temperature for an hour. The silver halides were removed by suction filtration and were washed carefully with cold water.

The combined filtrate and washings were stirred and cooled to +5°C, in an ice bath when a solution of 69 grams of sodium nitrite in 200 ml of water was added slowly at a rate such that the temperature remained between +5°C and 10°C. A solid separated from the reaction mixture during the diazotization procedure.

The suspension was brought to pH 8 by the careful addition of aqueous potassium hydroxide solution, warmed to 35°C, and then allowed to cool slowly to room temperature. After chilling the suspension thoroughly in an ice bath the crude 1-phenyl-5-aminotetrazole was collected on a filter, washed with cold water and dried at 100°C. Recrystallization of the material from 50% aqueous isopropyl alcohol, using charcoal to decolorize the solution, gave 113 grams of pure product that melted at 159° to 160°C, and on continued heating resolidified and then remelted at 199° to 200°C.

Example 2: 1-m-Tolyl-5-Aminotetrazole — A solution of 21.4 grams of m-toluidine in 120 ml of 95% ethanol was placed in a 1 liter, 3 necked flask equipped with a stirrer, a dropping funnel, a thermometer and an exhaust tube. With external cooling to keep the temperature of the reaction mixture below +10°C, a solution of 21.2 grams of cyanogen bromide in 80 ml of 50% ethanol was added dropwise with continuous stirring.

Under the same conditions a solution of 8 grams of sodium hydroxide in 20 ml of water was added, followed by a solution of 15 grams of sodium azide in 55 ml of water and 18 ml of concentrated hydrochloric acid diluted with 18 ml of water. The reaction flask was equipped with a reflux condenser, transferred to a steam bath and the mixture boiled under reflux for 6 hours. At the end of the reaction period about 120 ml of solvent was removed by distillation on the steam bath. The residual aqueous alcoholic solution was chilled thoroughly in an ice bath. The product crystallized and was removed by filtration and washed with cold 30% ethanol before drying at 100°C.

The crude 1-m-tolyl-5-aminotetrazole so obtained was recrystallized from 50% isopropyl alcohol from which it separated as lustrous needles melting at 165°

to 166°C. The yield of pure product was 21 grams.

Example 3: 1-m-Chlorophenyl-5-Aminotetrazole – To a solution of 12.7 grams of m-chloroaniline in 100 ml of 95% ethanol was added slowly, with cooling, a solution of 10.6 grams of cyanogen bromide in 50 ml of 95% ethanol. The resulting solution was warmed on the steam bath under a reflux condenser for 20 minutes after which hydrogen bromide was neutralized by addition of 6.9 grams of anhydrous potassium carbonate and the solvent was removed by distillation under reduced pressure from a warm water bath. The residue was extracted with 50 ml of warm benzene. The benzene solution of m-chlorophenylcyanamide so formed was transferred to a pressure bottle, cooled in an ice bath and then treated with 40 ml of a 15% solution of hydrazoic acid in benzene.

The pressure bottle was sealed and heated for 7 hours in a well shielded steam bath. 1-m-chlorophenyl-5-aminotetrazole separated from the benzene during the course of the reaction. After cooling the reaction mixture in an ice bath, the solid product was collected on a filter and washed with benzene. The air dried product was recrystallized from 90% isopropyl alcohol from which is separated as colorless leaflets that melted at 173° to 174°C, resolidified on continued heating and then remelted at 198° to 200°C.

Example 4: 1-m-Methoxyphenyl-5-Aminotetrazole – To a cooled solution of 12.3 grams of m-anisidine in ether in which was suspended 6.9 grams of anhydrous potassium carbonate was added dropwise with stirring a solution of 10.6 grams of cyanogen bromide in 50 ml of ether. The mixture was stirred at room temperature for several hours after which the supernatant was decanted into 100 ml of ether containing 5 grams of hydrazoic acid. The ether solution was boiled gently under reflux for about ten hours during which considerable solid separated from the mixture. The solid was separated by filtration, after cooling the ether suspension thoroughly in an ice bath, and recrystallized from 30% isopropyl alcohol from which it separated as glistening crystals that melted at 140° to 141°C.

Example 5: 1-β-Naphthyl-5-Aminotetrazole – A solution of 20.4 grams of β-naphthylamine in 150 ml of 95% ethanol was prepared in a 1 liter, 3 necked flask equipped with a stirrer, a dropping funnel, a thermometer and an exhaust tube. While keeping the temperature below 25°C, by occasional external cooling with an ice bath, a solution of 15 grams of cyanogen bromide in 80 ml of 95% ethanol was added dropwise with continuous stirring.

This was followed immediately by a solution of 5.7 grams of sodium hydroxide in 20 ml of water, added to the reaction mixture under similar conditions. The solid that separated during the addition of sodium hydroxide was redissolved by warming the reaction mixture to 50°C for 15 minutes.

After cooling the mixture to room temperature again, solutions of 11.3 grams of sodium azide in 35 ml of water and 8 grams of concentrated sulfuric acid in 30 ml of water were added successively. The resulting mixture was placed on a steam bath, the flask equipped with a reflux condenser and the mixture boiled under reflux for 8 hours. About 100 ml of solvent was then removed by distillation at atmospheric pressure.

The product separated from the residual solution on cooling as an almost colorless, crystalline solid. The crude product was contaminated with a small amount of β-naphthylamine that was removed by washing the crude material with warm ethylene chloride. Recrystallization of the remaining solid from 95% ethanol gave 14.5 grams of pure 1-β-naphthyl-5-aminotetrazole as a colorless crystalline solid melting at 191° to 192°C, and resolidifying on continued heating to re-melt again at 211°C.

Example 6: 5-Phenylaminotetrazole – A suspension of 122 grams of 1-phenyl-5-aminotetrazole in 800 ml of xylene was boiled under reflux with continuous stirring for one and a half hours. The original solid dissolved almost completely in the boiling xylene after which the product began to separate rapidly as heating was continued.

The resulting suspension was cooled thoroughly in an ice bath, the product was filtered by suction, pressed as free of xylene as possible, washed with benzene and air dried. The crude product weighed 119 grams and melted at 205° to 206°C. The material was recrystallized from 70% isopropyl alcohol using a little charcoal to decolorize the hot solution. The recrystallized material melted at 209°C, when heated rapidly to the melting point, at 206°C, on slower heating. The yield of recrystallized product was 113 grams.

Example 7: Calcium Salt of 5-Phenylaminotetrazole – A suspension of 1.61 grams of 5-phenylaminotetrazole and 0.37 gram of calcium hydroxide in 100 ml of water was warmed and shaken until complete solution was effected. A faint turbidity was removed by filtration and the clear, intensely sweet filtrate was evaporated to dryness.

The colorless, crystalline residue of calcium salt was recrystallized by dissolving it in the minimum amount of hot 65% ethanol and diluting the clear, hot solution with four volumes of acetone. The calcium salt darkened but did not melt below 300°C.

Example 8: Sodium Salt of 5-m-Chlorophenylaminotetrazole – A solution of 0.98 gram of 5-m-chlorophenylaminotetrazole in 10 ml of 0.5 N aqueous sodium hydroxide was evaporated to dryness. This sodium salt remained as a colorless, crystalline solid that failed to melt below 250°C. The sodium salt is readily soluble in cold water; its aqueous solutions exhibit an intensely sweet taste.

Example 9: Calcium Salt of 5-m-Chlorophenylaminotetrazole – A suspension of 0.98 gram of 5-m-chlorophenylaminotetrazole and 0.19 gram of calcium hydroxide in 50 ml of warm water was shaken until the organic compound dissolved. A slight turbidity was removed by filtration. The intensely sweet clear filtrate was evaporated to dryness leaving the calcium salt as a colorless, crystalline solid, readily soluble in water which darkened on heating to 300°C.

Example 10: Sodium Salt of 5-m-Tolylaminotetrazole – To a solution of 0.53 gram of anhydrous sodium carbonate in 50 ml of water was added 0.88 gram of 5-m-tolylaminotetrazole. The suspension was warmed and stirred until a clear solution resulted. The intensely sweet solution was evaporated to dryness leaving the sodium salt as a colorless, crystalline residue that was freely soluble in water.

8,9-EPOXYPERILLARTINE SWEETENERS

Perillartine is described in the literature as being approximately 2,000 times, sweeter than sucrose. However, this sweet taste is accompanied not only by the somewhat bitter aftertaste which is present to a greater or lesser degree with all synthetic sweeteners, but also by a pronounced licorice-like, or gingery-licorice flavor which has ruled out any usage of the product in this country. Perillartine (all optical isomers) has a maximum water solubility, at 25°C, of about 0.0003 M. This concentration is sufficient to provide a pleasantly sweet aqueous solution.

A derivative of perillartine, namely 8,9-epoxyperillartine is prepared by *E.M. Acton, M.A. Leaffer and H. Stone; U.S. Patent 3,699,132; October 17, 1972; assigned to Stanford Research Institute.* 8,9-Epoxyperillartine optical isomers have the formula:

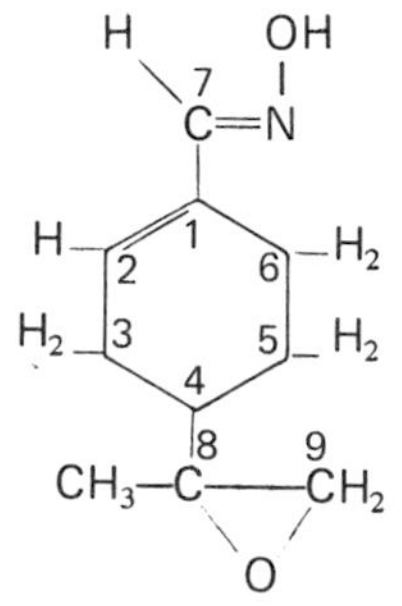

The numbering system used in the foregoing formula is that used with terpenes such as menthene. By another system of nomenclature, the optical isomers corresponding to the above formula can be designated as 4-(1,2-epoxy-2-propyl)-1-cyclohexene-1-carboxaldehyde, syn-oximes.

The optical isomer compounds of this synthesis which include the (+), or (d) isomer, the (–), or (l) isomer and the racemic (±), or (dl) isomer mixture are designated as 8,9-epoxyperillartines. In this connection, it may be noted that perillartine has the same formula as that shown above except that no epoxy group is present, and the number 8 and number 9 carbons are joined by an olefinic double bond.

The 8,9-epoxyperillartine compounds are white crystalline materials having a sweetness ranging variously from about 20 to 50 times that of sucrose, and a solubility in water at 25°C of about 0.003 M, or roughly 10 times that of perillartine. They are highly soluble in ethanol, benzene, acetone and other organic solvents. Further, these compounds have bitter aftertaste characteristics ranging from none to moderate and, quite unexpectedly, they are free of the undesirable gingery-licorice taste which characterizes the corresponding perillartine optical isomers. Aqueous solutions of the 8,9-epoxyperillartine, at concentrations of approximately 0.003 M, are pleasantly sweet and highly palatable.

In the dry state the compounds have an excellent shelf life, and aqueous solutions are stable under weakly basic conditions. They give no evidence of toxicity and are well adapted to be used, in the dry pellet or powder form, as synthetic sweetening agents either alone or in conjunction with other sweetening agents such as sucrose, dextrose, saccharin. They are also well adapted to be mixed, in dry form, with other food ingredients such as citric acid, flavoring, spices, starches, preservatives and dehydrated foodstuffs.

8,9-Epoxyperillartines can be prepared by the oxidation of the corresponding perillartine optical isomer, or isomer mixture, with m-chloroperbenzoic acid. The reaction is carried out by dissolving the perillartine in an organic solvent such as benzene, and then slowly adding m-chloroperbenzoic acid, or a solution thereof in the solvent. A 10 to 30% molar excess of the peracid is used. The ensuing reaction, which is somewhat exothermic, proceeds readily at room or moderately elevated temperatures.

Stirring is continued after all the peracid has been added while m-chlorobenzoic acid precipitates out. The system is then filtered, and the filtrate is washed with sodium bicarbonate solution to destroy excess peracid, followed by washing with water. After being filtered and dried with a suitable solid desiccant, the desired 8,9-epoxyperillartine isomer present in the solution can be recovered in good yield by crystallizing from solution or by stripping off the solvent in vacuo. The product can be purified by conventional methods such as recrystallization from appropriate organic solvents such as hexane, benzene or absolute ethanol.

During the reaction between the perillartine starting material and the peracid oxidant, it has been found that neither the 1,2-double bond in the ring nor the oxime functional group is affected. Further, the attack of the 8,9-double bond is highly stereoselective in that but one of the two possible stereomeric forms of the epoxide is produced.

Example 1: To a stirred solution of 2 grams (0.0121 mol) of racemic perillartine in 80 ml benzene at 25°C is added, in portions over a 10 minute period, 3.2 grams (0.0158 mol) of 85% m-chloroperbenzoic acid (a powdered dry solid) during which period the temperature rises to 30°C. The resulting solution is stirred for 12 hours while m-chlorobenzoic acid is precipitated. The system is filtered and the filtrate is washed first with two 50 ml portions of saturated aqueous sodium bicarbonate, and then with 50 ml of water.

After being dried with magnesium sulfate, the remaining liquid is concentrated in vacuo at 25°C, until crystals start to form. These crystals are then dissolved with heating, following which the solution is cooled to induce crystallization of the 8,9-epoxyperillartine product. On being recrystallized from benzene, the product is recovered as a white solid, MP 125° to 126°C, which is slightly soluble (0.003 M) in water at 25°C and has good solubility in hexane, benzene, alcohol and other organic solvents.

This material is found to be homogeneous by thin-layer chromatography in chloroform-ethyl acetate (1:1) on silica gel, it having R_f 0.52 compared to the (±) perillartine starting material having R_f 0.62. (These properties are the same whether racemic or enantiomeric perillartine is employed as the starting material.) The product is identified as (±) 8,9-epoxyperillartine by infrared and nuclear magnetic resonance spectra. This is confirmed by elemental analysis which shows the compound to have carbon, hydrogen and nitrogen contents of 66.47, 8.49 and 7.75%, respectively, as against theoretical values for these elements of 66.27, 8.34 and 7.73%, respectively.

Example 2: The foregoing Example 1 is repeated under essentially the same conditions except that (-) perillartine rather than (±) perillartine is reacted with the m-chloroperbenzoic acid. Here the (-) 8,9-epoxyperillartine obtained as product is also identified by infrared and nuclear magnetic resonance methods of analysis.

Elemental analysis shows it to have carbon, hydrogen and nitrogen contents of 66.08, 8.31 and 7.67%, respectively, as compared with theoretical values for these elements of 66.27, 8.34 and 7.73%, respectively. This compound has a MP of 116° to 117°C and an optical activity $[\alpha]_D^{24}$ of -105.6° at 1% concentration in ethanol. Its solubility characteristics are the same as those for the racemic isomer mixture of Example 1.

Example 3: Again following the procedure of Example 1, but substituting (+) perillartine for the (±) perillartine starting material, there is obtained (+) 8,9-epoxyperillartine as a white solid having a MW of 165.24. This compound has somewhat reduced sweetness and a higher bitter aftertaste than either the racemic isomer mixture of the (-) isomer of the foregoing examples. Its solubility characteristics are the same, however, as those of the other isomer materials. This compound, it will be noted, forms approximately one-half of the isomer mixture of Example 1.

Example 4: Taste tests by a panel consisting of 6 subjects are conducted to evaluate the relative sweetness characteristics of the (±) and the (-) 8,9-epoxyperillartine of Examples 1 and 2, respectively. In these tests each subject matches the sweetness of a given aqueous solution of either the (±) or the (-) compound with that of one of a series of sucrose solutions, each of a different concentration. It is found that the racemic (±) isomer mixture at a level in water of 0.054% (0.003 M) tastes relatively sweet and is equal in this respect to a 3.1% (0.09 M) aqueous solution of sucrose, its relative sweetness thus being at least 30 times that of the reference sucrose material.

Similarly, the (-) 8,9-epoxyperillartine at the same concentration (0.003 M) is found to be equal in sweetness to a 4.1% (0.12 M) sucrose solution on this direct comparison basis. Its relative sweetness is thus about 40 or more times that of sucrose. Further, the 0.003 M solution of this (-) isomer, while pleasantly sweet, is moderately sweeter than the 0.003 M solution of the (±) isomer.

Example 5: In another taste test by the panel of six subjects, the relative qualities of sweetness, bitterness and other flavors are evaluated for the compounds of Examples 1 and 2. In this test, each compound, in aqueous solution, is tasted a total of 12 to 36 times by the members of the panel. The data obtained by the panel show that, out of a total taste score of 100%, the racemic 8,9-epoxyperillartine isomer mixture of Example 1, at concentrations of both 0.002 and 0.003 M, has a sweetness of 70 to 72% and a bitterness of 15%. The balance (13 to 15%) is attributable to other taste qualities, none of which is objectionable to the palate. In particular, there is no characteristic perillartine flavor.

In tests with the (-) 8,9-epoxyperillartine isomer of Example 2, the solutions of both 0.002 and 0.003 M show a sweetness of 92 to 94%, no bitterness, and 6 to

8% of other taste qualities of an unobjectionable, nonperillartine character.

5-IMINO-4,4-DIMETHYL-2-IMIDAZOLIDINONE

A sweetening agent described by *R.J. Windgassen; U.S. Patent 3,340,070; September 5, 1967; assigned to Shell Oil Company* has the formula

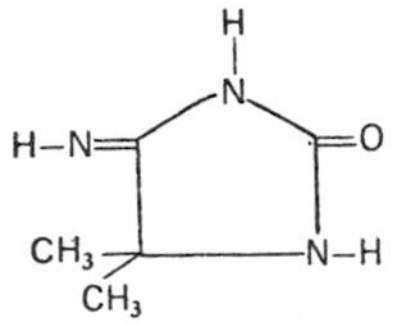

It melts at 280° to 283°C (with decomposition), and also can be named as 5-imino-4,4-dimethyl-2-imidazolidinone. It has also been found that sweetness is a unique characteristic of this compound, inasmuch as closely related compounds such as the following have been found not to be sweet.

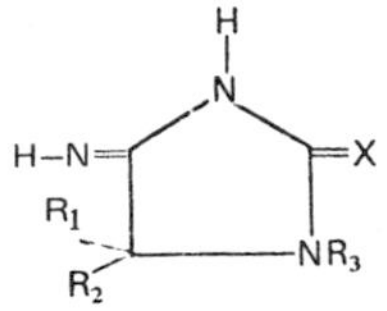

(1) R_1, R_2, R_3 = CH_3; X = O;
(2) R_1 = CH_3; R_2 = C_2H_5; R_3 = H; X = O
(3) R_1, R_2 = $-(CH_2)_5-$; R_3 = H; X = O;
(4) R_1, R_2 = CH_3; R_3 = H; X = S.

Another related compound (R_1 = phenyl; R_2, R_3 = H; X = O) was found to be definitely bitter. The 5,5-dimethyl-4-iminohydantoin is readily prepared by reacting alpha-ureidoisobutyronitrile (1-cyanoisopropylurea) with hydrogen in liquid ammonia using Raney nickel-chromium catalyst.

The sweetness of this compound was established as follows: A panel of four men was used. Solutions of 0.1, 0.01, 0.001 and 0.0001% of 5,5-dimethyl-4-iminohydantoin, in water, were tested against 2% sucrose solution. The sweetness of the sucrose solution was found to lie between the sweetness of the 0.1 and 0.01% solutions of the sweetener, indicating that the sweetener had between 20 and 200 times the sweetness of sucrose (weight basis). It was the conclusion of the panel that the sweetener probably was from 50 to 100 times as sweet as sucrose. No undesirable aftertaste was found.

5,5-dimethyl-4-iminohydantoin is readily soluble in water (12.5 grams dissolves in 100 grams of water at room temperature, 22°C), and is stable, with apparently adequate stability even in aqueous solution under acid conditions such as would exist in carbonated beverages.

The sweetener can be incorporated in the material to be sweetened in any of the usual ways: by simply mixing the sweetener with the material, or by first dissolving the sweetener in a suitable solvent or suspending it in a suitable liquid medium, then mixing the solution or suspension with the material to be sweetened, or by mixing the sweetener with a suitable solid carrier.

Since but a small amount of the sweetener ordinarily is required to impart the desired sweetness, it often will be most convenient to first formulate the sweetener with the suitable carrier, then mix the formulation with the material to be sweetened. As diluent there may be used any of the usual liquid and solid pharmaceutical carriers, including water, glycerol, starch, sorbitol, salt, sugar, citric acid, vegetable oils, and the like, which is nontoxic and compatible with the material which is to be sweetened.

Since the sweetener is intensely sweet, very small amounts of it are needed to impart the desired sweetness. Ordinarily, the desired sweetness is imparted by the addition, for example, from about 0.01 to about 0.25% of the sweetener. As little as 0.001% of the sweetener (same basis) often will be sufficient.

KYNURENINE DERIVATIVES

The chemistry and sweetness characteristics of kynurenine derivatives are given by *J.W. Finley; U.S. Patent 3,702,255; November 7, 1972; assigned to the U.S. Secretary of Agriculture.* Kynurenine derivatives are related to either of the following formulas

(1) benzene ring bearing, ortho to each other: $-C(=O)-CH_2-CH(COOH)-NH_2$ and $-NH-C(=O)R$

(2) benzene ring bearing, ortho to each other: $-C(=O)-CH_2-CH(COOH)-NH-C(=O)R$ and $-NH-C(=O)R$

It was found that these compounds exhibit a distinguishing characteristic in that they are about 25 to 50 times sweeter than sucrose. Sweetness is not a general characteristic of kynurenines as a class. Indeed, compounds which differ by the nature of the substituents on the basic kynurenine structure are tasteless or very sour (astringent). For example, kynurenine itself, has an astringent taste.

Because of their extraordinary sweetness, the kynurenine derivatives are useful as sweetening agents for foods and edible products of all kinds. It has been found, moreover, that for this sweetness to subsist, various elements in the structure are critical. One item is that the two side chains must be in ortho relationship. Another is that the amine group directly linked to the benzene ring must be acylated. Thus, kynurenine which contains a nonacylated amino group attached to the benzene ring is not sweet but has an astringent taste.

The compounds have certain structural features in common with tryptophane. Tryptophane exists in two optical forms, D and L. It is known that D-tryptophane is about 35 times sweeter than sucrose. On the other hand, L-tryptophane is rather bitter and in the racemic mixture of the two forms of tryptophane this bitter flavor is dominant so that DL-tryptophane is unsuitable as a sweetener. Although kynurenine and its derivatives can exist in different optical forms, the compounds discussed in this process in the racemic (DL) form exhibit the high degree of sweetness as mentioned and therefore are useful in such optically inactive form.

The compounds can be prepared by known methods. In one known technique DL-tryptophane is oxidized with H_2O_2 to yield N-formyl kynurenine. The kynurenine derivatives are soluble in water and stable, even in aqueous solution. As a result, they are useful for sweetening all types of materials which are intended for consumption or at least for contact with the mouth of the user.

Example 1: Synthesis of N′-Formyl Kynurenine –

O COOH
C–CH_2–CH–NH_2
O
NH–CH

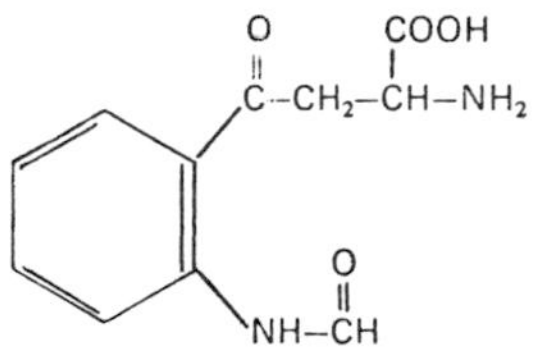

25 grams of DL-tryptophane were dissolved in water (about 1,000 ml), the solution adjusted to pH 2.0 by addition of hydrochloric acid, and filtered. An equimolar quantity of hydrogen peroxide (13.8 grams of 30% H_2O_2 solution) was added to the filtered solution. After standing for 30 minutes at room temperature, concentrated NH_4OH was added untii complete precipitation of the product was obtained. The precipitate was collected by centrifugation, then washed thoroughly with cold distilled water. The washed precipitate was then dissolved in hydrochloric acid, and concentrated NH_4OH added to the solution until complete precipitation of the product was obtained.

The precipitate was collected, washed with cold water, then dissolved in concentrated NH_4OH at pH 9.0. This solution was frozen and dried under vacuum. After drying, the material was washed three times with 200 ml portions of diethyl ether, yielding about 20 grams of N-formyl kynurenine.

Aqueous solution of N'-formyl kynurenine at different concentrations and aqueous solutions of sucrose at different concentrations were assessed for sweetness by a panel of 12 persons. The panel came to the conclusion that a 0.15% solution of N'-formyl kynurenine was equivalent in sweetness to a 5% solution of sucrose.

In another series of similar tests it was found that solutions of N'-formyl kynurenine at concentrations of 1.6%, 0.8%, and 0.4% were judged to be sweeter than a 10% solution of sucrose.

Example 2: Preparation of N'-Acetyl Kynurenine –

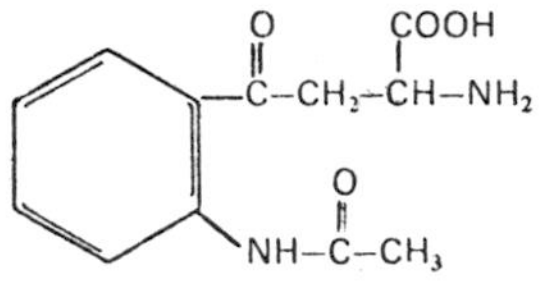

N'-formyl kynurenine was dissolved in 90% acetic acid and to this solution was added 2 volumes of acetic anhydride. After standing for about 2 hours, the product was precipitated by addition of ether, the precipitate being collected and washed with ether, yielding the desired N'-acetyl kynurenine. It was found that this compound was approximately 50 times sweeter than sucrose.

Example 3: Preparation of Kynurenine – The following reaction scheme was carried out in this synthesis:

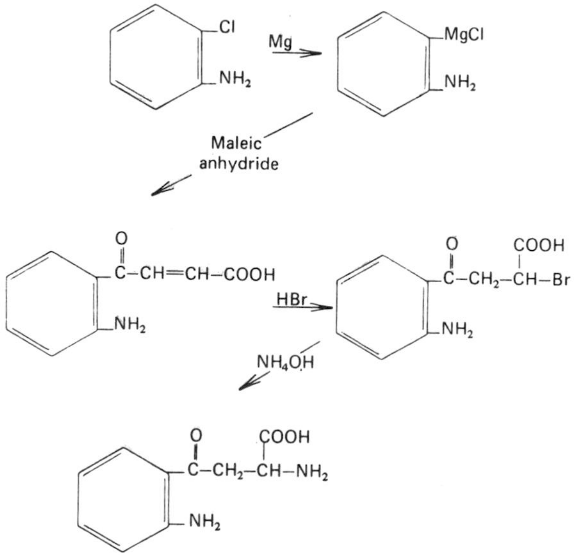

The following ingredients were heated under reflux for about 2 hours to produce the Grignard reagent:

	Grams
o-Chloroaniline	40
Tetrahydrofuran	40
Bromoethane	3
Magnesium	8

In the next step, 40 grams of maleic anhydride was added to the reaction mixture containing the Grignard reagent, and the system refluxed for 30 minutes. The reaction mixture was then filtered. The filtrate was diluted with 250 ml of water and extracted with an equal volume of ether. The aqueous solution was then treated with 2 N NaOH (about 50 ml) to precipitate the maleyl derivative which was collected and rinsed with distilled water.

In the next step the maleyl derivative was dissolved in 1 liter of 4% HBr and allowed to stand at room temperature for 30 minutes to form the bromo-succinyl derivative.

One liter of concentrated NH_4OH was added to the reaction mixture from the previous step and the reaction system allowed to stand overnight at room temperature. The next day the precipitate was separated and reprecipitated from 50% formic acid by addition of ether. The final yield of kynurenine was 48.5% (31.5 grams).

HELIOTROPYL NITRILE

It has been found that compositions which contain heliotropyl nitrile possess highly desirable sweetness and flavor characteristics with or without the addition of optional sweeteners and/or flavorants. *J.S. Neely; U.S. Patent 3,778,517; December 11, 1973; assigned to The Procter & Gamble Company* states that heliotropyl nitrile acts as a potentiator for sweetening agents resulting in compositions having enhanced sweetness characteristics.

It also has been found that compositions containing heliotropyl nitrile mixed or codissolved with other known flavorants results in a flavoring agent having a greatly enhanced flavor. Broadly, such flavorants include liquid and powdered extracts. Examples of such extracts include: oil of sweet birch, oil of spearmint, oil of wintergreen, anise oil, dill oil, celery seed oil, various citrus oils including lemon, orange, lime, tangerine and grapefruit oils, clove oil, peppermint oil, cassia, carrot seed oil, cola concentrate, ginger oil, angelica oil and vanillin.

When it is desired to have a liquid composition containing heliotropyl nitrile, it is important to utilize a suitable ingestible solvent in which to dissolve and/or suspend the heliotropyl nitrile to form the desired flavor and sweetening composition. It is equally as important to use a suitable solvent to codissolve and/or suspend the heliotropyl nitrile and other sweetening and/or flavoring agent which may be used.

For example, heliotropyl nitrile can be dissolved in ethyl alcohol and then diluted with an appropriate amount of water to yield the desired concentration of heliotropyl nitrile, which composition is suitable for flavoring and sweetening foods and beverages. In like fashion, ethyl acetate can be used to dissolve heliotropyl nitrile and hesperetin dihydrochalcone and can be subsequently diluted with water to yield a sweetening and flavor composition in a desired concentration.

Alternatively, heliotropyl nitrile or combinations can be dissolved in pure ethyl alcohol in a desired concentration and the solution employed as a flavoring and sweetening composition. Thus, it is seen that dissolution of the heliotropyl nitrile and previously disclosed combinations in polar, organic liquids, in mixtures and in mixtures with water, results in solutions suitable for use as artificial flavoring and sweetening agents.

Dry concentrated compositions wherein heliotropyl nitrile is admixed with a nontoxic, ingestible diluent are possible. In such concentrated compositions, heliotropyl nitrile is present in amounts of from 0.10 to about 5.0%. In the process, any of the commonly known diluents may be used. Nonlimiting examples of such diluents include: dry starch, powdered sucrose, lactose, kaolin, mannitol, dicalcium lactate, magnesium carbonate, magnesium oxide, calcium phosphate, powdered glycyrrhiza, or other commonly used diluents. Such dry compositions can then be used to enhance the flavor of and sweeten oral and labial compositions by addition in amounts to achieve a flavor and sweetness desirable to the user.

A typical soft drink contains carbonated water, flavoring and sweetening agents. A variety of flavoring agents are suitable for use in soft drinks. Generally, however, citrus oils are most preferred. Examples of suitable flavorants include heliotropyl nitrile, lemon, orange, lime, tangerine, and grapefruit oils, cola, root beer, grape, and the like. Such flavorants are generally present in amounts of from about 0.001 to about 0.10%.

Examples of suitable sweetening agents include heliotropyl nitrile either alone or in combination with saccharin, hesperetin dihydrochalcone, sucrose, levulose, and/or 6-(trifluoromethyl)tryptophane. Heliotropyl nitrile is most preferred as the flavoring and sweetening agent for soft drinks in amounts of from about 0.004 to about 0.75%.

Example 1: 20 grams of heliotropyl nitrile are thoroughly mixed with 80 grams of powdered lactose. The result is a dry sweetening and flavoring composition suitable for use without further treatment.

Compositions are prepared as in Example 1 except that dry starch, kaolin, mannitol, dicalcium lactate, magnesium carbonate, magnesium oxide, calcium phosphate, and powdered glycyrrhiza, respectively, are used in place of the powdered lactose. In each instance, the result is a desirable dry sweetening and flavoring composition.

Example 2: 1.0 gram of heliotropyl nitrile is dissolved in 100 grams of ethyl alcohol and 1,000 grams of water. The resulting solution is suitable for use as a sweetening and flavoring composition without further treatment.

Example 3: 1.0 gram of heliotropyl nitrile and 0.05 gram of hesperetin dihydrochalcone are dissolved in 20 grams of 1,2-dihydroxypropane with gentle warming and 1,000 grams of water is added thereto. The resulting solution is suitable for use as a sweetening and flavoring composition without further treatment.

Sweetening and flavoring compositions are prepared substantially as disclosed in Example 3 except that the hesperetin dihydrochalcone is, respectively, replaced by saccharin, glycerin, sorbitol, glucose, sucrose, dextrose, 6-(trifluoromethyl)-tryptophane, and levulose. Additional sweetening and flavoring compositions are prepared as in Example 3 except that hesperetin dihydrochalcone is replaced by 1:1 mixtures of hesperetin dihydrochalcone, respectively, with saccharin, glycerin, sorbitol, glucose, sucrose, and dextrose. Additional compositions are prepared as in Example 3 except that the hesperetin dihydrochalcone therein is replaced by a 1:1:1 mixture of hesperetin hydrochalcone, saccharin and sucrose. In each case, the result is a highly desirable sweetening and flavoring composition.

Example 4: One-tenth part of heliotropyl nitrile is dissolved in a synthetic pineapple oil (corresponding to winter fruit) consisting of 2.91 parts ethyl acetate, 0.61 part acetaldehyde, 0.45 part methyl n-valerate, 0.60 part methyl isovalerate, 1.40 parts methyl isocaproate and 0.75 part methyl caprylate to provide a sweetening composition and enhanced flavor composition.

2-(3-BROMOPROPOXY)-5-NITROANILINE

The synthesis of the sweetening agent 2-(3-bromopropoxy)-5-nitroaniline is given by *G.A. Crosby and G.C. Peters; U.S. Patent 3,845,225; October 29, 1974; assigned to Dynapol*. The steps are described below. The preparation of sodium-1-acetamido-4-nitrophenoxide is as follows:

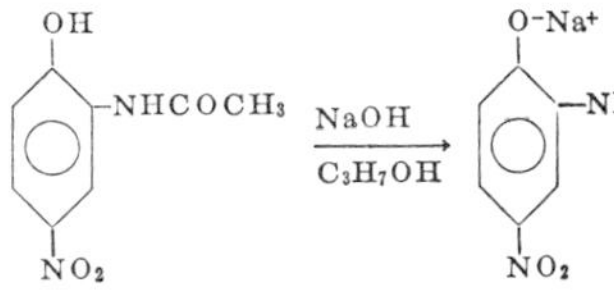

The sodium salt of 2-acetamido-4-nitrophenol is prepared by heating 8.32 grams of the starting material as shown above in 850 ml of n-propanol to near boiling for 30 to 60 minutes to effect solution, followed by cooling to about 50°C. Next, 44 ml of 1 N sodium hydroxide in n-propanol is added and the solution stirred to insure mixing and reaction of the reactants. After cooling and stirring for 1 hour the orange-red solution is evaporated to about ⅕ volume. Then, 500 ml of ether is added with rapid stirring to precipitate an orange solid that is collected on a filter and air dried to give 9.61 grams of the sodium salt (I).

The preparation of 1-(3-bromopropoxy)-2-acetamido-4-nitrobenzene (II) follows.

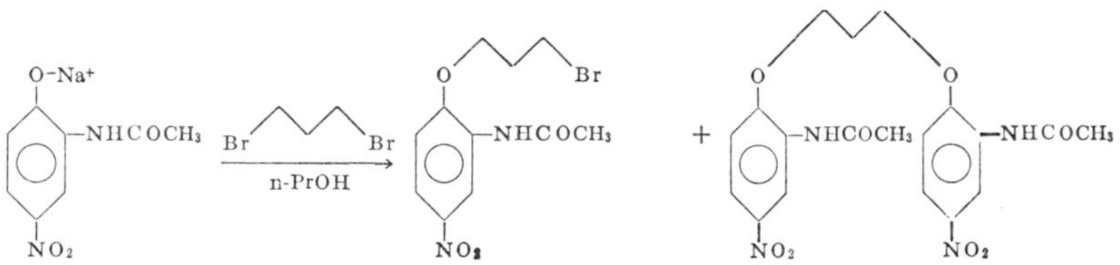

To the sodium salt (I; 23.8 mmol) in 150 ml of 1-propanol, there is added with stirring at reflux 1,3-dibromopropane (49.6 mmol). After stirring and refluxing for 10 hours, the reaction was diluted with ethyl acetate (250 ml), washed with saturated sodium bicarbonate solution until no more color was extracted, dried over anhydrous magnesium sulfate and evaporated to yield 1.50 grams of crude product. Chromatography with benzene solvent of the product on silica gel (70 grams; elution with 2, 4 and 6% ethylacetate in benzene) gave 420 mg of II and lesser amounts of III.

The preparation of 2-(3-bromopropoxy)-5-nitroaniline starting with compound II can be done via the following two routes:

1. 1 N HCl, reflux
2. 1 N NaOH

→ 2-(3-bromopropoxy)-5-nitroaniline (O–$CH_2CH_2CH_2Br$, NH_2, NO_2)

IV

A mixture of 1-(3-bromopropoxy)-2-acetamido-4-nitrobenzene (II; 1.0 mmol) in 1 N hydrochloric acid (20 ml) was stirred and refluxed for 2 hours and then diluted with distilled water. Then, 1 N sodium hydroxide was added until a precipitate formed which was extracted twice with ethyl acetate. The latter was dried over anhydrous magnesium sulfate and evaporated to yield 2-(3-bromopropoxy)-5-nitroaniline (IV) which has a sweet taste.

The halogenated sweeteners are mixed with the materials to be sweetened in an amount required to attain the desired level of sweetness. Usually, because of their high degree of sweetness, they are used in lesser amounts than sucrose is correspondingly used in like materials. For example, they are generally used in concentrations of about 0.005%. The addition can be on a volume or weight basis. Of course, higher or lower amounts can be used to attain a desired sweetness appropriate to the material. The technique of sweetening materials with the compounds is carried out by standard means. The sweeteners can be added directly to the material, or they can be incorporated with a diluent to increase their bulk so that smaller amounts of the compound may be metered into the diluent. As diluents, conventional diluents such as liquids or solids can be used. These include starch, sorbitol, sugar, citric acid or other like compatible materials.

3-AMINO-4-n-PROPOXYBENZYL ALCOHOL

The nonnutritive sweetener described by *G.A. Crosby and P.M. Saffron; U.S. Patents 3,876,814; April 8, 1975; and 3,952,058; April 20, 1976; both assigned to Dynapol Corporation* has the general formula of a propoxybenzyl alcohol as follows:

$O\text{-}n\text{-}C_3H_7$, NH_2, CH_2OH

This compound may be prepared by the five-step process shown on the following page.

(1) Esterifying 3-nitro-4-hydroxybenzoic acid to yield a first ester product.

(2) n-Propoxylating the 4-hydroxy group of the first ester product to yield 3-nitro-4-propoxybenzoic acid ester.

(3) Reducing the 3-nitro group of the 3-nitro-4-n-propoxybenzoic acid ester in acetic acid to an acetamido group.

(4) Hydrolyzing the acetamido group and the 1-benzoic acid ester functionality to yield 3-amino-4-n-propoxybenzoic acid, and

(5) Reducing the benzoic acid product of step 4 to the corresponding 3-amino-4-n-propoxybenzyl alcohol.

As a general rule, the sweetener may be used in any application where a sweet taste is desired. The sweetener may be used alone or in combination with other sweeteners, nutritive or nonnutritive. Also, if desired, binders or diluents may be added to the sweetener. This is not usually necessary, however, as the sweetener is a solid having excellent handling properties. This makes mixing the sweetener with an edible substance a simple conventional operation. The sweetener may be mixed with the edible substance as a solid or as a solution.

When this material is added to gelatin in combination with a fruit flavor and color to yield a dessert product, the 3-amino-4-n-propoxybenzyl alcohol imparts a sweet flavor to the product. Likewise, when 3-amino-4-n-propoxybenzyl alcohol is added to cola beverages and to chewing gum it imparts desired sweet flavors.

p-METHOXYCINNAMALDEHYDE (PMCA)

According to *J.S. Neely and J.A. Thompson; U.S. Patents 3,867,557; Feb. 18, 1975; and 3,928,560; December 23, 1975; both assigned to The Procter & Gamble Company* p-methoxycinnamaldehyde (PMCA) is not a new compound because it occurs naturally in several essential oils such as basil and estragon. However, it was never suggested as a sweetening and flavoring agent. PMCA has the structural formula:

CH=CH—CHO

$O—CH_3$

Compositions can contain PMCA alone or in combination with other sweeteners and/or flavorants. It has been discovered that when PMCA is mixed or codissolved with other known natural or synthetic sweetening agents, the resulting composition has enhanced sweetness characteristics. Examples of such other suitable known synthetic sweetening agents include: saccharin, 6-(trifluoromethyl)tryptophane, 5-(3-hydroxyphenoxy)-1H-tetrazole, and certain aglyconic dihydrochalcones.

PMCA may also be combined with other sweetening agents which may be naturally occurring or prepared from natural compounds. Examples include: sucrose, lactose, dextrose, levulose, maltose and the like.

Most preferred is PMCA mixed or codissolved with an aglyconic dihydrochalcone, preferably hesperetin dihydrochalcone. Such a composition possesses desirable enhanced flavor and sweetness characteristics.

It has also been discovered that compositions containing PMCA mixed or codissolved with other known flavorants results in a flavoring agent having enhanced flavor. Broadly, such flavorants include liquid and powdered extracts. Examples of such extracts include: oil of sweet birch, oil of spearmint, oil of wintergreen, anise oil, dill oil, celery seed oil, various citrus oils including lemon, orange, lime, tangerine and grapefruit oils, clove oil, peppermint oil, cassia, carrot seed oil, cola concentrate, ginger oil, angelica oil and vanillin. These flavors are obtained from the appropriate plant sources by extraction or may be synthetically made by known methods.

The solvents suitable for solubilizing the PMCA and other combinations containing PMCA are any of the polar, organic liquids and water containing polar, organic liquids. Of course, when it is desired to prepare artificial sweeteners suitable for prolonged or repeated ingestion by humans, it is necessary to use as a solvent for the PMCA a polar, organic liquid which is toxicologically acceptable.

Example 1: 20 grams of PMCA are thoroughly mixed with 80 grams of powdered lactose. The result is a dry sweetening and flavoring composition suitable for use without further treatment.

Compositions are prepared as in Example 1 except that dry starch, kaolin, mannitol, dicalcium lactate, magnesium carbonate, magnesium oxide, calcium phosphate, and powdered glycyrrhiza, respectively, are used in place of the powdered lactose. In each instance, the result is a desirable dry sweetening and flavoring composition.

Example 2: 1.0 gram of PMCA is dissolved in 100 grams of ethyl alcohol and 1,000 grams of water is added. The resulting solution is suitable for use as a sweetening and flavoring composition without further treatment.

Example 3: 1.0 gram of PMCA and 0.05 gram of hesperetin dihydrochalcone are dissolved in 100 grams of ethyl alcohol with gentle warming and 1,000 grams of water is added thereto. The resulting solution is suitable for use as a sweetening and flavoring composition without further treatment.

Sweetening and flavoring compositions are prepared except that the hesperetin dihydrochalcone is, respectively, replaced by saccharin, 5-(3-hydroxyphenoxy)-1H-tetrazole, glycerin, sorbitol, glucose, sucrose, dextrose, 6-(trifluoromethyl)-tryptophane, and levulose. Additional sweetening and flavoring compositions are prepared except that hesperetin dihydrochalcone is replaced by 1:1 mixtures of hesperetin dihydrochalcone, respectively, with saccharin, glycerin, sorbitol, glucose, sucrose, and dextrose. Additional compositions are prepared except that the hesperetin dihydrochalcone is replaced by a 1:1:1 mixture of hesperetin dihydrochalcone, saccharin and sucrose. In each case, the result is a highly desirable sweetening and flavoring composition.

Example 4: A concentrated, nonaqueous sweetening and flavoring composition having an intense sweetness is prepared in the following manner: 4 grams of PMCA is dissolved in 100 grams of ethyl alcohol. The resulting solution is suitable, without further treatment, for use as a highly concentrated flavoring and sweetening composition. The above composition is prepared and in addition 0.20 gram of hesperetin dihydrochalcone is dissolved. The resulting solution is a highly desirable flavoring and sweetening solution.

Example 5: 1.0 gram of PMCA is dissolved in a mixture of 1,000 grams of water and 50 grams of sorbitan monooleate polyoxyethylene with gentle warming. The resulting solution is suitable for use as a flavoring sweetening composition without further treatment.

Example 6: A vanilla-flavored sweetening composition suitable for simultaneously sweetening and flavoring is prepared as follows: 1 part PMCA and 0.5 part vanillin are dissolved in 50 parts ethyl alcohol. The result is an enhanced vanilla-flavored sweetening composition having a prolonged flavor impact but lacking the bitter aftertaste frequently associated with vanillin.

Example 7: 5 parts of PMCA and 0.25 part of hesperetin dihydrochalcone are dissolved in 100 parts lemon oil. The resulting solution possesses, without further treatment, an enhanced lemon flavor possessing desirable sweetness characteristics.

The lemon oil is replaced by an equivalent amount of oil of sweet birch, oil of wintergreen, anise oil, dill oil, celery seed oil, bitter almond oil, orange oil, lime oil, clove oil, peppermint oil, tangerine oil, cassia, carrot seed oil, angelica oil, cola concentrate, and ginger oil and mixtures, respectively, such as 1:1 mixtures of lemon and orange oil, lemon and lime oil, and cola concentrate and ginger oil. Mixtures in a ratio of about 1:1:1 are also prepared replacing lemon oil. An example of such a mixture is lemon oil, lime oil and orange oil. In each instance the result is a sweetened flavor composition having an enhanced flavor and sweetness.

SPECIAL OXIMES

Unsaturated Aldoximes

The oximes described by *E.M. Acton, M.A. Leaffer and H. Stone; U.S. Patent 3,780,194; December 18, 1973; assigned to Stanford Research Institute* have the structure similar to the known sweetening agent perillartine. The three compounds of interest are: 1-cyclopentene-1-carboxaldehyde syn-oxime (I), 1-cyclohexene-1-carboxaldehyde syn-oxime (II) and tiglaldehyde syn-oxime (III). These compounds have relatively good water solubilities ranging from about 0.02 to 0.15 molar at saturation.

The taste intensity of the oxime compounds in solution is 40 to 50 times that of sucrose, when calculated on a mol for mol basis. These are much smaller ratios than that (2,000) calculated for perillartine, but because of the enhanced solubilities as compared with that of perillartine, it is feasible to prepare solutions of these oximes with up to 8, 2 and 10 times, respectively, the taste intensity of a saturated perillartine solution. While sweetness in these oximes is accompanied in varying degree by other minor taste qualities, described as bitter, menthol, mint or coconut, each has the potential for use in combination with sucrose, as a partial substitute in food and drink. Such products would be low in caloric value and reduced in cariogenicity.

Example 1: A 0.005 molar solution of the oxime (I) had a taste intensity approximately equivalent to the taste intensity of a 0.25 molar sucrose solution. The tastes perceived in solutions of the oxime (I) were identified as sweet (estimated as 54% of the total taste), bitter (23%), and menthol or coconut and mint (20%).

Example 2: A 0.004 molar solution of 1-cyclohexene-1-carboxaldehyde syn-oxime (II), MP 99° to 100°C, had a taste intensity approximately equivalent to the taste intensity of a 0.25 molar sucrose solution. The tastes perceived in solutions of this oxime were identified as sweet (estimated as 52% of the total taste), menthol and mint (22%), and bitter (8%).

Example 3: A 0.005 molar solution of tiglaldehyde syn-oxime (III), MP 35° to 38°C, had a taste intensity approximately equivalent to the taste intensity of a 0.25 molar sucrose solution. The tastes perceived in solutions of this oxime were identified as sweet (estimated as 37% of the total taste), bitter (20%), and phenolic (10%).

The oxime compounds I, II and III which take the form of white to off-white crystalline solids are well adapted to be used, in dry pellet or powder form, or in a solution in water or in ethanol or other appropriate organic solvent, as synthetic sweetening agents either alone or in conjunction with other sweetening agents such as sucrose, dextrose, saccharin, or the like. They are also well adapted to be mixed with other food ingredients such as citric acid, flavoring, spices, starches, preservatives and dehydrated foodstuffs. They are particularly well adapted, for example, to be used in various dietary foods and beverages as a replacement for approximately one-half of the sucrose which would otherwise be used in the product.

1,4-Cyclohexadiene-1-Carboxaldehyde syn-Oxime

E.M. Acton, M.W. Lerom and H. Stone; U.S. Patents 3,919,318; November 11, 1975; and 3,952,114; April 20, 1976; both assigned to Stanford Research Institute report that the compounds 4-methyl-1,4-cyclohexadiene-1-carboxaldehyde syn-oxime, 4-methoxymethyl-1,4-cyclohexadiene-1-carboxaldehyde syn-oxime, and 4-(1-methoxyethyl)-1,4-cyclohexadiene-1-carboxaldehyde syn-oxime, as well as the compound 1,4-cyclohexadiene-1-carboxaldehyde syn-oxime have a high degree of sweetness which is accompanied by very little off-taste. The unsubstituted oxime compounds have the formulas:

(I)

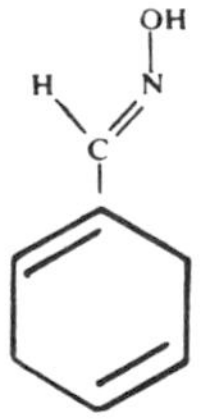

OH
N
H
C

1,4-Cyclohexadiene-1-Carboxaldehyde syn-Oxime

(II)

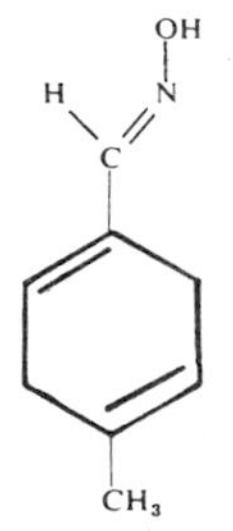

4-Methyl-1,4-Cyclohexadiene-1-Carboxaldehyde syn-Oxime

(III)

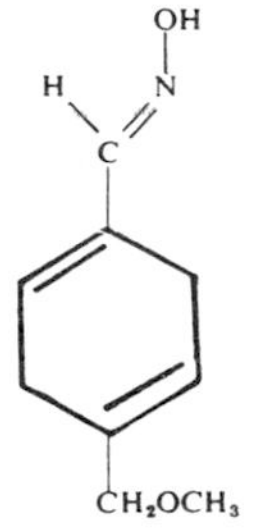

4-Methoxymethyl-1,4-Cyclohexadiene-1-Carboxaldehyde syn-Oxime

(IV)

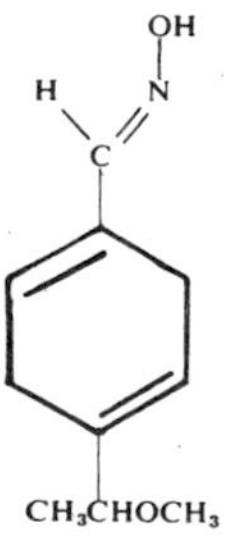

4-(1-Methoxyethyl)-1,4-Cyclohexadiene-1-Carboxaldehyde syn-Oxime

Using tasting panels of five or six experienced people both the intensity of the taste and its quality were evaluated for each of the oximes. In the first of these tests, the taste intensity of oxime solutions of varying concentration was evaluated against an 0.25 M (8.55%) sucrose solution taken as the standard of unit intensity. In this fashion the concentration of each oxime compound having essentially the same taste intensity of that of the sugar was determined. This concentration worked out to be 0.0012 M (0.015%) for (I), 0.0005 M (0.007%) for (II), 0.001 M (0.017%) for (III) and 0.0008 M for (IV).

Using solutions of each oxime having intensity levels between about 0.4 and 1.5 or 2.0, the panel members estimated the percent of the taste quality which was identifiable as sweet, as well as the percentage of off-taste characteristics which was identified as bitter, other, etc.

The table given below presents information as to the melting point, optical isomerism and solubility of the various oximes, together with data derived from the findings of the tasting panels.

	(I)	(II)	(III)	(IV)
Melting point, °C	103.5°-104.0°	106.5°-107.0°	91.0°-91.5°	108° sublimes
Optical isomerism	None	None	None	dl pair
Concentration of saturated solution	0.19% 0.015 M	0.008% 0.0006 M	0.34% 0.02 M	0.07% 0.0039 M
Total taste intensity of saturated solution (compared to 0.25 M sucrose as 1)	(~12)*	1.3	(~18)*	(~5)*
Taste intensity of oxime compared to sucrose				
Molar basis	200X	500X	225X	300X
Weight basis	560X	1250X	460X	570X
Taste characteristics (% of total taste)				
Sweet	70	83	89	94
Bitter	3	2	2	1
Menthol, Coconut, Licorice	9	11	4**	2

*Extrapolated from values at lower concentrations near unit intensity
**Some fruit, berry off-tastes

STEVIOSIDE EXTRACTION

An improved method for extraction of stevioside from the leaves of *Stevia rebaudiana* is described by *G.J. Persinos; U.S. Patent 3,723,410; March 27, 1973; assigned to The Amazon Natural Drug Company.*

Stevioside is a compound that is 300 times as sweet as sugar and has been found to be nontoxic. This property, in the light of the present day activity toward the production of a sugar substitute, has resulted in considerable interest being shown in the compound and its production on a commercial basis.

The method of producing stevioside according to this process is carried out by grinding the leaves of the *Stevia rebaudiana* Bertoni (Compositae) plant to a coarsely milled state. The milled leaf mixture is thereafter subjected to an initial defatting treatment to defat the plant material prior to any subsequent extracting treatment. The defatting may be carried out by boiling the ground plant material mixture with an organic extraction solvent for the plant fat, the solvent employed being one in which stevioside is substantially insoluble, for a period of time sufficient to effectively defat the plant material.

An example of a suitable organic solvent is chloroform in which the ground plant mixture may be treated on a continuous basis for a period of time sufficient to achieve the desired defatting. The cycling of the organic solvent may be carried out for up to 150 hours or more, or the cycling may be continued until the fresh extractive solvent is no more than a pale yellow color. The defatting of the ground mixture may be carried out in a Soxhlet extraction apparatus or a Lloyd-type extractor, both of which operate on a continuous basis.

After the ground leaf mixture has been defatted, the resulting mixture is then air dried for a suitable period of time and then mixed with a base material, such as, for example, calcium carbonate, to render the ground mixture basic when it is placed in a liquid suspension. The ground mixture combined with the basic compound is to be subjected to further extractive treatment to effect the removal of stevioside from the ground leaf mass. This extractive step may be carried out in one of two ways, each of which is described in the following examples.

Example 1: 8.0 kg of coarsely milled leaves of *Stevia rebaudiana* Bertoni (Compositae) were continuously extracted with chloroform as a solvent in a Lloyd-type extractor to defat the mixture. The cycling of solvent was continued until the fresh chloroform extractive was no more than a pale yellow color. After this treatment the ground plant material was removed from the extractor and air dried.

400 grams of calcium carbonate powder and 32 liters of dioxane were thereafter added to the dried ground leaf material. This mixture was heated to boiling with continuous stirring and was maintained in this state for 2 hours. The mixture was then filtered through a Buchner funnel, with the plant material remaining on the filter pad being washed with an additional 8 liters of boiling dioxane. This procedure was repeated three times and the three dioxane filtrates were combined and reduced to a syrup in vacuo at 50°C. An equal volume of methanol was then added to the syrup and the resulting solution was set aside overnight to allow crystallization to proceed.

The crystals were formed overnight and were collected using a Buchner funnel, and washed thoroughly with ice cold methanol. The mother liquor, together with the methanol washings, were reduced to a syrup in vacuo at 50°C, and an equal volume of methanol was again added, and the mixture set aside overnight to crystallize. The crystals formed were removed by filtration using a Buchner funnel and the resulting crystals were washed with ice cold methanol.

Both crops of crystals were combined, dissolved in 3 liters of boiling dioxane, filtered, and the solution was brought to room temperature (25° to 27°C). An equal volume of methanol was added and the solution was set aside overnight. Crystals that formed were removed by filtration using a Buchner funnel and the crystals were washed with ice cold methanol. The crystals were dried in vacuo at 100°C and weighed. The yield from 8.0 kg of air dried *Stevia rebaudiana* leaves was 520 grams (6.5%) of stevioside. Additional stevioside can be obtained from the mother liquors obtained above.

Example 2: One kilo of dried, coarsely ground leaves of *Stevia rebaudiana* Bertoni (Compositae) was extracted continuously with boiling chloroform in a Soxhlet apparatus for 150 hours. The plant material was then air dried, mixed with 50 grams of calcium carbonate, moistened with distilled water, and packed into a pharmaceutical percolator. After 1 hour the percolate was collected at a moderate rate. Enough distilled water was added so that 6.0 liters of percolate was collected. The percolate was then frozen and lyophilized.

To each 100 grams of lyophilized powder was added 800 ml of dioxane and this was heated on a steam bath (100°C) for 1 hour. The hot mixture was filtered through a Buchner funnel. This procedure was repeated twice and the filtrates were combined and reduced to a volume of 100 ml in vacuo at 40°C.

To this concentrated solution was added 100 ml of methanol and the solution was allowed to stand at 25° to 27°C for 24 hours. The crystals of crude stevioside that formed were then removed by filtration using a Buchner funnel, and washed with cold methanol.

The mother liquor was reduced to dryness in vacuo using a rotating evaporator at 40°C. A minimum amount of dioxane was added and heated to boiling on a steam bath until solution was effected. The volume was reduced to 75 ml in vacuo at 40°C on a rotating evaporator. The solution was transferred to a beaker, 75 ml of methanol was added, and the solution allowed to stand at room temperature for 24 hours. The crystals that formed were removed by filtration, and the procedure was repeated as above until no more stevioside was formed.

MALTITOL SWEETENER

Maltitol is used by *M. Mitsuhashi, M. Hirao and K. Sugimoto; U.S. Patent 3,741,776; June 26, 1973; assigned to Hayashibara Company, Japan* as a major sweetener for low or noncaloric foods, and at the same time it serves the functions of providing solids volume, increased viscosity, body, luster, moisture retention and stability to foods.

Maltitol is obtained by hydrogenation of pure maltose, in the form of a nonreducing substance as represented by the following formula:

```
    CH2OH        CH2OH
     |            |
     C----O       C----OH
  H /|     \ H  H/|
   |/ H     \|  |/ H
   C         C  C
 HO|\ OH  H /|_O_|\ OH  H  CH2OH
     \|    |/       \|  |/
      C----C         C--C
      |    |         |  |
      H    OH        H  OH
```

The process for producing maltitol is exemplified as follows. Pure maltose is dissolved in water to prepare a 50% aqueous solution. To the resulting solution is added 8% Raney nickel as a reduction catalyst. The mixture is gradually increased in temperature up to 90° to 125°C with constant stirring. By introducing hydrogen at a pressure of 20 to 100 kg/cm^2 the mixture is caused to absorb the hydrogen. After cooling, the reaction mixture is freed of the Raney nickel and is purified in the usual manner by the use of active carbon and ion exchange resin. The product upon concentration yields maltitol in a colorless, transparent and viscous state.

Maltitol has no caloric value because it is not digested or absorbed by digestive organs of higher animals. This was demonstrated by experiments with live rabbits. The intestines of test rabbits not fed for 24 hours beforehand were closed at both ends and were injected with 50 cc of a 20% aqueous solution of maltitol or an equilmolecular amount of a sucrose solution each. After the lapse of several hours, the sugar or sugar alcohol left in the intestines was estimated. It was then found that, while 90% of the sucrose intake had been lost due to absorption and digestion, maltitol had shown no loss, thus proving its impossibility of being absorbed and digested in the digestive organs. It was also found that maltitol has no harmful stimulus because the intestinal walls exposed to it showed no irregularity such as congestion.

Recent reports have disclosed that xylose and sorbitol, both known as noncaloric sweetening materials, are actually metabolized and cannot be as noncaloric as maltitol. Thus, maltitol has no energy value as food and, in addition, can improve the palatability of foods with sweetness and body.

Moisture Retention and Viscosity: The remarkable moisture-retaining property and viscosity of maltitol are naturally expected from its chemical structure. With the features the additive can also serve as a stabilizer for flavorings, colorants, etc. Maltitol is highly viscous, as shown by the viscosity values of a 70% solution of it at different temperatures as follows:

°C	Cp
22	274
30	167
40	94.5
50	60
59	40.3

Thermal Stability: Maltitol is highly stabilized against heat. When heated with direct fire, it undergoes no coloration at all up to 200°C and is colored only slightly above 200°C. Also it will not solidify fast on cooling.

Maltitol is useful for the preparation of various soft drinks, including carbonated drinks such as colas, lactic acid drinks such as Calpis, and artificial fruit juices, especially concentrated ones. Substitution of maltitol for starch syrup, dextrose or sucrose in ordinary sweet drinks causes replacement of the total amount of carbohydrates in the solid contents of the drinks by the low-caloric substance. This is not only valuable from medical and dietetic viewpoints but is desirable for improving the palatability, imparting suitable viscosity and maintaining flavors of the drinks.

Example 1: Preparation of Concentrated Syrup – A recipe for preparing 100 liters of a concentrated syrup containing orange juice is as given below.

	Liters
Concentrated (1/5) orange juice	7.90
60% maltitol solution	74.56
50% citric acid solution	1.97
Orange base	0.50
Orange essence	0.25
Water	14.80

Maltitol is particularly harmonizable with orange juice to give good sweetness and body. In addition, the overall concentration can be much increased by a remarkable reduction of the proportion of water because, unlike sugar, maltitol will not crystallize at high concentrations.

MALTITOL AND MALTOTRIITOL MIXTURE

M. Mitsuhashi, M. Hirao and K. Sugimoto; U.S. Patent 3,705,039; Dec. 5, 1972; assigned to Hayashibara Company, Japan give the background and descriptions for preparation of a low calorie sweetener mixture of maltitol and maltotriitol.

Common acid or enzyme conversion starch syrups contain branched oligosaccharides and dextrins, due to amylopectin of branched structure in the starch, and therefore have extremely high viscosities. By simultaneously using α-1,6-glucosidase, which debranches branched structures, to convert amylopectin into only linear chained molecules during the saccharification procedure, and by hydrolyzing with β-amylases, a mixture of sugars is obtained, which only contains linear chained molecules, such as maltose, maltotriose and a small amount of tetraose, etc.

Sugar alcohol, obtained by hydrogenating this mixture, has an unexpected high sweetness. As the material is a polyalcohol, it is heat-stable and its viscosity is lower than conventional starch syrup, due to its linear chained structure. Further, this viscosity is variable to a wide extent by the degree of β-amylolysis. That is, when β-amylolysis is carried out to nearly 100%, the resultant product is made up mostly of maltose, whereas when decomposition is suspended at an earlier stage, the product contains linear chained oligosaccharides and its sweetness somewhat decreases.

By using combinations of α-amylase, glucoamylase and acids besides β-amylase in producing the material, the rate of linear chained oligosaccharides excluding maltose can be changed. As the compositions of sugars and viscosities are variable, sugar alcohols with parallel properties are obtainable by hydrogenation.

Other processes give hydrogenation products of common malto syrups which contain less than 60% of maltose, produced by hydrolysis of starches with only β-amylases, and contain more than 50% of branched chain oligosaccharides and dextrins. Thus, the viscosities of these products are excessively high with no possible means of regulation. Organic solvent precipitation is the only means to remove the branched chained dextrins and to increase the purity of maltose. This requires much trouble and expense, and thus its commercialization is impossible. However, the sweetener of this process has an essentially different molecular structure and is not comparable with other sweeteners.

The sweetener of this process is a nonreducing substance containing maltitol and maltotriitol as main components and has the following formula.

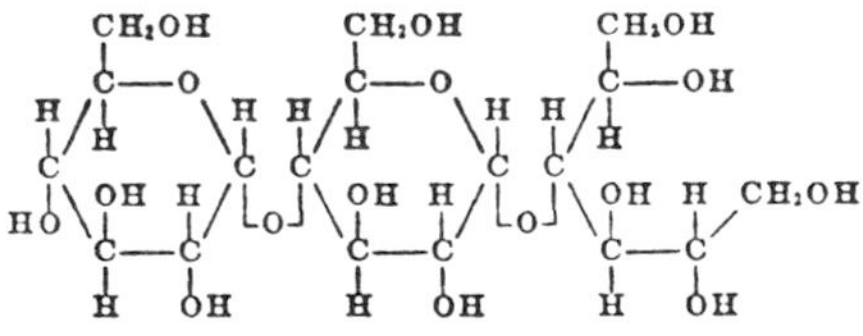

```
     CH2OH          CH2OH          CH2OH
      |              |              |
      C----O         C----O         C----OH
   H /|      \H   H /|      \H   H /|
   |/ H       |   |/ H       |   |/ H
   C          C   C          C   C
 HO\ OH  H  /└O┘\ OH  H  /└O┘\ OH  H  CH2OH
    |    |  /     |    |  /     |    | /
    C----C        C----C        C----C
    |    |        |    |        |    |
    H    OH       H    OH       H    OH
```

Maltotriitol

Maltitol

A process for production of this sweetener, which contains mostly maltitol and maltotriitol, is as follows: A 30% by weight suspension of starch is enzymatically liquefied at 90°C, pH 6.0, using a liquefying amylase of 0.2% on a dry basis to form a uniform liquefied solution. The solution is cooled rapidly to 50°C. 20 units of α-1,6-glucosidase, obtained from a culture broth of *Escherichia intermedia,* and 10 units of β-amylase extracted from wheat bran respectively per gram of starch, are added to the solution, and saccharified for 30 hours. Thus, a sugar solution which contains 85 to 95% maltose and 5 to 15% maltotriitol is obtained. After purification with active carbon and ion exchange resins a colorless and clear sugar solution is obtained.

After the sugar mixture, thus obtained and consisting of maltose and higher linear chained oligosaccharides, was concentrated to 40 to 50%, 8% of Raney nickel catalyst based on the starch was added to the solution and was heated in an autoclave with stirring to 90° to 125°C. During the heating procedure, hydrogen was charged at a pressure of 20 to 100 kg/cm^2. The catalyst used was removed from the mixture after absorption of hydrogen. A colorless, transparent and viscous mixture of maltitol, maltotriitol and higher linear chained oligosaccharide alcohols was obtained after purification with active carbon and ion exchange resins.

Mixtures of lower saccharification degrees contained a small amount of higher linear chained oligosaccharide alcohols, such as maltotetraitol, etc. Thus, the viscosities of hydrogenated products of lower conversion degree or with less maltose contents are slightly higher. However, the viscosities of these products are not as high as those of saccharification products obtained by only using common β-amylases. When combinations of α-1,6-glucosidase, β-amylase, α-amylase, or other enzymes or acids are used, starch syrups consisting only of various compositions of linear chained molecules are obtained. Results of studies on the characteristics of using this sweetener, which consists of maltitol with maltotriitol, for production of foods and drinks are as follows.

Sweetness: The results of panel tests on the degrees and qualities of sweetness of this sweetener show that this sweetener has a mild and well harmonized blandness that surpasses those of only maltitol. It has better aftertaste than maltitol. Its sweetness degree is higher than glucose, though less than cane sugar. It has a sweetness equivalent to 75% of that of cane sugar without any significant difference from maltitol. Panel tests conducted with groups of 30 persons gave the following results.

(1) Sweetness of cane sugar, maltitol, the maltitol-maltotriitol sweetener and glucose by a paired preference test: By significant difference determination,

there were obtained significant differences at 1 and 5%. At least 5 tests were repeated on each substance, at concentration degrees of 70, 35, 20 and 10%. From the results the following order was obtained.

Cane sugar > maltitol = maltitol-maltotriitol > glucose. This shows that the mixture has a sweetness that comes between that of cane sugar and glucose, though no significant difference was observed between maltitol and the maltitol-maltotriitol.

(2) Comparison of the sweetness degrees between the maltitol-maltotriitol sweetener and cane sugar: To reduce viscosity, 35% of an aqueous solution of the maltitol-maltotriitol sweetener was prepared and compared with cane sugar solutions of 5, 10, 15, 20, 25 and 30% concentration degrees. It was found that the 35% aqueous solution of the sweetener has a sweetness equivalent to 25% aqueous cane sugar solution.

Noncrystallization and Crystallization Preventive Properties: The maltitol-maltotriitol sweetener dissolved freely in water up to 100%. In cases of aqueous solutions with concentration degrees of 70 to 90%, no formation of crystallization was observed after a standing period. Further, to 70% aqueous solutions of cane sugar and glucose were added 10% of this sweetener and the resulting solutions were compared with the control at room temperature. The control solutions began to crystallize after 1 day, whereas the material with sweetener added showed no turbidity even after a period of 1 week.

Noncaloric Properties: Oligosaccharides, such as maltose, maltotriose, etc., are easily decomposed by various amylases, whereas the linear chained oligosaccharide alcohols are hardly decomposed. The results are shown in the following table.

Saccharification Rate with Various Enzymes

	Units/ml --- of --- MA	SA	Buffer Solution, pH	°C	Time of Reaction, hr	Hydrolysis of the Sweetener, %	Hydrolysis of Maltose, %
Rhizopus glucoamylase	5	10.0	5.0	40	5	7	100
Porcine pancreatic maltase	0.17	0	7.5	40	10	5	82
Yeast maltase	0.07	0	6.5	35	8	10	75

Note: Test samples contain 85% maltitol. MA = maltase activity; SA = saccharifying activity.

Reactions were performed with test solutions comprising 5 ml of substrate (concentration degree 1%), 4 ml of buffer solution and 1 ml of enzyme solution. The absorption test of the sweetener within digestive organs of higher animals showed that the absorption rate was zero and thus this substance was completely noncaloric.

Into both end-tied intestines of rabbits, after fasting for 24 hours, were charged 50 ml of 20% aqueous maltitol-maltotriitol sweetener and aqueous cane sugar solution of the same concentration as controls, of respectively equivalent amounts. After a few hours, the residual amount of sugars in the intestine were determined. 90% of the cane sugar was absorbed, while the sweetener according to the process hardly decreased, thus proving that this sweetener was not absorbed within the interior of the bodies. No unusual symptom, such as engorgement was observed.

Xylose, sorbitol, etc., which have conventionally been considered as noncaloried sweeteners, were absorbed in the bodies and therefore they cannot be defined as strictly noncaloried as is the case with the sweetener of the process. This noncaloried sweetener has a strong sweetness and imparts a heavy body to food products, and accordingly is an indispensable sweetener for the production of noncaloried foods and drinks.

Viscosity, Moisture and Flavor Retentive Properties: As maltitol-maltotriitol sweetener contains higher oligosaccharide alcohols, the viscosity of this sweetener is naturally higher than cane sugar or maltitol, etc. However, this sweetener consists only of linear chained molecules, thus no excessive increase of viscosity was observed. Also, the sweetener, which is a polyalcohol, has moisture and flavor retentive properties. Results of tests performed on moisture retentive properties of sponge cakes and flavor retentive properties of natural juice showed that in each case this sweetener was equal or even superior to cane sugar.

The viscosity value of a 70% concentration solution, containing 85% maltitol on a dry basis was higher than that of cane sugar. However, this viscosity value can be controlled by the extent of β-amylolysis. The results from the above performed tests are shown below. The regulation of viscosity is possible by simultaneously using α-amylases, acids, etc. These means are impossible in the cases of hydrogenated products made from conventional malt syrups.

Temperature, °C	Viscosity, cp
20	295
30	187
40	103
50	70
60	51

Heat Stability and Chemical Stability: Concentration of the maltitol-maltotriitol sweetener by direct heating (up to 200°C) to an almost anhydrous state causes no color-action or change of quality. When heated over 200°C, white smoke was observed coming from the sweetener and it gradually colored. The sweetener had difficulty in hardening instantly upon cooling. Heating the aqueous solution with 1% amino acid added did not cause discoloration.

This maltitol-maltotriitol sweetener has many superior characteristics as a sweetener source for foods and drinks, as well as a suitable additive for noncaloried foods. Accordingly, this sweetener is a suitable additive for carbonated beverages such as cola drinks, etc., lactated beverages, such as Calpis, concentrated fruit juice, etc.

Utilization of this sweetener as a substitute for glucose, cane sugar and other syrups makes even higher concentrated solutions possible, and all the carbohydrate contents may be converted into noncaloried substances, with resulting medical and beauty effects. Due to its ability to impart viscosity and flavor stability to the products, first rate drinks can be produced utilizing this sweetener, which has a decent taste and no undesirable aftertaste as in the cases of cane sugar, maltitol, starch syrups, etc.

Similarly, when it is used in sponge cakes, Japanese cakes and others, the calorie content can be maintained at a minimum and its moisture retentive property prolongs the shelf life greatly, and improves the product's texture, which is a vital problem for these products, and their flavor retentive property thus permits great improvement to be realized, eliminating the disadvantages of cane sugar such as drying and crystallizing.

This sweetener can be added to bakery products such as biscuits, cookies, etc., to produce nonsugar or low caloried foods. Heat resistance of the sweetener prevents over-baking caused by heating, cracking and deformation which are usually observed upon cooling the baked products, increases yield, prevents dispersion of flavor and increases the durability of foods.

When it is used in jellies, etc., the sweetener not only imparts sweetness to the products but make production of the products with noncaloried components possible. It prevents discoloration and crystallization of the products, imparts moisture and flavor retentive properties to the products and fully exhibits its efficiency of preserving the initial qualities of the products after prolonged periods.

Example 1: Process for Bottling Chestnut Syrup Preserve – Chestnuts, after removal of their astringent coats were soaked in water overnight. To the chestnuts was added 0.1 to 0.3% of aluminum potassium sulfate and the mixture was cooked in water to restrain smear and to tighten the sarcocarp. Water was drained. The chestnuts were soaked in 50% aqueous maltitol-maltotriitol sweetener solution, heated for 10 minutes at 80°C, and left to stand for a night.

The next day, to 130 grams of the chestnuts was added a dissolved solution comprising aqueous maltitol-maltotriitol sweetener solution (purity of maltitol 90%), 0.04% of saccharin were added to aqueous maltitol-maltotriitol sweetener solution (purity of 90%) and the thus obtained solution was diluted to a syrup containing 50 to 65% maltitol. 110 grams of this syrup per 130 grams chestnuts was added to the chestnuts, and the result was packed in bottles.

Products thus prepared had a light beautiful yellow tint and did not develop the coloration caused by reducing sugars such as the case when glucose was employed. Sarcocarps had suitable tightness with improved surface luster and desirable sweetness. Conventional chestnut syrup preserves, except those using cane sugar, were generally considered not applicable for the market, owing to their extreme coloration. However, the sugar alcohol sweetener, according to this process, imparts better results and more desirable coloration values than cane sugar. The noncaloric property of this sweetener enables the reduction of the total calorie value to half and thus provides low-caloried foods.

Example 2: Process for Canning Fruits – The sugar solution to be charged into canned oranges was prepared by dissolving 50 kg of the sweetener which contained 85% of maltitol (dry base), 25 grams of saccharin in water to a total volume of 100 kg. After the solution was charged to the peeled oranges, the products were canned and sterilized according to usual methods. The syrup produced according to this process, had a suitable viscosity, a mild sweetness that harmonized well with the acid taste of the oranges, retained its flavor satisfactorily and the syrup itself was noncaloried.

Example 3: Process for Production of Sponge Cakes – A popular formula for production of sponge cakes is shown below.

	Grams
Maltitol-maltotriitol sweetener (maltitol content 80%, dry base)	1,000
Egg	1,100
Flour	500
Honey	50

Dough was prepared according to the above formula and the usual method. The dough was poured into an iron pan, covered with paper, and baked in an oven appropriately heated at 180° to 190°C. The products had an appealing baked color and a soft, spongy and improved texture.

Delicious products with suitable moisture were available even after a prolonged storage period, due to the fact that their degradation or drying are delayed by the addition of the sweetener. Also a reduction of carbohydrates to half makes production of low calories foods possible.

LACTITOL SWEETENER

The use of lactitol as a special sweetening agent with other functional properties suitable for use in low calorie foods is described by *K. Hayashibara and K. Sugimoto; U.S. Patent 3,973,050; August 3, 1976; assigned to Hayashibara Company, Japan.*

A procedure for the production of lactitol is as follows. To pure lactose in the form of a 30% aqueous solution is added 8% Raney nickel as a reduction catalyst. The mixture is gradually heated to 100° to 130°C with constant stirring. By introducing hydrogen at a pressure of 50 to 100 kg/cm^2 the mixture is caused to absorb the hydrogen at a rate of 1 mol/mol of lactose. After cooling, the reaction mixture is freed of the Raney nickel and is purified in the usual manner by the use of active carbon and ion exchange resin. The product upon concentration yields a lactitol solution in a colorless, transparent and viscous state.

Lactitol obtained in this way showed upon analysis no trace of direct reducing sugar, thus indicating that it had become a polyhydric alcohol. It yielded 50% of whole sugar. This means that galactose equivalent to one-half molecule had been formed by the hydrolysis, and the result demonstrated that the product was pure lactitol.

Sweetness: A panel test on sweet taste showed that the sweetness of this substance is mild and refreshing and leaves no thick taste behind. As for the intensity of sweetness, lactitol is milder than sucrose and is comparable to dextrose. It is by far sweeter than lactose, the starting material. It is well harmonized with saccharin, and other artificial sweetening agents, and its sweetness can be freely adjusted thereby.

Noncrystallinity: Even at fairly high concentrations, lactitol scarcely crystallizes. When mixed with sucrose and dextrose, it also serves to avoid the crystallization of those sugars.

Noncaloric Value: Lactitol has no caloric value because it is not digested or absorbed by digestive organs of the higher animal. This was demonstrated by experiments with live rabbits. The intestines of test rabbits not fed for 24 hours beforehand were closed at both ends and were injected with a 20% aqueous solution of lactitol or an equimolecular amount of a sucrose solution each. After the lapse of several hours, the sugar or sugar alcohol left in the intestines was estimated.

It was then found that, while 85% of the sucrose intake had been lost due to absorption and digestion, lactitol had shown no loss, thus proving its impossibility of being absorbed and digested by the digestive organs. It was also found that lactitol has no harmful stimulus because the intestinal walls exposed to it showed no irregularity such as congestion. Recent reports have disclosed that xylose and sorbitol, both known as noncaloric sweetening materials, are actually metabolized and cannot be as noncaloric as lactitol.

Moisture Retention and Viscosity: The moisture-retaining property and viscosity of lactitol are naturally expected from its chemical structure. With a molecular weight of 360 which is comparable to that of sugar and by the breakage of the pyranose ring of glucose by reduction, lactitol has a branched, complicated steric structure. Moreover, because it has nine hydroxyl groups, lactitol displays a remarkable influence of hydrogen bonds and has great moisture-retaining and moisture-absorbing properties. It thus serves as a stabilizer generally for flavorings and colorants that have polarity. Further, the intermolecular action of lactitol combines with the steric structure of the compound to give luster to foods.

Thermal Stability: In lactitol the reducing group of lactose is stabilized by hydrogenation, and therefore lactitol is more stable against heat than lactose. Ordinarily lactose is very stable thermally as compared with glucose, sucrose, fructose, xylose, etc., and is decomposed at melt temperatures ranging from 200° to 260°C. Lactitol is even more stable. It is dehydrated while being vaporized at 250°C. By virtue of this extreme stability against heat, it can be used as an additive to many different foods and drinks with no danger of decomposition or coloration upon various heat treatments.

Example 1: Preparation of Carbonated Drinks — A carbonated lemon drink is prepared by the use of the standard recipe given in the *National Bottler Gazette,* March issue (1939). That is, the drink so prepared has saccharinity of 10.9, CO_2 gas volume of 3.5, and acidity adjusted with citric acid or the like to 0.145. In addition to suitable amounts of colorant and flavoring, lactitol is used as a sweetening material in the mixture in order to give body. Lactitol blends well with acid to provide a pleasant aftertaste and ensure thorough distribution of the flavor. Further, this sweetening agent which has no food value itself makes it possible to produce an ideal noncalorific drink for dietetic purpose. Other refreshing drinks such as colas can be made in the same way.

Since lactitol is quite compatible with other sweetenings, artificial or natural, the intensity of its sweetness may be adjusted by the use of another artificial agent, for example, to attain best palatability.

Example 2: Preparation of Concentrated Syrup — A recipe for preparing 100 liters of a concentrated syrup containing orange juice in accordance with the process is as shown in the table on the following page.

	Liters
Concentrated (1/3) orange juice	7.500
75% lactitol solution	75.960
50% citric acid solution	1.976
Orange base	0.500
Orange essence	0.250
Water	13.814

Lactitol blends particularly well with orange juice, and, unlike sugar, it will not crystallize at high concentrations and therefore can increase the overall concentration of the syrup with a sharp reduction of the proportion of water. It necessary, the sweetness can be suitably increased by the addition of saccharin.

Example 3: Preparation of Ice Cream – Ice cream is prepared by the following recipe:

Milk fat	10%
Skim milk powder	11%
Lactitol	20%
Stabilizer	0.2%
Artificial sweetening	Suitable amount
Flavoring	Suitable amount
Water	58.5%

When a mixture of the above formulation is processed in the usual manner into ice cream, the product has a smooth texture and good body. With very good overrun, the cream has quite refreshing, balanced flavor. Still it is a noncaloric food.

SOLUBLE MALTOSE POLYMERS

A special process is used by *H.H. Rennhard; U.S. Patent 3,766,165; October 16, 1973; assigned to Pfizer Inc.* to produce soluble glucose and maltose polymers which can be used to replace sugars in dietetic foods. Insoluble polymers can be used as a substitute for flour or other starches in dietetic foods.

The starting materials used in the melt polymerization process are maltose or glucose, although other simple sugars may be used as well. The sugars are supplied to the process as dry anhydrides or dry hydrated solids and are in powdered form.

The acids used as catalysts, crosslinking agents or polymerization activators may be any one of a series of relatively nonvolatile, edible, organic polycarboxylic acids. In particular it is preferred to use citric, fumaric, tartaric, succinic, adipic, itaconic or terephthalic acids.

The acid moieties are likely to serve as crosslinking agents between different polyglucose or polymaltose molecules in the insoluble polymers whereas, in the soluble polymers, each acid moiety is more likely to be esterified to only the polymer molecule.

The performance of this portion of the process involves the steps of combining the dry powdered glucose or maltose with the proper amount of acid; the heating and melting of the glucose or maltose and the acid under reduced pressure; the maintenance of the molten conditions in the absence of water until substantial polymerization occurs; and the separation of the individual polymeric product types.

The anhydrous melt polymerization must be carried out at a pressure below atmospheric pressure. The preferred pressures do not exceed about 300 mm, e.g., from about 10^{-5} to 100 to 300 mm Hg. The vacuum is required in order to exclude air from the polymerization and to remove the water of hydration and the water liberated in the polymerization reaction. Air should be excluded from the environment of the polymerizing mixture in order to minimize decomposition and discoloration of the polyglucoses or polymaltoses formed in the polymerization. A fine stream of nitrogen has also been found to be useful with this process as a method for excluding air and removing the waters of hydration and polymerization which are formed. Where the nitrogen purge is used, the vacuum requirements are lessened but pressures of 100 to 300 mm Hg or less are still preferred.

The thermal exposure (reaction time and temperature) used in the production of soluble polyglucoses or polymaltoses by melt polymerization should be as low as possible, since discoloration, caramelization and degradation increase with prolonged exposure to high temperature. Fortunately, however, as the temperature of the polymerization is increased, the time required to achieve substantially complete polymerization decreases.

The inclusion of a food acceptable polyol such as sorbitol in the saccharide-carboxylic acid reaction mixtures prior to polycondensation yields superior products. In most cases, 90% or more of the polyol cannot be isolated from the condensation product, demonstrating that it has been chemically incorporated in the polymer. These additives function as internal plasticizers to reduce viscosity, and also provide improved color and taste.

This is evident, for example, in the manufacture of hard candy from such condensation polymers, where the rheological properties of the melt are improved during processing, foaming is minimized, and a better tasting product of lighter color is obtained. In addition to sorbitol, other food-acceptable polyols include glycerol, erythritol, xylitol, mannitol and galactitol. Polyol concentrations of from about 8 to 12% by weight of the total reaction mixture provide such advantages. Chemical purification is not generally required for the products of this process. Where insoluble and soluble glucoses or maltoses are produced together, separation may be required.

Neutralization of the polyglucoses or polymaltoses may be desirable for certain applications, despite the very low levels of acid catalyst which are employed. For example, where the polyglucoses are to be used in dietetic foods containing whole milk, excess acid which may be present in the unneutralized polyglucoses will tend to curdle the milk. In the case of the soluble polyglucoses or polymaltoses, the solutions of the polyglucoses or polymaltoses are neutralized directly. This neutralization may be accomplished by adding carbonates of potassium, sodium, calcium or magnesium to the solutions of polyglucose or polymaltose.

Decolorization of the soluble and insoluble polyglucoses and polymaltoses is often desirable for certain uses, despite their inherently light color as produced. Soluble polyglucose or polymaltose may be decolorized by contacting the soluble polyglucose or polymaltose with activated carbon or charcoal, by slurrying or by passing the solution through a bed of the solid adsorbent. Soluble and insoluble polyglucoses and polymaltoses may be bleached with sodium chlorite, hydrogen peroxide or similar materials which are used for bleaching flour. The insoluble polyglucose is a yellow powder and often does not require bleaching at all.

Where insoluble polyglucose is to be used as a flour substitute in dietetic foods it may be ground or subdivided mechanically so that it manifests a consistency similar to that of wheat flour. Typically, 325 mesh material is used as the wheat flour substitute. The solutions of soluble polyglucose or polymaltose are almost tasteless and the insoluble polyglucose is a bland-tasting yellow powder.

The soluble polyglucoses and polymaltoses are useful for imparting the physical properties of natural foods, other than sweetness, to dietetic foods from which the natural sugars have been removed and replaced by artificial or other sweeteners. In baked goods, for example, the polysaccharides affect rheology and texture in a manner analogous to sugar and can replace sugar as a bulking agent.

Typical uses for the soluble polyglucoses are found in low calorie jellies, jams, preserves, marmalades, and fruit butters; in dietetic frozen food compositions including ice cream, iced milk, sherbet and water ices; in baked goods such as cakes, cookies, pastries and other foodstuffs containing wheat or other flour; in icings, candy and chewing gum; in beverages such as nonalcoholic soft drinks and root extracts; in syrups; in toppings, sauces and puddings; in salad dressings and as bulking agents for dry low calorie sweetener compositions containing saccharin.

Example 1: Citrated Soluble Polyglucose by Melt Polymerization – An intimate mixture of 500 grams of powdered anhydrous glucose and 12.8 grams of finely ground citric acid was placed in a stainless steel tray and heated in a vacuum oven at 160°C at a pressure of 0.1 mm Hg for 8 hours. The light yellow product was completely water-soluble and contained only a trace of unreacted glucose. The following data for the polymer were determined: Reducing value (RV) is 7.0 (Hagedorn-Jensen iodometric method); apparent number average MW ($a\overline{M}_n$) is 9,100; pH (5% aqueous solution) is 2.9; acid equivalent is 9.6 mg NaOH/g; optical rotation +63.6° (c is 1, water); intrinsic viscosity $[\eta]$ is 0.053 dl/g.

Example 2: Tartrated Polyglucose with Sorbitol, Melt Polymerization by Continuous Process – A preblend of 700 lb of Cerelose monohydrate, 78.6 lb of sorbitol monohydrate, and 7.2 lb of anhydrous tartaric acid was melted at 120°C in a steam-jacketed screw conveyor. The molten material was continuously fed to a vacuum-operated, 0.25 ft^3, continuous double-arm mixer and heated to 165° to 245°C at a pressure of 75 to 100 mm Hg at a feed rate adjusted to obtain a retention time of about 5 minutes. Data for the soluble polymer from a representative run were as follows:

Reducing value (Munson & Walker)	3.8
pH (5% aqueous solution)	3.2

(continued)

Acid equivalent	4.8
Residual glucose (%)	0.75
Residual sorbitol (%)	1.77
Nondialyzable fraction (%)	57.4

Example 3: Pudding – A pudding, incorporating the products of this process as ingredients, was made using the following proportions of ingredients and according to the following directions.

	Grams
Soluble polyglucose citrated	6.0
Cornstarch	2.5
Sodium saccharin (suitable amount)	
Imitation vanilla flavoring concentrate	0.02
Sodium chloride	0.05
Whole milk	45.0

Premix all the ingredients except the milk. Heat the milk gently while adding the mixed ingredients to it. Allow the mixture to boil gently for 10 to 15 minutes. Pour the mixture into pudding molds and allow to cool and solidify.

SACCHARIN COMBINATIONS AND SPECIAL FORMULATIONS

SACCHARIN AND DIPEPTIDE

Simple Combinations

Saccharin is commonly used as a sugar substitute but suffers the drawback of having an objectionable bitter, metallic aftertaste. It has been reported that from one-quarter to one-third of the population is "saccharin sensitive" and perceives an offtaste regardless of the saccharin concentration.

It has also been reported that the incidence of offtaste due to saccharin is a function of the concentration of the compound and that everyone can be expected to obtain an offtaste from saccharin at some concentration. It has been suggested to eliminate the aftertaste by combining saccharin with such ingredients as pectin or sorbitol, dextrose, maltose, etc., but such a combination has not overcome the problem.

On a weight-for-weight basis saccharin at its threshold level of sweetness is about 700 times as sweet as sucrose. This greatly enhanced sweetness of saccharin relative to sucrose decreases as the concentration of saccharin increases. At normal use levels saccharin is only 150 to 200 times as sweet as sucrose. Consequently, to obtain a given increase in sweetness level with saccharin it is necessary to employ a proportionately greater concentration of saccharin. This increased level of saccharin causes a larger segment of the population to perceive an objectionable aftertaste.

While individuals vary in the degree to which they find the aftertaste of saccharin objectionable, the occurrence of the objection is so widespread that considerable effort has been expended in devising formulations to overcome this problem. The problems caused by the aftertaste of saccharin are compounded in products such as preserves, jams and jellies which normally have sugar concentrations up to 60 to 75%. Substituting a quantity of saccharin which gives a sweetness equivalent to this quantity of sugar results in a noticeably undesirable aftertaste.

J.A. Hill and A.L. La Via; U.S. Patent 3,695,898; October 3, 1972; assigned to E.R. Squibb & Sons, Inc. found that the aftertaste of saccharin can be masked by adding a sweet-tasting dipeptide to saccharin. The dipeptide need be present in only very small quantities. Even when present in undetectable amounts, that is, amounts below their threshold level of taste, the dipeptides begin to manifest their effectiveness in overcoming the objectionable aftertastes due to saccharin.

The dipeptide sweetening agents are superior to sucrose in sweetness and possess a lingering sweetness which has been found by taste panel evaluation to be preferable to that of sucrose. The dipeptides have the ability of masking the bitterness and potentiating the sweet taste of saccharin.

The dipeptide sweetening agents have a fairly constant ratio of sweetness to that of sucrose over all sweetness levels. Saccharin, on the other hand, is found to be 700 times as sweet as sucrose when both are compared at their threshold levels of sweetness but only 150 to 200 times as sweet as sucrose at normal use levels. The dipeptides are L-aspartyl dipeptides of the formula:

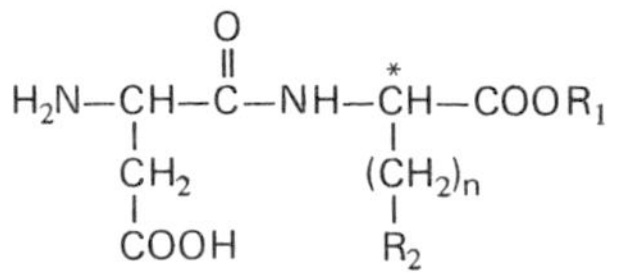

```
              O
              ||      *
H2N—CH—C—NH—CH—COOR1
     |             |
     CH2          (CH2)n
     |             |
     COOH          R2
```

where the carbon marked with an asterisk is of the L-configuration, n is an integer from 0 to 5, R_1 is an alkyl radical of up to 6 carbon atoms, an alkyl aryl or alicyclic radical of up to 10 carbon atoms, and R_2 is a mono- or di-unsaturated alicyclic radical of up to 8 carbon atoms, or when n is 0, a phenyl radical.

Some examples of some specific peptides are L-aspartyl-2,5-dihydro-L-phenylalanine, L-aspartyl-L-(1-cyclohexene-1)-alanine, L-aspartyl-L-phenylglycine and L-aspartyl-L-2,5-dihydrophenylglycine.

The dipeptides demonstrate an effect in masking the aftertaste of saccharin when present in an amount as little as 2% that of saccharin. The minimum amount needed will vary depending on the sweetness of individual dipeptides relative to sugar.

The taste of the composition improves as the amount of dipeptide is increased. The upper limit of dipeptide is determined not by taste consideration alone, but rather by other factors such as cost and solubility. The dipeptides are more expensive than saccharin and for many applications, e.g., soft drinks, a relatively expensive sweetening agent cannot be used. In such situations the quantity of dipeptide used will be at least that level or near that level at which masking of the saccharin aftertaste begins.

Another factor limiting the amount of dipeptide which may be used is that of solubility as there is little to be gained by employing the dipeptide at levels beyond its solubility. Generally, the dipeptides are used in an amount of about 1 part of dipeptide to from 2.5 parts to 12 parts of saccharin.

The compositions may be used to impart their sweetness to a variety of food products, liquids and pharmaceutical preparations.

Examples of the food products in which the compositions may be used are fruits, vegetables, juices, meat products such as ham or bacon, sweetened milk products, egg products, salad dressings, ice creams and sherbets, icings, syrups, cake mixes, pastry mixtures, gelatin, chewing gum and candy. Liquids include beverages such as carbonated and noncarbonated soft drinks, wines and liqueurs.

Examples of pharmaceutical preparations are orally administered suspensions of antibiotics, e.g., penicillin, ampicillin, nystatin or oxytetracycline; liquid vitamin preparations; and syrups, e.g., cough syrup; toothpaste; mouthwash, etc. The compositions are stable, water-soluble compositions which may be utilized under various physical forms, for example, as powders, liquids or pastes.

They may also be used with liquid or solid carriers such as water, glycerin, sorbitol, starch, salt, citric acid or other appropriate pharmaceutically acceptable nontoxic carriers. The saccharin may be used in the form of the pharmaceutically acceptable alkali metal salt, ammonium salt, or amine salt, e.g., the N-methyl-glucamine salt, or the alkaline earth metal salt, or as the free acid.

Example 1: Four 8-oz samples of black coffee are prepared and the samples sweetened, respectively, with the following sweetening agents.

	Sweetening Agents	Grams	Percent
A	Sucrose	19.2	8
B	Sodium saccharin	0.036	0.015
C	Sodium saccharin and	0.024	0.01
	L-aspartyl-L-phenylglycine methyl ester	0.144	0.06
D	Sodium saccharin and	0.024	0.01
	L-aspartyl-L-phenylgylcine methyl ester	0.072	0.03

The sample containing sucrose is taken as the standard for purposes of comparison. The taste of B sample sweetened solely with sodium saccharin is evaluated as poor due to the noticeable aberrant taste quality of the saccharin. The sample sweetened with mixture C of saccharin and dipeptide is found to be oversweet. The aberrant taste aspects due to saccharin, however, are not detected. The sample sweetened with mixture D has a sweetness comparable to that of the sucrose, but with the added advantage of a better quality of taste. The aberrant taste aspects due to saccharin are not detected.

Example 2: A solution is prepared containing 0.3 mg of sodium saccharin and 0.025 mg of L-aspartyl-L-phenylglycine methyl ester in 5 ml of water. The solution contains 0.006% sodium saccharin and 0.0005% of the dipeptide, the ratio of the two materials being 12:1. The concentration of the saccharin is equal to the threshold level at which the bitter aftertaste is detectable while the concentration of the dipeptide is one-tenth that of its threshold level of sweetness. Taste tests on this solution do not reveal any of the aftertaste normally found with solutions containing this quantity of saccharin.

Example 3: A solution is prepared containing 0.3 mg of sodium saccharin and 0.05 mg of L-aspartyl-L-phenylglycine methyl ester in 5 ml of water. The concentration of saccharin (0.006%) is equal to the threshold level for its bitter aftertaste, while the concentration of the dipeptide (0.001%) is about one-fifth that of its threshold level of sweetness. With this ratio (6:1) of saccharin to dipeptide there is effective masking of the bitter aftertaste of saccharin. The sweetness quality of the solution is evaluated as better than that of Example 2.

Example 4: A solution is prepared containing 0.3 mg of sodium saccharin and 0.1 mg of L-aspartyl-L-phenylglycine methyl ester in 5 ml of water. At this ratio (3:1) of sodium saccharin to dipeptide there is effective masking of the saccharin aftertaste. The sweetness quality of the solution is evaluated as better than that of Example 3 and as better than sucrose alone.

Example 5: A solution is prepared containing 0.3 mg of sodium saccharin and 0.00625 mg of L-aspartyl-2,5-dihydro-L-phenylalanine methyl ester in 5 ml of water. The solution contains 0.006% of sodium saccharin and 0.000125% of the dipeptide, the ratio of the two materials being 48:1. The concentration of the dipeptide is one-tenth that of its threshold level of sweetness. Taste tests on this solution do not reveal any of the aftertaste normally found with solutions containing this quantity of saccharin.

Example 6: A dry mix is prepared by blending 5.5 grams of the potassium salt of penicillin G, 0.5 gram of sodium saccharin, 0.25 gram of the methyl ester of L-aspartyl-L-phenylglycine, and 58.0 grams of fine granulated sugar. A syrup for oral administration is then prepared by adding to the dry mix sufficient water to make 100 ml. The reconstituted syrup has an improved and enhanced sweetness character due to the dipeptide.

Example 7: A concentrated liquid sweetener is prepared containing the following ingredients.

	Grams
Sodium saccharin	18.0
L-aspartyl-2,5-dihydro-L-phenylalanine methyl ester	1.8
Povidone (polyvinylpyrrolidone)	0.1
Glycerin	10.0
Sodium benzoate	0.1
Methylparaben	0.1
Distilled water	68.9

Example 8: A dry mix is prepared containing the following ingredients.

	Grams
Sodium saccharin	5.0
L-aspartyl-L-phenylglycine methyl ester	10.0
Calcium lactate	25.0
Acacia powder	60.0

The foregoing ingredients are mixed, wet-granulated with a liquid containing 50% water and 50% alcohol SD-3A, and dried to give a granular free-flowing powder.

Example 9: A bulk sweetener is prepared by mixing, warming and spray drying the following ingredients.

	Grams
Sodium saccharin	0.8
L-aspartyl-L-phenylglycine methyl ester	2.0
Hydrolyzed corn syrup solids	54.0
Distilled water	43.2

Synergistic Combinations

D. Scott; U.S. Patent 3,780,189; December 18, 1973; assigned to G.D. Searle & Company developed synergistic formulations containing dipeptide and saccharin. Special methodology is described for the measurement of the synergistic effects.

The sweetening potency of saccharin is, depending upon the edible material to which it is added, up to 500 times that of sucrose. It has been discovered that when saccharin is mixed with one of the dipeptides, the sweetening potency is greatly enhanced. The addition of L-aspartyl-L-phenylalanine methyl ester, for example, to saccharin in coffee raises its potency to about 1,000. It has been found that this synergistic effect is manifest over a range of ratios of saccharin to dipeptide of between about 1:15 and 15:1, with the range of ratios of saccharin to dipeptide of between about 1:15 and 5:1 being particularly preferred.

Edible materials to which these compositions can be added include fruits, vegetables, juices, meat products such as ham, bacon, and sausage, egg products, fruit concentrates, salad dressings, milk products such as ice cream and sherbet, icings, syrups, corn, wheat and rice products such as bread, cereals, and cake mixes, fish, cheese and cheese products, nut products, beverages such as coffee, tea, carbonated soft drinks, beers and wines, and confections such as candy. An example of a suitable tablet adapted for addition to edible materials, e.g., coffee, is as follows.

	Milligrams
L-aspartyl-L-phenylalanine methyl ester	25
Saccharin	5
Mannitol	200
Sorbitol	5
Magnesium stearate	1

Specific examples of the synergistic effect resulting from the mixture of the dipeptides with known sweetening agents are given below. The concentrations in the following formulations are given for the purpose of illustration only and it is recognized that individual taste preferences for degrees of sweetness vary so substantially that it is difficult to delineate well-defined limits for useable ranges of concentration.

Example 1: A noncarbonated imitation strawberry drink is prepared using four different sweetening compositions: Formulation 1, sucrose alone; Formulation 2, L-aspartyl-L-phenylalanine methyl ester alone; Formulation 3, saccharin alone and Formulation 4, saccharin and L-aspartyl-L-phenylalanine methyl ester together. The compositions of each mixture, amounting to a total volume of 300 ml, are shown on the following page.

	Grams
Formulation 1:	
Sucrose	27
Dextrose	2.61
Citric acid	0.15
Imitation strawberry flavor	0.12
Lactose	1.827
Water	268
Formulation 2:	
L-aspartyl-L-phenylalanine methyl ester	0.216
Citric acid	0.15
Imitation strawberry flavor	0.12
Lactose	1.827
Water	299
Formulation 3:	
Saccharin	0.054
Citric acid	0.15
Lactose	1.827
Imitation strawberry flavor	0.12
Water	299
Formulation 4:	
L-aspartyl-L-phenylalanine methyl ester	0.108
Citric acid	0.15
Lactose	1.827
Saccharin	0.0165
Imitation strawberry flavor	0.12

It was determined by a taste panel that each of the latter preparations was equivalent in sweetness.

Example 2: Approximately 3,000 ml of instant coffee is prepared by dissolving 16.5 grams of instant coffee in 3,000 ml of water. 600 ml portions of the instant coffee are taken and to each was added one of the following sweetening preparations: (1) 24 grams of sucrose, (2) 0.16 gram of L-aspartyl-L-phenylalanine methyl ester, (3) 0.048 gram of saccharin and (4) 0.012 gram of saccharin with 0.08 gram of L-aspartyl-L-phenylalanine methyl ester. It was determined by a taste panel that each of the latter sweetened coffees has an equivalent sweet taste.

Example 3: Aqueous samples of L-aspartyl-L-phenylalanine methyl ester (APM), sodium saccharin (SS) and compositions composed of mixtures of the two were analyzed by a flavor panel comprised of five experienced tasters and each sample was evaluated at least three times.

Samples were examined at about 22°C and each panel member consumed 3 to 5 milliliters of solution. The test solutions were compared with a 10% sucrose solution to make certain that they were iso-sweet. Solutions were defined as iso-sweet when the flavor panel assessed that the solutions were equal in sweetness. The precision of sweetness determination is limited to about 10% of the sweetness level.

Stock solutions were prepared for each sweetener using 1.0 gram of material diluted to 100 milliliters volume using distilled water. Serial dilutions were prepared from the stock solutions for samples to be used in the flavor analysis. The concentrations of APM and SS necessary to result in predetermined ratios at an equivalent sweetness to 10% sucrose are shown in the table below.

. . . . . Ratio.		. Actual Weights (g/100 ml) .	
APM	SS	APM	SS
1	0	0.075	-
0	1	-	0.05
1	1	0.0166	0.0166
1	2	0.010	0.020
1	2½	0.0083	0.0208
1	3	0.0072	0.0216
1	5	0.0058	0.0290
1	10	0.0035	0.0350
1	15	0.0031	0.0465
5	1	0.0275	0.0055
15	1	0.0450	0.003

Potency values as a function of concentration were determined by comparing sucrose, APM and SS at various concentrations at which they were iso-sweet.

Sucrose (%)	APM (%)	P_1	SS (%)	P_2
0.34	0.007-0.001	400	0.0047	723
2.0	-	-	0.0036	556
3.0	-	-	0.0061	492
4.0	-	-	0.0096	417
4.3	0.02	215	-	-
5.0	-	-	0.0142	353
6.0	-	-	0.0204	294
7.0	-	-	0.0273	257
8.0	-	-	0.0359	223
9.0	-	-	0.0444	203
10.0	0.075	133	0.053 (0.050)	194 (200)
15.0	0.15	100	-	-

Potency is based on sucrose and is defined as the ratio of concentration of sucrose in solution to concentration of test compound in solution at which the two solutions are determined to be iso-sweet. Interpolated values of potency were determined from the table above.

The determination of theoretical percent sucrose equivalent (percent S) (i.e., that value of sucrose equivalence which would be expected in a mixture with no synergistic effect) is complicated by the concentration dependence of potency. The potency of a compound in a mixture is a function not only of its concentration but also the presence of other sweeteners. It is possible, however, to calculate high and low limits for theoretical percent sucrose equivalent, thereby bracketing the true value.

The percent S is calculated from the general formula (which neglects any interaction): Percent S = $P_1 \times C_1 + P_2 \times C_2$ wherein percent S is the theoretical percent sucrose equivalent; P_1 is the potency of APM; P_2 is the potency of SS; C_1 is the concentration of APM; and C_2 is the concentration of SS.

The lower bound on theoretical percent sucrose equivalent is determined by using potencies which are equal to the potencies that each component would have if each contributed all of the sweetness to the mixture. The potencies are evaluated at the concentration of APM and SS where each is iso-sweet with a 10% sucrose solution. These are the lowest values of potency which the components would have in solution when iso-sweet with 10% sucrose.

The upper bound on the theoretical percent sucrose equivalent is determined using potencies at the concentration of each component independently of the other. These are the largest values for potencies which the components would have in a mixture iso-sweet with a 10% sucrose solution, thus yielding the highest value for percent S. The lower bound is calculated from the following formula: Percent S = $(P_{1\,at\,10\%} \times C_1) + (P_{2\,at\,10\%} \times C_2)$ where percent S is the theoretical percent sucrose equivalent; $P_{1\,at\,10\%}$ is the potency of APM evaluated at the concentration of APM iso-sweet with a 10% sucrose solution (constant at 133); $P_{2\,at\,10\%}$ is the potency of SS evaluated at the concentration of SS iso-sweet with a 10% sucrose solution (constant at 200); C_1 is the concentration of APM; and C_2 is the concentration of SS and the upper bound is calculated from the formula: Percent S = $(P_{1\,at\,C_1} \times C_1) + (P_{2\,at\,C_2} \times C_2)$ where percent S is the theoretical percent sucrose equivalent; $P_{1\,at\,C_1}$ is the potency of APM evaluated at the concentration of APM; $P_{2\,at\,C_2}$ is the potency of SS evaluated at the concentration of SS; C_1 is the concentration of APM; and C_2 is the concentration of SS. The results of these two calculations are presented in the table below.

Ratio		Percent S (lower bound)			Percent S (upper bound)		
APM	SS	APM	SS	Total	APM	SS	Total
1	0	10	----------	10	10	----------	10
0	1	----------	10	10	----------	10	10
1	1	2.2	3.3	5.5	3.9	5.4	9.3
1	2	1.3	4.0	5.3	2.9	5.8	8.7
1	2½	1.1	4.2	5.3	2.5	6.0	8.5
1	3	0.96	4.3	5.3	2.3	6.2	8.5
1	5	0.77	5.8	6.6	1.9	7.2	9.1
1	10	0.47	7.0	7.5	1.3	7.9	9.2
1	15	0.41	9.3	9.7	1.1	9.3	10
5	1	3.7	1.1	4.8	5.2	2.8	8.0
15	1	6.0	0.60	6.6	6.9	1.7	8.6

Since all of the above solutions were iso-sweet with a 10% sucrose solution, synergism is present when the calculated values of total percent S are less than 10. It is apparent from the table, that this occurs over a range of ratios of SS to APM of between about 1:15 and 15:1, with the largest synergistic effect being observed at a ratio of SS to APM of between about 1:1 and 1:15.

SACCHARIN AND GLUCONO-DELTA-LACTONE

Blend with Buffer

Several benefits are obtained by *P. Kracauer; U.S. Patent 3,285,751; November 15, 1966; assigned to Cumberland Packing Company* with a special formulation containing saccharin, glucono-delta-lactone in combination with a buffer salt.

When these three ingredients are used in combination, the resulting composition, although it is substantially free of calories, has sufficient bulk to be easily handled, and particularly to be packed in individual unit dose packets, and can be used for the sweetening of any food to a degree closely approximating the taste of natural sugar without leaving any bitter aftertaste.

It is preferred to use per amount of artificial sweetener having the sweetening power of approximately 10 grams of sugar between about 0.3 and 1 gram of glucono-delta-lactone and between 0.3 and 1 gram of sodium citrate or potassium citrate. It should be noted that it is essential to use both the glucono-delta-lactone and the sodium citrate or potassium citrate in combination.

If glucono-delta-lactone is used alone as the filler, while it will not alter the original taste of the artificial sweetener with which it is mixed, it will eventually cause milk to curdle. Consequently, such a composition cannot be used for the sweetening of coffee, for example, for baking with milk, or for the sweetening of milk products.

On the other hand, if sodium citrate or potassium citrate is used alone as the filler with the artificial sweetener, it cannot be used for the sweetening of drinks containing tannic acid and/or caffeine, such as coffee, tea, mocha, etc., because of a chemical reaction which results in discoloration (darkening) of the coffee, tea or the like. When glucono-delta-lactone is mixed with the sodium citrate and/or potassium citrate the resulting composition can be used for the sweetening of milk or milk products without any danger of curdling, and it also can be used for the sweetening of coffee, tea or any other drink containing tannic acid and/or caffeine, without causing any discoloration.

The special composition can be simply formed by blending the artificial sweetener or sweeteners with the glucono-delta-lactone and the sodium citrate and/or potassium citrate in the desired proportions. A preferred method of manufacture consists of mixing the components with an aqueous-alcohol mixture and then carefully drying at low temperature, in accordance with a normal granulation method, to form an extremely uniform composition which dissolves very quickly and consists of uniform granules of optimum size.

Example: A uniform mixture is made of 300 grams of glucono-delta-lactone, 1,000 grams of sodium citrate and 10 grams of calcium saccharin. Each 1.3 grams of this mixture has the approximate sweetening power of 10 grams of sugar.

Potassium Bitartrate Buffer

Another buffer is used by *M.E. Eisenstadt; U.S. Patent 3,647,483; March 7, 1972; assigned to Cumberland Packing Corporation* to reduce or eliminate the aftertaste due to saccharin. The saccharin artificial sweetener is mixed with cream of tartar powder and with the glucono-delta-lactone and sodium gluconate and/or potassium gluconate in a ratio of 1 part of saccharin artificial sweetener to ⅛ to 5 parts of cream of tartar powder (most preferably ⅙ to 2 parts), with 2 to 10 parts of a physiologically acceptable, soluble gluconic acid salt, such as sodium gluconate and potassium gluconate (most preferably 3 to 7 parts) and with 3 to 15 parts glucono-delta-lactone (most preferably 5 to 10 parts).

When these components are used in these proportions, the most desired effect of sweetness approaching that of natural sugar, without any bitter aftertaste and without any undesired effect on any food or beverage to which the sweetening composition is used, is obtained.

It should be noted that all of the components of the composition must be used in combination in order to achieve the desired results. Thus, using glucono-delta-lactone alone or even glucono-delta-lactone plus the gluconate, e.g., sodium gluconate and/or potassium gluconate, with the saccharin artificial sweetener, results in a composition that cannot be used for the sweetening of milk or milk products because of a danger of curdling. Furthermore, the sweetening of coffee, tea or any other drink containing tannic acid and/or caffeine cannot be accomplished with such composition without causing discoloration.

On the other hand, if cream of tartar powder alone is used with the saccharin artificial sweetener, it is not possible to mask the bitter aftertaste of the saccharin. It is only by using the cream of tartar powder together with the gluconate and the glucono-delta-lactone in the proportions indicated above, that it is possible to obtain a complete masking of the bitter aftertaste of the saccharin without adversely affecting the taste of any food or beverage to which the composition is applied while still obtaining a sweetening composition which is entirely free of calories.

Example 1: 15 pounds of glucono-delta-lactone, 5 pounds of sodium gluconate, 1 pound of soluble saccharin and 2 pounds of potassium bitartrate are thoroughly and uniformly mixed. The resulting mixture is many times as sweet as natural sugar so that a small amount thereof can be used in place of sugar to give a sweetening effect without providing any calories.

Approximately 1 gram of the composition will give the sweetening effect of two teaspoons of sugar. This composition can be used to sweeten beverages or in cooking, in all quantities even to highly sweeten beverages, without causing any bitter aftertaste and without adversely affecting the taste of the food or beverage to which it is applied.

Example 2: A sweetening composition is prepared as in Example 1, however using 5 pounds of potassium gluconate, 5 pounds of glucono-delta-lactone, 1 pound of saccharin and ½ pound of potassium bitartrate. Approximately 1 gram of the above product gives the sweetening power approximating that of two teaspoons of sugar.

Example 3: A composition is prepared as in Example 1, however, using 6 pounds of glucono-delta-lactone, 4 pounds of sodium gluconate, 10 ounces of sodium saccharin and 2 ounces of potassium bitartrate. Each gram of the above product gives approximately the sweetening power of two teaspoons of sugar. The product contains no calories.

Example 4: A sweetening composition is prepared as in Example 1, however, using 3 pounds of glucono-delta-lactone, 8 pounds of sodium gluconate, ½ pound of saccharin and 1 pound of potassium bitartrate. Approximately 0.5 gram of the above product is used to obtain the sweetening of two teaspoons of ordinary sugar.

Example 5: A sweetening composition is prepared as in Example 1, however using 300 pounds of glucono-delta-lactone, 200 pounds of potassium gluconate, 36 pounds of calcium saccharin and 6 pounds 4 ounces of potassium bitartrate. This composition contains no carbohydrates, no calories, no cyclamates and no sodium.

Approximately 0.024 ounce of the above product is used to obtain the sweetening equivalence of two teaspoons of ordinary sugar. Any of the above compositions may be used for the sweetening of beverages, in baking, preparing cooked fruits, in cooking, in making candies, etc.

Sodium Gluconate Buffer

In the process of *J.J. Liggett; U.S. Patent 3,684,529; August 15, 1972* there is found an improvement in taste quality of saccharin by use of sodium gluconate in combination with glucono-delta-lactone. Broadly, the sweetening mix consists essentially of at least 50% by weight of an edible salt of saccharin in mixture within the range of from 5 to 25% by weight of glucono-delta-lactone and from 5 to 25% by weight of sodium gluconate.

The sweetener can be combined with a solid pharmaceutical carrier and compressed into tablet form. Any of the customarily used fillers and adjuvants such as dextrans and whey solids of high bulk and low density, lactose and starch can be added as a bulking agent, inert extender or carrier. Binders such as carboxymethylcellulose and acacia can also be used.

In addition, it may be desirable to include mixtures of carbonates, and organic acids in the manner well known in tablet technology. Preferably such tablets should be proportioned in weight to provide sweetness equivalent to from ½ to 2 teaspoonsful of sucrose, i.e., from 2½ to 10 grams of sucrose.

The sweetening composition can be combined with a conventional bulking agent such as lactose or starch, and blended into granulated form. Preferably, the granulated composition will have a sweetening power approximately equivalent to sugar on a weight for weight basis so that 1 teaspoonful of the granulated composition will be equivalent in sweetness to about ½ to 2 teaspoonsful and more, preferably 1 teaspoonful of sucrose.

The sweetening composition can also be extended with water so that aqueous solutions can be manufactured simply by dissolving the sweetening composition in water to the desired concentration. A preferable concentration is one in which 2 drops of the solution are equal in sweetness to from ½ to 2 teaspoonsful of sucrose.

The sweetener can be formed simply by blending powdered solution saccharin with powdered glucono-delta-lactone and powdered sodium gluconate. The ingredients in the desired proportions are mixed thoroughly and sifted to provide a uniform mixture. It may be desirable to mix the components of the composition with an aqueous-alcohol mixture and then carefully dry at a low temperature in accordance with normal granulation procedures to thereby form an extremely uniform composition which dissolves very quickly and consists of uniform granules of optimum size.

Example 1: 75 pounds of powdered sodium saccharin are thoroughly blended with 12.5 pounds of powdered sodium gluconate and 12.5 pounds of glucono-delta-lactone to provide a uniform mixture. The resulting mixture is several hundreds of times sweeter than natural sugar and can be used in place of the sugar to give a sweetening effect with essentially no caloric intake in all food and beverage products.

Throughout the range of concentrations required to obtain conventional sweetness levels including highly sweetened beverages, the resulting sweetening effect is achieved with no bitter aftertaste. 1 gram of the above sweetening composition is equivalent to about 1.5 grams of sodium saccharin alone and to about 450 grams of natural sugar (sucrose) on a weight-for-weight basis in concentrations giving a sweetness actually equivalent to 2.5% sucrose.

A cup of coffee sweetened with 22 mg of the composition of Example 1, tastes as if it had been sweetened with 2 teaspoons (approximately 10 grams) of ordinary sugar.

Example 2: A sweetening composition is prepared in the manner of Example 1 using 90 pounds of sodium saccharin, 5 pounds of sodium gluconate and 5 pounds of glucono-delta-lactone.

Example 3: A sweetening composition is prepared in the manner of Example 1 using 50 pounds of sodium saccharin, 25 pounds of sodium gluconate and 25 pounds of glucono-delta-lactone.

Example 4: A sweetening composition is prepared in the manner of Example 1 using 70 pounds of sodium saccharin, 10 pounds of sodium gluconate and 20 pounds of glucono-delta-lactone.

Example 5: A sweetening composition is prepared in the manner of Example 1, using 70 pounds of sodium saccharin, 20 pounds of sodium gluconate and 10 pounds of glucono-delta-lactone.

Example 6: A sweetening composition is prepared in the manner of Example 1 using 60 pounds of sodium saccharin, 20 pounds of sodium gluconate and 20 pounds of glucono-delta-lactone.

Example 7: A sweetening composition is prepared in the manner of Example 1 using 80 pounds of sodium saccharin, 10 pounds of sodium gluconate and 10 pounds of glucono-delta-lactone.

Example 8: A granulated composition having a sweetening power approximately equivalent to sugar on a weight-for-weight basis is prepared by blending 8 pounds of sodium saccharin, 1 pound of sodium gluconate and 1 pound of glucono-delta-lactone with 449 pounds of lactose which serves as a bulking agent or extender. The powdered ingredients are thoroughly blended using a conventional mixing apparatus to provide a uniform mixture 5 grams (1 teaspoonful) of which is equivalent in sweetness to 5 grams of sugar.

Example 9; 222 grams of the sweetening composition of Example 1, are dissolved in distilled water, q.s. to 1,000 ml, to which is added approximately

1 gram each of benzoic acid, USP, and methyl-p-hydroxybenzoate as preservatives. 1 drop (0.05 ml) of the resultant solution is equivalent to 5 grams (1 teaspoonful) of sugar.

Example 10: Tablets are made by blending a solution of 25 grams of water and 5 grams of acacia with 100 grams of the powdered sweetening composition of Example 1. 5 grams of sodium benzoate and 65 grams of an inert extender, e.g., lactose, are blended with the dried mixture which is then compressed into tablets. A tablet weighing 50 mg is equivalent in sweetness to approximately 2 teaspoonsful of sugar. The tablet does not give an offtaste even though it contains a high concentration of saccharin.

From the above examples, it is seen that a sweetening composition consisting essentially of at least 50% of an edible salt of saccharin in mixture within the range of 5 to 25% by weight of glucono-delta-lactone and from 5 to 25% by weight of sodium gluconate, provides sweetness levels which are higher than those obtained with sodium saccharin alone.

Bicarbonate Buffer

Formulations containing saccharin, lactose and cream of tartar are satisfactory for sweetness without bitter taste. However, such a product contains calories because of the presence of lactose and in fact a sweetening composition which is the equivalent in sweetness to 2 teaspoons of sugar contains about 3 calories.

While this is of course extremely low as far as caloric content is concerned, 3 calories are undesirable, particularly in the case of diabetics who might use large amounts of sweeteners and who would not wish to upset their carbohydrate intake in any way. In addition, lactose is also known as milk sugar, and therefore persons of orthodox Jewish faith cannot use this composition in conjunction with meat products.

M.E. Eisenstadt; U.S. Patent 3,946,121; March 23, 1976; assigned to Cumberland Packing Corporation uses an edible bicarbonate in a formulation which is calorie-free, has no lactose and will not cause curdling of milk in coffee. The use of the bicarbonate alone with the saccharin artificial sweetener has no effect whatsoever in masking the bitter aftertaste of the saccharin. It is only by combining the bicarbonate with the glucono-delta-lactone that it is possible to completely mask the bitter aftertaste of the saccharin while at the same time avoiding disadvantages of curdling of milk, sourness, etc.

Example 1: 25.5 pounds of glucono-delta-lactone, 1.8 pounds of calcium saccharin and 2.1 pounds of potassium bicarbonate are thoroughly and uniformly mixed. The resulting mixture is many times as sweet as natural sugar so that a small amount can be used in place of sugar to give a sweetening effect without providing any colories.

Approximately 0.6 gram of the composition will give the sweetening effect of two teaspoons of sugar. This composition can be used to sweeten beverages or in cooking, in all quantities even to highly sweeten beverages without causing any bitter aftertaste, without sourness, and without adversely affecting the taste of the food or beverage to which it is applied. In addition, the above composition may be freely used by persons with high blood pressure because it is totally sodium free.

Gluconate and Fructose

In the combination of gluconate, fructose and saccharin *M.E. Eisenstadt; U.S. Patent 3,743,518; July 3, 1973; assigned to Cumberland Packing Corp.* states that the amount of fructose should be about 5 times the amount of the saccharin artificial sweetener, and the amount of the gluconate should be about 12 times the amount of the saccharin artificial sweetener.

The term gluconate is meant to refer to glucono-delta-lactone, gluconic acid, and salts of gluconic acid such as sodium gluconate, potassium gluconate and calcium gluconate.

Example 1: 5.5 pounds of calcium saccharin, 27.8 pounds of fructose and 66.7 pounds of sodium gluconate are thoroughly mixed to provide a uniform mixture. The resulting mixture is many times as sweet as natural sugar and in fact 1 gram is equivalent in sweetness to 2 teaspoons of natural sugar. 1 gram of the composition provides a single calorie, and although the fructose is a carbohydrate, the composition can nevertheless be used by diabetics because the fructose is converted into glycogen even without insulin.

This composition can be used to sweeten beverages or in cooking in all quantities, even to highly sweeten beverages, without causing any bitter aftertaste and without adversely affecting the taste of the food or beverage to which it is applied.

Example 2: A sweetening composition is prepared as in Example 1, however using 5 pounds of sodium saccharin, 18 pounds of fructose and 74 pounds of calcium gluconate. Each gram of the above product gives approximately the sweetening power of two teaspoons of sugar.

Example 3: A sweetening composition is prepared as in Example 2, however using 34 pounds of glucono-delta-lactone and 40 pounds of sodium gluconate in place of the calcium gluconate.

SACCHARIN AND CALCIUM GLUCONATE

It is reported in another process by *P. Kracauer; U.S. Patent 3,489,572; January 13, 1970* that the offtaste of saccharin can be eliminated by addition of small amounts of organic calcium or magnesium salts.

Suitable salts are of glycerino phosphoric acid and of gluconic acid, namely calcium or magnesium glycerino phosphate and calcium or magnesium gluconate (food grades). The required amounts to be added might differ to obtain the optimum effect, depending on the sweetening composition. Generally, sweetening agent plus between 20 to 60% of a calcium or magnesium salt of glycerino phosphoric acid or gluconic acid will be required.

In a liquid sweetening composition there should be 5 to 25% of a soluble saccharin and 2 to 7% of a calcium or magnesium salt of gluconic acid or 2 to 5% of a calcium or magnesium salt of glycerino phosphoric acid. The granulated compositions contain 0.8 to 2% of saccharin, 0.3 to 1% of a calcium or magnesium salt of gluconic acid (or 0.2 to 1% of a calcium or magnesium salt of glycerino phosphoric acid) and the balance of a suitable inert carrier, e.g., lactose. Sweetening tablets contain about 7 to 15 mg of saccharin and 5 to 10 mg of a calcium or magnesium salt of gluconic acid, in an effervescent base consisting of a soluble carbonate and edible organic fruit acid, or 7 to 15 mg of saccharin plus 4 to 8 mg of a calcium or magnesium salt of glycerino phosphoric acid, in an effervescent base consisting of a soluble carbonate and an edible fruit acid.

Example 1: Liquid Sweetener –

	Grams
Soluble saccharin	13
Calcium gluconate	4
Benzoic acid, USP	0.1
Tegosept M (methyl-p-hydroxybenzoate), a preservative	0.1
Distilled water, q.s. to 100 cc	

Dissolve in hot water (approximately 80°C) the benzoic acid and Tegosept M. Add and dissolve the calcium gluconate. After cooling off add and dissolve the soluble saccharin. Filter and bottle.

In place of calcium gluconate in this example, 4 grams of magnesium gluconate or 2.5 grams of magnesium or calcium glycerino phosphate may be used. Two drops are equivalent to about 1 teaspoon of sugar in sweetening power.

Example 2: Granulated Sweetening Composition –

	Grams
Soluble saccharin	1.5
Calcium gluconate	0.7
Lactose, USP	97.9
	100.1

Blend thoroughly the above ingredients and sift. Pack, preferably in envelopes, 1 gram of the above mixture. This serving is equivalent to about 5 grams of sugar in sweetening power. In place of calcium gluconate, the same amount of magnesium gluconate or 0.5 gram of magnesium or calcium glycerino phosphate may be used.

Example 3: Sweetening Tablets (Effervescent) –

	Grams
Saccharin sodium	1.4
Sodium bicarbonate	14.6
Tartaric acid	12.0
Calcium gluconate	0.7
Carbowax (6000)	1.3
	30.0

Blend thoroughly saccharin with sodium bicarbonate. Incorporate one-half of the Carbowax amount. Blend separately tartaric acid with calcium gluconate and the other half of the Carbowax amount. Blend thoroughly both portions.

If necessary (preferably) compress to large tablets, slug, granulate through Fitzmill.

Compress at medium pressure. Two tablets weighing 0.15 gram each are equivalent to about 1 teaspoon of sugar in sweetening power (approximately 5 grams). In place of calcium gluconate the same amount of magnesium gluconate or calcium or magnesium glycerino phosphate may be used.

SACCHARIN AND CITRATE BUFFER

P. Kracauer; U.S. Patent 3,780,190; December 18, 1973; assigned to American Sweetener Corp. uses a soluble citrate as an extender and debittering agent in special sweetener formulations containing saccharin.

It has been found that the addition of minute amounts of d-galactose to the composition may assist in further elimination of the bitter aftertaste of saccharin, as well as the enhancement of flavor. In addition, the inclusion of minor amounts of alkaline chloride, particularly sodium or potassium chloride has been found to assist in the creation of a more sugar-like taste in foods and drinks.

The preferred method of formulation is a conventional wet granulation technique. The wet granulation technique results in a fluffier composition which offers more volume and consequently is more convenient to dispense. In the typical wet granulation technique, all of the ingredients are carefully mixed and wet granulated using about 1 to 2% of a 50% alcohol solution. The wet granulated mixture is then dried at low temperature and sifted to obtain the desired uniform crystal granulate.

Example 1: The following ingredients are skillfully mixed and uniformly blended:

	Grams
Soluble saccharin	3.2
Sodium citrate	50
Fumaric acid	0.8
Potassium chloride	0.2
Sodium chloride	0.1

Approximately 0.55 gram of this mixture is equivalent in sweetening power to about 2 teaspoons of sugar. It is completely and instantly soluble in cold and hot foods and drinks and does not have a bitter aftertaste.

Example 2: The following ingredients were uniformly mixed and blended:

	Grams
Sodium citrate	45
Soluble saccharin	3.3
Calcium monophosphate	0.1
Sodium chloride	0.1
d-Galactose	0.5

Approximately 0.5 gram of this mixture is equivalent in sweetening power to about 2 teaspoons of sugar.

Example 3: The following ingredients were uniformly mixed and blended:

	Grams
Sodium citrate	45
Soluble saccharin	3.6
Fumaric acid	0.7
Calcium monophosphate	0.1
d-Galactose	0.2

Approximately 0.5 gram of this compound is equal in sweetening power to about 2 teaspoons of sugar.

SACCHARIN AND LACTOSE

Critical Lactose Level

A low-calorie mix containing saccharin and lactose was found by *B. Eisenstadt; U.S. Patent 3,259,506; July 5, 1966; assigned to Cumberland Packing Company* to give an improved sweetener for food products. Using saccharin alone as the artificial sweetener, the components of the mixture are present in a ratio of 75 to 125 pounds of lactose per each 3 to 10 pounds of saccharin artificial sweetener.

The compositions can be used for the making of candy, in place of sugar, to provide low-calorie candies which are as sweet as though the same had been sweetened with cane sugar, without adversely affecting the taste of the candy. The same holds true for the use of compositions for meat products which might have to be sweetened, and for fruits, for example for canned fruits and the like to make the products of relatively low caloric value, while being sweetened without adversely affecting the taste of the food product.

The compositions can also be used in the place of sugar, for the making of baked goods, for example, bread, muffins, and the like, provided that the amount of sugar required in the particular recipe is not so great that the bulking effect of the sugar is necessary for the making of the desired baked goods.

The term "saccharin artificial sweetener" is meant to refer to saccharin itself and the salts such as sodium saccharin, potassium saccharin, etc. It has been found that lactose has the effect, when used in sufficient quantity, and when used with a saccharin artificial sweetener to depress or completely destroy the bitter aftertaste of the saccharin artificial sweetener.

However, it has further been found that if too great an amount of the lactose is used together with the artificial sweetener, then the lactose not only depresses or removes the bitter aftertaste of the artificial sweetener, but in addition, the lactose has the undesired effect of changing the taste of the food or beverage to which it is applied.

Thus, for example, while up to about 1 gram of lactose will not have any pronounced adverse effect on the taste of a cup of coffee, any amount substantially greater than 1 gram of lactose per cup of coffee will change the taste of the coffee.

It should be noted that although lactose is chemically a sugar, it is not generally used for sweetening. It apparently has a leaching effect on other tastes than sweetness so that it leaches out or masks the bitter aftertaste of the artificial sweeteners when used in sufficient quantities and if used in too great a quantity it will also leach out or mask the taste of the food or beverage to which it is applied.

Example 1: A sweetening composition is prepared by thoroughly mixing 100 pounds of lactose (powdered) with 8 pounds of sodium saccharin to provide a uniform mixture.

Example 2: Tongue Polanaise is prepared by boiling a steer tongue, cutting the same into cubes, and adding the cubes to a sauce prepared from the following: 1½ ounces raisins, ¾ cup of a sweet wine, 2 onions, 1 can of tomato sauce, ⅛ cup of lemon juice and 7½ grams of the composition of Example 1. The resulting Tongue Polanaise has an excellent flavor, no bitterness and tastes as though the same had been cooked with pure sugar.

Cream of Tartar Addition

In a later process *M.E. Eisenstadt; U.S. Patent 3,625,711; December 7, 1971; assigned to Cumberland Packing Corporation* uses cream of tartar in combination with lactose or lactose/dextrose mixtures to reduce undesirable bitter aftertastes.

It should be noted that although lactose is chemically sugar (sometimes called milk sugar) it is not generally used for sweetening. It apparently has a leaching effect on other tastes than sweetness so that it leaches out or masks the bitter taste of the artificial sweeteners when used in sufficient quantities. However, if used in too great a quantity it will also leach out or mask the taste of the food or beverage to which it is applied.

For this reason, if lactose alone was used in order to obtain a sufficient amount of lactose with the saccharin to mask the bitter aftertaste of the saccharin, the amount of lactose would be so great that it would also leach out or mask the taste of the food or beverage to which the composition is applied.

On the other hand, dextrose is actually a sweetening agent and dextrose alone does not have the effect of sufficiently masking the bitter aftertaste of the artificial sweetener. Therefore, with the dextrose it is necessary to use the cream of tartar powder in order to achieve the desired masking of the bitter aftertaste.

Furthermore, if cream of tartar powder alone were used with a saccharin artificial sweetener, there is no possibility at all of masking the bitter aftertaste of the saccharin.

It is only by using the cream of tartar powder with the lactose and/or dextrose, that it is possible to obtain a complete masking of the bitter aftertaste of the saccharin without adversely affecting the taste of any food or beverage to which the composition is applied.

Example 1: 50 pounds of lactose (powdered), 4 pounds of cream of tartar powder and 1.0 pound of calcium saccharin are thoroughly mixed to provide a uniform mixture. The resulting mixture is many times as sweet as natural sugar so that a small amount can be used in place of the sugar to give a sweetening effect with low caloric intake. This composition can be used to sweeten beverages or in cooking in all quantities, even to highly sweeten beverages, without causing any bitter aftertaste and without adversely affecting the taste of the food or beverage to which it is applied.

Example 2: A sweetening composition is prepared as in Example 1, however using 48 pounds of lactose, 1.75 pounds of powdered cream of tartar and 1.5 pounds of sodium saccharin. A cup of coffee sweetened with 1 gram of the composition of this example tastes as though it had been sweetened with 2 teaspoons (approximately 10 grams) of ordinary sugar. The above compositions may be used in baking, in preparing cooked fruits, in cooking, in the making of candies, etc.

Example 3: 50 pounds of powdered dextrose, 4 pounds of cream of tartar powder and 1.0 pound of calcium saccharin are thoroughly mixed to provide a uniform mixture. Each approximately 1 gram of the above product gives substantially the sweetening power of 2 teaspoons of sugar. The taste is akin to that of natural sugar and does not have any bitter aftertaste.

SACCHARIN AND GALACTOSE

D-galactose in minor amounts was found to reduce the aftertaste of saccharin according to *P. Kracauer; U.S. Patent 3,667,969; June 6, 1972; assigned to American Sweetener Corp.* Preferably, 20 to 50 parts of d-galactose is mixed with 100 parts of saccharin in order to produce the nonbitter, low calorie, artificial sweetening composition.

However, somewhat smaller or larger amounts of d-galactose may be used without detracting from the effectiveness of the resulting composition. It has been found that the utilization of amounts of d-galactose in excess of the amount of saccharin may not only eliminate the aftertaste normally associated with saccharin but also could effectively reduce or eliminate the sweetness of this compound.

While the d-galactose-saccharin composition may be used alone, it has been found that selected additional ingredients may be used either alone or in combination to produce a sweetening composition having an enhanced flavor, i.e., a more sugar-like taste, in foods and drinks.

Thus, for example, it has been found that the addition of 10 to 40 parts of sodium chloride for each 100 parts of saccharin creates a more sugar-like taste in foods and drinks. It has also been found that the flavor of the sweetening composition may be enhanced by incorporating 1 to 5 parts of calcium hydroxide in the composition for each 100 parts of saccharin.

The utilization of bulking agents is desirable since the sweetening power of saccharin is quite concentrated, i.e., 1 part of saccharin is equal to 350 to 400 parts of sugar, and in the absence of a bulking agent normal servings of the artificial sweetener would be quite small. Any edible, soluble material which will not adversely affect the basic sweetening compound of the food or drink to be sweetened may be used as a bulking agent.

Typical bulking agents include carbohydrates such as sucrose, dextrose, sorbitol. Normally 0.7 to 1.0 gram of the bulking agent will be added to the sweetening composition in order to create an individual portion equal in sweetening power to about 2 teaspoons of sugar.

Although the utilization of such bulking agents will add a small amount of calories to the composition, the amounts utilized on an individual serving basis will add only about 3 calories as compared to the 35 to 40 calories which would be present if sugar were used as the sweetener.

Although the composition may be formulated by blending the materials into a uniform mix, the preferred method of formulation is a conventional wet granulation technique. The wet granulation technique results in a fluffier composition which offers more volume and consequently is more convenient to dispense.

In the typical wet granulation technique all of the ingredients are carefully mixed and thereafter are wet granulated using 1 to 2% of a 50% alcohol solution. The wet granulated mixture is then dried at low temperature and sifted to obtain the desired uniform crystal granulate.

Example 1: The following ingredients are mixed and uniformly blended:

	Grams
Soluble saccharin	3,000
d-Galactose	1,000
Sodium chloride	800
Calcium hydroxide	60

Approximately 0.049 gram of this mixture is equivalent in sweetening power to about 10 grams of sugar. It is completely and instantly soluble in cold and hot foods and drinks.

Example 2: The following ingredients were uniformly mixed and blended:

	Grams
Soluble saccharin	310
d-Galactose	110
Sodium chloride	90
Sugar	7,500

About 0.8 gram of this mixture has the approximate sweetening power of two teaspoons of sugar.

Example 3: The following ingredients were uniformly mixed.

	Grams
Soluble saccharin	3,000
d-Galactose	900
Sodium chloride	800
Calcium hydroxide	70
Sugar	90,000

The above mix was wet granulated using 1 to 2% of a 50% alcohol solution. The mixture was thereafter dried at a low temperature and sifted to obtain a uniform crystal granulate. Approximately 0.94 gram of the above very light composition is equivalent in sweetening power to about 9 to 11 grams of sugar.

Example 4: The following ingredients were uniformly mixed.

	Grams
Soluble saccharin	3,200
d-Galactose	1,400
Calcium hydroxide	80

When about 0.5 gram is dissolved in a 12-ounce serving of a carbonated, flavored drink, the drink will be as sweet as it would have been if sweetened with about 10 grams of sugar.

Example 5: The following ingredients were uniformly mixed.

	Grams
Soluble saccharin	3,000
d-Galactose	110
Dextrose	85,000

Approximately 0.841 gram of this mixture is equivalent in sweetness to about 10 grams of sugar.

Example 6: The following ingredients were uniformly mixed.

	Grams
Soluble saccharin	3,100
d-Galactose	1,000
Sodium chloride	900
Calcium hydroxide	84
Sorbitol	75,000

When 0.812 gram of the above blend is dissolved in a 4-ounce orange drink its sweetness is equivalent to a sweetened orange drink containing 10 grams of sugar.

SACCHARIN AND RIBONUCLEOTIDES

Use of ribonucleotides, ribonucleosides and their deoxy analogs with saccharin to reduce the aftertaste is described by *M.H. Yueh; U.S. Patent 3,647,482; March 7, 1972; assigned to General Mills, Inc.* The ribonucleotides and their deoxy analogs would include the free acid form of the phosphate moiety or, as is frequently done in the case of the flavor enhancers, in the form of the sodium or other salts of the phosphate moiety. The base moiety of the nucleotides and the nucleosides is preferably the purine base although pyrimidine bases may be used.

In addition, it is also possible to use the deoxy form of the ribonucleosides and nucleotides such as the 2'- or 3'-deoxynucleosides and deoxynucleotides. Typical flavor modifiers which may be used include 5'-inosinic acid and 5'-guanylic acid, 5'-cytidylic acid and 5'-uridylic acid as well as the alkali metal salts of the same. Similarly, the ribonucleosides, andenosine, guanosine, inosine, thymidine and cytidine may be used. Suitable deoxy compounds include 2'-deoxyadenosine, 2'-deoxycytidine, 3'-deoxyadenosine, etc., as well as the deoxy forms of the ribonucleotides, such as deoxyguanylic acid, deoxyadenylic acid, deoxyinosinic acid, etc.

In use it is necessary to include one or more of the above flavor modifiers either in the artificial sweetening composition or in the sweetened food product. The amount of the flavor modifier used can be varied considerably. Even minute quantities have some effect upon reducing the bitter aftertaste of artificial sweeteners but in general it is preferable to use a quantity of at least 0.001% based on the weight of the food or beverage product.

The quantity can be increased up to a level of about 0.1% and even higher. However, commercial considerations usually would dictate that the flavor modifier not be used in a concentration in excess of 0.1% based on the weight of the ultimate food or beverage. Where the product is the artificial sweetener per se, as for example a dry solid saccharin, the amount of modifier is generally in the range of 0.02 to 1 part per part of the saccharin solids. The preferred range is 0.1 to 0.5 part.

It has been found that substantially improved results can be obtained if the saccharin and flavor modifier are heat-treated. In the method of heat treatment the saccharin and flavor modifier are added to an open vessel of water and the water is boiled 30 to 60 minutes. A long period of boiling, however, is not believed to be detrimental so long as sufficient water remains to prevent burning of the saccharin and flavor modifier. For example, satisfactory results have been obtained by boiling for 3 hours.

It should be recognized that boiling as used herein, means open boiling and not refluxing. The boiled mixture of saccharin, nucleotide and water may be used as a sweetening composition. If desired, the water may be removed by vacuum evaporation or spray drying. The dried sweetener then may be ground if desired. The dried sweetener may be dissolved in a suitable solvent such as water and can be used as a liquid sweetener.

Example 1: A series of three beverages, Samples 1A through 1C, were prepared using orange-flavored Kool Aid as the beverage base. Sample 1A was a preferred embodiment of the process and included 0.940 gram of saccharin, 0.188 gram of a nucleotide composition, 4.3 grams of orange-flavored Kool Aid and ⅞ gallon of water. The saccharin and nucleotide composition had been previously heat-treated according to the process.

The heat-modified saccharin and nucleotide composition was prepared by dry blending by weight, 5 parts of saccharin and 1 part of nucleotide. The nucleotide was a 1 to 1 mixture of disodium guanylate and disodium inosinate. The dry blend was dissolved in 50 parts of water. The solution was boiled for 30 minutes and then vacuum dried. The dry material was finely ground.

Sample 1B was prepared identically to Sample 1A except the saccharin and nucleotides were not heat-treated. Sample 1C was a control and included the type of sweetener used in diet beverages such as artificially sweetened cola beverages. This sweetener comprises a mixture of sugar and saccharin. Sample 1C was prepared by mixing 4.3 grams of orange-flavored Kool Aid, ⅞ gallon of water, 19.5 grams of sucrose, 19.5 grams dextrose and 0.740 gram saccharin.

Samples 1A through 1C were taste-tested by a panel of five persons. All five persons found that Sample 1C was less bitter than Sample 1B. One person found that Sample 1A was less bitter than Sample 1C, three persons found no difference in bitterness between Samples 1A and 1C and one person found that Sample 1C was less bitter than Sample 1A.

The panel results thus show a substantial improvement in the process when the saccharin and nucleotide are heat-treated as compared to the saccharin and nucleotide without heat-treatment. The panel results also indicate that the heat-treated saccharin and nucleotide sweetener is equal to the sugar and saccharin mixture as regards level of bitterness. Moreover, the beverage prepared using the heat-treated saccharin and nucleotide mixture has substantially less carbohydrate content than the beverage prepared using sugar-saccharin mixture.

Example 2: A carbonated cola drink was prepared according to the process by mixing 4 ml of a cola flavoring, 1.2 ml of an 85% aqueous solution of H_3PO_4, 2.28 grams of the ground heat-modified saccharin and nucleotide composition of Example 1, 0.79 gram of caramel color and 1 gallon of water. The resulting solution was carbonated by passing carbon dioxide gas therethrough. The drink may be carbonated using conventional carbonation techniques. The carbonated cola may then be bottled.

Example 3: A liquid form of artificial sweetening agent was prepared by dissolving 10 grams of saccharin and 2 grams of a nucleotide mixture in 100 ml of water. The nucleotide mixture included 1 gram of disodium guanylate and 1 gram of disodium inosinate. The solution was boiled for 30 minutes. The boiled solution, which had a volume of about 50 ml was diluted with water to a concentration of about 2% saccharin and about 0.4% nucleotides. The final solution contained about 0.1% benzoic acid and 0.05% propylparaben as preservatives.

SODIUM CHLORIDE INSOLUBLE FRACTION

According to *M. Yoshida and M. Yoshikawa; U.S. Patent 3,295,993; January 3, 1967* the bitterness and aftertaste of saccharin is removed by treatment of solutions with sodium chloride and with alkaline compounds. Sodium bicarbonate and sodium salts of amino acids are typical of the sodium salts of weak acids that are suitable. Sodium glutamate is the amino acid salt most readily available at the lowest cost.

Merely adding common salt and sodium bicarbonate or sodium glutamate to an aqueous solution of the synthetic sweetening agent is not sufficient for removing the bitter taste component. While the reagents, when added in relatively large amounts, may completely alter the taste of the solution, this is not the effect to be achieved. The reagents are to be used in amounts sufficiently small that their presence cannot be detected by their taste. The exact nature of the reaction taking place is not fully understood but it was found that small amounts of sodium chloride and of alkaline materials, when added to a solution of saccharin cause the formation of an insoluble precipitate upon heating to a temperature at or near the boiling point of water. The precipitate may be separated from the remainder of the solution which is then found to have a taste better than the original solution of the artificial sweetening agent, and better than the mixture of the original solution with reagents.

The improvement is greatest when the solution is rapidly cooled from the boiling range to a temperature near room temperature or lower; and an easily filtered precipitate of crystalline appearance is produced by such chilling. The precipitate has a distinct bitter taste. The clear filtrate when evaporated at ambient pressure or in a vacuum to a concentration greater than the solubility of the sweetening agent at room temperature yields crystals of the sweetening agent which are not visibly different from the starting material, but which yield aqueous solutions of a taste much closer to that of sucrose solutions. The sweetening strength of the treated material is unaffected.

It is difficult to separate the bitterness-forming precipitate from a hot aqueous solution immediately after its formation. The boiling solution first becomes turbid and milky, and the precipitated particles appear to be extremely small. They grow with time, and particularly upon cooling. Rapid chilling not only improves the filtration behavior of the precipitate but also appears to cause more complete separation of the bitter constituent from the liquid phase. When cooling is slow, the improvement achieved is not as great as that available with cooling at a high rate.

Example 1: Soluble saccharin (saccharin sodium) is dissolved in water; 0.1 part sodium chloride and 0.02 part sodium bicarbonate were then added with stirring, and the clear solution produced was heated to a temperature at or near the boiling point for 20 minutes. It was then cooled to room temperature within a few minutes by immersion of the container in running cold water.

The precipitate formed was removed from the liquid phase by filtration and the filtrate was evaporated until crystals formed. More than 90% of the original weight of the sweetening agents was recovered. The formed crystals yielded solutions free from the slightly bitter taste perceptible in the original solution both before and after the addition of the sodium chloride and the sodium bicarbonate.

The combined amount of chloride and bicarbonate should not normally exceed 10% of the amount of the sweetening agent if it is not desired that the reagents significantly affect the taste of the product. The quantity of the reagents added to the sweetening agent may be reduced below the quantities indicated with a corresponding decrease in effectiveness. Since the sensitivity to bitterness varies greatly from one person to another, a generally valid lower limit of effectiveness of the reagent mixture cannot be established.

The concentration of the sweetening agent in the treated original solution should be sufficiently low to prevent precipitation of the desirable material with the bitter component upon heating and chilling. The solubilities of saccharin and soluble saccharin are well known. Sodium bicarbonate may be replaced by other weakly alkaline materials, more specifically by the sodium salt of another physiologically tolerated weak acid, as illustrated by the following example.

Example 2: Soluble saccharin was dissolved at room temperature in sufficient water to make an almost saturated solution, i.e., about 20 to 35% water. 0.4 part sodium chloride and 0.7 part sodium glutamate were added, and the resulting solution was heated to boiling for about 30 minutes. The container holding the hot solution was then plunged into ice-water, and the temperature of the solution was lowered with vigorous agitation to 5°C within less than 2 minutes.

A crystalline precipitate formed, and the remainder of the solution became turbid. The cold solution was filtered and the clear filtrate had lost the bitter component noticeable prior to boiling. Crystallization of the sweetening agents from the partly evaporated filtrate yielded a solid product which was as effective a sweetening agent as the original mixture, but lacked bitterness.

Example 3: One part soluble saccharin (saccharin sodium), 0.24 part sodium chloride and 0.004 part sodium bicarbonate were dissolved in enough water to make a solution almost saturated at room temperature, i.e., about 20 to 35% water. The solution was boiled 20 minutes, and then quickly cooled to room temperature by tap water externally applied to the container. Filtration of the cooled solution yielded a clear filtrate free from the bitter taste noticeable prior to boiling.

No unfavorable effects resulted from an extension of boiling time to 60 minutes, regardless of the concentration of sweetening agents or reagents. While the rate of cooling had a perceptible effect on the elimination of the bitter component, the lowest temperature reached upon rapid cooling was not found to have a significant effect if the minimum temperature was at or below room temperature (approximately 0° to 30°C).

CALCIUM CHLORIDE-STARCH HYDROLYSATE COMBINATIONS

It has been found by *J.B. Bliznak; U.S. Patent 3,773,526; November 20, 1973; assigned to Alberto-Culver Company* that the bitter aftertaste of water-soluble saccharin salts is reduced by mixing therewith 0.5 to 2 parts calcium chloride per 1 part saccharin salt. If the water-soluble saccharin salts and the calcium chloride are incorporated into a major proportion of a starch hydrolysate having a low DE and the mixture dried, as by spray drying, the overcoming of the aftertaste is enhanced.

While various starches can be used to make the starting starch hydrolysate, as a practical proposition corn and sorghum starches are most desirable, particularly waxy maize and waxy milo starches. The starch hydrolysate solids content of the aqueous starch hydrolysate solution can be varied but, in general, it is desirably in the range of about 55 to 58%, by weight. Lower solids concentrations produced finished dried sweeteners having lower bulk densities than those obtained with the higher solids concentrations.

The mixture of the starch hydrolysate solution, the water-soluble saccharin salt and the calcium chloride, in an aqueous solution, are fed into the preheater of the spray drier, and then fed into the spray drier.

In those cases in which a starch hydrolysate is incorporated with the water-soluble saccharin salt and the calcium chloride, the starch hydrolysate constitutes a distinctly major proportion, and the saccharin salt and calcium chloride constitute a distinctly minor proportion of the total solids of the final sweetener composition. The proportions of the saccharin salt and calcium chloride utilized, although always distinctly minor in the starch hydrolysate sweetener compositions, are variable, depending upon the degree of sweetness desired in the final sweetener composition. The following are examples of such compositions. All parts stated are by weight.

Examples 1 through 6:

Example	Calcium Saccharin	Calcium Chloride	Milo Starch Hydrolysate*	Cornstarch Hydrolysate**
1	10	5	-	-
2	10	20	-	-
3	10	15	-	-
4	4.5	3.3	92.2	-
5	4	2.5	-	93.5
6	5	4	91	-

*DE, 12 to 13
**DE, 5 to 6

SACCHARIN AND ADIPIC ACID

E. Polya; U.S. Patent 2,971,848; February 14, 1961; assigned to General Foods Corporation reports that when adipic acid is used in combination with artificial sweeteners to produce a high sweetness level and tart character in food products, there is no objectionable bitter or medicinal character noticeable in the product. It appears adipic acid, by virtue of its organoleptic properties, is capable of masking any bitter or medicinal aftertaste caused by such artificial sweeteners and in addition enhances the sweetness effect accomplished by using them.

In the case of gelatin jelly desserts the virtues of adipic acid in combination with saccharin are highly significant. In general, gelatin jelly dessert products which are fruit-flavored, should have a very sweet as well as a tart taste. Tartness is usually imparted by such edible food acids as citric acid.

It has been found that when adipic acid is used in combination with the artificial sweetener, saccharin, the sweetness contribution of the adipic acid in combination with the saccharin offers an improved fruit-like tart flavor of increased sweetness and no bitter or medicinal aftertaste is noticed.

SACCHARIN AND MANNITOL-GUM

Mannitol and a hydrophilic colloid or thickening agent is used by *H.W. Block, J.M. Adams and T.P. Finucane; U.S. Patent 3,476,571; November 4, 1969; assigned to General Foods Corporation* for an artificially sweetened fruit-flavored beverage mix.

The use of mannitol and the hydrophilic colloid blends the taste impact of the saccharin to give a more acceptable sweetening effect while at the same time, delaying the dispersion or impact of the artificial sweeteners on the taste buds. In a formulation having fruit flavors, fruit color and a suitable food acid such as adipic, citric, malic, tartaric or fumaric acid, the sweetener appears to also enhance the desirable tartness of the reconstituted formulation. The tartness impact is particularly evident when the sweetener is combined with adipic acid as the acidulent.

The ingredients may be merely dry-blended with any conventional fruit-flavored formulation. The usual amount of mannitol added will be 8 to 20 parts by weight, for each 1 part of the total weight of the artificial sweetener, and the edible thickener will range in content depending on the thickening effect desired.

However, in beverage formulations the level of thickener will usually be 0.8 to 1.2 parts, by weight, for each 1 part of the total weight of the artificial sweeteners. At these ratios the sweetener may be substituted for sugar while obtaining an improved flowability and a desirable organoleptic impact.

The conventional fruit-flavored product may contain any of the ingredients compatible with a fruit-flavored dessert or beverage such as buffer salts, clouding agents, anticaking agents, carbonating agents, acidifying salts, flavors and color.

When the fruit-flavored product is packaged in a container adapted to provide multiservings of the reconstituted product, and the product does not include a high level of a bland hydrophilic colloid or edible thickener as gelatin or pectin, it may be desirable to avoid separation of the sweetener from the remaining components and thereby obtain uniformity in product quality at all servings of the product by granulating the artificial sweeteners together with the mannitol and the bland edible thickener.

This may be done by known techniques such as mixing the artificial sweetener, that is, saccharin, mannitol and edible thickener with acetone or other volatile water-soluble solvent and water, the acetone and water being blended in equal parts. About 10 to 30 parts by weight of the water-solvent are used for each 100 parts by weight of dry ingredients. The mixture is mixed until homogeneous and then dried and granulated.

Alternatively, the dry ingredients may merely be wet by spraying a small amount of water (5 to 10%) onto the dry ingredients to thereby form moistened granules which are then dried under mild drying conditions, e.g., 70° to 120°F. In still another method, the ingredients may be dissolved in excess solvent and/or water, dried and then granulated. The granules of artificial sweetener, mannitol and edible thickener can then be incorporated in any fruit-flavored dry mix to obtain an easily flowing mixture of improved appearance which does not separate into a nonuniform distribution of ingredients during subsequent storage.

Aside from the organoleptic properties of the mannitol and bland hydrophilic colloid or edible thickener in masking the bitter or medicinal aftertaste of the artificial sweetener, the nonhydgroscopic character of the mannitol serves to enhance the stability of all the ingredients and inhibits the tendency of the formulation to cake, brown, or degrade due to moisture pick-up from the atmosphere. The relatively large bulk of the mannitol used in the composition enables the formulation to retain stable character despite the use of some hygroscopic ingredients.

Example 1: The sweetener was prepared by dry blending the following ingredients.

Ingredients	Parts by Weight
Mannitol	10.0
Gum arabic	1.1
Sodium saccharin	0.4

The sweetener was then used in a low-calorie gelatin dessert product of the following formulation.

Ingredients	Parts by Weight
Sweetener	39.6
Gelatin	51.2
Citric acid	6.0
Trisodium citrate	1.4
Lemon flavor	1.0
Color	0.8

About 6 to 7 grams of the above dry mix was then stirred into 237 ml (1 cup) of hot (180°F) water until a clear solution was obtained. The gelatin solution was then allowed to gel and when tasted had a sweet, tart taste with no residual bitter or medicinal aftertaste being discernible in the mouth after ingestion. A control sample containing saccharin at the same levels as above but without the mannitol and gum arabic gave a distinctly bitter, medicinal and metallic aftertaste when the sample was consumed.

Example 2: The sweetener of Example 1 had carboxymethylcellulose substituted for the gum arabic and was blended well and then sprayed with water to give a granular product having a moisture content of 5 to 10%. The moistened granules were then dried in air at a temperature of 110°F in several minutes. The sweetener was then used in a low-calorie beverage formulation as follows.

Ingredients	Parts by Weight
Sweetener	12.0
Citric acid	11.2
Trisodium citrate	2.4
Tricalcium phosphate	0.8
Orange flavor	0.1
Color	0.1

A teaspoon of the above beverage formulation was stirred into 6 ounces of chilled water until a clear solution was obtained. The beverage when consumed gave a desirable sweetness and tart orange flavor with no discernible offtaste. The beverage left no residual medicinal or bitter aftertaste on ingestion when compared to a control sample not having the mannitol-carboxymethylcellulose modifiers.

The dry beverage formulation had a desirable bulk density (about one-half that of a naturally sweetened product), and a flowability and appearance similar to a naturally sweetened formulation. The product was stable when stored in moisture pervious envelopes at an atmospheric temperature of 80°F and a relative humidity of 40% over a 3- to 4-month period.

Examples 3 and 4: The procedures of Examples 1 and 2 were followed with the exception that adipic acid was used in the formulations instead of citric acid. It was noted on ingestion that the taste of the samples containing adipic acid as the acidulent gave an enhanced tartness and sweetness above that of the Example 1 and 2 samples.

Example 5: The procedure of Example 2 was followed with the exception that about 1 part of cornstarch was used as a clouding agent. The reconstituted product of this example had the same desirable properties of the Example 2 formulation with the exception that the beverage resembled the appearance of orange juice more closely.

In addition to the above example, the sweetener was used in other fruit-flavored beverage and dessert formulations including cherry, grape, strawberry and raspberry with the same desirable results, i.e., increased sweetness and tartness with no residual medicinal aftertaste in the reconstituted product as well as increased flowability, stability and desirable appearance in the dry form.

SACCHARIN AND PECTIN

C.A. Weast; U.S. Patent 2,536,970; January 2, 1951 discusses the fact that pectin inhibits the characteristic bitter taste of the saccharin without affecting the sweet taste. The conjoint incorporation of pectin and saccharin in foodstuffs (a) minimizes or eliminates the characteristic by-taste which some people encounter from saccharin; (b) permits the use of saccharin by most individuals in concentrations greater than about 0.01% and as high as 0.05% without unpleasant aftertaste; and (c) stabilizes the artificially sweetened food so that it does not become bitter upon cooking or standing.

In canned fruits, flavor stabilization is apparent when the pectin content of the canned fruit is 15 times the saccharin. Generally speaking, the best flavor stabilization is obtained when the ratio of pectin to saccharin in the finished canned fruit is on the order of 30:1 to 40:1; but edible and desirable formulations may be obtained with higher ratios, such as 50:1.

Example 1: Canned Apricots — Mature apricots are pitted and halved, graded for size, washed and sorted, in accordance with standard commercial practice. The fruit is filled into cans, an average of 19¾ oz avoir being placed in each No. 2½ size can. The can is then filled with liquid medium, which requires 10 to 12 fluid ounces, depending upon the size of the fruit. A suitable liquid medium is made by dissolving 15 pounds of 150 grade citrus pectin and 13½ oz saccharin in water to make 100 gallons. This gives a liquid medium analyzing 1.8% pectin and 0.1% saccharin.

The filled cans are exhausted by heating in accordance with conventional canning practice until a closing center temperature of 160° to 170°F is attained; the container is then hermetically sealed, and processed by heating typically for 20 to 30 minutes at 212°F in a continuous agitating cooker.

The finished container will show from 10 to 15 inches vacuum at room temperature, and the fruit possesses the color and texture typical of ordinary commercial canned apricots. The product corresponds in sweetness to ordinary canned apricots packed in light syrup, but the viscosity of the liquid is more than twice that of the liquid in ordinary apricots canned in heavy syrup. To persons of normal taste response, the product is free of bitter by-taste and bitter aftertaste, and remains so even when stored in the sealed container for periods of many months. A typical run of dietetic canned apricots prepared as above described analyzed as follows.

TABLE 1: ANALYSIS OF DIETETIC CANNED APRICOTS

Pectin	1.3%
Saccharin	0.034%
Carbohydrate	8.6%
Protein	0.66%
Fiber	0.4%
Ash	0.4%
Water	88.5%
Calories per 100 grams	35
Calories per pound	160

It is to be noted that the pectin and saccharin content of the finished product is quite different from that of the liquid medium in which the raw fruit was packed, because during processing the saccharin and pectin become distributed through both the solid and the liquid portions of the canned fruit. Of the 1.3% total pectin shown by analysis in the finished product, 0.6% is supplied by the pectin added with the liquid medium, and 0.7% represents pectin naturally present in the apricots.

Example 2: Dietetic Canned Peaches – Peeled sliced peaches are filled into cans, one of the improved sweetening tablets is added thereto, the can is completely filled with water, and then exhausted, heat processed and sealed in a manner similar to that described in conjunction with Example 1.

The tablet which is added may be made in an ordinary tablet press. One tablet containing 0.2 avoir ounce pectin and 0.011 avoir ounce saccharin is sufficient to sweeten one No. 2½ can of peaches. The ratio of pectin to saccharin in the tablets may be varied between 15:1 and 50:1, and the weight of the tablet predetermined to properly sweeten any preselected unit quantity of food.

Example 3: Dietetic Canned Kadota Figs – A suitable liquid medium is prepared by dissolving 15 pounds of 150 grade pectin and 12 ounces of soluble saccharin in enough water to make 100 gallons. If desired, 1 gallon of lemon juice may be added for flavoring purposes. This liquid medium analyzes 1.8% pectin and 0.09% saccharin.

The figs are graded, sorted, washed and blanched. The fruit is filled into cans, the average filling weight amounting to 18¼ oz avoir per No 2½ can. The remainder of the can is then filled with the liquid medium and the cans are exhausted, heated and sealed in the manner described in Example 1, except that the heating is continued for 30 to 40 minutes at 212°F.

The dietetic product thus obtained corresponds in sweetness to figs packed in light syrup, but the liquid portion is far more viscous, the viscosity being more than twice that of the liquid obtained from ordinary canned figs in heavy syrup. To persons of normal taste response, the product is free of bitter by-taste and bitter aftertaste, and remains so even when stored in the sealed container for periods of many months.

Of the 1.3% total pectin shown by analysis in the finished product, 0.76% is supplied by the pectin added with the liquid medium, and 0.54% represents pectin naturally present in figs.

SACCHARIN AND MALTOL

Maltol, also known as 2-methyl-3-hydroxy-4-pyrone, is a gamma-pyrone of rapidly increasing acceptance for enhancement of odors and flavors of many products. For example, maltol has been found to enhance the apparent sweetness of natural sugars. Thus, part of the natural sugar in many products may be replaced with relatively very much smaller amounts of maltol.

In marked contrast, it has been found that maltol does not exhibit a sweetness-enhancing effect with nonnutritive sweetening agents as is observed with natural sugars such as sucrose, dextrose and glucose. Furthermore, while not enhancing the sweetness of nonnutritive sweetening agents, maltol has been found to mask the bitter and metallic aftertaste commonly associated with the use of saccharin.

In addition to aftertastes associated with the use of saccharin and its salts, it is noted that such aftertastes are observed following the use of other nonnutritive sweetening agents such as arabitol, also known as 1,2,3,4,5-pentanepentol; xylitol, an isomer of arabitol and stevioside, an isolate of the plant, *Stevia rebaudiana,* and many others.

The details on the use of maltol for this improvement in the flavor quality of nonnutritive sweetening is given by *J.M. Griffin; U.S. Patent 3,296,079; January 3, 1967; assigned to Chas. Pfizer & Co., Inc.*

The effective amount of maltol required to mask the bitter aftertaste can be computed, based on the weight of the product sweetened with nonnutritive sweetening agent; generally speaking, maltol is found effectively to mask the aftertaste when it is present in an amount to provide from 5 to 400 ppm based on the sweetened product. The aftertaste-masking amount of maltol necessary may be found by adjusting the concentration within the stated ranges. The proper amount will depend on the nature of the product and the amount of sweetener used.

Higher amounts of maltol are required if a particularly bitter ingredient such as stannous fluoride, commonly used in toothpastes, is to be masked with the sweetening agent. Lesser amounts of maltol are required if only small amounts of nonnutritive sweetening agents are added to products, such as canned fruits, which contain natural sugars. Generally, 30 ppm of maltol has been found effective to mask the bitter flavor of a commercial sweetening preparation, for table use, which contains an artificial sweetener; this represents an amount of maltol to provide about 0.05% by weight based on the sweetener.

Dietetic syrup such as cherry-cola, grape, orange and raspberry, sweetened with nonnutritive sweetening agents are freed of aftertaste by adding 25 to 50 ppm of maltol; since these syrups typically contain about 0.4% by weight of sweeteners, this represents an amount of maltol sufficient to provide from 0.62 to 1.25% by weight based on sweetener. Low-calorie carbonated beverages are freed of aftertaste by adding from 5 to 30 ppm of maltol; since these ordinarily contain 0.25% by weight of sweeteners, this represents an amount of maltol sufficient to provide from 0.2 to 1.2% by weight based on the sweetener.

In addition to maltol, an amount of from 0.2 to 15 parts based on each part of the sweetening agent, of a polyhydric alcohol such as propylene glycol, glycerol or sorbitol is added. Products with very desirable mildness and freedom from aftertaste and bitterness are thus obtained by combining nonnutritive sweetening agents, polyhydric alcohols and maltol; these are especially useful for table use when it is desired to sweeten, for example, coffee and tea or grapefruit.

Example 1: Sodium saccharin is added to water in an amount to provide 0.1% by weight. The solution has a disagreeable metallic aftertaste. Maltol is added in increments to provide 5, 10, 20, 30, 40, 100, 200, 300 and 400 ppm, respectively. The solution is tasted after each addition and it is found to be free of aftertaste when it contains between 5 and 400 ppm of maltol. Maltol has thus been added effectively in an amount to provide from 0.5 to 4% by weight based on saccharin.

The procedure is repeated substituting a 0.166% aqueous solution of stevioside, a nonnutritive sweetening agent derived from a plant, for the 0.1% saccharin solution. Stevioside is about 300 times as sweet as sucrose. The aftertaste of this agent is masked by the addition of maltol in an amount to provide from 5 to 400 ppm by weight based on the aqueous solution; this is an amount of maltol sufficient to provide from 0.3 to 2.4% by weight based on the sweetener.

The procedure is repeated substituting for the 0.1% solution of saccharin, respectively, the following aqueous solutions: 0.025% perillartine, also known as 1-perillaldehyde-α-antioxime, which is 2,000 times as sweet as sugar, and 15% solutions of xylitol and of arabitol. In all cases the aftertastes of these nonnutritive sweetening agents are masked by from 5 to 400 ppm of maltol. The amount of maltol used with perillartine is from 2 to 160% by weight based on the sweetener; the amount of maltol used with xylitol and arabitol is from 0.003 to 0.26% by weight based on the sweetener.

Example 2: A raspberry-flavored dietetic syrup simulating a 40% sugar-containing syrup, but containing no sugar, is formulated with soluble saccharin, carboxymethylcellulose, raspberry flavoring and water to make a total volume of 100 ml.

The syrup is characterized by a bitter and metallic aftertaste. Maltol in an amount to provide 25 ppm, based on the syrup masks the aftertaste. Similarly, cherry-cola, grape and orange-flavored syrups are prepared. The aftertaste is effectively masked with 25 and 50 ppm of maltol in the cherry-cola flavored syrup, with 50 ppm of maltol in the grape-flavored syrup and with 50 ppm of maltol in the orange-flavored syrup.

SACCHARIN AND PIPERAZINE

Piperazine hydrate is used to reduce the aftertaste of saccharin and also to enhance the solubility of saccharin according to *C.A. Long; British Patent 1,104,251; December 22, 1965; assigned to The Sanitas Company Ltd., England.*

The basic organic compound is piperazine, but other bases, for example, monoethanolamine, diethanolamine, lysine or other amino acids, are suitable. The sweetening composition may be prepared in any one of a variety of forms, for example, in aqueous solution or dissolved in a mixture of water and a di- or polyhydric alcohol, e.g., propylene glycol or glycerol, in tablet form or in granular form. When tablets are required, conventional lubricants and pelleting aids, for example stearic acid, may be incorporated.

The basic organic compound may be introduced into the saccharin in any conventional manner; direct mixing of the two components is in many cases all that is required. It may be advisable, particularly,in applications where inorganic metallic ions are present, to incorporate within the sweetening composition a sequestering agent, whereby the offtaste may be retained at a minimal level. A preferred sequestering agent is ethylenediaminetetraacetic acid (EDTA).

Example 1: This composition gave a sweet taste with no offtaste subsequently.

Saccharin	60 mg
Piperazine hydrate ($6H_2O$)	65 mg
EDTA	60 mg
Distilled water to	120 ml

Example 2: This composition gave a warmer sweet taste than Example 1 with no offtaste subsequently.

Saccharin	60 mg
Piperazine hydrate ($6H_2O$)	120 mg
EDTA	60 mg
Distilled water to	120 ml

Example 3:

Saccharin	60 mg
Monoethanolamine	40 mg
EDTA	60 mg
Distilled water to	120 ml

A similar composition was prepared in which the EDTA was omitted. Both compositions were successful in attaining sweetness without residual offtaste.

Example 4: This composition again gave good sweetness without offtaste.

Saccharin	60 mg
Monoethanolamine	40 mg
Lysine hydrochloride	60 mg
Distilled water to	120 ml

Example 5:

Saccharin	360 mg
Diethanolamine	210 mg

The two components were melted together and solidified. Good sweetness was attained after an interval, possibly due to a time lag wherein the components of the saccharin composition in solid form are selectively dissolved.

Example 6:

Saccharin	60 mg
Piperazine hydrate ($6H_2O$)	65 mg
EDTA	60 mg

The components were melted together, and on solidification yielded granules which could be pressed into tablet form.

Example 7:

Saccharin	120 mg
Piperazine hydrate ($6H_2O$)	65 mg
Distilled water to	240 ml

Example 8: A preferred, commercial composition contains saccharin, piperazine hydrate and EDTA in solution in a mixture of water and glycerol or propylene glycol in the following proportions.

Saccharin	3.25 g
Piperazine hydrate	6.5 g
EDTA	4.5 g
Glycerol or propylene glycol	50 ml
Distilled water to	100 ml

SACCHARIN AND TRYPTOPHANE

A combination of tryptophane and saccharin is prepared by *T. Nonomiya, T. Ojima, S. Yamaguchi and M. Ito; U.S. Patent 3,717,477; February 20, 1973; assigned to Ajinomoto Company, Inc., Japan*. The intensity of sweetness is remarkably strengthened because of a synergistic effect, and at the same time the quality of taste shows a very significant improvement with the elimination of the unpleasant aftertaste.

Experimental results on the synergistic effect and quality improvement mentioned above are reported below. The panel used in the experiments consisted of 100 persons, both male and female, who were screened from approximately 1,000 people based on the sensitivity of their taste.

Example 1: The potency of an aqueous solution of each individual sweetener was measured in comparison with sucrose, by repeatedly obtaining the points of subjective equality between the sweeteners and sucrose with the participation of 25 to 50 panel members. In the example, a sweetener solution was paired with several concentrations (5 to 7 grades) of sucrose solution and pairs of samples were presented at random. Then, the panelists were asked to indicate which one in each pair had a stronger sweetness. The data obtained were analyzed by the probit method and the sweetness of the test sample was represented by the concentration of sucrose required to attain the sweetness equivalent to the test sample.

TABLE 1: THE POTENCY OF SWEETNESS OF SACCHARIN AND TRYPTOPHANE

Sweet Substance	Equivalent Concentration (sweeteners and sucrose, g/dl)				
Sodium saccharin	0.005	0.01	0.02	0.04	0.08
Sucrose	2.2	3.5	5.5	7.4	9.2
D-tryptophane	0.05	0.1	0.2	-	-
Sucrose	3.0	5.1	8.5	-	-
DL-tryptophane	0.05	0.1	0.2	-	-
Sucrose	1.5	2.8	4.9	-	-

In a comparison with sucrose the sweetening potency of both saccharin and tryptophane falls as their concentration increases. On the other hand the bitterness increases when the concentration becomes higher. The potency of DL-tryptophane is about one-half that of D-tryptophane.

Example 2: Test solutions containing different concentrations of sodium saccharin, DL-tryptophane or D-tryptophane as shown in Table 2 were prepared, and the degree of bitterness or unfavorable taste of each test solution was checked by an organoleptic test by 15 panel members. The organoleptic test was performed by requesting the panel to point out the degree of bitterness shown hereunder (ranking method). A (no bitter taste), 0; B (slightly bitter taste), 1; C (bitter taste), 2; D (strongly bitter taste), 3.

TABLE 2: POTENCY OF BITTERNESS OR UNFAVORABLE TASTE OF SODIUM SACCHARIN AND D- OR DL-TRYPTOPHANE

	Sample								
	Sodium Saccharin				DL-Tryptophane			D-Tryptophane	
Concentration (g/dl)	0.01	0.02	0.04	0.08	0.05	0.1	0.2	0.05	0.1
Judgement:									
A	8	6	1	1	9	6	2	10	7
B	5	8	7	3	3	4	3	4	4
C	2	1	5	5	2	3	4	0	2
D	0	0	2	6	1	2	6	1	2
Average mark	0.60	0.67	1.53	2.07	0.67	1.07	1.93	0.47	0.93

SACCHARIN AND OXAZOLIDINONES

W.E. Walles; U.S. Patents 3,449,339; June 10, 1969 and 3,642,704; February 15, 1972; both assigned to The Dow Chemical Company describes a mixture of saccharin and substituted oxazolidinone compounds to produce a complex of improved properties as a sweetening agent.

Complex is used to designate a substance composed of separate molecules bonded together by forces not so strong as conventional covalent bonds but stronger than hydrogen bonding, and strong enough that the complexes are stable under ordinary conditions of preparation and storage.

Example 1: Complex of Saccharin with a Copolymer of 5-Methyl-3-Vinyl-Oxazolidin-2-one and Styrene – A copolymer is made from 15 molar proportions of styrene and 85 molar proportions of 5-methyl-3-vinyloxazolidin-2-one. The complectant is a white, finely divided, granular material readily soluble in ethanol. The average molecular weight of the polymer for each cyclic carbamate unit present is 123.6.

A dry mixture containing 124 grams of the copolymer of 5-methyl-3-vinyloxazolidin-2-one with styrene and 103 grams (½ mol) saccharin is added with mixing and stirring to 500 milliliters of warm 95% ethanol. These solids promptly dissolve and disappear in the ethanol with the resulting formation of the desired complex. Formation of the complex is indicated by an ultraviolet fluorescence spectrum different from the combined fluorescence spectra of the starting materials.

Example 2: Artificially Sweetened Alcoholic Cordial – A flavoring essence derived primarily from seeds, stems, and roots of umbelliferous plants, principally anise (*Pimpinella anisum*) and dill (*Anethum graveolens*) together with minor amounts of oil of coriander (*Coriandrum sativum*) and stem and root of angelica (*Angelica archangelica*) is combined with a syrup containing a sweetening syrup of which the sweet flavor is derived from a complex of saccharin and polymeric 3-vinyl-5-methyloxazolidin-2-one, slightly thickened with gum acacia.

The resultant intensely sweet highly flavored concentrate syrup is diluted with alcohol and water and thereafter artificially colored by the addition of small amounts of soluble chlorophyll to obtain an artificially sweetened, sugar-free herb-flavored alcoholic cordial. The water and alcohol are added in such amounts as to provide a cordial containing alcohol in the amount of approximately 30% by weight of alcohol-water mixture.

METHODS FOR INCREASING BULK OF MIXES

Since saccharin is approximately 300 times sweeter than sucrose, there is the problem of handling small amounts in typical uses such as household replacement for sugar. In this chapter the most important methods are presented for increasing the bulk or particle size of powders containing saccharin. The processes mainly involve agglomeration and spray-drying. More complete information on agglomeration can be found in another publication, *Agglomeration Processes in Food Manufacture,* Noyes Data Corporation, Park Ridge, N.J. (1972).

PILLSBURY LOW CALORIE DRINK MIX

Basic Process

Conventional fruit flavored beverage mixes are unagglomerated mechanical mixtures which contain an edible acid, a sweetener, flavoring (either natural or artificial) and coloring and are usually dissolved directly in cold water. Some of these mixes utilize natural sugars such as sucrose or dextrose or some combination as the sweetening agent, whereas other mixes use a synthetic noncaloric sweetener to provide a low calorie dietetic type mix and beverage.

Conventional unagglomerated mixtures of this type are not readily wettable and dispersible when added to a liquid, particularly a cold liquid, and tend to lump up in difficult to disperse and wet masses when attempting to form the desired beverage therefrom and usually require vigorous stirring to get them into solution. These products tend to cake and lump up in storage, particularly in humid climates, and have poor flow characteristics. These conventional mixes also require the use of a nonfunctional inert bulking agent or diluent such as mannitol which diluent usually comprises approximately half of the product. This diluent is expensive and materially increases the cost of the product, and in the case of a low calorie type mix, increases the caloric content thereof since there is no known noncaloric bulking agent suitable for use in this type of mix.

In the conventional dry mix, the color of the mix is very pale compared with the natural color of the fruit or its flavoring and the color of the beverage when reconstituted in a liquid, the coloring of the dry mix being so pale as to not be properly suggestive of the fruit with which the flavoring corresponds. Conventional mixes are difficult to measure with any degree of speed and accuracy, particularly when the mix is being packaged. As a result, it is difficult to obtain a uniform amount of product in each package, and the handling of the material is more difficult and expensive than is desirable. The conventional mixes are quite dusty with the usual undesirable results associated therewith. Since these mixes contain coloring, when the dust disperses and settles anywhere but in the container where it belongs, the dusty particles stain the object upon which they come to rest when they acquire moisture in any way.

An improved process and products is given by *W.L. Ganske, J.A. Stein, R.G. Gidlow and F.H. McCarron; U.S. Patent 3,433,644; March 18, 1969; assigned to The Pillsbury Company* which requires agglomeration of the mix.

The low calorie type drink mixes consist primarily of citric acid and a noncaloric artificial sweetener with minor quantities of flavoring and coloring added. Some of the acidity may be provided by other acidulants such as fumaric or tartaric acid. However, enough citric acid is provided to provide a sufficient amount of bonding agent for the product.

The flavoring may be of any type desired and is usually a fruit flavoring such as raspberry, orange, grape, lime, strawberry, cherry and the like. The coloring is synonymous with the type of flavoring in each particular instance, such as yellow for lemon, purple for grape, etc. In some instances, it is desirable to have a buffer such as sodium citrate present. If a buffer is used, less acidulant and more sweetener is used than when there is no buffer.

Broadly speaking, the ingredients of the beverage mix are reduced to fine pulverulent form (as by grinding) and thoroughly mixing these ingredients to provide a uniform homogeneous mixture. The mixing may be done before or after reduction to the desired size. However, if the flavoring is in the form of tiny capsules, these capsules are not subjected to any grinding or other reducing step to avoid breaking, and are mixed with the other ingredients when they are of the proper size. The majority of the particles are fine enough to be capable of passing through a 150 sieve. The particle size is controlled in part by the way in which they are moisturized during agglomeration and the kind of citric acid used, as described in more detail later.

Once the ingredients have been uniformly mixed, they are wetted with enough liquid (preferably water) to wet the citric acid particles to a degree sufficient to cause the surfaces to become tacky and adhesive. The liquid is a solvent which dissolves a portion of the particles on which it is deposited and forms an adhesive solution on the surface thereof which serves to bond the particles together in the form of porous agglomerates.

The wetted particles are then brought into random contact with one another whereby they stick together in clusters and form the desired porous agglomerated product, with the citric acid serving as the primary bonding agent in the low calorie mixes.

The most successful agglomeration of this low calorie beverage mix is accomplished by limiting the amount of moisture increase or the amount added to a maximum limit of about 3.5%. If the material is over-moisturized, too much goes into solution, making it difficult if not impossible to form porous agglomerates therefrom. Excess solution also tends to foul the agglomerating equipment and to fill the spaces between whatever particles remain, thereby reducing or destroying the porosity of the product. The increase in moisture content falls within a preferred range of 0.5 to 3.5%, with the optimum increase being in the range of 1.5 to 2.5%.

Thus, it is desirable to retain as much as possible the physical particulate entity of the citric acid during wetting and agglomeration. The moisturization is achieved while the particles are in a densely dispersed agitated condition such as is provided on the apparatus to be described, with the moisture and particles being thoroughly intermixed so that the particles of the mixture are uniformly moistened.

The moisturization is accomplished by condensation, as this method provides the best manner of controlling the degree and uniformity of moisturization. Best results are obtained by carefully controlling the condensation conditions, which can be expressed by the relation between the temperature of the dry material fed to the humid atmosphere of the agglomerator and the wet and dry bulb temperatures of the humid atmosphere from where the agglomerating liquid comes.

It is desirable to limit the feed temperature of the material to a maximum of about 120°F, with the temperature not higher than 100°F. The humid atmosphere has a dry bulb temperature not lower than 255°F, with a range of 255° to 275°F considered desirable. The wet bulb temperature of the humid air should not exceed 170°F, with a wet bulb temperature of 160° to 170°F being a desirable range when the feed temperature of the mixture is about 100°F. Lower wet bulb temperatures can be used with lower feed temperatures. Any combination of feed temperature and wet and dry bulb temperatures which will produce the same results and degree of moisturization as that produced by a feed temperature of 100°F, a dry bulb temperature of 255° to 275°F, and a wet bulb temperature of 160° to 170°F is considered to be within the scope of this process.

When the wet agglomerates have been formed, they are almost immediately subjected to a drying operation to remove all or a major portion of the added moisture to improve the agglomerated product and make it better capable of withstanding the rigors of subsequent handling. The starting material usually contains not more than 4% total moisture, and preferably only 2 to 3%, and the agglomerated product is dried back to within 0.5% plus or minus of the original moisture content.

One set of conditions involves dispersing the pulverulent particles to be agglomerated at a temperature of about 100°F in a humid atmosphere having a wet bulb temperature of 160° to 170°F and a dry bulb temperature of 255° to 275°F, condensing moisture on the particles and increasing the moisture content thereof by not more than 3.5% and within the range of 1.5 to 2.5%, the condensed moisture forming adhesive surfaces on the citric acid particles, bringing the wetted particles into random contact whereby they form porous agglomerates, and then drying the wet agglomerates to remove substantially all of the added moisture therefrom.

The process will be explained in greater detail with reference to the accompanying diagrammatic illustration of the apparatus used to carry out the method. The agglomerating apparatus illustrated in Figure 6.1 includes a feed hopper **10** which serves as a source of dry beverage mix starting material to be agglomerated, which material is of the requisite degree of fineness.

FIGURE 6.1: BASIC PROCESS FOR DRINK MIX

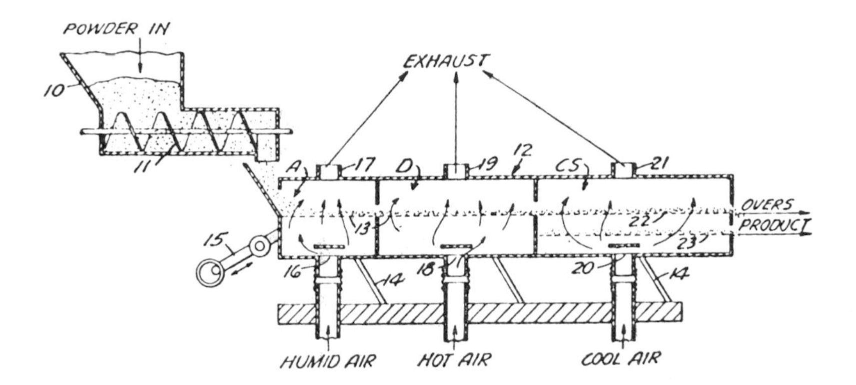

Source: U.S. Patent 3,433,644

The material to be agglomerated is drawn from the hopper **10** into the screw type conveyor **11** which continuously feeds the fine particulate material into the elongate agglomerator structure **12**. The agglomerator has a perforated or foraminous screen **13** extending longitudinally upon which the starting material is initially deposited and across which it travels from the inlet towards the discharge end and in the course of the travel is converted into the desired agglomerated end product.

The agglomerator, in the form shown, is divided into a longitudinal series of adjacent operational compartments successively labeled **A**, **D** and **CS** and in which the agglomerating, drying and cooling and sizing respectively take place. The agglomerator illustrated is also provided with means for transporting the bed of material through the agglomerator structure **12** to provide the generally forward motion of the bed of material passing therethrough, which vibrating means includes the supporting rocker arms **14** and the eccentric drive pitmans **15** and suitable drive means therefor.

As the material travels over the perforated screen **13**, it is maintained in a dispersed, agitated substantially fluidized condition by a stream of humid air or water vapor such as steam which is constantly injected into the lower portion of the first compartment or agglomerating section **A** through the port **16** and passes upwardly through the screen **13** and the bed of material at a velocity sufficient to maintain the particulate bed in a substantially fluidized state with a high particle-vapor ratio, and the particles in close, relatively dense, agitated proximity, and after passing through the bed of material is removed through a suitable discharge or exhaust port **17** provided on top of the agglomerator.

To avoid excessive fouling of the screen **13** by the wetted adhesive particles, the screen may be of the endless continually moving type which continuously leaves and returns to the agglomerating section A, with means provided for cleaning that portion of the screen removed from the agglomerating compartment between its removal and its reentry thereto, or the fluidizing vapor or gas may be superheated so that the particles adjacent to the screen absorb the superheat without condensation and without becoming sticky, while the particles in the upper strata of the bed of material in compartment A and out of contact with the screen have the moisture condensed thereon.

After the agglomeration of the material in the agglomerating compartment A due to the formation of adhesive surfaces on the particles by the condensation of the water vapor on the particles and the bringing together of the adhesive particles by the general agitation and mild turbulence of the bed provided by its fluidized nature and the vibration of the agglomerator, the agglomerated material passes to the next stage or compartment **D** in which the added moisture is removed and the material is dried by means of a current of warm air passing constantly through the screen **13** in the bed of agglomerates, the stream of hot air entering the bottom of compartment **D** through duct **18** and exhausting therefrom through the exhaust opening **19** in the roof. Drying air having a temperature of about 250°F is desirable. However, considerable latitude is permissible in the drying temperature.

The drying of the agglomerates strengthens and rigidifies them and removes the stickiness to enable them to be handled, packaged, stored and transported. After drying, the warm dried agglomerates pass to the next or final stage or compartment **CS** in which the agglomerates are optionally subjected to a stream of cooling air entering from duct **20** and exhausting through escape openings **21** in the roof to cool the agglomerates to room temperature and place them in a rigid state.

In the last compartment **CS** the agglomerates are also classified during cooling. The hot dried agglomerates pass from the screen **13** onto a coarser classifying screen **22**, the products falling through the screen **22** onto the finer collection screen **23**, the oversized material remaining on the screen **22**, the product and oversized material being discharged as indicated.

Thus, the material to be agglomerated is introduced into a warmer atmosphere of humid air or steam which is partially or nearly saturated with water vapor and therefore condensation occurs on the surface of the cooler particles. Solution occurs on the surface of the citric acid particles forming a sticky tacky adhesive surface thereon. Due to the agitation and proximity of the particles in the agglomerating chamber, the particles comprising the mixture stick to one another when making contact with each other to progressively build up soft agglomerates which are characterized by the proportionately large number of voids and interstices extending therethrough providing a liquid with ready access to substantially all of the particles forming the agglomerates to quickly disperse and dissolve them in a liquid. The agglomerates are then subjected to conditions which cause drying and the removal of the added moisture resulting in hardening or firming of the agglomerates.

The low calorie beverage agglomerates are very porous and characterized by the multitude of fine particles in each agglomerate which are randomly bonded together at their interfaces by the citric acid which functions as the bonding agent. These particles define a large number of voids or interstices which form a network of passages through the agglomerate to enable a liquid to penetrate quickly the agglomerate interior and rapidly reach all of the particles comprising the agglomerates. This action thus breaks the bond holding individual particles together and disperses them in the liquid for quick (virtually instant) dispersion and dissolution without any lumping into difficult-to-penetrate balls or masses in the liquid. The agglomerates are further characterized by being free-flowing and noncaking, even under severe atmospheric conditions involving significant changes in humidity. Thus, by grouping large numbers of these very fine pulverized particles together into larger porous agglomerates, the advantages of fine individual particle size for maximum rate of solution and of the larger size agglomerates which provide the free-flowing noncaking characteristics are obtained.

The agglomerated product has a bulk density which is lower than that of the starting material, is substantially dustless and the agglomerates are friable yet strong enough to resist breaking down and pulverizing or powdering when subjected to the rigors of handling, shipping and storing.

The lower bulk density of the agglomerated product is of particular advantage in this type of product. The dry mix is usually packaged in amounts sufficient to make one or two quarts of the desired beverage. Since only a very small amount of mix is needed to make one or two quarts of beverage (5 to 7 grams for 2 quarts), only a minute-sized package is actually needed to house the volume of mix. However, for convenience of handling, display, etc., a package considerably larger than the minimum required is actually used, resulting in a considerable amount of excess space in each package. The lower bulk density product enables this small amount of mix to occupy more space and thereby better fill these packets.

Also, the agglomerated form better assures complete emptying of the contents from the package. Portions of conventional unagglomerated powdered mix tend to remain in the corners of the package and get thrown away, rather than used, or at least make the complete emptying of the package more difficult.

The unagglomerated dry mixture from which the agglomerated product is formed is a pale color which is only vaguely reminiscent of the natural fruit which corresponds to the flavoring. This is because in the unagglomerated form, the coloring in the physical admixture of ingredients is of such a small amount that it is diluted and disguised by the much larger quantities of the other ingredients.

In contrast, the agglomerated product has a deep intense color which is much darker and deeper than the unagglomerated material. The color of the agglomerated material is very similar in color to and much more suggestive of the natural fruit of the flavoring of the mix and to the actual color that the drink will be when reconstituted in water than the unagglomerated material. This more uniform and intense coloring of the material takes place when the particles are wetted during the agglomeration operation. The wetting of the coloring and the intimate relationship of the agitated wetted particles causes the coloring to be spread uniformly over virtually all the particles comprising the mix so that the coloring is more uniformly distributed throughout the entire mixture. The agglomerated product is easy to measure at high speed and with a high degree of accuracy, which is particularly important during the packaging. It disperses and dissolves almost instantly when added to a liquid.

Mix with Fumaric Acid

The basic process for agglomeration to produce a cold-water-soluble low calorie fruit-flavored drink mix is given in the previous patent description (U.S. Patent 3,433,644). There are important advantages to the use of fumaric acid in combination with citric acid in the formula for the drink mix. *F.H. McCarron; U.S. Patent 3,506,453; April 14, 1970; assigned to The Pillsbury Company* gives the essential steps for treating fumaric acid in the mix to achieve cold water solubility.

The aggregates are prepared by forming a mixture of fumaric acid particles having a particle size of 75 μ and coated with between about 0.001 and 0.01 part by weight per part by weight fumaric acid of a dialkyl ester of a sodium sulfosuccinate, citric acid particles in an amount of 3 parts by weight and an artificial sweetener, and converting the mixture to the form of porous aggregate in which the artificial sweetener is homogenously dispersed.

The porous aggregates containing fumaric and citric acid particles are firmly bonded together in random fashion. The aggregates are inherently resistant to mechanical segregation into layers of particles of differing sizes and densities during storage and other handling. Moreover, the aggregates are free-flowing and substantially nonhygroscopic in nature. Upon reconstitution of the aggregates in an aqueous medium, the aggregates readily disperse to provide a homogenous beverage mix. Necessity for additional bonding agents which would increase the caloric content such as starch, sugars, corn syrup solids and dextrin gums is obviated by this process.

Since the fumaric acid particles, including the dialkyl esters of sodium sulfosuccinate fumaric acid particles, are substantially nonagglomerable, it was believed that the addition of the fumaric acid particles to citric acid particles would decrease the agglomerability in proportion to the amount of added fumaric acid particles. It was observed that a combination of citric acid and fumaric acid particles agglomerated more readily than when only citric acid was employed. In comparison to agglomerates where citric acid particles were used, the resultant combination of ingredients provided a superior agglomerate.

Suitable fumaric particles and the manner of preparing the particles are given in U.S. Patent 3,151,986. Typical dialkyl esters of sodium sulfosuccinate coatings include di-(2-ethylhexyl)-sodium sulfosuccinate, di-(n-dodecyl)-sodium sulfosuccinate, dimyristyl-sodium sulfosuccinate and di-(n-heptyl)-sodium sulfosuccinate. Average particle size of the fumaric acid particles ranges from 20 to 75 μ.

The amount of fumaric acid based upon 3 parts by weight citric acid ranges from 1 to 9 parts by weight. It has been found that greater amounts of fumaric acid are relatively unagglomerable with concomitant mechanical segregation of the particulate ingredients resulting therefrom. Lesser amounts of fumaric acid result in screen fouling during processing and greater hygroscopicity and segregation in the resultant product. From about 4 to 6 parts by weight fumaric acid particulates per 5 parts by weight citric acid will provide excellent agglomeration and aggregate characteristics.

All of the ingredients excepting the flavoring and fumaric acid were physically mixed and then ground to a fine pulverulent form in a Micro Pulverizer Model TH. The micro-pulverized particles, coated fumaric acid and flavoring agent were then blended in a ribbon mixer.

The anhydrous citric acid particle size was such that substantially all the citric acid particles were less than 80 μ with at least a major portion being less than 10 μ. Fumaric acid average particle size was within the range of 20 to 75 μ.

The blended ingredients were then agglomerated on a vibratory agglomerator. Humid air was passed through the agglomerating section at a rate of 200 feet vertical superficial velocity, this air having a wet bulb temperature of 165°F and a dry bulb temperature of 275°F.

The material was fed at 80°F into the agglomerator at a free feed rate of 300 pounds per hour. The wet agglomerates were dried in the drying section with drying air having a dry bulb temperature of 250°F, which air was passed through the drying section at a rate of 100 feet vertical superficial velocity.

The agglomerated mixes had a bulk density of about 18 to 22 pounds per cubic foot. In contrast, the powdered starting material from which the agglomerated product was formed has a bulk density of about 37 pounds per cubic foot. The same material in conventional unagglomerated form which includes a bulking agent (mannitol) has a bulk density of about 34 pounds per cubic foot. Thus, it will be appreciated that the agglomerated product has bulk volume almost twice that of the conventional unagglomerated mix.

The agglomerated low calorie mixes dissolve in a matter of a few seconds (virtually instantly) upon addition to water, even cold water, without requiring any mixing to obtain the desired dispersion and dissolution. In contrast, the conventional unagglomerated mixes require vigorous stirring to break up the lumps and masses which naturally form and get the material into solution. Even with vigorous stirring, it usually takes a minute or more to dissolve all the conventional material.

The resulting aggregates were firm and resistant toward mechanical segregation when subjected to a testing via a conventional jolting apparatus. Hygroscopicity of the resulting aggregates was less than that of aggregate compositions containing citric acid as the sole acid ingredient.

PARTIALLY SYNTHETIC SWEETENERS

Foam System with Sucrose

M. Pader and J.J. Miles, Jr.; U.S. Patent 3,170,800; February 23, 1965; assigned to Lever Brothers Company state that a sugar substitute can be prepared by dissolving sucrose, saccharin and a whipping agent, whipping the solution to provide a foam, drying the foam and subdividing the dried foam to obtain a material having a particle size similar to that of granulated sucrose. In order to insure complete solution of this product, it has also been found advantageous to coat the surfaces of the particles with an edible surface-active agent.

The product obtained has a bulk density of 0.2 to 0.4 g/cc, and on a volume basis a caloric value of approximately 20 to 40% of household granulated sugar. The material can be used on an equal volume basis for sweetening coffee, tea, grapefruit, etc., and the material has the general appearance of household granulated sugar.

The artificial sweetening agent is used at a level sufficient to bring the sweetening power of the composition to the sweetening power of an identical volume of sucrose. The combination of sucrose with saccharin reduces the bitterness associated with the use of nonnutritive sweetener, a marked advantage over other low calorie sweetening products containing bulk extenders to allow the use of high volumes. Thus, a minimum of bitterness was detected when the product was tasted directly, whereas a pronounced bitterness was apparent when a mixture of saccharin and nonsucrose materials (e.g., 24 Dextrose Equivalent corn syrup) containing the same weight percentage of saccharin was tasted.

The whipping agent is a partially hydrolyzed casein material. Other materials, such as partially hydrolyzed soy protein, gelatin and various gums, such as gum arabic, guar gum and alginates may be used. The important consideration with regard to the selection of an appropriate whipping agent is the ability of the agent to maintain the porous structure of the foam during drying.

While gums and caseinates are functionally suitable for the whipping operation, they are less suitable overall to partially hydrolyzed casein. For example, 1 to 2% gelatin in conjunction with gums yields a satisfactory foam, but the final product is not as readily soluble as one made with a casein hydrolysate. Sodium caseinate (0.3 to 2%) also is satisfactory for whipping, particularly in the presence of 0.1 to 0.5% algin, but the final product leaves a substantial surface foam when it is dissolved, necessitating the use of relatively large amounts of edible surface-active agents. Foams stabilized with algins alone may tend to have excessively large gas pockets resulting in a dried foam which, when comminuted to the desired particle size, has too high a bulk density because the particles do not contain a sufficient amount of finely entrapped air.

The foam has an overrun of 300 to 350% calculated as percent occluded gas per weight of foam based on a formula in which the volume of occluded gas per weight of foam multiplied by one hundred divided by the weight of foam is equal to the percent overrun. Its structure must be fine-celled, like whipped cream, rather than coarse-celled like a sponge, to ensure foam stability during dehydration and proper structure after dehydration.

The introduction of finely divided bubbles of a gas which does not react with the components of the mixture may be used to facilitate foam formation. A Hobart mixer fitted with a wire whip is satisfactory, as are foaming devices such as those used for the continuous production of cake batters.

It may be seen that it is important that the whipping agent be used at a minimum level in order to reduce the amount of extraneous material present in the composition to as low a level as possible. Certain whipping agents contribute to the formation of a stable foam on the surfaces of food in which the composition may be dissolved. While this foam is considerably reduced and in most cases avoided by the use of a suitable surface-active agent, it is preferable that the problem be minimized by the use of as low a level of the whipping agent as possible.

The drying of the foam must be carefully controlled. The conditions must never be such that caramelization of the sugar occurs. On the other hand, foam layer thickness, temperature, pressure and other factors may be controlled to yield optimal rates of drying.

While the optimal conditions will vary depending on drying facilities and foam properties, they must be such that when the water is removed the sugar is preferably in a crystalline state. Thus, the final product under the microscope appears to be a tightly bound agglomerate of sugar crystals in contact with each other at only a few points, the major volume of each particle being occupied by air. Conditions conducive to dehydration to an amorphous product, e.g., rapid drying at temperatures over about 300°F, are preferably avoided, although such amorphous products eventually can become crystalline on storage.

Any convenient means may be used to comminute the dried foam to the desired particle size. Of course, the extent of comminution determines in part the bulk density of the final sugar substitute, and means should be used which provide a maximum percentage of particles in the desired size range as well as a minimum of very fine particles. A gentle mill in conjunction with a particle classifying device is quite suitable. Excessively large particles are returned to the mill for further comminution while particles of the proper size are drawn off for further processing. The very fine material can be returned to the next batch of ingredients prior to converting it to a foam.

Edible surface-active agents reduce the tendency of the composition to float on the surface of liquids to which the composition has been added, and also prevent the formation of a foam or scum on the surface of food attributed to the whipping agent. It is preferred that the surface-active agent be applied to the surface of the dry foam granules because of its greater effectiveness in this location.

A preferred edible surface-active agent is a water-dispersible mono-diglyceride mixture. Any animal or vegetable fat or oil may be used as the source of the mono-diglyceride mixture. Other food-grade surface-active agents which are well known in the art may also be employed.

As the finished product sometimes exhibits a tendency to cake during storage, it has been found helpful to include an edible anticaking agent in the product. One such agent which has been found to be suitable is sodium alumino silicate.

Example: A sugar substitute was prepared having the following composition:

	Percent
Sugar	99.095
Casein hydrolysate (partial)	0.200
Saccharin	0.600
Mono-diglyceride mixture derived from cottonseed oil	0.005
Sodium alumino silicate	0.100

The composition was prepared by dissolving the sugar, casein hydrolysate and saccharin in an amount of water sufficient to provide a nearly saturated sugar solution. The solution was then mechanically whipped to an overrun of about

250 to 350%. The resulting foam was spread on trays to a depth of ½ inch and dried in a hot-air oven at about 140°F.

The dried foam was comminuted and classified according to particle sizes. That portion of the material having a particle size similar to that of household granulated sugar was retained. The finer material was set aside for reprocessing in a subsequent batch. This utilization of fines is advantageous in reducing the cost of the entire operation.

The mono-diglyceride mixture was then uniformly sprayed on the surfaces of the granulated material, and sodium alumino silicate was then evenly dispersed throughout the product and the product was packaged. The product was found to have a bulk density of 0.23 and on a volume basis had about 25% of the caloric value of household granulated sugar.

Foam System with Lactose

In another process lactose is used in place of sucrose in the foam system as described by *J.P. McNaught; U.S. Patent 3,170,801; February 23, 1965; assigned to Lever Brothers Co.* The product consists essentially of lactose, and is modified by containing a water-dispersible edible protein whipping agent and an artificial sweetener. It is preferably in the form of hollow spheres and has an alpha-monohydrate crystalline lactose content ranging from about 25 to 98%, a bulk density of from about 0.20 to 0.40 g/cm^3, a total moisture content of about 3 to 8% and on a volume basis, a caloric value of approximately 20 to 40% of household granulated sugar. The material has the general appearance of granulated household sugar.

In order to prepare a free-flowing spray-dried composition, it is important that the product has a total moisture content of about 3 to 8% by weight and that a maximum amount of the lactose content of the product be in the α-monohydrate crystalline form. When in this state, lactose particles maintain their free-flowing characteristics under adverse storage conditions to a greater degree than other sugars such as dextrose, corn syrup solids, finely divided sucrose, etc. Since the caramelization temperature of lactose is higher than that of sucrose, the use of the former material is also advantageous as an effective means of avoiding or minimizing the presence of caramelized specks in the product.

A spray-dried product having the desired bulk density (0.20 to 0.40 g/cm^3) and being as resistant to caking as regular granulated sucrose is obtained by drying the particles to a total moisture content preferably within the range of about 4 to 6%. This range can be extended to include levels of about 3 to 8% if an anticaking agent is added. At these moisture levels which include water present as water of hydration in addition to the free moisture in the product, the lactose monohydrate (alpha) crystalline content ranges from a minimum of about 25% to a maximum of about 98% although a minimum of about 45% is preferred in the absence of an anticaking agent.

For example, one product which was prepared with a moisture content in the desired range had a monohydrate lactose content of about 65%. It can be seen that it is extremely important to spray-dry the lactose solution under conditions whereby the moisture level of the product is within the desired range since the result in terms of the required α-monohydrate crystalline lactose content leads to the production of particles having the superior properties set forth above.

The whipping agent is a water-dispersible edible protein such as a hydrolyzed caseinate. A preferred caseinate for use is commercially available under the designation Sheftene. This material is a powdered agent obtained by partial hydrolysis of casein and is composed of the chloride salts of sodium and calcium in addition to the hydrolyzed casein.

The introduction of a soluble gas into the lactose solution prior to drying is used to obtain free-flowing particles having a low bulk density. Although it is best to use carbon dioxide, any gas which is soluble in the lactose solution and which is not detrimental to the product may be utilized. Suitable gases include nitrogen and air in addition to carbon dioxide.

Example: A sweetener when prepared had the following composition.

	Percent
Lactose	93.5
Water	5.5
Sodium saccharin	0.795
Partially hydrolyzed calcium caseinate (Sheftene)	0.195
Mono-diglyceride derived from cottonseed oil	0.01

The composition was prepared by adding the lactose (alpha hydrate), casein hydrolysate and saccharin to an amount of cold water (68°F) sufficient to provide a feedstock containing 50% solids. This batch was spray-dried to a bulk density of about 0.22 to 0.25 gram per cubic centimeter and a moisture content of about 5 to 6%. A high pressure pump was used to feed the slurry to the spray nozzle at 150 to 200 psig and CO_2 was introduced between the pump and the nozzle. Tower air flow was countercurrent with an inlet temperature of 450°F. A Spraying Systems nozzle having a number 3 orifice and a number 2 chamber was used. The spray dryer was of pilot plant size, 29 feet high and 9 feet in diameter.

The temperature of the water used in preparing the aqueous mixture of this example was about room temperature. However, water having a temperature as high as about 95°F can be employed with comparable results. Furthermore, the solids content of the aqueous solution can vary between about 40 to 50%. The pressure in the pump used to feed the aqueous mixture to the spray nozzle can range from about 125 to 350 psig. The air temperature in the spray-drying tower may be varied from about 300° to 450°F.

The spray-dried product obtained in the example was classified according to particle size. The portion of the material having a particle size similar to that of household granulated sugar was retained while the finer material was set aside for reprocessing in a subsequent batch.

When the spray-dried product was used in water or coffee, some surface foam remained after the lactose dissolved in the liquid. However, when 0.01% of the mono-diglyceride was added to the particles, the product left practically no surface film when dissolved in water or coffee. The mono-diglyceride was incorporated in the product by first preparing a 1% mono-diglyceride/99% lactose mixture and adding this material at a 1% level to the spray-dried particles. The product contained about 65% monohydrate and was equal to granulated sugar in caking characteristics.

No significant flavor difference could be observed between this product and granulated sugar (sucrose) when tested on an equal volume basis in coffee. Since the product was about four times lighter than sugar (0.22 vs 0.85 g/cm^3 for sugar), it contained approximately one-fourth as many calories.

Maltodextrin Extender

A sweetening agent which has the particulate form and has the volume sweetening power equal to that of granulated sugar while the calories are two-thirds that of table sugar is described by *K. Newton and A.J.H. Sale; U.S. Patent 3,950,549; April 13, 1976; assigned to Lever Brothers Company.*

The extender may, for example, be a starch hydrolysate of low dextrose equivalent of 15 to 30, for example maltodextrin, or a waxy starch hydrolysate having a dextrose equivalent of 5 to 15. However, owing to the friability of the particles of such a product, a considerable proportion of it may be reduced to fines in the period between manufacture and retail sale. This spoils the appearance of the product at the table, and particularly its ability to glisten or reflect light irregularly in the way that ordinary table granulated sugar does.

The process provides a less friable particulate product, whose particles comprise solid sugar occluded in the glassy matrix formed by the water-soluble bland carbohydrate (polysaccharide) extender containing nonnutritive sweetener dissolved in it.

The size of table granulated sugar particles varies widely from country to country and even from one area to another, and accordingly no hard and fast dimensions can be set for the particles of the product. The product is provided in a form passing a sieve of aperture size 1.5 mm and retained on a sieve of aperture size 0.25 mm. The particles of solid sugar present embedded in the polysaccharide/nonnutritive sweetener matrix will, of course, be correspondingly smaller. The artificial sweetener used to form the product may be saccharin.

Example: Maltodextrin (dextrose equivalent 20; 21.4 kg) and saccharin (sodium salt; 0.2 kg) were put into the bowl of a mixer equipped with a planetary stirrer, and water (7.2 kg) was added rapidly with stirring. Stirring was continued until a uniform solution of maltodextrin and saccharin had formed.

To the stirred solution there was then added fine (caster) sugar (18 kg; all passing 30 BS sieve, aperture size 0.5 mm; 60% retained by BS 44 sieve, aperture size 0.355 mm). Some, though very little, of the sugar dissolved, and there was obtained a slurry consisting essentially of particles of caster sugar suspended in a solution of maltodextrin and saccharin.

The slurry was then dried at 95°C in thin layers in steam-jacketed trays in a chamber at 50 mm mercury absolute pressure. After 2 hours the brittle coarse glass-like form (moisture content less than 3% by weight) that had formed was scraped from the trays and crushed between rollers, and the coarse particles thus obtained were further reduced in a mill. The milled material was sieved, the fraction passing BS 12 (aperture size 1.4 mm) and retained on BS 60

(0.25 mm aperture) being collected as the product. (A typical yield is 70%.) The particulate product consisted essentially of particles of caster sugar embedded in a glassy matrix of maltodextrin and saccharin. Its composition in terms of solids content was approximately 55% sugar; 44.5% maltodextrin; and 0.5% saccharin. The bulk density of the uncompacted, free-flowing product was 0.35 g/cm^3. It had an appearance resembling that of ordinary table granulated sugar in its translucency and its glistening aspect, and was relatively nonfriable. It had not more than half the calorie content of table granulated sugar when used at the same volume.

The fines obtained during the earlier operations of milling and sieving consisted essentially of fragments of glassy matrix (i.e., material containing very little sugar). These can be mixed with starting material (maltodextrin, sugar and saccharin in appropriate proportions) and recycled.

Agglomerated Spray-Dried Process

The expanded sweetener product of *N. Wookey, A.F. Jackson, M.C. Collins and B. Francis; U.S. Patent 3,930,048; December 30, 1975; assigned to Ranks Hovis McDougall Limited, England* is based on a process which produces agglomerated spray-dried particles with a bulk density in the range of 0.2 to 0.75 g/cc, as compared to 0.9 for granulated sugar.

A method of producing such a sweetening material includes the steps of feeding to a spray drier a mixture containing dextrose and a synthetic sweetening agent, such mixture being the liquid phase in the form of a water-based syrup and in the solid phase in the form of powder; contacting the syrup and powder as it passes through the spray drier whereby all or most of the powder particles will pick up some of the syrup and will become sticky, and will agglomerate while drying and screening the spray-dried material so as to separate any unagglomerated powder, together with the agglomerates which are above and below a predetermined range of screen sizes, from the intermediate-sized remainder, which constitutes the end product.

For commercial production such method is continuous, in which case the oversized and undersized agglomerates are pulverized after separation to reduce them to powder which, together with any unagglomerated powder, may be recycled through the spray drier and will then constitute all or part of the powder supply.

Alternatively, all or part of the separated material (comprising the oversized and undersized agglomerates and any unagglomerated powder) may be mixed with the water to form syrup and then recycled through the spray drier, so as to constitute all or part of the syrup supply thereto.

In a modification of the above-described methods, some of the material supplied to the spray drier may be dextrose unmixed with synthetic sweetening agent. Such unmixed dextrose material may be in powder form, or in the form of water based syrup or both.

The following two examples will explain in greater detail two ways of carrying the process into effect. In both cases the apparatus used was a Niro spray drier having recycle equipment attached which comprises a recycle fan and ducting to return powder to the top of the spray drier. Screening was effected by a Locker screen using 8 and 23 British Standard Industrial Wire Mesh screen sizes (medium class)—British Standard Specification 481:Part 1:1971.

Example 1: A continuous supply of dextrose syrup of 94 DE mixed with 0.45% (by weight) of saccharin is fed into the upper part of the spray drier. At the start, to seed agglomeration, a quantity of dextrose powder of 94 DE is injected for a period of about 30 minutes while spraying with saccharin solution to bring the circulating powder up to the 0.45% concentration of saccharin. During the spray drying porous agglomerates will form, resulting from individual particles becoming sticky and cohering with other particles. The rate of feed of syrup was about 0.5 ton per hour and the rate of feed of seeding powder was about 1.5 tons per hour.

As the agglomerates emerge from the spray drier, they are first passed through a coarse screen, which removes oversized agglomerates above 7 mesh (BS 410: 1969) and then passed through a second finer screen which collects intermediate sized agglomerates in the range of 7 to 18 mesh size (BS 410:1969). These will form the end product. Undersized agglomerates, below 18 mesh size (BS 410: 1969), together with any unagglomerated powder pass through the finer screen.

The oversized and undersized agglomerates and any unagglomerated particles are collected and fed into the recycle circuit where they pass through the impeller (recycle fan) which reduces their size down to powder. This powder is continuously fed into the top of the spray drier where it constitutes a continuous supply of seeding material.

The intermediate-sized porous agglomerates consist of cohering particles each comprising a mixture of dextrose and saccharin, and the product has a bulk density in the order of 0.30 g/cc.

Example 2: The procedure of Example 1 was followed except that the syrup input to the spray drier was dextrose only without any synthetic sweetening agent. The intermediate-sized agglomerates thus consisted of cohering particles of anhydrous dextrose. These agglomerates were sprayed with a saccharin solution and were then dried on a fluidized bed drier.

The resultant product was a mass of porous anhydrous dextrose agglomerates having at least a partial coating of saccharin, some of which had penetrated into some of the interstices in each agglomerate. The bulk density was given in the order of 0.40 g/cc.

Pulverized Sugar Combination

A sponge-like material made up of fine particles of sugar and of synthetic sweetener agglomerated into granules was developed by *W.M. Grosvenor, Jr.; U.S. Patent 3,011,897; December 5, 1961; assigned to The American Sugar Refining Company.*

The finely divided sugar which is used in making the composition may be ordinary powdered sugar such as 6X or 10X powdered sugar. Such powdered sugar is made by grinding granular sugar to the desired fineness. The synthetic sweetener can be separately pulverized or ground to the desired size if it is not already in a sufficiently finely divided state. Or the synthetic sweetener can be mixed with the granular sugar before grinding and the mixture ground together. Alternatively and advantageously, the synthetic sweetener can be mixed with a small part of the granular sugar as a preliminary step of the process, and the resulting preliminary mixture then mixed with the larger amount of granular sugar.

The synthetic sweetener which is used in making these compositions is soluble saccharin. Where the saccharin is in the form of a coarse powder, it is separately ground before mixture with the pulverized sugar or is mixed with the granular sugar and ground at the same time.

The proportions of sugar and synthetic sweetener will in general be such that the resulting granular porous sponge-like product will have a sweetness similar to or comparable with that of a similar volume of granular sugar. With a bulk density of the porous sponge-like product equal to about one-half that of granular sugar, the amount of sugar by weight in the porous granular product will be approximately one-half that of the granular sugar, and the amount of synthetic sweetener will be sufficient to impart an overall sweetness comparable with that of granular sugar. With a bulk density of one-third that of granular sugar, the amount of synthetic sweetener will supply about two-thirds of the sweetness; with a bulk density of about 60% that of granular sugar the amount of synthetic sweetener will supply about 40% of the sweetness.

The conversion of the finely divided sugar and synthetic sweetener into the form of an agglomerated product is accomplished by moistening the surfaces of the particles and agitating the particles to cause agglomeration of the fine, moistened particles to form agglomerates of a sponge-like character and low bulk density, and by drying the agglomerates, and with screening, if necessary, to separate oversize granules and fines.

The examples illustrate the use of different sugars, in which the amount of synthetic sweetener is approximately equal, in sweetening power, to the amount of sugar with which it is used, so that, with the production of a porous, sponge-like granular product having half the bulk density of granular sugar, the composition will have a sweetness comparable with that of an equal volume of granular sugar. As the bulk density of the product is varied to a somewhat higher or lower bulk density, the sweetening power of the composition can be similarly varied to one which is somewhat greater than or somewhat less than that of a corresponding volume of granular sugar.

The following examples are on the basis of approximately 20 pounds of sugar with an amount of synthetic sweetener approximately equal in sweetening power to that of the sugar. In each case, the mixture of sugar and synthetic sweetener is ground, in a suitable mill such as a micropulverizer or Bauermeister U-Z-U-00 mill, to a fineness corresponding to around 10X sugar.

Example 1: A mixture was made of 20 pounds of fine, granular sugar and 30 grams of sodium saccharin, the mixture was ground, and the finely powdered mixture subjected to agglomeration. Then the agglomerates were dried to form a porous sponge-like granular product of low bulk density, approximating one-half that of the granular sugar.

Example 2: 20 pounds of anhydrous dextrose was used with 45.5 grams of saccharin sodium, this being the approximate amount to increase the sweetness of the resulting porous granules to a sweetness comparable with a corresponding amount of granular sucrose. The mixture was ground and agglomerated in the manner above described.

Agglomerated Sugar plus Sweetener

It was known that partially synthetic sweetening agents could be prepared by intermixing a natural sweetening agent (e.g., a sugar such as sucrose and lactose) with an artificial sweetener (e.g., saccharin) and agglomerating the mixture. It was also known that such agents could be prepared by intermixing an agglomerate of a natural sweetening agent with an artificial sweetener in the dry state. This latter procedure, however, suffered the disadvantage that a nonsegregating, uniform product was difficult to prepare and maintain.

The uniform mixing of two dry products, one of which (the agglomerate) is highly friable, is extremely difficult since to assure equal distribution rather drastic mixing methods are necessary which tend to disintegrate the agglomerate and thereby render the entire mixture unsatisfactory. Moreover, even if such a substantially uniform product is obtained, it tends to separate on standing so that when shipped in individual containers, the product is no longer uniform throughout the depth of the container.

According to *R.W. Walton; U.S. Patent 3,795,746; March 5, 1974; assigned to E.R. Squibb & Sons, Inc.*, the disadvantages can be avoided in the process of mixing an agglomerated sugar with a liquid solution or suspension of an artificial sweetener. Among the artificial sweetening agents that can be used are saccharins, such as sodium saccharin and magnesium saccharin, and the dipeptides.

After the natural sweetening agent has been agglomerated, but preferably before it has been fully dried, the agglomerate is contacted with a spray of the liquid solution or suspension of the artificial sweetening agent. The amount of spray used depends on a number of factors, namely, the degree of agglomeration of the natural sweetening agent, the concentration of the artificial sweetening agent in the spray, the rate of passage of the agglomerate through the spray, the intrinsic sweetness of the natural sweetening agent and the artificial sweetening agent, and the desired sweetness of the product.

In general, the solution or suspension of artificial sweetening agent is sprayed at high pressure of from about 660 to about 900 psig from an orifice having a diameter of about 0.009 inch to about 0.011 inch. Under these conditions the saccharin solution or suspension is mechanically atomized to substantially a molecular dispersion, that is to a particle size from about 10 to about 20 microns. At the same time that the saccharin solution or suspension is sprayed over the agglomerated natural sweetening agent, the latter is subjected to a hot air stream to remove any residual moisture from the agglomerate and to dry the artificial sweetening agent immediately after it contacts the agglomerate. In this manner the artificial sweetening agent is adsorbed and absorbed onto and into the agglomerated sugar and becomes bonded and does not separate from the agglomerated sweetening agent upon standing or storage, or when subjected to normal vibration.

Thus, if the natural sweetening agent is sucrose and the desired sweetness of the final product is that of sucrose and the sucrose is agglomerated so that its bulk density is halved, then sufficient artificial sweetener must be added to account for the loss of sweetness per unit volume due to the fact that only one-half as much sucrose is actually present as is in the corresponding volume of granulated sucrose. If sodium saccharin is used as the artificial sweetener, this means the addition of about 3 grams of saccharin per kilogram of sucrose, since sodium saccharin is more than three hundred times as sweet as sucrose.

After the agglomerate has been sprayed with the artificial sweetener, the mixture is dried immediately to give the final product. As the spraying is done before the agglomerate has been fully dried, the final drying step not only removes the solvent used in forming the suspension or solution of artificial sweetener, but also removes any residual water present from the agglomeration step.

The product contains agglomerated sucrose having a density of from about 18 pounds/cubic foot to about 30 pounds/cubic foot (optimally about 20 pounds per cubic foot to about 23 pounds per cubic foot), mixed with sodium saccharin in the weight proportion of about 20 grams to about 40 grams (optimally about 27 grams to about 33 grams) of sodium saccharin per kilogram of sucrose. Such a product has a sweetness equal to about 1 to 1.2 times that of sucrose.

Example 1: Finely powdered sucrose is fed into an agglomerator. The sucrose is contacted with hot wet air and the agglomerates are formed and carried on a vibrating bed to the second area in which hot dry air sets the agglomerate and removes the moisture. The agglomerates formed in this manner have a density from about 19.0 to about 22.0 pounds per cubic foot. The sucrose is supplied at a rate of about 3,200 kg/hr. 1.2 parts of sodium saccharin is dissolved in 1 part of very hot water in a mixing tank. The solution is pumped at a pressure of about 900 psig to a spray gun and through a full flat fan nozzle having a diameter of 0.011 inch to cover the moving bed of agglomerate from the second area.

At the same time a secondary hot air stream is introduced from under the agglomerate bed to remove the water introduced and to lock the sodium saccharin into the agglomerate. The solution is continuously sprayed at a rate of about 13.55 liters per hour to add about 9.55 kilograms of sodium saccharin to about 3,200 kilograms of the agglomerate. The normal agglomerator fines recycling system requires five minutes to reach equilibrium prior to discharging a uniform product.

Example 2: The procedure of Example 1 is followed except that a solution of 1.95 parts of water, 26.00 parts of methanol and 7.00 parts of sodium saccharin is used instead of the aqueous sodium saccharin solution. The solution is continuously sprayed at a rate of 51.7 liters per hour to add 9.55 kilograms of sodium saccharin to 3,200 kilograms of the agglomerate.

Example 3: The agglomerated sucrose obtained by following the procedure of Example 1 but without adding the aqueous sodium solution is conveyed by a device having a vibrating screw feeder into the top side inlet of a verticle chamber. As the agglomerate falls by gravity a solution of 1.2 parts of sodium saccharin in 1 part of very hot water is pumped at high pressure to a spray gun and out a full flat fan nozzle to cover the falling stream of agglomerate. Hot air is forced into the bottom of the chamber and flows upward past the agglomerate to dry out the water.

DRUM DRYING WITH STARCH

Numerous approaches have been made to the production of sucrose substitutes which have the bulk and general appearance of sucrose, as, for instance, those of ordinary granulated sugar, but which are characterized by a caloric content very substantially less than that of sucrose. Various of such earlier approaches are discussed in U.S. Patents 3,320,074 and 3,325,296. In addition, these patents describe certain procedures for the production of sucrose substitutes.

Thus, in the first of these two patents, an aqueous solution of a water-soluble dextrin, having a dextrose equivalent of about zero, is subjected to pressure, aerated and then spray-dried, the final product having a bulk density not exceeding about 0.15 g/cm^3 and having a caloric content in the range of about 2.7 to less than 5.5 calories per level teaspoon.

In the second of the patents, an aqueous solution of a water-soluble starch hydrolysate, having a dextrose equivalent in excess of 13 but not more than 28, and a noncaloric artificial sweetener is subjected to vacuum drying as, for instance, on a vacuum drum dryer. The resulting products, after milling and classification, have caloric contents as low as about 3 calories per level teaspoon.

Sweetening compositions of the type disclosed in U.S. Patent 3,320,074, prepared as they are by a spray-drying operation, tend to have an undesirable chalky look and feel and are readily distinguishable from sucrose. On the other hand, while sweetening compositions of the type described in U.S. Patent 3,325,296 bear a reasonably good resemblance to sucrose and have been and are being commercially marketed, their method of manufacture, requiring as it does, a vacuum drying procedure, notably the utilization of vacuum drum drying equipment, involves very substantial costs in such equipment, the drying operations are cumbersome and the economics of the drying operation leave much to be desired.

Furthermore, commercially marketed sweetening compositions made by the vacuum drying procedure of U.S. Patent 3,325,296 generally have bulk densities of 0.18 to 0.19 g/cm^3 and a caloric content of not less than 3 calories per level teaspoon. As reported in the patent, bulk densities in the range of about 43 to 78% of that of sucrose are obtained. In this connection, it may be noted that sucrose, in the form of ordinary granulated sugar, has a bulk density of about 0.9 g/cm^3 (which is 4.5 grams per level teaspoon of about 5 cm^3).

In the process of *W.H. Schmitt and R.A. Lukey; U.S. Patent 3,653,922; April 4, 1972; assigned to Alberto-Culver Company* pulverulent or granular free-flowing, water-soluble, low calorie sweetening compositions having the general appearance of sucrose, as, for instance, that of ordinary granulated sugar, are produced, without aeration and without resort to vacuum drying, in a form having exceptionally low bulk densities and exceptionally low caloric contents per level teaspoon. These sweeteners are obtainable with bulk densities in the range of 0.09 to 0.12 gram per cubic centimeter, and caloric contents from about 1 or 1½ to 2 calories per level teaspoon.

A certain type of water-soluble starch hydrolysate and an essentially noncaloric artificial sweetener are used to prepare an aqueous solution, the total solids in solution ranging from 72 to 78%. The proportions of noncaloric artificial sweetener utilized, although always distinctly minor, are variable, depending upon the sweetness characteristics of the noncaloric artificial sweetener utilized, and on the degree of sweetness desired in the final sweetening composition. In general, the noncaloric artificial sweetener will usually fall within the range of about 3 to 15%, by weight, of the starch hydrolysate.

The viscosities of the aqueous compositions or solutions, fed into the pool prior to being picked up by the dryer drums, are variable. The viscosities depend, of course, not only on the concentration of the solids in the aqueous composition or solution, for a particular starch hydrolysate and a particular noncaloric artificial sweetener, but also on the temperature of such aqueous composition or solution. Thus, in an illustrative case of an aqueous composition or solution containing 76% total solids, in which, on the total solids basis, the starch hydrolysate constitutes 91% and the essentially noncaloric artificial sweetener constitutes 9%, the viscosity is about 25,000 cp at 37°C, about 11,200 cp at 50°C and about 8,800 cp at 55°C.

It has been found that starch hydrolysates having a dextrose equivalent in excess of 25 produce finished dried sweeteners which are far too hygroscopic to be packaged and handled satisfactorily. Furthermore, it has also been found that dried sweetener compositions can be prepared from starch hydrolysates which have dextrose equivalents which are of the order of about zero. Hence, those water-soluble starch hydrolysates which are useful for the production of the sweetener compositions are those which have dextrose equivalents in the range up to 25, particularly desirably in the range of 6 to 15.

The accompanying figures show illustrative forms of a double drum drying apparatus setup, operating at atmospheric pressure, in which the drying of the aqueous solutions of the starch hydrolysate and noncaloric artificial sweetener can be carried out.

FIGURE 6.2: DRUM DRYING WITH STARCH

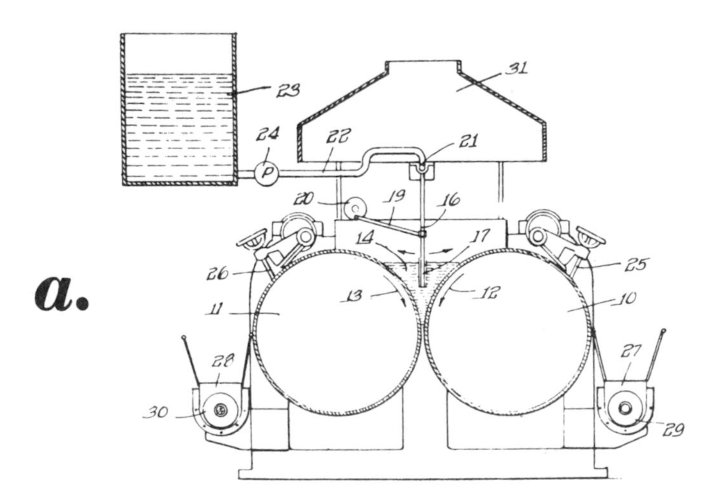

(continued)

FIGURE 6.2: (continued)

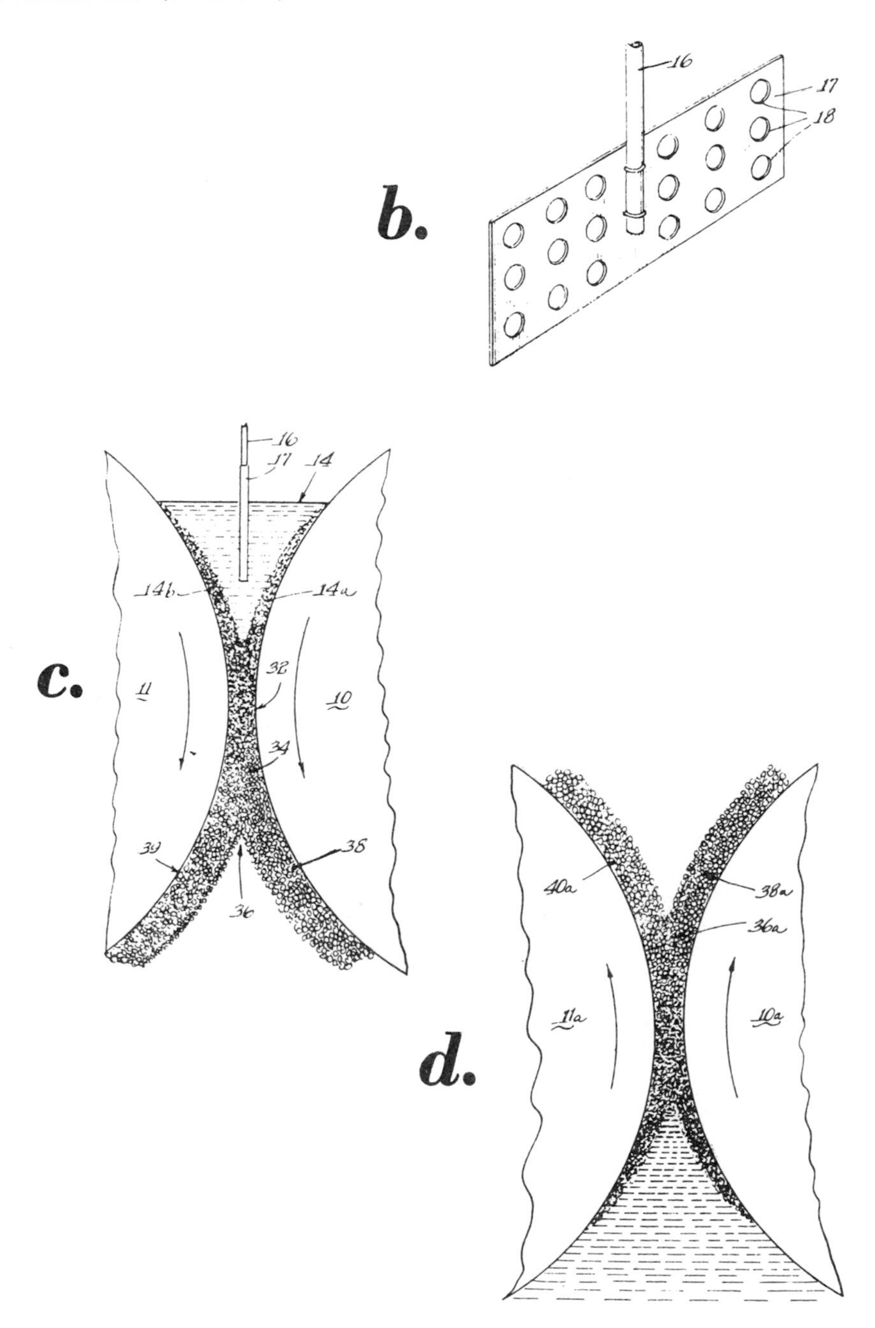

Source: U.S. Patent 3,653,922

As shown in Figure 6.2a the dryer has a pair of internally steam heated cylindrical drums **10** and **11** supported for rotation towards each other. A pool **14** of the aqueous starch hydrolysate-essentially noncaloric artificial sweetener composition is maintained in the crease or space between the drums and the pool is maintained by feeding into it from the feed pipe **16**. The feed pipe dips into the pool and, as shown in Figure 6.2b, is attached to an agitator vane **17** having a plurality of perforations **18**, the feed pipe-agitator vane assembly being adapted to be oscillated or moved in the pool by any suitable means such as shown in Figure 6.2a, by a link **19** connected to the feed pipe and operated by an eccentric **20** so as to perform the additional function of agitating or mixing the aqueous composition sufficiently to inhibit formation of a skin on the surface of the aqueous composition and/or to break such skin as may form.

The thin films of the aqueous composition which are picked up as a coating by each of the drums become confluent in the pinch between the drums and the resulting confluent bead is pulled apart by the motion of the counter-rotating drums in their normal paths. This forms a disrupted hot film on each drum which is carried along and the water is evaporated from them thereby producing a dried film which is removed by the scrapers or doctor knives **25** and **26** which ride on the respective drums **10** and **11**.

The dried sweetener composition is removed from the drums by the scrapers in the general form of flakes having a moisture content which can be controlled within desired limits but is preferably about 1 to 4%. The dried flakes fall into troughs **27** and **28** and are conveyed through them by suitable screw conveyors **29** and **30** and discharged into a mill or grinder where the dried flakes are milled or ground to a desired particle size, preferably granules of a size of the order of ordinary granulated sugar. A conventional vapor hood **31** is provided to remove the vapors from the drying operation.

Referring now to Figure 6.2c, it may first be pointed out that the liquid level of the pool is in constant agitation from the movement of the agitator vane and from the boiling action caused by vaporization of the solution which is heated by the drums. The solution tends to start to dry in a somewhat V-shaped area shown above the center line of the closest proximity (the pinch) between the two drums. As the partially dried composition **14a** and **14b** approaches the midpoint **32**, the temperature of the solution increases to maximum due to the proximity of the drums (the gap dimension at approximately 0.010 inch). There is then substantially no temperature gradient and the temperature is at or above that required for drying.

As the material gets into approximately the area **34**, cellular structure is being formed and it is aided by the tearing action, at **36**, caused by the adherence of the film to the drum surfaces **38** and **39**. The tearing action at **36** causes a mechanical pulling of the partially expanded cellular material radially outwardly from the drums, discouraging the formation of any film and encouraging the exposure and expansion of the cellular material.

Referring now to Figure 6.2d, where the drums are rotated in opposite directions and where the aqueous composition is applied to the drum surfaces from the bottom, the composition is applied to drums **10a** and **11a** by suitable pans, applicator rolls, spray or other means not shown. The disruption of the partially dried cellular material at **36a** results from the fact that the cellular material at

38a and **40a** adheres to the adjacent drums.

It may be noted that the finished dried products obtained as shown in Figure 6.2d generally do not give quite as low bulk densities as in the case of Figure 6.2c, although the bulk densities obtained in the Figure 6.2d are substantially lower than are obtained on a single drum dryer or by vacuum drum drying techniques.

It is important that there be a mechanical tearing action as shown at **36** and at **36a** concomitant with the simultaneous application of heat. While this is best achieved by means of a double drum dryer arrangement, it can also be accomplished by other means as, for instance, by a belt drying arrangement modified to bring about a compression of the dried film and the tearing or disrupting in a drying environment in which the starch hydrolysate-substantially noncaloric artificial sweetener aqueous composition is applied to a drying surface so as to cause a substantially uniform temperature throughout the material being dried, such temperature being equal to or above that required to effect drying and providing a means for mechanically disrupting the interior of the drying material.

The bulk density of the finished dried sweetener compositions is affected primarily by the concentration of solids in the aqueous starch hydrolysate-essentially noncaloric artificial sweetener composition to be dried, the manner in which the drying is carried out, the temperature of drying, and the speed of the drying drums.

As pointed out above, the solids concentration of the aqueous composition to be dried should fall within the range of 60 to 80%. The lower solids concentrations, in this range, produce finished dried sweetener compositions having lower bulk densities than those obtained with the higher solids concentrations. However, even with solids concentrations of about 76% in the aqueous compositions to be dried, it is common to obtain finished dried sweetener compositions with a bulk density of the order of 0.11 g/cc, and with caloric contents of the order of 1½ calories per level teaspoon.

It has been found that the drums should be heated and maintained at a temperature in the range of 120° to 180°C or, in terms of steam pressure, in the range of about 20 to 125 psig, especially satisfactory being steam pressures corresponding to a temperature in the range of 160° to 170°C. Drum speeds are variable, but usually being in the range of 8 to 20 rpm. The drums are spaced apart, at their closest line of contact along their longitudinally extending surfaces, of from about 0.008 to 0.012 inch.

When aqueous compositions or solutions containing a distinctly major proportion of a starch hydrolysate of the general type shown in U.S. Patent 3,235,296 and a distinctly minor proportion of an essentially noncaloric artificial sweetener, having a solids content of, say 80%, are dried on a single drum dryer operating at atmospheric pressure, and the resulting dried composition as removed from the dryer is then milled or ground to a pulverulent or granular composition similar to ordinary granulated sugar, the bulk density of such composition is very decidedly greater than the bulk density of such a product when dried on a double drum dryer operating at atmospheric pressure under the conditions which have been described in accordance with this process; and this is the case even where the same aqueous composition, the same drying temperature, the same drum speed, and the same method of particle reductions are utilized.

The bulk density of the dried sweetener compositions is of the order of 0.11 gram per cubic centimeter and such that a level teaspoon contains only about 1½ calories.

CODRIED DIPEPTIDE-ORGANIC ACID

M. Glicksman and B.N. Wankier; U.S. Patents 3,761,288; September 25, 1973 and 3,922,369; November 25, 1975; both assigned to General Foods Corporation claim that the solubility of dipeptide sweetening compounds can be significantly increased by codrying with organic acids. In addition, use of certain bulking agents in these compositions will produce a product with low bulk density which has the appearance of table sugar. An especially suitable bulking agent is dextrin material.

The dextrin material used for the production of these sugar substitutes must dissolve easily in water to produce a clear solution and must be nonhygroscopic. In this regard the DE (dextrose equivalent) of the dextrin material has been found to be a critical parameter. The dextrin material must have a sufficiently low molecular weight to be easily soluble in water and to produce a clear solution so that the final product will have the essential reflecting surfaces in order to give the appearance of a crystalline product. On the other hand the molecular weight of the dextrin material must be high enough so that hygroscopicity is avoided.

Accordingly, it has been determined that for the production of table sugar substitutes the dextrin material should have a DE in the range of about 5 to 10. Additionally it has been found that the best results are obtained if the dextrin material contains little or no monosaccharide (i.e., glucose) and contains an irregular distribution of the other lower (one to eight saccharide units) saccharides with a preponderance of the hexamer and heptamer. Such corn syrup dextrins have been produced by means of enzymatic hydrolysis of starch and are typified by the products available under the name Mor-Rex.

Example 1: 3 grams of citric acid and 1 gram of L-aspartyl-L-phenylalanine methyl ester are dissolved in 50 ml of water with stirring. The resulting solution is spread on a stainless steel tray (2.1 ft^2) and allowed to dry at ambient conditions for about 2 days. The dry material was then scraped from the tray and ground with a mortar and pestle. One-half gram samples of this ground material were added with stirring, to beakers containing 200 ml of water at 40°F. The material completely dissolved in an average time of about 55 seconds yielding solutions which were sweet with a slight acid taste.

Example 2: A solution was prepared containing 800 grams of water (80°F), 241 grams of 5 DE Mor-Rex and 5.95 grams of L-aspartyl-L-phenylalanine methyl ester. This solution was placed in a tray at a 1.5 inch thickness and freeze-dried in a Stokes Freeze Drier for 48 hours. The material was then ground to a fine powder using a Waring Blendor at a high speed.

Example 3: A solution was prepared according to the method of Example 2 and this solution was drum dried at a temperature of 130°C on a drier operating at 25 lb/in^2 and 6.25 rpm.

Example 4: A solution containing 384.05 grams of water, 241 grams of 5 DE Mor-Rex and 5.95 grams of L-aspartyl-L-phenylalanine methyl ester was prepared. This solution was then spray dried in a Niro Spray Dryer at an air pressure of 5.2 kg/cm^2, and air inlet temperature of 160°C, an air outlet temperature of 75°C and a rate of solution flow of 15 cc/min.

Equal weight samples of the sweetening compositions of Examples 2, 3 and 4 were dissolved in coffee samples and were organoleptically determined to have substantially equivalent sweetness levels. This sweetness level is not found to significantly differ from control coffee samples containing an equal amount of the untreated dipeptide material, thus indicating the absence of any degradation of the dipeptide material during the drying operations.

The solubility rate of the powders from Examples 2, 3 and 4 was evaluated by recording the times required for complete solution of 1.5-gram samples of these powders (containing about 0.036 gram of sweetener) into 170 ml of water at a temperature of 40°F, with stirring. The results are summarized in the table below.

	Seconds
Example 2	62
Example 3	40
Example 4	60

When 0.036 gram-samples of L-aspartyl-L-phenylalanine methyl ester are sought to be dissolved in 170 ml of water at 40°F, with stirring, average times for complete solution run about 30 minutes.

Additional tests have shown that varying the level of dipeptide in the sweetening compositions up to the level of about 1 part dipeptide per part of bulking agent, does not have any appreciable or predictable effect on the rate of solution. All samples were found to dissolve in water as cold as 40°F in less than 2 minutes, whereas complete solution of equivalent amounts of the dipeptide material taken alone requires a times of about 30 minutes.

The bulk density of the final sweetening composition can be controlled by varying the solids concentration of the solution prior to drying. The bulk density may also be controlled by changing the method of drying, by varying the rate of drying or by varying the conditions of pressure or vacuum under which the solution is dried. Bulk densities ranging as low as about 0.04 g/cc can be obtained by the process.

LOW BULK DENSITY PROCESSES

Anticaking Lactose Combination

The particular objective of the process of *C.J. Endicott, P.W. Brown and L.S. Andrews; U.S. Patent 3,746,554; July 17, 1973; assigned to Abbott Laboratories* utilizing lactose is to produce a low bulk density noncaking powder equivalent to the sweetness of an equal amount of sucrose.

In general, a mixture of lactose and saccharin or a saccharin salt, for example,

98.8% lactose and 1.2% saccharin is dissolved in water and concentrated to about 50% solids content. The solution is then pumped to a conventional spray dryer where, prior to introduction of the solution into the dryer, a nontoxic gas such as nitrogen or carbon dioxide is injected into the solution. The solution-gas mixture is then immediately spray dried to produce hollow spherical particles of low bulk density and having an outer appearance similar to granular sugar. While the product resulting from the spray drying process will have a satisfactory low bulk density (0.10 to 0.20 g/ml) and hence low caloric value on a volumetric basis, it will lack stability. This instability is exhibited in caking, increase of bulk density and decrease of solubility upon exposure to humid conditions.

It was found that the lactose in the spray-dried product is amorphous and present in predominantly the β-form. It is believed that exposure to high humidity causes hydration of the lactose to crystalline α-lactose monohydrate. The large amount of water, about 11 to 12%, necessary for crystallization causes fusion and deformation of the spray-dried hollow spherical particles.

After crystallization the unbound water is evaporated, leaving a hard cake. The caking results in not only loss of granularity but also a decrease in solubility due to destruction of the thin wall particle structure and the change in the lactose from a highly soluble β-form to a much less soluble α-hydrate. Serious caking of the product results at 40 to 50% relative humidity at room temperature with resultant fusion and breakdown of the hollow spherical particles. This results in an undesirable increase in bulk density. Since α-lactose monohydrate is relatively insoluble, particularly in comparison to β-lactose, this may explain the decrease in solubility of the spray-dried product after exposure to humid conditions.

It is apparent that for a product designed to provide equivalent sweetness as sugar on a volumetric basis and adapted to be stored under normal conditions of temperature and humidity, such instability resulting in an increase in bulk density, caking and decrease in solubility cannot be tolerated.

As a means of stabilizing a low bulk density product, it has been found that treatment of the spray-dried product with an organic solvent such as a water-soluble alcohol or acetone is effective. By contacting the spray-dried product with the organic compounds, for example by spraying or vapor treatment, and then drying to remove the compound, a uniform, free-flowing product results with only a slight increase in bulk density.

Subsequent exposure of the treated product to relative humidities ranging from 50 to 90% at room temperature for 40 hours showed that effective stabilization of the spray-dried product was obtained. The treated product is free-flowing after extended exposure to high humidity with no evidence of caking or decrease of solubility. It is believed that treatment of the spray-dried product with the noted solvents results in crystallization of the amorphous material without hydration of the lactose. Treatment can be effected by spraying the amorphous product with the selected compound, by exposing the product to vapors of the selected compound, by utilizing a fluidized bed system of vapor treatment and drying, or by any suitable means effective to expose the amorphous product to the desired organic compound.

Example 1: A sweetener containing 98.8% lactose and 1.2% calcium saccharin is prepared by mixing the ingredients in water having a temperature of about 190°F and sufficient to obtain approximately a 56% solids solution. The feed solution is fed to a spray drier by means of a high pressure pump operated at a pressure of about 1,000 psi. Carbon dioxide is injected into the feed solution at a point on the high pressure side of the pump and prior to introduction of the solution into the drier. The CO_2-injected feed solution is sprayed into an air stream having a temperature of about 300°F to obtain a spray-dried product having a bulk density of about 0.10 g/ml.

Since it is desired that the final product be equivalent to sugar in sweetness on a volumetric basis, a definite relation must be maintained between product composition and bulk density. For example, a finished product found by taste tests to be about 10% too sweet at a bulk density of 0.18 g/ml can be lowered in sweetness level by reducing the bulk density to 0.16 g/ml. If it is desired to maintain the higher bulk density in such a case, then the composition can be modified accordingly, reducing the saccharin content and increasing the lactose content. This, of course, would have a negative effect on the caloric content of the product.

With sucrose having a bulk density of close to 1.0 g/ml, it is apparent that a product having a bulk density of about 0.20 g/ml would have approximately 20% the caloric value of an equivalent volume of sucrose. At a bulk density of about 0.30 g/ml, the product would have a caloric value of approximately 30% that of an equivalent volume of sucrose.

Depending on the caloric content desired, bulk density can be varied from about 0.10 to 0.40 g/ml. In turn, the lactose content of the product can be varied from about 75% to essentially 100% with the remainder of the composition being made up of saccharin and its salts. The amount of saccharin used depends to some extent upon the end use of the product. If intended solely as a coffee sweetener, more saccharin may be required since coffee is somewhat bitter. Likewise, citrate soft drinks may require more saccharin in comparison to cola drinks. Additionally the relative sweetness of saccharin diminishes at higher concentrations. As a consequence, the concentration of saccharin used can vary considerably. A composition range of from 97 to 99.25% lactose and 0.75 to 3% saccharin is suitable for most purposes with a composition of 98.8% lactose and 1.2% saccharin being preferred.

Example 2: Product produced with the process illustrated in Example 1 is stabilized by treatment with ethanol, applying ethanol vapors to the product in a vacuum blender. Ethanol vapor is admitted to a Pfaudler glass-lined blender (3 cubic feet capacity) containing 9 kg of spray-dried product. Treated product shows a high degree of crystallinity with lactose predominantly in the β-form, and no α-lactose monohydrate being apparent. Bulk density was 0.21 g/ml and the product was free-flowing after 40 hours' exposure to 75% relative humidity at room temperature.

For adequate crystal conversion, addition of 4 to 6 ml of ethanol per 100 grams of product and a 20 to 30 minute vapor treatment time is sufficient. Below these limits, the ethanol remains bound to the partially amorphous material and prolonged drying is required, even at a drying temperature of 70° to 80°C. A vapor treatment and drying process cycle of about 1½ hours is satisfactory.

Peebles Process

Lactose is used in combination with an artificial sweetener in the process of *D.D. Peebles and C.A. Kempf; U.S. Patent 3,014,803; December 26, 1961; assigned to Foremost Dairies, Inc.* to prepare a free-flowing powder with a low bulk density. Figure 6.3a is a flow sheet of the process and Figure 6.3b is a side view of the special apparatus used for agglomeration.

FIGURE 6.3: PEEBLES PROCESS

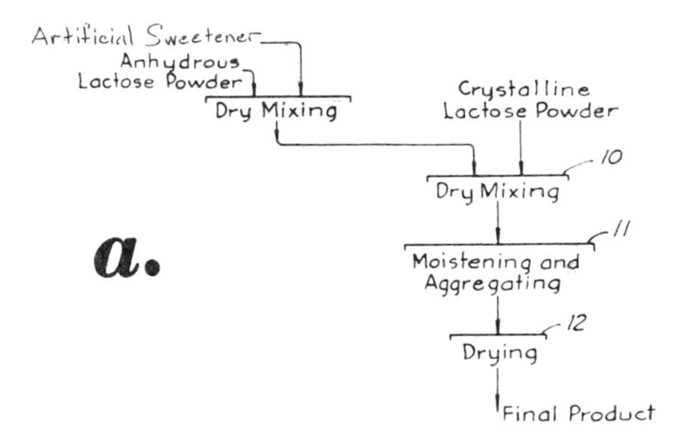

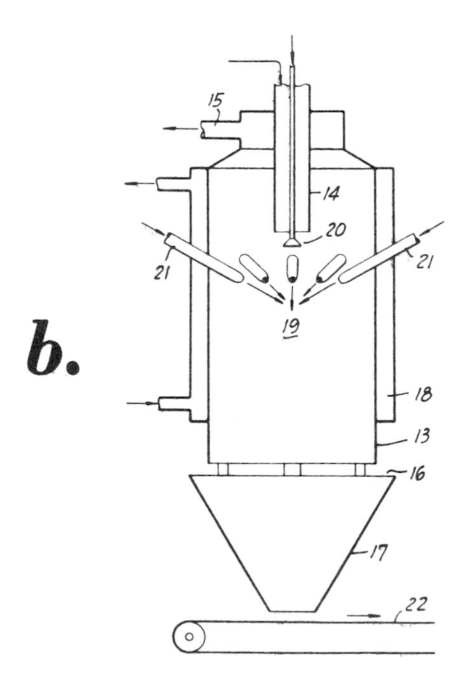

Source: U.S. Patent 3,014,803

The procedure illustrated in Figure 6.3a is as follows: crystalline lactose powder (i.e., alpha lactose monohydrate), of edible grade, is applied to the dry mixing operation **10** where it is intermixed with a measured amount of dry artificial sweetener powder. The artificial sweetener has been premixed with a measured amount of anhydrous lactose powder having the major part of its lactose in the amorphous form. In the dry mixing operation **10**, the crystalline and anhydrous powders are homogeneously dispersed throughout the lactose. In step **11** the dry powder mix is treated to convert the same to the form of moist porous aggregates.

In effect, this step is a moistening and aggregating operation in which the particles of powder are moistened whereby particularly the particles of anhydrous lactose powder become sticky. The powder particles are then caused to be brought into repeated random contact, whereby they cling together in the form of moist composite porous aggregates. This moistening and aggregating operation is carried out in such a manner that the particles of crystalline lactose, and likewise the particles of sweetener, are not materially affected.

The moist composite aggregates leaving step **11** are subjected to drying **12** without crushing or grinding, to produce a final product having a free moisture content of less than 1%, and a total moisture content of about 3.5%.

The crystalline lactose powder supplied to the method has a relatively small particle size, as for example a size such that the majority of the material passes through a 100 mesh screen. Such a powder can be made by spray drying a lactose concentrate containing very fine lactose crystals, or by grinding a coarser crystalline material. The moisture content of such a crystalline product can be of the order of 5.2 to 6.0% (total). The anhydrous lactose powder can be made by spray drying a concentrated lactose solution, whereby the lactose content in the powder produced is in the amorphous form.

The equipment shown in Figure 6.3b for carrying out the moistening and aggregating step may consist of a vertical chamber **13** having a downwardly extending inlet conduit **14** for receiving the dry powdered mix as conveyed pneumatically from a suitable supply source. Some air is removed from the chamber through **15**, whereby air is drawn into the chamber through the lower opening **16** between the main part of the chamber and the discharge hopper **17**. The sides of the chamber can be kept warm by circulating warm air through the jacket **18**, thereby preventing moisture from condensing upon the inner surfaces. The dispersed powder passes through zone **19** where it is comingled with finely atomized water discharged from nozzle **20**. Also some saturated steam is introduced by way of nozzle **21**. The falling powder acquires moisture from the water droplets and water vapor whereby the material discharged upon the conveyor **22** has a predetermined moisture content.

Good results have been secured when the material discharging from hopper **17** has a total moisture content of the order of about 8 to 16%, from 10 to 12% being deemed optimum. The temperature level within the zone **19** can be adjusted by adjusting the proportioning between the water and the steam introduced. Good results have been secured by maintaining an average temperature level within zone **19** of the order of from 100° to 150°F, about 130°F being optimum.

As the particles of powdered material are comingled with the vapor and atomized water in zone **19**, moisture distributes itself upon the surfaces of the particles. The time period and treatment temperature within the zone is such that the lactose and artificial sweetener crystals are not materially affected. However, the particles of anhydrous lactose become relatively sticky whereby they serve as a sticky medium to cause adherence between all of the particles, when they are brought into random physical contact. Sufficient comingling takes place within the treatment zone whereby the particles are brought into repeated direct contact with the result that the particles are caused to adhere together in the form of moist porous random aggregates. The time period of treatment in the equipment may range from 10 to 60 seconds.

Immediately after the aggregates are formed and before they have been discharged from hopper **17**, some of the sticky lactose commences to crystallize, thereby making the aggregates stronger and less susceptible to breakage. Such crystallization is accompanied by reduction in surface stickiness, whereby the aggregates discharged upon conveyor **22** do not tend to adhere together in the form of a cake.

On conveyor **22**, the aggregates are held in a quiescent and compact mass, and during this holding period, which may range from about 30 to 90 seconds, there is a reduction in temperature (e.g., from 100° to 90°F) and a further reduction in surface stickiness with an increase in the strength of the aggregates. The conveyor **22** delivers the material in free-flowing form to the equipment used for final drying. This equipment is such as to avoid a substantial amount of crushing or grinding of the aggregates.

Suitable equipment for this purpose employs a screen provided with small perforations and which is vibrated to cause the material to progress from the feed to the discharge end of the same and to apply vertical motion to maintain the powder as a loose working layer. Warm dry air is delivered upwardly through the screen to pass upwardly through the layer of powder. The number and size of the openings in the screen are so chosen that the product moving along the screen is fluffed or levitated to form a layer several times the thickness it would normally have if at rest. By this method the product is caused to progress along the screen and is at least partially supported by the cushion of air intermingled therewith. This provides drying without rough or mechanical handling which might break up the powder aggregates.

The final product from the dryers may be subjected to screening and sizing operations to produce a material of relatively uniform size requirements. A typical product has a particle size such that the bulk of the material passes through a 20 mesh screen, but remains on an 80 mesh screen. Undersize material passing through an 80 mesh screen can be rejected or returned to the method, as for example back to the mixing operation **10**.

The anhydrous lactose powder is a minor part of the total amount of lactose. For example, good results have been obtained when the anhydrous powder comprises from 20 to 30% of the total lactose powder. When too small an amount of anhydrous powder is used, the aggregates formed are relatively weak. An excessive amount of anhydrous powder causes excessive stickiness, thus interfering with proper continuous operation of the apparatus.

Bulk density of the product is dependent somewhat upon the specific procedures used for forming the aggregates, but may vary from 250 to 350 g/l. The major part of the lactose present is in the hydrate or crystalline form. The material is free-flowing and is not hygroscopic with respect to atmospheric moisture. It can be packaged in ordinary cartons provided with a pouring spout.

When a quantity of the product is deposited upon a quantity of cold water in a tumbler, the mass of aggregates is quickly wetted and sinks to the bottom of the tumbler. It quickly dissolves with simple stirring. High wettability is attributed to the fact that water freely penetrates a mass of the material, and also penetrates the pores of the individual aggregates, whereby the rate of solution is governed by the size of the individual lactose and sweetener crystals, rather than the size of the aggregates. In hot water the product wets and dissolves almost instantly.

By way of example, in one particular instance the product was made as follows. An edible lactose powder was prepared by spray drying a lactose concentrate containing fine seed crystals of lactose, to produce a crystalline lactose powder having such particle fineness that 100% of the material passes through a 100 mesh screen, and 40% through a 200 mesh screen. This material had a moisture content (total) of about 5.8%. About 66 parts of such crystalline lactose powder was dry mixed with 24 parts (by weight) of anhydrous lactose powder. There had been previously mixed with the anhydrous powder an amount of dry powdered artificial sweetener to produce the desired sweetness.

The anhydrous lactose powder had a particle fineness such that 100% of the material passed through a 100 mesh screen and 60% through a 200 mesh screen. It had a moisture content (total) of about 1.5%. The artificial sweetener powder had a particle fineness such that 100% of the material passed through a 100 mesh screen and 50% through a 200 mesh screen. It had a moisture content (total) of about 2%.

The dry mix of the three ingredients was then supplied continuously to the apparatus of Figure 6.3b, with the atomized water and steam being introduced into the zone **19** to maintain this zone at an average temperature of about 130°F. The moist porous aggregates being delivered from the hopper **17** had a moisture content (total) of about 10%. The holding time during transit upon the conveyor **22** was about 40 seconds. The drying apparatus employed was of the type previously described with an inlet temperature for the drying air of about 275°F. A screen analysis of the final product was as follows:

	Percent
Through 20 mesh screen	0
On 50 mesh screen	78.4
Through 50 on 80 mesh screen	18.8
Through 80 on 100 mesh screen	2.8

The final product had a moisture content (total) of about 1.5%, and a bulk density of about 300 g/l. It was a white free-flowing material having about the same sweetness as granulated sugar, and it was readily wettable and soluble in either hot or cold water.

Foremost-McKesson Process

Another process is given in *British Patent 1,152,610; May 21, 1969; assigned to Foremost-McKesson, Inc.* for the production of a low density, free-flowing artificial sweetener powder. The process consists in preparing a concentrated lactose syrup having a substantial part of its lactose content in the form of fine lactose seed crystals. After adding the desired amount of sweetening agent, the concentrate is subjected to beating together with a whipping agent and a gas such as air to form a relatively stiff foam.

The foam is fed to a spray drying operation utilizing a centrifugal atomizer of the type which does not use shattering impacts and which is capable of producing discrete or atomized particles without serious disruption of the foam cells. The particles discharging from the atomizer are dispersed in a hot drying gas whereby they are converted to a discrete dry material. The apparatus is especially adapted to carry out the foregoing steps.

The resulting discrete material may be used as a final product, or this product may be further processed by passing the same through an aggregating operation. After treatment in the aggregating operation, the discrete material may be subjected to final drying. The final product has a density within the range of from 80 to 200 g/l. A substantial part of the lactose content is crystallized in the form of alpha-lactose monohydrate, whereby the product is relatively nonhygroscopic.

The discrete particles produced by the spray drying operation are identifiable in the aggregated product. They are spheroidal shaped, and with interior and exterior walls that are porous and within which fine lactose crystals are dispersed. Such a product is not sticky or dusty, it has high wettability, and it readily disperses in either cold or hot water with simple stirring.

The steps of the method as illustrated in Figure 6.4a consist in supplying concentrated lactose syrup containing fine lactose seed crystals to the mixing operation **10**, where it is mixed with an artificial sweetener as indicated. Instead of introducing the sweetening agent at this point, it may be introduced in the subsequent beating operation. In step **11** the concentrate is subjected to mechanical beating together with a gas and a whipping agent to form a foam. Before processing in step **11** it is desirable to heat the material to an elevated temperature such as from 140° to 180°F.

The lactose syrup as supplied to step **11** may have a concentration of from 35 to 60% solids and the fine crystals present may amount to from 5 to 25% of the total lactose content. The whipping agent may be a suitable protein hydrolysate having whipping properties, such as soybean protein hydrolysate. The gas supplied to step **11** may be air, or other gases may be used, such as nitrogen, carbon dioxide and the like. The amount of gas employed may range from about 1.5 to 15 cubic feet (at 125°F) per gallon of the liquid mix.

The foam produced in step **11** is then subjected to a spray drying operation which involves atomizing the foam and causing the resulting discrete particles to be dispersed in a drying atmosphere, such as hot air. With respect to atomizing the foam, it was found that conventional centrifugal atomizers which utilize shattering impacts serve to break down the foam by disrupting the foam cells, whereby the desired end product is not obtained.

FIGURE 6.4: FOREMOST-McKESSON PROCESS

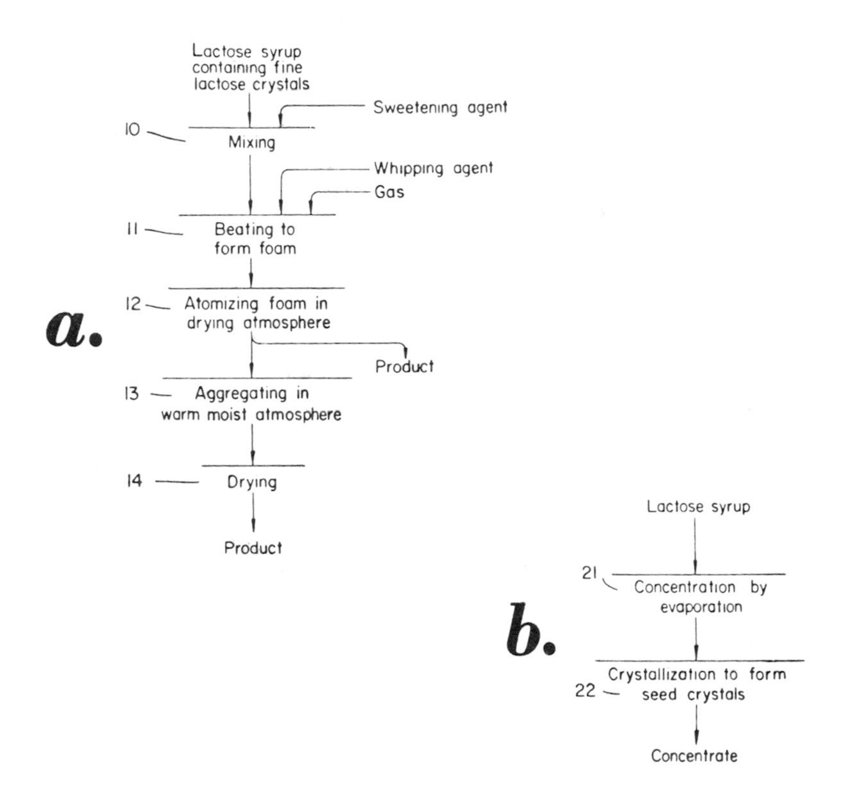

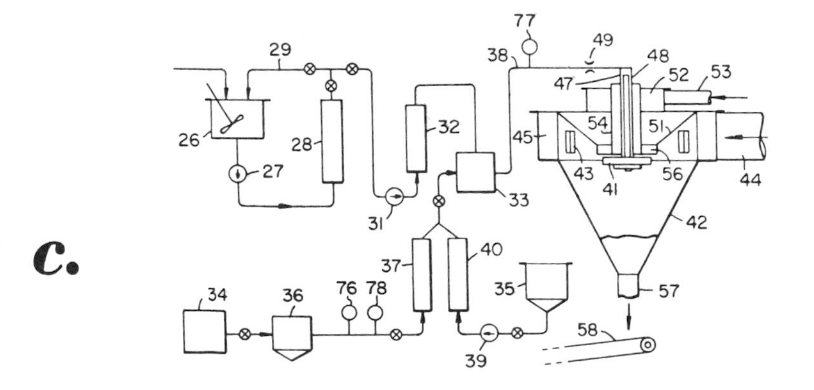

(continued)

FIGURE 6.4: (continued)

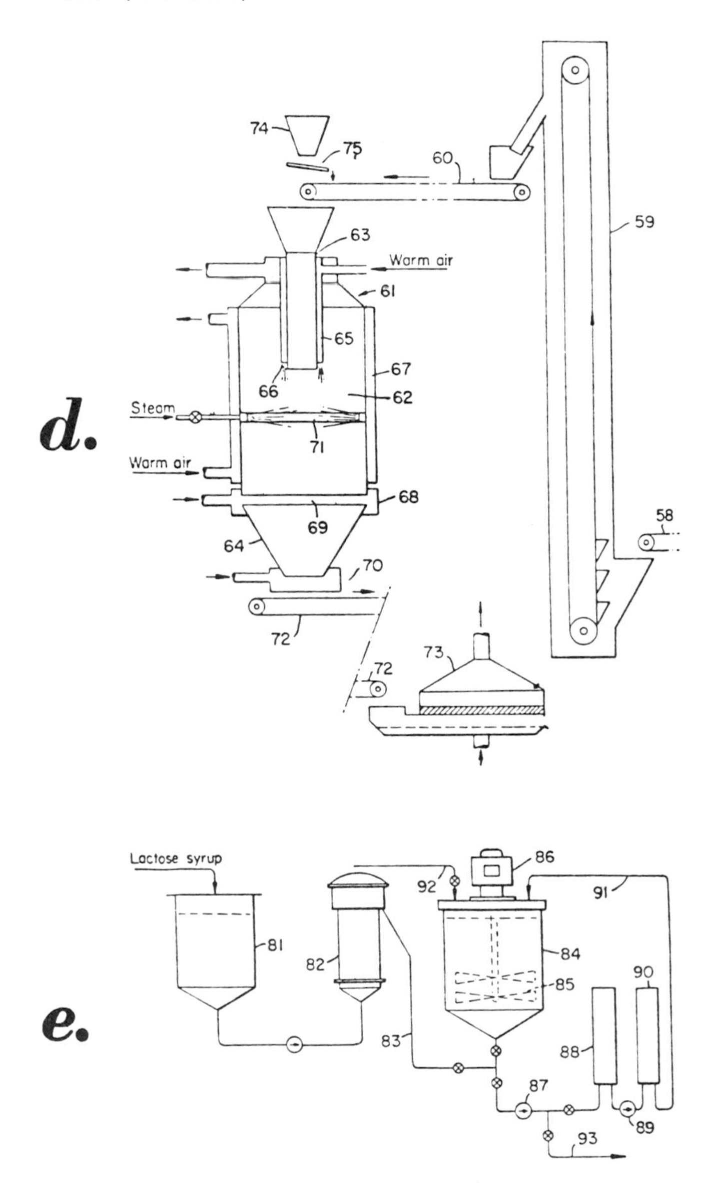

Source: British Patent 1,152,610

It has been discovered that when a stacked disk centrifugal atomizer is used, the foam cells are not seriously disrupted, and the dry discrete material produced is altogether different in that the dry particles are spheroidal, with the exterior walls of the particles being porous. The interiors of the bulk of the particles have numerous cell spaces or cavities separated by porous septa that are likewise porous with fine lactose crystals dispersed therein.

A centrifugal atomizer causes the foam to be broken up into discrete particles by discharge from the peripheral edges of rotating disks, without disruptive impacts. This serves to form discrete particles which retain the numerous foam cells, and it is such particles that produce a discrete dry product according to the process.

During spray drying some crystallization of lactose occurs due to the seeding action of the crystals present in the feed concentrate. The extent of such crystallization depends on control of the drying conditions as will be presently explained.

The spray-dried material from step **12** may be used without further processing. Its moisture content may vary from 1.2 to 4.5%, depending on the drying conditions maintained. As will be presently explained in greater detail, the drying conditions in the spray drying operation may be such as to produce a product which has the major part of its lactose in crystalline form.

The product obtained from spray drying has a bulk density within the range of from 80 to 200 g/l. The particles are of a size ranging from about 30 to 500 microns, with the major part of the material ranging from 80 to 250 microns.

In many instances it is desirable to subject the spray-dried product to further processing to promote more complete crystallization of lactose and to increase the average particle size. Thus, the spray-dried material is shown being subjected to the aggregating operation **13**. The dry material from the spray drying operation, which may have a total moisture content of from 1.2 to 4.5%, is passed through a chamber where it is subjected to a warm moist atmosphere. The surfaces of the particles are made sticky, and the particles are caused to be brought into random contacts whereby the particles cling together as aggregates.

These aggregates, which may have a total moisture content ranging from 3.5 to 10%, are subjected to final drying **14** to produce a final product which may have a total moisture content of from 1.5 to 5.5%. The drying operation is carried out with a minimum amount of attrition whereby the aggregates formed remain substantially intact. The aggregated product has a slightly higher bulk density than the product produced by the spray drying operation. The aggregated product has improved flow characteristics, the particle size is increased, and there is an absence of fine dust particles.

When the product is subjected to the aggregating operation, this step is carried out in such a manner as to cause further crystallization of lactose, whereby the amount of lactose in crystallized form is increased in the final product. Preferably a holding period is interposed between aggregating and final drying during which lactose crystallization is promoted.

Figure 6.4b illustrates a suitable procedure for preparing the concentrated lactose syrup. Thus a refined lactose syrup, which may contain from 35 to 45% solids, is further concentrated by vacuum evaporation in step **21** to produce a concentrated syrup which may contain from 50 to 63% solids. This syrup is then supplied to the crystallizing operation **22**. As supplied to the crystallizing operation, the concentrate may be at a temperature of from 80° to 140°F. In the crystallizing operation, which is carried out in the presence of continuous agitation, fine dispersed crystals are formed whereby at the end of the crystallizing operation from 15 to 35% of the lactose is in the form of the seed crystals.

Referring to the equipment of Figures **6.4c** and **6.4d**, the concentrate is supplied to the feed tank **26**, which is equipped with a suitable agitator. This tank is connected by piping to a reheating circuit including the pump **27**, and heater **28**. Thus the concentrate can be continuously recirculated through a bypass line **29** to bring it to a suitable temperature level such as 150° to 210°F. A pump **31** that is preferably of the positive displacement type takes the liquid material from the reheater circuit and delivers it through the flow meter **32** to the beater device **33**. This device also receives compressed air from the compressor **34**.

The compressor delivers air through the filter **36** and the flow meter **37**. The whipping agent is also supplied to device **33** from tank **35**. The agent, in liquid form, is delivered by positive displacement pump **39** through the flow meter **40**. The beater consists of a closed chamber supplied with high speed agitating means capable of effective beating or whipping action.

From the beater device **33** the foam is delivered through the pipe **38** to the centrifugal atomizer **41** of the spray drying equipment. This equipment consists of a spray drying chamber **42** having its upper portion arranged to receive tangentially directed hot air. Thus the upper portion of the chamber is shown provided with openings **43** which are louvered to direct the incoming air tangentially. Hot air is supplied through the connecting ducts **44** and **45**. Within the drying chamber **42** there is the centrifugal atomizer **41** of the stacked disc type.

The annular feed manifold within the atomizer has two feed pipe connections **47** and **48** with the pipe **38**, through the restricting orifice **49**. The orifice serves to maintain a desired pressure within beater device **33** and the pipe line **38**, as for example, a pressure of the order of from 40 to 120 psig passage. On the discharge side of the orifice and in the feed lines **47** and **48**, the pressure is substantially atmospheric. Therefore the foam as it is delivered to the interior of the atomizer is at substantially atmospheric pressure. Because of the release of pressure as the foam passes through the orifice, there is a great increase in foam volume.

It is desirable to provide a down draft of hot air in the area immediately surrounding the periphery of the atomizer. Thus the upper part of the chamber is provided with the conical partition wall **51** which receives hot air from the exterior box **52** and duct **53**. A cylindrical shroud **54** is shown disposed about the drive shaft of the atomizer. The central part of the partition **51** forms an annular opening **56** about the shroud whereby air is directed downwardly in a region immediately surrounding the periphery of the atomizer. The discharge conduit **57** at the lower end of the chamber is shown supplying the dried material to the conveying means **58**, which may be of the endless belt type.

In instances where the product from the spray dryer is used without further processing, any excess moisture present can be removed by a suitable dryer, such as one of the shaker type. Assuming, however, that the spray-dried material is to be subjected to aggregating, there can be used the additional equipment shown in Figure 6.4d. Thus conveyor **58** delivers the material to the elevator **59**, which together with conveyor **60** delivers the material to the hydrating and aggregating apparatus **61**. This consists of a jacketed chamber **62** having a downwardly extending inlet conduit **63** and a lower collecting cone **64**. The interior portion of conduit **63** is provided with jacket **65** that receives warm air.

Conveyor **72** receives moist aggregates and delivers them to the dryer **73**, which may be one of the shaker type in which the aggregates are contacted with warm air to remove excess moisture.

Suitable sizing equipment can be used to size the final dried material, with return of undersized fines to the process by way of hopper **74** and vibrated feeder **75**. The operation of the apparatus can be shown in the following example.

Lactose concentrate containing fine crystals of lactose is delivered continuously to the mixing tank and is continuously recirculated through the heater and bypass pipe. Concentrate at 150° to 210°F is continuously withdrawn through the pump and supplied to the beater. In the closed chamber of this beater the concentrate is subjected to intense mechanical beating action together with air and the whipping agent. The beating or whipping taking place at this point is carried out at a pressure of 40 to 120 psi. This pressure is determined by the pressure of the incoming air indicated by gauge **76**.

Also, as previously explained, pressure is maintained in the discharge line **38** from beater device **33** as indicated by gauge **77**. The temperature of the compressed air should be maintained constant and is indicated by thermometer **78**.

The foam formed under pressure is caused continuously to flow through the line and the orifice, after which it expands in volume in the feed lines to the atomizer. The expanded foam is delivered to the interior of the atomizer and passes through the stacked disks of the atomizer to be discharged from its periphery in the form of foam particles which are dispersed in the hot dry air within the dryer chamber thereby effecting flash drying with delivery of discrete material through the conduit to the conveyor.

The conveyor delivers the material from the spray drying equipment to the elevating conveyor and from there to the belt conveyor. It is caused to drop from the discharge end of the conveyor through the conduit, during which time it is dispersed in the air. It falls in dispersed condition from the lower open end of the conduit and is contacted by the warm moist atmosphere maintained by steam discharging from the pipe.

Contact with the warm moist atmosphere causes the particles to become moist and sticky and the particles during downward movement are caused to be brought into random contacts to form moist aggregates. These aggregates are delivered by the collecting cone **64** to the conveyor **72**. This in turn delivers the moist aggregates to the dryer **73** where excess moisture is removed without substantial breakage or crushing of the aggregates. Assuming that a sizing operation is carried out after the aggregates have been dried, reject fines can be returned to the hopper **74** from which they are delivered to the conveyor **60** and fed to the aggregating apparatus **61**.

The conveyor provided between the aggregating apparatus and the final dryer **73** provides a holding period during which further lactose crystallization takes place. This may be a period of the order of from 1 to 5 minutes.

Figure 6.4e illustrates suitable equipment which can be used to prepare the lactose concentrate. It consists of a tank **81** for receiving the refined lactose syrup and which supplies the syrup to the concentrator **82**. The concentrator should be of the type capable of producing concentrates of the order of 50 to 63% solids without material heat deterioration. The discharge line **83** from the concentrator serves to deliver material to the crystallizing tank **84**.

This crystallizer is of the batch type and is provided with an agitator **85** driven by motor **86**. It is connected to an external cooling circuit including the pump **87**, heat exchange cooler **88**, pump **89**, heat exchange cooler **90** and the return line **91**. A line **92** is shown for the introduction of a synthetic or artificial sweetener. Valve controlled line **93** serves to discharge the concentrate from the outlet side of pump **87** to the apparatus of Figure 6.4c.

In the batch operation of the apparatus shown in Figure 64.e, a quantity of refined lactose syrup is supplied to the tank, and this syrup is then concentrated by passing the same through the concentrator. The concentrator supplies the concentrate to the crystallizing tank. When a sufficient amount of material is in the crystallizing tank, a crystallizing cycle can be carried out as follows. With the agitator in operation, the pump is placed in operation and the valves operated to cause the concentrate to be circulated through the cooling stages **88** and **90**. Assuming that the concentrate is initially at a temperature of 125° to 165°F, at the beginning of the cycle, cooling can be carried out to rapidly lower the temperature level to about 80° to 120°F.

During this time a substantial number of seed crystals of lactose are formed, and these crystals are relatively quite fine due to shock cooling and to the presence of continuous agitation. By properly controlling the time and temperature conditions and the rate of cooling, a predetermined amount of the lactose can be crystallized to provide a proper material for supplying to the apparatus of Figure 6.4c. The sweetening agent can be added in dry form at the end of the crystallizing cycle.

Previous reference has been made to using the product discharging from the spray dryer. When the dryer is operated to produce a product having a low moisture content of from 1.2 to 4.5%, little if any of the particles are in the form of aggregates. Such a product can be used as a filler with other discrete materials (e.g., in premixes containing cereal flour and the like). When the dryer is operated to increase the moisture content of the discharging material, an increased amount of hydration takes place during spray drying, and an increased percentage of the material is in the form of aggregates. Operation to provide a moisture content of from 6 to 10% with some retention on the walls of the dryer serves to aggregate the product to the extent that it appears to be granular.

When the spray dryer is operated to discharge material having a moisture content in excess of about 4.5%, excess moisture should be removed by final drying before the material is packaged or otherwise employed. A hold period of from 1 to 5 minutes can be interposed before final drying to promote further crystallization.

In actual practice, the moisture content of the material discharging from the spray dryer has been maintained within the range of from 1.2 to 4.5%, where the material is further processed by aggregating. However, the moisture content may be maintained within a higher range of from 4.5 to 8% to increase the capacity of the dryer and to promote crystallization within the spray dryer and during transit to the aggregating apparatus.

Example 1: A refined lactose syrup containing 40% lactose solids was concentrated and precrystallized by apparatus such as shown in Figure 6.4e. The syrup was concentrated to 63% solids in the evaporator **82**, and supplied to the crystallizing tank **84** at about 105°F. The agitator **85** of the crystallizer tank was operated continuously at about 56 rpm. A crystallizing cycle was carried out by continuously recirculating the concentrate through pump **87** and the coolers **88** and **90**. The cooling was such as to reduce the temperature of the concentrate from 150°F at the beginning of the cycle to about 105°F over a period of about 1 minute. The concentrate was then held at the lower temperature level for a period sufficient to reduce the dissolved solids to about 40%. This crystallizing cycle served to crystallize 55% of the lactose in the form of fine seed crystals of alpha-lactose monohydrate. A dry sweetening agent was then added.

The sweetened concentrate produced as just described was then processed by the equipment shown in Figure 6.4c. The material introduced into the tank **26** of Figure 6.4c was continuously agitated and was recirculated through pump **27** and heater **28** to elevate the temperature to about 170°F. The material at this temperature was then supplied by pump **31** to the beater **33**. The beater was supplied with air under pressure at 125°F and at a pressure of about 80 psi.

The quantity of air supplied was about 10 cubic feet of air for each gallon of concentrate processed. The chamber of the beater **33** was also continuously supplied with a liquid whipping agent, the flow of whipping agent being metered whereby a substantially constant proportionate amount was added to the concentrate. The whipping agent employed was soybean protein hydrolysate. The rate of supply was such that 0.7% (dry solids basis) of such protein hydrolysate was contained in the liquid content of the foam being delivered through line **38**. The pressure maintained in line **38** was about 70 psi. This foam was continuously supplied through the orifice **49** to the atomizer **41** of the spray dryer **42**.

The spray dryer was operated with inlet air temperatures through conduits **44** and **53** of about 330°F and with an outlet temperature of about 215°F. The centrifugal atomizer **41** had an outside diameter of 11.5 inches and was operated at a speed of about 9,000 rpm. The discrete material delivered through the conduit **57** of the spray dryer had a total moisture content of 1.9%.

This material was delivered by the conveyors **58, 59** and **60** to the aggregating apparatus **61**. The aggregating apparatus was supplied with saturated steam at a pressure of 35 psi. The discrete material delivered into the upper part of the chamber **62** fell downwardly by gravity and was collected in the cone **64** in the form of moist aggregates. The jacket **67** was maintained at a temperature of 120°F. The humidity and temperature of air entering the rings **68** and **70** was 35% RH and 45°F. The material leaving cone **64** was at 65°F. The product remained on conveyor **72** for a holding period of 3 minutes, during which time the temperature raised to 90°F. The total moisture content of this material was

7.5%. This moist material was then subjected to drying in the dryer **73** to produce a final product having a total moisture content of 4.0%.

The product produced had a particle size ranging from about 30 to 1,500 μ, with the bulk of the material ranging from 300 to 1,000 μ. When viewed by microscope, most of the particles were aggregates. The particles bonded together to make up the aggregates were spheroidal shaped, with fine lactose crystals dispersed in the walls.

The crystals were bonded together in such a manner as to form connected voids therebetween whereby the walls of the particles were porous. Some such crystals were found to be present in the material leaving the spray dryer **42**, but a substantial part of the lactose of such particles was in the glass state. According to our observations, some crystallization of lactose took place within the spray dryer due to the presence of seed crystals of lactose, and substantial additional crystallization took place during processing after spray drying, including treatment within the aggregating apparatus **61**, retention of the material on the conveyor **72**, and during the initial stages of the final drying. The porous walls included the septa between the numerous cells or cavities forming the interiors of the spray-dried particles.

As withdrawn from the spray dryer the product produced had a bulk density of about 150 g/l and the final dry product taken from the dryer **73** had a bulk density of 185 g/l. When a quantity of the final product was deposited upon the surface of cold water, it wet and sank in a few seconds. It dissolved immediately upon simple stirring with a spoon.

Example 2: The procedure and apparatus used were the same as in Example 1, up to the point of forming a foam and supplying the same under a pressure of about 70 psi through line **38** and from there to the atomizer **41**. Drying conditions within the dryer chamber **42** were adjusted by reducing the air outlet temperature to provide discharging discrete material having about 6% total moisture. After a holding period of about 8 minutes, this material was passed through a dryer (like dryer **73**) to produce a product which was less hygroscopic than the spray-dried material produced in Example 1, due to further hydration during and immediately following spray drying. This product had a bulk density of about 165 g/l and was noticeably more granular than the product produced by spray drying in Example 1, due to the presence of some aggregates.

Example 3: The product produced by Example 2, before final drying, was subjected to aggregating and final drying by substantially the same procedure and equipment as in Example 1. The moisture content of the material leaving the aggregating apparatus was about 9% and the moisture content of the final dried material was 4%. The bulk density of the final product was 185 g/l, and the bulk of the particles had a size ranging from 300 to 1,000 μ. This product was nonhygroscopic, being superior in this respect to the product of Example 2. Also it had substantially the same flow, wettability and dispersibility properties as the final product of Example 1.

Example 4: The procedure and apparatus used were the same as in Example 1, up to the point of forming a foam and supplying the same under a pressure of about 70 psi through line **38** and from there to the atomizer **41**. However, drying conditions within chamber **42** were adjusted by reducing the outlet air temperature to provide discharging discrete material with a moisture content of 10%.

The discrete material was permitted to lodge on the side walls of chamber **42** for an average period of retention, which served to promote crystallization and formation of aggregates. The moist discharging material after holding for about 2 minutes was passed through a final dryer (like dryer **73**) to reduce the moisture content to about 4.2%.

The material was granular in appearance and the bulk of the particles were aggregates formed of hollow spheroidal shaped particles as previously described. The bulk density was 180 g/l and the particle size ranged from 30 to 1,500 μ. This product was relatively nonhygroscopic when exposed to the atmosphere and had substantially the same free-flow, wettability and dispersibility properties as the material produced by Example 1.

OTHER BULKING PROCESSES

In this section three processes are described for the preparation of dried material with increased particle size. These particular processes involve spray drying or agglomeration and are frequently cited for the preparation of low calorie dry mixes containing artificial sweeteners in place of sucrose.

Although the products discussed in these processes are dry milk, coffee, tea and other food products, the principles pertaining to equipment, material handling and operational procedures can be applied to mixes containing artificial sweeteners where the objective is to increase particle size or bulk.

Spray Drying Foamed Material

In spray drying the size of the dried particles depends on the size of the sprayed droplets. Methods of controlling the atomization to produce a coarse spray are well known. Other factors being constant, however, large droplets must require more time to dry than smaller ones. In spray drying there is always a limited amount of time available depending on the size and design of the dryer. Particles striking the walls or floor of the dryer before being sufficiently dry will adhere.

If the air velocity is low, so that the particles fall freely, the time available depends on the height of the dryer and is less for larger particles than for smaller ones. If the air velocity is great enough to carry the particles in the exhaust air, the time the particles are in the dryer is roughly directly proportional to the dryer volume and inversely proportional to the air flow rate. With excessive air flow the particles may be removed from the drying chamber and collected before completion of drying. Increasing the air temperature will increase the rate of drying but temperature increase must be limited because of its effect on product quality and because of danger of fire. Thus the production of spray dried products of large particle size presents serious problems, particularly where it is necessary to limit the size of the dryer and hence limit the time available for drying.

Another problem in spray drying is control of bulk density of the dried material. This is especially important in consumer products where the product may have to be used by the spoonful. In packaging in a container of predetermined volume, slack fill and excessive overfill must be avoided. For particular drying equipment, the bulk density can normally be varied only slightly by small changes in conditions such as air temperature. Spray dryers, then, are normally considered to be quite inflexible, particularly in regard to control of particle size and bulk density.

I.M. Reich and W.R. Johnston; U.S. Patent 2,788,276; April 9, 1957; assigned to Standard Brands Incorporated converts the solution to be dried into a foam which is fed continuously to the dryer in the usual manner and spray dried.

The foam is a dispersion of small gas bubbles in the liquid, the liquid being the continuous phase and the gas the discontinuous phase. There is no apparent chemical reaction with the liquid extract although there may be some solution of gas in the extract.

It has been found that the bulk density of the dry product (weight per unit volume in the final packaged condition) is correlated with the density of the foamed extract which is spray dried. The foam density is in turn dependent upon the amount of gas incorporated in the foam; the more gas, the lower the foam density. The gas incorporated in the foam serves to increase the puffing effect which usually accompanies the atomization of liquids into an atmosphere of heated gas.

The bulk density of porous particulate materials must depend upon the intrinsic density of the solid material, the amount of porosity within the individual particles, and also upon the shape and size distribution of the particles. Control of the amount of gas used to form the foam controls the bulk density by determining the porosity within the particles. As explained below, a reduction in the foam density permits more rapid drying of large particles.

Typical spray-dried hollow particles are characterized by a more or less spherical shape with a single empty space surrounded by a thick wall. In contrast, the process produces particles which have many internal spaces each enclosed by a thin wall. To a large extent the size and number of internal particle spaces can be easily controlled in practice.

It has been found that strong shearing forces applied to the foam just before it is dried serve to reduce the size and increase the number of individual bubbles in the foam. This results in a larger number of smaller spaces within the dried particles with correspondingly thinner walls surrounding each space although the particle size may be large. As a result of this state of increased subdivision of the dry material within the particles, light transmitted and reflected by them is scattered to a greater extent. This condition has the effect of causing the product to appear lighter in shade just as the comminution of materials which transmit and reflect light generally makes them appear to be of a paler cast. Thus, by varying the bubbles size in the foam, the intensity of color in the dry product may be controlled.

Another factor which affects the shade of the dry particles is the nature of the gas used to prepare the foam. For example, it has been demonstrated that carbon dioxide foam produces larger spaces in the particles and consequently a darker color than does nitrogen foam used in the same manner. Foams made with nitrogen are more stable than those made with carbon dioxide. Thus, under similar conditions, a carbon dioxide foam has a coarser structure and hence yields a product with a thicker internal wall structure.

In order to increase the particle size of spray-dried particles without increasing the size of the dryer it is necessary to minimize the time required for drying, which time of course, depends on the rate of drying and the amount of water which must be removed from the particles. The rate of drying of sprayed droplets is generally understood to be dependent upon the rate of heat transfer from the surrounding heated air to the droplets. This rate depends on the area available for drying, the air temperature and the coefficient of heat transfer. If the diameter of the sprayed droplets is increased, the area available for drying increases in proportion to the square of the diameter but the amount of water to be evaporated increases as the cube of the diameter because of the increased weight of liquid in the larger particles. It has been determined empirically and reported in the literature that the coefficient of heat transfer is inversely proportional to the diameter of droplets.

The time required to dry a particle of given size will also depend upon the density of the particle which is proportional to the product bulk density if the particle shape and size distribution are not changed. Thus, in summary, the required drying time is proportional to the square of the particle diameter, to the ratio of water to solids in the extract and to the bulk density of the dry product. For example, should it be desired to double the average particle diameter without increasing the drying time, temperature being unchanged, this could be accomplished by adjusting the foam density to produce a product of one-half the bulk density and by halving the ratio of water-to-solids in the solution or suspension. The process provides a convenient means for adjusting the bulk density.

Any gas which will not react to the detriment of the product, for instance, air, nitrogen and carbon dioxide in solid or gaseous form, may be used to produce the foam. In the case of aqueous coffee extracts the coffee solids concentration is preferably in the range of 40 to 70%. In the case of aqueous tea extracts the tea solids concentration is preferably in the range of 45 to 75%. The density of the foam may vary within wide limits. In the case of aqueous solutions it may be, for example, from 0.05 to 1.0 g/ml; for aqueous coffee extracts, 0.4 to 0.7 gram per milliliter; for aqueous tea extracts, 0.1 to 0.4 g/ml.

The table below presents data on gas (carbon dioxide) to liquid ratios at 70°C, by weight and by volume for various foam densities. The volumetric ratios are calculated at 1 atmosphere pressure, at which the foam density is measured, and at 275 atmospheres (4,040 psia), which corresponds to a typical spray pressure. The calculations are also based on a liquid extract having a density of 1.24 g/ml at 70°C.

Foam Density at 1 atm., g/ml	Volumetric Ratio, 1 atm.	Volumetric Ratio, 275 atm.*	Weight Ratio
1.0	0.23	0.00047	0.00029
0.1	11.6	0.0237	0.0147
0.05	24.6	0.0502	0.0311

* Neglecting possible solubility of gas in extract.

Example 1: A pure aqueous coffee extract containing 55% solids was spray dried in a ten-foot spray dryer of the type shown in Figure 40, page 842, *Chemical Engineers' Handbook,* edited by John H. Perry, 3rd edition, McGraw-Hill (1950). This dryer is of mixed flow design, i.e., the flow of spray and drying gas is partly parallel and partly countercurrent, and employs a high pressure atomizing system. It is unusually efficient in terms of production rate for its size. Calculation of the overall average contact time available by dividing the dryer volume by the air flow rate indicates a time of approximately 4 seconds.

The coffee extract was pumped to a controlled pressure of 40 psi and a controlled flow of carbon dioxide gas was introduced into the extract at this pressure. Since it was desired in this instance to produce a dark colored product, carbon dioxide was used rather than air and the foam was not subjected to a great deal of shearing force. The combined flow of gas and liquid extract in the form of a foam was then raised to a pressure of 150 psi with a booster pump which fed a high pressure reciprocating pump in which the foam was compressed and raised to the 4,000 psi pressure employed for spraying.

By regulation of the amount of carbon dioxide introduced the density of the foam at the high pressure pump measured at atmospheric pressure was adjusted to 0.5 g/ml. The bulk density of the dry product obtained was 0.25 g/ml. The product was relatively dark in color, of very coarse particle size and had excellent solubility in hot and cold water.

Example 2: A pure aqueous tea extract containing 65% solids was converted into a foam and spray dried in the manner described in Example 1. The tea extract was pumped to a controlled pressure of 650 psi and a controlled flow of carbon dioxide gas was introduced into the extract at this pressure. The combined flow of gas and liquid extract, in the form of a foam, was then fed directly to a high pressure reciprocating pump in which the foam was compressed and raised to the 4,500 psi pressure employed for spraying. The density of a foam sample withdrawn and measured at atmospheric pressure was about 0.25 gram per milliliter. The bulk density of the dry product obtained was about 0.13 g/ml. The product obtained was like that obtained in Example 1 in that it was relatively dark in color, was of very coarse particle size and had excellent solubility in hot water.

Granular Powders

To produce agglomerates of the desired density it is essential that the moisture added to the powdered particles be carefully controlled. Skim milk powder, which has great affinity for water, is difficult to wet uniformly, while whole milk powder, a fat-bearing material, is quite uniformly surface wetted by spraying. Uniform wetting is essential to the production of uniform agglomerates under impact.

Using skim milk powder as an example, fine dense agglomerates or clusters are produced if the moisture is added in the range of 7 to 8% and the product is subjected to impact quickly after wetting. The dense small particles so produced can be rewetted, as for example when the milk is being reconstituted, only by pouring water on the dry material before stirring. These dense particles also retain all of the normal hygroscopic characteristics of regular skim milk powder. These dense small agglomerates may or may not be desired in a skim milk solid, although such agglomerates may be highly desirable for products other than milk.

A low density, agglomerated product can be made by the process of *E.C. Scott; U.S. Patent 2,900,256; August 18, 1959.* The process is especially suitable for treating protein-containing powdered materials. For purposes of illustration, the starting material may be ordinary dried milk powder derived from previous desiccation. The production of a relatively low density, spongy skim milk product requires uniform wetting to the extent of 10 to 14% by weight, and allowing the wetted protein to swell and the milk sugar to hydrate and absorb its water of crystallization requirements, thereby expanding the individual particles of powder.

Preferably, the milk powder is gravity fed at a uniform rate between two fan-shaped sprays of liquid, for example, water or skim milk. The liquid may be maintained at a temperature of about 160°F. Condensed milk at this temperature has proved to be an excellent wetting agent. It promotes good cohesion and also provides additional solids and increased yield.

Spray nozzles are disposed to emit a fine spray of the hot liquid downwardly in the direction of the falling airborne powder. If the spray were directed upwardly, the flow path of the particles would become distorted, and some particles would be hit repeatedly by the spray. It is preferable that the wetting be as uniform as possible.

Before the wetted material is subjected to impact to produce agglomerated clusters, it is desirable to permit the liquid to penetrate through the powder particles to completely hydrate the same. This may be accomplished by permitting a short time, say 15 to 20 seconds, to elapse between the spray step and the impacting step. Preferably during this time period the wetted mass is subjected to vigorous agitation to promote thorough mixing and uniform distribution of the moisture. A ribbon mixer equipped with prongs serves well for this purpose.

The spongy mass discharged from the ribbon mixer may contain some rather large sized chunks. The material is then introduced into a hammer mill operating at high speed in which the rotating blades break up the chunks and simultaneously effect the necessary concussion or impact to agglomerate the particles into clusters or granules of the desired size. The size of the clusters is controlled by the degree of wetting, as previously indicated, the speed of the blades in the mill, the taper of the blades and the size of the screen in the discharge duct of the mill. The tapered or angularly disposed mill blades strike the particles tangentially and cause them to collide repeatedly with each other. A flat or square blade tends to carry the particles directly through the mill, which minimizes the opportunity for agglomeration.

The clusters must then be dried without causing disintegration or further agglomeration. The preferred method for drying is by depositing the clusters on a porous conveyor belt in a uniform layer and advancing the layer through an oven in which hot air at about 160° to 240°F is circulated upwardly through the material.

Other suitable means for drying may also be employed; for example, the clusters may be dropped into a countercurrent stream of hot air which is moving at a rate sufficient to buoy the clusters so that they fall slowly to the bottom of the vessel. Drying is effected during the period of descent. Drying time is approximately 30 seconds.

The clusters may be sized by passing them between slightly spaced rolls and then cooling to about 80°F. Cooling may be carried out by advancing the clusters on a porous belt through an enclosed space through which cold air is circulated upwardly through the belt, or by other suitable means. Fines may be removed by screening.

The resultant product consists of relatively coarse, porous clusters of convenient size for handling. They are readily wettable by water so that they dissolve easily and quickly and do not pack in the same manner as fine powders when they are mixed with water. The clusters or agglomerates formed can be mixed into a liquid and dissolved very quickly with a minimum of agitation; they are readily flowable in dry form and will not cake in storage, even though they are readily water-soluble.

Reference is made to Figure 6.5. Dried milk powder, or other dried material which is to be formed into clusters in accordance with the process, is charged into the hopper **10**. A screw conveyor **12** driven by a motor **14** connects to the bottom discharge opening in the hopper and serves to convey powdered material at a uniform rate from the hopper into the cylindrical hydration vessel **16** through a screen **18** in the outlet opening leading from the screw conveyor.

Mounted at diametrically opposite points in the wall of the hydration vessel are nozzles **20** through which wetting liquid is directed downwardly in the form of a fine fan-shaped spray. Powder passing through the screen **18** is subjected to the fan-shaped spray patterns of liquid emitted from the nozzles. The wetted particles fall to the bottom of the vessel and are discharged into a ribbon mixer **22**. The ribbon mixer has thin spiral tape **24** and a series of radial prongs **25** disposed around a central shaft **26**. The shaft is driven so that wetted material entering the mixer **22** becomes thoroughly agitated as it is advanced to the end of the mixer from which it is discharged into the hammer mill **28**.

The tapered rotor blades **30** within the hammer mill rotate at a relatively high speed (1,700 to 2,300 rpm) and cause the particles to be subjected to concussion and impact which brings about formation of the clusters by agglomeration. The agglomerated clusters pass through the screen **32** in the bottom of the mill, which serves to provide control over the size of the clusters. The clusters fall from the mill onto the conveyor **34** which has a porous belt for supporting the clusters. The clusters are carried by the conveyor through the oven **36** in which hot air rises from the air inlets **38** through the clusters on the belt. The dried clusters discharged from the end of the conveyor pass through a pair of sizing rolls **40** which are spaced so as to break up any clusters that may have bunched together during the drying step.

To decrease the length of the oven, several belts may be disposed, one above the other, within the oven to carry the clusters back and forth until sufficient time has elapsed to complete the drying operation. From the sizing rolls the sized clusters drop through chute **41** to another conveyor **42** within cooling zone **44** supplied with cold air through ducts **46**. The finished product is discharged from the left end of the conveyor **42** through chute **45**.

A pilot plant model built in accordance with the drawing was operated in the production of milk agglomerates from dried skim milk powder. The hopper was adapted to hold about 20 pounds of skim milk powder. The powder was continuously and uniformly discharged through the 14 mesh screen **18**, which was about ¾ inch wide and 6 inches long. The screen insures discharge into the vessel continuously and at a uniform rate.

FIGURE 6.5: APPARATUS FOR PRODUCING GRANULATED FOOD PRODUCTS

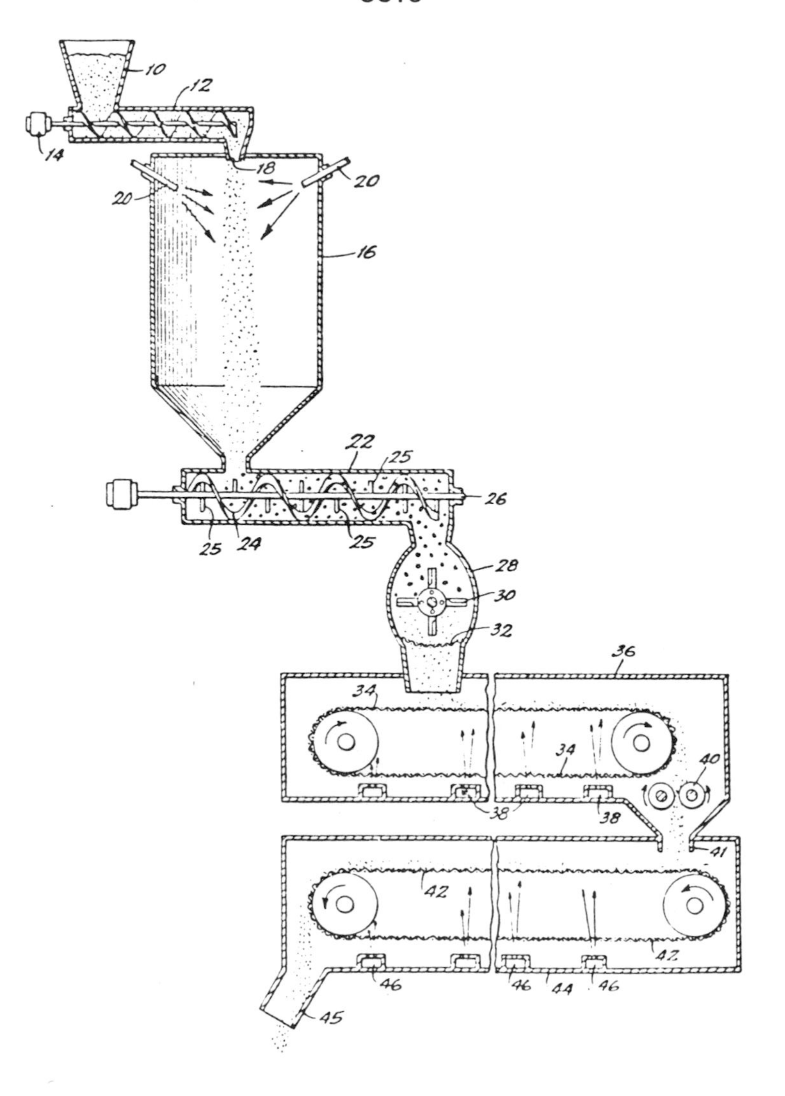

Source: U.S. Patent 2,900,256

Skim milk at 160°F was introduced through the nozzles **20**. The powder flowing through the hydration vessel **16** was delivered at the rate of 395 pounds per hour and sufficient milk was introduced through the nozzles to effect 12% wetting

(60 pounds per hour of 9% skim milk). The wetted powder was advanced through the ribbon mixer which was operated at a speed of about 75 rpm. The time elapsing did not exceed about 10 seconds. The hammer mill **28** was equipped with a bevelled blade rotor and operated at 2,200 rpm using a ½ inch perforated screen **32**. The resultant product had the properties described above.

In a modified form, a conveyor belt may be disposed between the hydration vessel and the hammer mill to supplement the ribbon mixer, the object being to lengthen the time of hydration of the particles by the liquid prior to the agglomerating step.

Blaw-Knox Process

H.L. Griffin; U.S. Patent 2,893,871; July 7, 1959; assigned to Blaw-Knox Company gives a process and apparatus for the conversion of powdered materials into agglomerated products with low bulk density.

Referring to Figure 6.6a, the powder to be treated is fed continuously from an overhead hopper **10** to a horizontal vibrating feeder **11** which spills from its forward edge into the agglomerating zone **12** below. The vibrator is electrically driven in known fashion, its motion being horizontal. The powder falls in a thin, straight line, sheet-like stream.

In one form successfully used, the straight line feeder edge from which the powder spilled was 6 inches wide and was provided with a vertically adjustable gate **11a** to control the size of the orifice above the spill-over edge, thereby to control the thickness of the free-falling sheet.

The agglomerating zone **12** requires no housing provided the ambient air is sufficiently clean and quiet. Just above it and equally spaced on either side of the plane in which the powder falls are two parallel steam pipes **13** and **14**, horizontally disposed and parallel to the plane, having longitudinal slit orifices and forming nozzles **13a** and **14a**. These nozzles receive low pressure steam or other moistening fluid from a common source by way of lines shown diagrammatically in Figure 6.6b. They deliver two thin sheet-like jets **A** and **B** of the moistening vapor downwardly at an angle from opposite sides of the vertical plane **C** in which the powder falls. These nozzles are shown in enlarged view in Figure 6.6a and are further shown in Figure 6.6b as seen from beneath to disclose the slit orifices.

Ordinarily, the nozzles are from 2 to 4 inches apart, on centers, and are formed of ½ inch (nominal) standard pipe, with steam supplied preferably at both ends by lines **D** and **D'** to get a more uniform discharge over the length of the slit orifice.

The two converging jets **A** and **B** collide at the vertical plane **C** or, in other words, the powder falls into the apex of the V-shaped space or trough between these opposed but converging and colliding jets. After passing through this zone of collision, the solid product falls a short distance through the air at substantially room temperature before entering a top opening in a drying unit below.

The drying unit has its top opening or inlet **15** directly beneath the jets, and the opening is large enough to receive the falling solids, being some 12 inches

FIGURE 6.6: SPECIAL AGGLOMERATION PROCESS

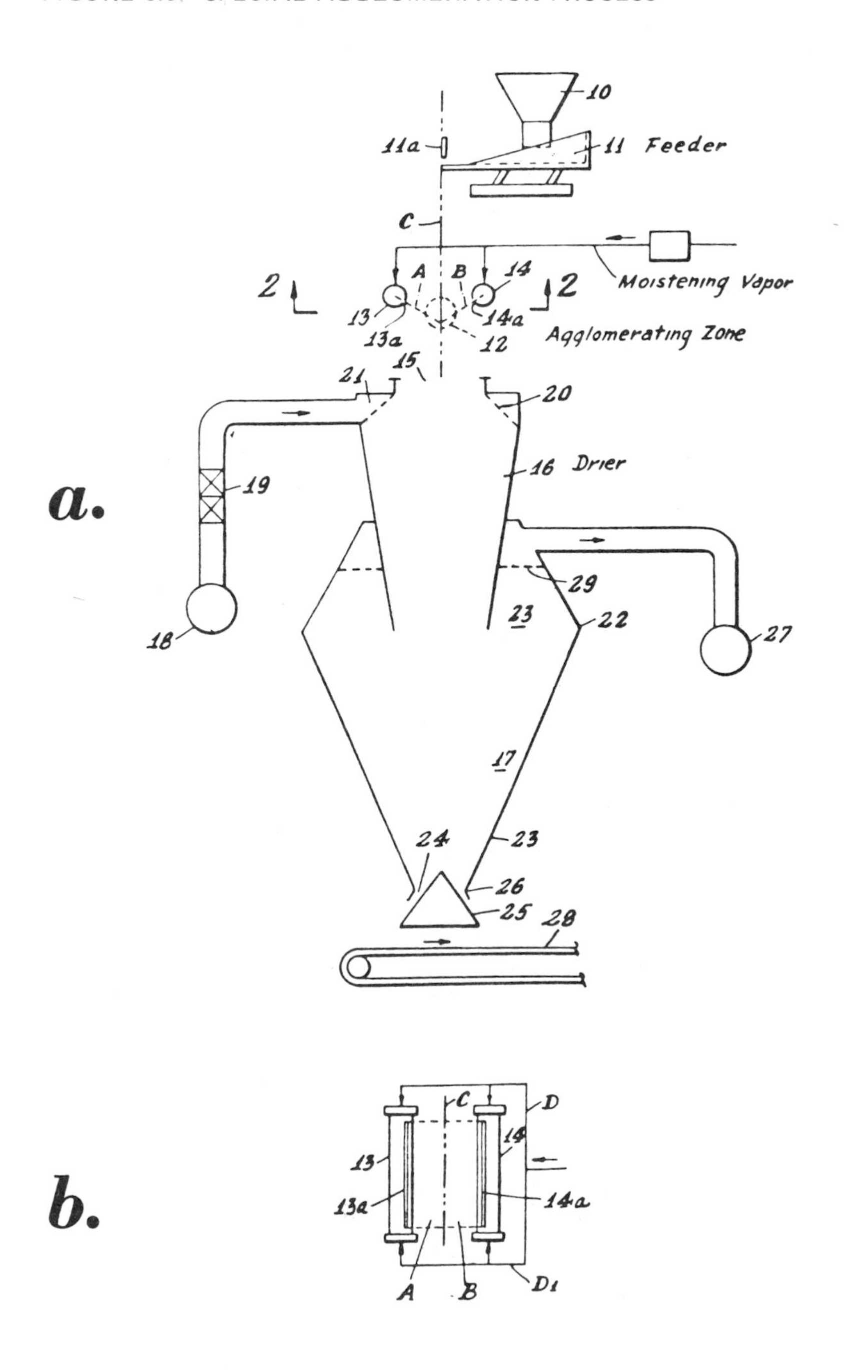

Source: U.S. Patent 2,893,871

in diameter when the powder is fed in a stream of the order of 6 inches in width. The inlet **15** is spaced several inches below the jets so that there is a short but clear zone through which the solids fall before reaching the dryer, after leaving the zone of collision of the jets.

The drying unit consists of a generally cylindrical, slightly converging, dryer proper **16** together with a surrounding housing **17**, both being arranged on a vertical axis. Warm air is delivered to the dryer at the top of a blower **18**. A steam or other suitable heater **19** is located in the supply duct beyond the blower **18**. An inclined inner perforated plate **20** of annular form is mounted at the top of the dryer and forms an annular distribution chamber **21** into and around which the warm air flows and from which it passes down through the perforations into the dryer proper, enveloping the solid material entering through the top opening **15**. There are no baffles or obstructions in the dryer **16**, and the solids pass through as a freely falling body.

Surrounding the lower half or so of the dryer unit, and extending below it, is a housing **17** having an enlarged section **22** opposite the lower end of the dryer, and having a converging or hopper bottom **23** with an annular solids outlet **24** formed between the steep inverted cone **25** and the rim of the opening **26** at the hopper bottom. The upper part of the chamber **23** thus formed around the dryer is connected to an exhaust blower **27**. Warm air issuing from the bottom of the dryer, and containing the moisture taken up from the falling solids, is drawn gently to the exhaust, while solids similarly issuing fall to the outlet at the bottom which delivers to any suitable collection means such as a bin, car or conveyor belt **28**.

There is a perforated plate **29** across the annular chamber **23**, just below the air outlet, to provide better distribution of the air. Entrainment is minimized by the fact that the enlarged flow space provided by the housing **22**, as compared with the dryer, causes a sharp reduction of velocity and thus promotes separation of solids from air. Any residual solids that remain in the effluent air may be removed by a cyclone separator, bag filter or other known means.

Dried skim milk powder, conventionally prepared by spray drying in the manner of a "low heat" product for drinking, and having a moisture content of the order of 2.5 to 3.5%, is fed from the feeder at the rate of about 750 pounds per hour, in a thin sheet-like stream about 6 inches wide and about one-eighth inch in thickness. The steam nozzles **13** and **14** are on centers 2 to 4 inches apart, and their slit orifices, about 7 inches long and 0.006 inch wide, are directed downwardly at an angle of about 40° from the central vertical plane. The two jets thus are about 80° apart.

The milk powder falling into the zone of turbulence where the two sheet-like jets collide is subjected to two actions. The particles are superficially moistened by condensation of a portion of the steam, as well as by contained moisture in the steam; and at the same time they are tumbled about, while remaining largely in a confined zone affording intimate contact, with the result that they adhere together in clusters or agglomerates. All but an inappreciable portion of the particles are thus agglomerated.

The agglomerates fall through the air space, some 6 to 10 inches, between the line of collision and the top of the dryer. The precise action occurring in this stage is not fully known, but is believed to involve a cooling and setting or further agglutination of the agglomerates prior to their admission to the warm air of the dryer. The production of good agglomerates involves a time factor, and this stage between the zone of turbulence and the drying provides that factor, supplementing the brief dwell in the zone of turbulence.

The drying is done under mild conditions of temperature and flow rate, and with the agglomerates falling freely without baffling. The inlet air in the illustrative case of treating dried milk is from about 120° to 185°F, being less warm when less moist, and the outlet air is about 100° to 160°F. The usage of air is from 400 to 500 cubic feet per minute, where the dryer **16** has a diameter of about 3 feet and a length of about 4 feet. The supply and exhaust blowers are operated to maintain a pressure approximately equal to atmospheric at the solids inlet **15** of the dryer, so that there is little or no flow of warm air outward or of cool air inward. There is inflow of air at the solids outlet **24**, which cools the effluent agglomerates.

The process variables require no extensive change when the process is used with other materials which it is capable of agglomerizing by reason of their capacity for becoming self-adherent when superficially moistened. Its key step is the combined moistening and tumbling in a zone of turbulence which confines the powder particles so as to afford intimate contact in their turbulent state, that zone being created when a powdered material is fed into the bottom of the trough formed by two converging and colliding low pressure jets of moistening fluid.

The width of the jets (about 7 inches in the example above) can be increased or decreased to vary the throughput of the unit, since the action is the same in any unit of length beyond the minimum which is economically feasible. The sheet-like form of the stream or jet is best. Moistened warm air or other compatible fluid, capable of releasing its moisture upon contact with the cooler powder, can be used in place of low pressure steam having that same capability; but the process is more readily carried out with steam because of the relative ease with which steam is made available and is held uniform in quality. Water jets can also be used, but in such cases the jets are much thinner through use of narrower nozzle slits.

The optimum feed rate of the powder and of the moistening vapor, as well as the optimum angle of incidence of the jets, vary somewhat with different powders according to their differences in density, moisture absorption and requirement of moisture for good adhesion. A slower feed of powder or a higher supply of moisture is used where more moisture absorption is needed or where absorption is slower. The angle of incidence is variable from about 15° to 45° in terms of angular departure from the central plane along which the powder is fed.

Variation of angle or of the feed rates varies the bulk density of the product. The optimum condition is one with which the powder is tumbled sufficiently well and long to get the desired agglomeration, and is recognizable as one with which there is neither upward tossing of the particles to such extent that they are freed from the zone of collision and are dispersed, as occurs when the angle is too great, nor direct passage of any substantial number of particles through the zone of collision, without sufficient moistening or tumbling, as occurs when the angle is too small.

The significance of the angular relation of the jets will be plainer upon consideration of two extremes which bring about no useful amount of agglomeration. If the jets are 180° apart, so that they collide head on, there is strong turbulence but particles fed into the zone of turbulence are largely dispersed upwardly and outwardly, and do not return to the zone to receive enough moisture and close enough contact to effect agglomeration. On the other hand, if the angle between the jets is too small, so that they are nearly parallel, there is too little turbulence when they come together and the particles are swept through without being moistened and tumbled in close relation so as to be predominantly converted into agglomerates.

At intermediate angles the particles are sufficiently moistened superficially, and are tumbled about sufficiently in a confined zone of turbulence affording intimacy of contact, so that a predominant proportion of them are agglomerated to yield a bulk product having improved properties derived from its predominantly agglomerate form. This desirable condition of predominant agglomeration can be effected by the use of a low pressure moistening fluid, the jets of which supply the small amount of moisture needed while also providing, through their collision at these intermediate angles, the confined zone of turbulence affording also the intimacy of contact that causes the moistened particles to adhere to one another in clusters to a predominant extent.

The condition yielding good results is one in which the particles appear as if boiling, without boiling over, in the zone where the sheet-like jets collide. The steam jets themselves are scarcely visible as they emerge but show some condensation from passage through the air. There is little or no visible vapor escaping from this zone as uncondensed or unused steam.

The nozzles **13** and **14** can be mounted so as to be horizontally adjustable to vary their spacing, angularly adjustable to vary the angle of incidence of the jets, and vertically adjustable to permit variation of the space between the nozzles and the dryer inlet **15**. In general, the gap between the line of collision and the dryer top is 6 to 10 inches, and an optimum spacing for each set of conditions is readily found.

The solid material takes up very little moisture in the agglomerating step, something on the order of half a percent and only superficially; but it usually is desirable to remove it by a current of air or compatible gas having the appropriate capacity for moisture absorption. Under some conditions, this air current need not be heated. The product removed from the dryer has substantially the same moisture content as the starting powder. Where the added moisture is unobjectionable, the air-current drying may be omitted; but in such case the free-fall of the agglomerates through ambient air is desirably made considerably longer than the gap above the dryer, this being to permit them to become well set and to lose any surface tackiness by passage through the air. In such case a housing is desirable.

The process has a further capability and advantage in treating powdered materials consisting of two or more component substances which are desirably put into intimate admixture. An example is a mixture of sugar and cocoa. The two components may be fed separately into the plane **C** from two opposed feeders, or may be coarsely mixed and fed as a single body. In either case, when the falling sheet reaches the zone of collision of the sheet-like jets the resulting tumbling of the particles effects a very intimate mixing.

In other cases of mixed materials, as in the case of malted milk, it is possible to effect the mixture while one or more components are in the liquid state and before it is dried to a powder form. A malted milk powder so prepared has been agglomerated with notable success by this process using the apparatus described above and essentially the same feed rates and jet angle described for the case of skim milk powder. The mixing of malt and milk powder may, however, be effected in the agglomerating step itself, by the action of the colliding jets. The easy and rapid solution or dispersion of an agglomerated product composed of malt and milk powder makes it possible to prepare a malted milk beverage at the point of use by simple addition of water to reconstitute the liquid milk.

Except for the agglomeration, the milk or other material thus treated undergoes no material change of either a physical or a chemical nature so far as has been determinable. The heating in the dryer is so brief and mild as to have no appreciable effect other than the removal of the added moisture, and there is no material net change in moisture content in the overall process.

When reconstituted with water, the agglomerated milk product presents striking advantages over the conventional powdered milk and also over the special products now marketed. The product disperses quickly in water. If not stirred, it forms a thin custard-like mass at the bottom of the container, but this mass is readily penetrated by water so that complete solution occurs in a matter of minutes, while the comparison products, forming a sticky mass which often remains dry in the interior, are much more slowly penetrated and do not dissolve without stirring within any feasible time. If stirred, either immediately or after a time, the product of this treatment goes into solution more rapidly than the comparison products, and so rapidly as to qualify as an "instant" product, requiring only brief stirring.

By making certain modifications or additions at the feed or the agglomerating steps, or both, the product can be made even better in respect of the mechanical strength of the aggregates and the percentage of relatively large aggregates. Further, by making possible the production of larger and firmer aggregates as the immediate product of the agglomeration, which are oversize relative to the desired maximum size in the end product, there is a readier control of the aggregate size and resulting bulk density of the end product through mechanical breakup of the intermediates into smaller aggregates in the course of finishing operations, which may include after-drying and screening. The basic apparatus and process already described are still employed.

One of these modifications and apparatus attachments causes the addition of somewhat more moisture to the product in total, and therefore makes it preferable in many instances to remove moisture in an after-dryer in order to assure against caking; but this increment of moisture can be kept small, raising the total moisture to no more than about 5 to 5.5%.

SPRAY-DRIED WITH WHIPPING AGENT

Unilever Limited; British Patent 999,073; July 21, 1965 provides a process for the preparation of a low calorie sweetening composition, in which a slurry comprising lactose, a water-dispersible edible protein whipping agent and an artificial sweetening agent is spray dried to produce a granular product having a density of 0.2 to 0.4 g/ml, a moisture content of 3 to 8% by weight, and containing at least 25% by weight of lactose in the alpha monohydrate crystalline form.

The amount of artificial sweetening agent included in the composition is preferably sufficient to bring the total sweetening power of the composition approximately to that of an equal volume of household sugar, although if desired larger or smaller proportions of the artificial sweetening agent may be used.

The whipping agent used is a water-dispersible edible protein, for example a hydrolyzed caseinate, and a particularly satisfactory caseinate for this purpose is one obtained by the partial hydrolysis of casein, and which comprises, in addition to the hydrolyzed casein, sodium and/or calcium chloride. The amount of whipping agent needed is not large; in the case of the preferred hydrolyzed casein, for example, an amount of whipping agent forming 0.1 to 2% by weight of the composition, is sufficient to give a good product.

The moisture range of 3 to 8% represents the water present as water of hydration as well as free moisture in the product. The moisture content of the sweetening composition is preferably from 4 to 6% if an anticaking agent is not incorporated.

In addition to the essential ingredients already stated, other additives may be used to improve the characteristics of the sweetening composition. Thus, since the sweetening composition is of low density, it may show a tendency to float on the surface of liquids; this tendency can be reduced, and the rate of dissolution of the sweetening composition consequently increased, by incorporating into the slurry or mixing with the spray-dried granular product an edible surface-active agent. This surface-active agent is preferably a water-dispersible mono-diglyceride mixture although any food-grade surface-active agent may be employed.

It is desirable that the product remain free-flowing, and to attain this end, if no anticaking agent is added, the composition should comprise at least 45% by weight of lactose in the alpha monohydrate crystalline form. However, if an appropriate amount of anticaking agent is added, the level of alpha monohydrate lactose can be as low as 25% by weight of the product. The presence of an anticaking agent does, of course, reduce the tendency of the finished product to cake during storage and can advantageously be used whatever the alpha lactose monohydrate content of the composition may be. Sodium aluminum silicate has been found to be a particularly suitable anticaking agent for these sweetening compositions.

If desired, the composition may also contain conventional edible preservatives such as benzoic acid, sorbic acid or the salts of these acids. The production of particles having a low bulk density is facilitated by the introduction of a soluble gas into the slurry prior to drying it. Carbon dioxide is a preferred gas, but any gas which is soluble in the slurry and which is not detrimental to the product may be used, for example nitrogen or air. Preferably the particles obtained are hollow.

Example: A low calorie sweetening composition was prepared having the following ingredients, amounts being given in percent by weight.

Lactose	93.5
Water	5.5
Sodium saccharin	0.795
Partially hydrolyzed calcium caseinate	0.195
Mono-diglyceride derived from cottonseed oil	0.01
	100.000

This composition was prepared by adding the alpha lactose monohydrate, casein hydrolysate and saccharin to sufficient water (at room temperature) to form a slurry containing 50% solids. The slurry was pumped to a spray nozzle at 10 to 14 kg/cm^2 gauge (150 to 200 psig), carbon dioxide gas being introduced between the pump and the nozzle. Air flow in the tower was countercurrent, with an inlet temperature at 232°C. The slurry was spray dried to a bulk density of about 0.22 g/ml and a moisture content of about 5%. The product contained about 65% of alpha lactose monohydrate.

The spray-dried product was classified according to particle size; the portion of the material having a particle size similar to that of household granulated sugar was retained and the finer material was reprocessed.

The mono-diglyceride was incorporated into the product by mixing one part of a 1% mono-diglyceride per 99% lactose mixture with 99 parts by weight of the spray-dried particles. The product obtained was equal to granulated sugar in caking characteristics. No significant flavor difference was observed when the sweetening composition and granulated sugar were tested on an equal volume basis in coffee, but, since the product was about 7 times lighter than sugar, it contained only approximately one-seventh as many calories.

SPRAY-DRIED WITH DEXTRIN

The compositions specified by *H.T. Gebhardt; U.S. Patent 3,320,074; May 16, 1967; assigned to Afico SA, Switzerland* consist essentially of dextrin having a bulk density below about 0.15 g/cc and a dextrose equivalent of about zero, and from about 1 to 12%, preferably from about 4 to 9%, based upon the weight of the dextrin, of at least one noncaloric artificial sweetener. In addition a small amount of a noncaloric flow conditioner, up to about 0.25% by weight of the dextrin and sweetener, may if desired be included in the sweetening composition.

Dextrins are carbohydrates commonly derived from starch by dilute acid hydrolysis, enzyme action or dry heating. Vegetable, cereal and root starches may be converted to dextrin, and dextrin prepared from tapioca root is particularly preferred because of its white color where the product is to be used as a table sugar substitute. The color of the dextrin is of lesser importance where the product is to be mixed with flavorants or other colored substances. The dextrin may be decolorized or otherwise whitened where color is important.

Desirably, the dextrin used in this process should be of a purity such that significant quantities of mono-, di- and trisaccharides are not present. Such dextrins are accordingly essentially free from natural sweetness. By virtue of its freedom from sugars, the product has a Dextrose Equivalent (percent reducing sugar calculated as dextrose) of zero or very close thereto, not exceeding about 1. Likewise, the dextrin gives no reaction or only a slight reaction with Fehling's solution or with iodine. It is water-soluble and precipitated from aqueous solution by ethanol.

The products are made by a specific process which provides bulk extenders having bulk densities substantially below those heretofore available without the use of emulsifiers, gums, stabilizers or the like. Moreover, the process provides a product which will not separate or segregate during handling and packaging, the artificial sweetener remaining intimately dispersed among the particles of the bulk extenders and thus ensuring that all portions of the product are substantially uniform in composition and will remain so in packaging and even after long periods of storage. In contrast, products made by earlier processes frequently separate during storage and packaging and, therefore, are not uniform throughout.

An aqueous solution containing between about 30 to 65% by weight of dextrin having the characteristics heretofore set forth and up to about 12%, suitably from about 4 to 9% (based on the weight of the dextrin), of at least one noncaloric sweetener is subjected to a pressure between about 200 and 2,000 psi. A nonreactive, noninflammable gas is introduced to aerate the solution, and the aerated solution is then spray-dried to produce a granular free-flowing solid. The product may have a moisture content of about 10% by weight or less, preferably about 6% or less. After spray-drying, the flow conditioner, if any, is mixed with the granular product by conventional techniques.

Virtually any nonreactive, noninflammable gas that does not react with the constituents of the solution and which is a gas at the operational temperature and pressure can be used. Air, nitrogen, carbon dioxide, helium, carbon tetrafluoride and the like are illustrative satisfactory gases. The amount of aeration gas added should be sufficient to provide a product of the bulk density desired, since bulk density of product is dependent upon the amount of aeration gas. Generally, from about 0.05 to 0.50 scf of aeration gas per pound of dissolved dextrin is suitable. A solution temperature of from about 40° to 180°F has been found satisfactory.

The bulk density of the product is readily controlled by varying solution concentration, amount of aeration gas added, and solution temperature and pressure within the ranges specified. Solution concentration may vary between about 30 and 65% total solids, but a concentration in excess of about 40% is preferred. Concentrations below about 30% generally are insufficient to obtain the required density, while solutions having a concentration in excess of about 65% become too viscous to give a product of the desired appearance and free-flowing characteristics. Solutions having concentrations between about 40 and 55% by weight of total solids are most preferred.

The quantity of aeration gas may be adjusted to give the required bulk density in accordance with density measurements made as the spray-drying operation proceeds. Increasing the aeration gas concentration decreases the product density, and conversely. Normally, a concentration between about 0.13 and 0.25 standard cubic feet per pound of dextrin is adequate to achieve the desired density.

Solution pressure may be varied from about 200 to 2,000 psi, the exact value chosen depending primarily upon the diameter of the atomizing nozzle, velocity of air flow and other characteristics of dryer design. With nozzle diameters of about 0.025 inch to 0.030 inch, pressures between about 450 and 550 psig are preferred and give a satisfactory product.

Optionally, the aqueous dextrin solution or the water used to prepare the dextrin solution may be pasteurized before spray-drying by any conventional manner such as by maintaining the solution at an elevated temperature for a substantial period of time. For example, it was found that adequate pasteurization takes place when the solution is maintained at a temperature between 150° and 200°F for from 10 to 20 minutes, though other conditions known to the art may as readily be used. Pasteurization may be particularly advantageous when substantial time may elapse between the preparation of the solution and the spray-drying operation.

The spray-drying apparatus and its operating conditions are generally conventional, it being important that the spray-drying conditions enable production of a product having a moisture content below about 10% and preferably below 6% by weight. Higher product moisture contents or higher dryer temperatures may result in caking of the product. Choice of drying temperatures and drying gas temperature depends upon the characteristics of the spray dryer used.

DRYING WITH MALTO-DEXTRIN

W.C. Braaten; U.S. Patent 3,325,296; June 13, 1967; assigned to Norse Chemical Corp. states that attempts to produce a sucrose substitute have not resulted in a commercially acceptable product, primarily because they were unable to yield a product which had the crystalline appearance and the behavior-in-use characteristics of ordinary granulated sucrose.

An improved sugar substitute is prepared by combining an artificial sweetener with starch hydrolysate and drying the solution in a particular way. One of the suitable starch hydrolysates is a corn syrup solid identified by the Corn Industries Research Foundation as "malto-dextrin." This is a conventional starch degradation product having a dextrose equivalent (DE) in excess of 13% but not more than 28%. It is available in large supply and at relatively low cost, and it has virtually no discernible sweetness so that the sweetness level of the finished product can be controlled by regulating the proportion of the artificial sweetener used.

An important aspect of this process is that in the production of the product the selected sweetener and the starch hydrolysate are combined in an aqueous solution. As a result, the sweetener and the hydrolysate become intimately and uniformly mixed. This solution is then dried and particulated. By exercising certain controls over the drying, the nature of the end product can be varied and predetermined. Thus it is possible to convert aqueous solution of starch hydrolysate and artificial sweetener either into a product having substantially the same density and crystalline appearance as sucrose, or into a product which has the same crystalline appearance and behavior-in-use characteristics as sucrose, but a greatly decreased bulk density.

To be more specific, if the starch hydrolysate-artificial sweetener solution is dried on an ordinary vacuum drum dryer, and the resulting flakes which are scraped from the drum are milled and classified, i.e., graded for particle size, the product will be practically indistinguishable in appearance, taste and behavior-in-use characteristics from granulated sucrose, provided of course that the correct amount of sweetener is used. Its density and caloric value will be, or can be, made substantially the equivalent of sucrose.

On the other hand, if the solution is dried in a way which produced a cellular or porous structure so that each particle of the milled product has an irregular shape, and perhaps also some porosity, the bulk density of the composition will be far less than that of sucrose, with the result that the caloric value per unit of bulk or volume will be less than that of sucrose, though its sweetness strength is equal to that of sucrose. Controlled low temperature vacuum drying will produce this cellular or porous structure. One way of getting the needed control is to use the drying method known as lyophilization. In lyophilization the solution is placed in a container and frozen at a very low temperature into a cake, which is then transferred to a vacuum chamber where it is heated and dried.

If desired, additives such as gum acacia and gelatin and conventional stabilizers and preservatives may be incorporated in the solution before it is dried, to give the composition the benefits expected from their addition. The incorporation of a small amount of gum acacia and gelatin seems to improve the structural strength of the particulate product and hence minimizes attrition due to shipping and handling, an advantage which is especially important where the product is dried to a cellular or porous form with a view towards reducing its bulk density. Gum acacia and gelatin also tend to make the product less hygroscopic and enhance its crystalline appearance.

Example 1: 30 parts of malto-dextrin and 1 part of artificial sweetener, both in fine pulverulent form, were dissolved in 20 parts of water, by heating. The resulting syrupy solution was then dried by spreading it in a thin film onto the surface of a steam heated vacuum drum dryer, in which the pressure was maintained at about 100 mm Hg. This produced dry flakes which were scraped from the drum and particulated. The resulting product had the crystalline appearance and flowability of granulated sucrose, contained less than 5% moisture, and had approximately the same bulk density, caloric value and sweetness as granulated sucrose. When used as a sweetener for coffee or tea, either hot or cold, it gave no evidence of floating, but on the contrary quickly dissolved. Heat did not affect the product any more than it affects ordinary granulated sucrose.

Example 2: 40 parts of malto-dextrin and 2 parts of artificial sweetener, both in fine pulverulent form, were dissolved in 5 parts of water by heating. This formed a syrupy solution. One-half part of gum acacia was dissolved in 10 parts of boiling water. The two solutions were combined and thoroughly mixed, and then dried in a vacuum dryer, as in Example 1. The resulting product had the desired crystalline appearance of granulated sucrose, contained less than 5% moisture, and had approximately the same bulk density and caloric value as sucrose, but considerably greater sweetness.

Example 3: 40 parts of malto-dextrin and 2 parts of artificial sweetener, both in fine pulverulent form, were dissolved in 25 parts of water by heating. The resulting syrupy solution was placed in a container to a depth of about one-half inch, and frozen at -40°C. The frozen cake was then transferred to a vacuum chamber and exposed to a pressure of less than 0.1 mm Hg and radiant heat of 70° to 80°F for 3 hours. This reduced the moisture content of the material to less than 5% and gave it a cellular or porous structure.

The cellular-porous cake was then particulated, i.e., granulated, to a grain or particle size comparable to that of ordinary granulated sucrose. The resulting product appeared crystalline and because of the irregular shape of its individual particles, and perhaps also such porosity as the particles may have had, the product had a bulk density of approximately 60% that of ordinary granulated sucrose. On a volume basis, its sweetness strength was equivalent to that of sucrose and hence the caloric value of a given volume of the product was considerably less than an equal volume of sucrose.

Example 4: 100 parts of malto-dextrin in fine pulverulent form were dissolved in 50 parts of water, and mixed with a solution of 4½ parts of artificial sweetener in fine pulverulent form in 10 parts of water. The resulting solution was then lyophilized, as in Example 2, and particulated. This produced a product that had the crystalline appearance of granulated sucrose, a bulk density 75% that of granulated sucrose, a caloric value approximately 75% that of sucrose, and a sweetness on a volume basis substantially equal to that of sucrose.

Example 5: 1,000 parts of malto-dextrin in fine pulverulent form were dissolved in 500 parts of water to form a first solution. 60 parts of artificial sweetener also in fine pulverulent form were dissolved in 100 parts of water to form a second solution. A third solution was formed of 10 parts of gum acacia in 40 parts of water, and 4³⁄₁₀ parts of gelatin were dissolved in 40 parts of water to form a fourth solution. These four solutions were combined and thoroughly mixed. The resulting mixture was then placed in trays in a vacuum chamber. The air was withdrawn from the chamber and at the same time the contents of the chamber were heated to about 50°C. This caused a foaming and a puffing of the material as the moisture was drawn from it. The degree of vacuum to which the chamber was evacuated was about 0.50 mm Hg and the time involved was 3 hours. The material was then removed from the vacuum chamber and particulated by passing it through a 20 mesh screen.

The resulting particulate product had approximately a 5% moisture content, a bulk density one-half that of granulated sucrose, and a sweetness approximately the same as that of sucrose on a volume basis. Like all the products of all of the other examples, this product also had a crystalline appearance and the easy flowability of granulated sucrose.

Example 6: 5 pounds of corn syrup solids were dissolved in 1,000 cc of water. The corn syrup solids, like the malto-dextrin used in the preceding examples, was a conventional starch hydrolysate derived from simple hydrolysis of cornstarch, but having a DE of about 40%.

Another solution was prepared by dissolving 125 grams of artificial sweetener in 150 cc of water. A third solution was prepared in which 10 grams of acacia and 5 grams of gelatin were dissolved in 100 cc of water. The three solutions were combined and thoroughly mixed, cooled to room temperature and dried in a vacuum chamber. The dried material was particulated and had the appearance of granulated sucrose, but its density was only 43.3% that of sucrose. This product was more hygroscopic than the products of the aforesaid examples, and hence had to be kept in dry storage, or otherwise protected against humid conditions.

Example 7: 100 grams of corn syrup of 80% solids concentration and having a dextrose equivalent of 26 was heated to 70°C. A solution of 5 grams of artificial sweetener and 10 ml of water was added and mixed in thoroughly. The resulting syrupy solution was cooled and dried by heating in a high vacuum oven and particulated. The particulated material had the crystalline appearance of granulated sucrose and a density 52% that of sucrose.

Example 8: 100 grams of malto-dextrin was dissolved in 50 ml of water by heating to 85°C. Six-tenths of a gram of saccharin sodium hydrate was dissolved in 4 ml of water, and a third solution consisting of four-tenths of a gram of acacia and one-tenth of a gram of gelatin was dissolved in 10 ml of boiling water. These individual solutions were combined, mixed and cooled to room temperature; and then low temperature vacuum dried, and particulated. The resulting product had the crystalline appearance of granulated sucrose and a density 78% that of sucrose.

GRANULAR CLUSTERS

A. Schapiro; U.S. Patent 3,100,909; August 20, 1963; assigned to Roto-Dry Corporation found that it is possible to produce a variety of water-soluble and/or water-dispersible food products by forming water-soluble, water-absorptive or water-dispersible food solids in the presence of an aqueous or hydrophilic, tacky bonding agent derived from the other ingredients by solution in a dispersing liquid and mixing until a granular mass of agglomerates or clusters is formed.

In more detail, an initially slightly tacky, moist granular mass is formed which is then subjected to a low temperature thermal conditioning treatment. In this thermal conditioning treatment the granules are subjected to heat while being tumbled to expose fresh surfaces to the action of the heat. This thermal conditioning step removes most but not all of the moisture. The phrase "thermal conditioning step" is used advisedly because this important step is not a mere drying step. Surface moisture is evaporated but some of the surface moisture migrates into the body of the granules and is held absorptively and/or adsorptively and/or as water of crystallization in such a way that a friable product results. In this friable product any lumps are easily broken up into small granules.

The moisture content of the granules is sufficiently low that the product is seemingly dry, friable and can be poured easily, yet the moisture content (virtually all of which is present as internal moisture) is sufficient to render the product quickly soluble and nondusting. Too much moisture impairs the desired free-flowing qualities and too little moisture renders the granules too hard and too slow to dissolve in water. Generally a total moisture content of about 1 to 6% based on the weight of product is preferred. This moisture, which can be called internal or bound moisture, aids to disperse the product, and also serves with the soluble ingredients as a binder to hold together small particles of the several ingredients in the form of clusters or agglomerates, each cluster or agglomerate containing all of the components of the mixture.

Glycerin or other hydrophilic liquids may replace all or part of the water in the bonding agent. The granular mixture is free-flowing, yet it does not readily form a dust, nor does it pick up moisture readily from the air. This material can be used in high speed packaging machines and also in dispensing machines where small unit quantities are dispensed. The product is also readily soluble or dispersible in water, either in hot or in cold, depending upon the specific ingredients. The initial blending and mixing operation of the ingredients may be carried out in several ways, such as the following.

Example 1: Solid food ingredient A and an aqueous solution or dispersion of solid food ingredient B are mixed, as by adding A first to a blender and then adding the aqueous solution or dispersion of B. The proportion will be such that after thorough blending, a moist and tacky but granular mix results.

The solid food ingredient A may be any one or a mixture of the following: a starch product including unmodified potato starch, unmodified tapioca starch, unmodified cornstarch, unmodified sago starch; modified potato, tapioca, corn and sago starches which dissolve or thicken at lower temperatures than the unmodified starches or whose solutions have lower viscosities; pregelatinized starch, dextrins and amylose; a sugar including dextrose, sucrose, lactose, corn syrup solids; a proteinaceous food derivative having when moistened adhesive, binding or thickening properties, including wheat gluten, pectin, amylopectin, casein, casein salts, gelatin.

Examples of aqueous solutions or dispersions of solid food ingredients B are tomato paste, orange concentrate, lemon concentrate, pureed fruits and vegetables, concentrated milk products including condensed whole milk, condensed skim milk, condensed cream, single-strength cream, half-and-half milk and cream, and condensed whey. It will be understood also that part or all of the water in these solutions or dispersions of food solids may be replaced by glycerin, propylene glycol or syrups of sorbitol, glucose, invert sugars, molasses, etc.

Example 2: Solid food ingredient C and solid food ingredient D are added to a blender and water is added in quantity sufficient to form, upon thorough blending, a granular, moist tacky mixture. Instead of water, a water-miscible solvent such as glycerin, propylene glycol, ethyl alcohol or ethyl acetate may be used, either alone or in admixture with water. Also mixtures of these solvents may be used.

Food ingredients C and D in Example 2 are any of the solids listed under A above and any of the food solids listed as dispersions under B above.

The blended, moist, tacky granular mix produced as described under either Example 1 or Example 2 is then further subjected to a thermal conditioning step which consists of subjecting the material to surface drying while the mass is being constantly stirred and tumbled to bring up fresh moist surfaces of the agglomerated particles or clusters for drying off the surface moisture. It is preferred to use radiant heat as the source of heat, and to subject the blend to a gentle tumbling action of a character which does not disintegrate the clusters or agglomerates, but which continuously brings fresh portions of the mixture into contact with the heat.

A low velocity current of air is passed over the mixture during agitation and heating. For this purpose it is preferred to use an upwardly inclined trough type screw conveyor which has interrupted flights of a character such that each screw segment will lift a portion of the mass a distance forwardly and upwardly, then allow the mass to rest or to slip back to be picked up by another flight, to commingle with a later portion of the mass and to move forwardly. A trough is provided under and partly around the screws, and infrared lamps are disposed above the screws. Alternatively, the thermal conditioning step may be carried out by tumbling in a vacuum. The temperature of the food material is held low to avoid loss or change of flavor, and in general ranges from 75° to 180°F.

In Examples 1 and 2 a batch operation is described in which the granular mass is transferred after blending to a heating-tumbling mixer (such as the preferred screw conveyor trough) for the final heat conditioning step. The whole operation may be made continuous, particularly when using the screw conveyor trough arrangement, so that proportioning feeders may be used to continuously feed into the lower end of the trough the constituent food materials, where the revolving screw mixes and blends continuously, and conveys the blend into the radiant heat zone of the conveyor where the surface moisture is evaporated from the ever-changing surfaces of the granules or clusters which are brought into the heat rays. All of the following specific examples may be carried out as continuous operations, as here described, with the final dry granular clustered or agglomerated products discharged from the upper end of the conveyor trough. Other tumbling means may be used, with applied radiant heat and air circulation.

Example 3: Nondietetic Carbonated Soft Drink Beverage — 80 to 85 parts by weight of sucrose of quickly soluble, baker's special type and 10 to 15 parts of citric acid were added to a muller. Operation of the muller was commenced and 3 to 5 parts of a refined corn syrup were added. Orange concentrate may be substituted for part of the corn syrup. Coloring matter was added to the syrup, for example, any of the standard food colors such as FDC Yellow No. 5 or No. 6. Unless otherwise indicated, parts are by weight.

A flavoring material such as oil of orange or orange concentrate may also be dispersed in the syrup but preferably it is added at a later stage. An edible emulsifying agent such as sorbitan-mono-oleate or fatty acid glycerides may also be added. The emulsifying agent functions to emulsify the flavoring agent. Mixing was continued until the citric acid and sucrose crystals were completely and uniformly coated with the syrup. Then 10 to 12 parts of granular or powdered sodium bicarbonate were added and mixing continued. Within a short time, usually about 5 to 10 minutes, the mass fluffed up. Mixing was continued until the fluffed-up mass subsided. Generally, subsidence occurred about 12 to 15 minutes after the mass first fluffed up.

The mass was then thermally conditioned as above described. The mass in the muller was transferred to the screw conveyor. The screw conveyor moves the mass along and tumbles it as it moves, thereby exposing it uniformly to the heat. During this heat treatment and tumbling action in the conveyor, a flavoring material may be added. If the flavoring material is rather volatile, it is added near the outlet end of the conveyor. Alternatively, the flavoring material may be added to the mass after completion of the heat treatment and mixed with the solid material in a simple mixer.

The material thus produced may be stored in fiberboard drums having steel lids closed with ordinary tightness. Prior to packaging the material in retail packages, it is screened or run through a comminuting device to break up lumps. The material, if properly prepared, has no tackiness at temperatures up to 100°F, but like many powdered or granular materials, it forms soft, easily broken, friable lumps which can be easily broken up prior to packaging.

Example 4: Dietetic Carbonated Soft Drink Beverage — The procedure of Example 3 was followed but the mixture was modified as follows to provide a dietetic beverage. 100 parts of citric acid were placed in a muller followed by 4 to 10 parts of 70% sorbitol syrup in which a suitable coloring agent was dissolved. Artificial sweetener was added in an amount for desired sweetness. 5 parts of a powdered, partially converted dextrin gum were also added to improve tackiness. The sorbitol syrup used in this beverage is not as efficient as a tackiness agent as the corn syrup of Example 3. Mulling is commenced and, when a uniform coating of the solid particles is achieved, 90 parts of granular or powdered sodium bicarbonate are added. Mulling is then continued until fluffing up and subsidence have occurred and the mass is then subjected to heat treatment as in Example 3.

SUGAR SUBSTITUTES AND SPECIALTY INGREDIENTS

This chapter is a continuation of the previous one and it contains a group of patents for processes where the objective was to create a sugar substitute or a product with reduced calories. In some instances this is accomplished by use of enhancers and sweetness potentiators in combination with sucrose.

SUGAR SUBSTITUTES AND EXTENDERS

Methyl Glucoside Sweetener

The use of artificial or synthetic sweeteners to replace sugar in all types of products has expanded enormously, particularly the use in the so-called "one calorie" beverages. These sweeteners, and products using them are favored by diabetics who must limit their carbohydrate intake. The concentrated nature of these synthetic sweeteners is such that very small amounts are required, and for convenience they are all often extended with other materials, such as lactose, low-density corn syrup solids, casein, gum arabic, glycine, and the like which are essentially calorigenic themselves.

More significantly, the synthetic sweeteners in their extended form cannot replace sugar in many foods, since they lack the other functional properties of sugar—bulk, humectancy, water binding, and plasticizing. Thus, additional non-caloric ingredients are required if any significant reduction in calorie content is to be realized.

The bulk extenders, which have been used with the nonnutritive sweetener approved for human consumption, i.e., saccharin, are relatively few in number and many of the things which have heretofore been used have serious disadvantages. Some are nutritive and the use of any large amount defeats the purpose of using a nonnutritive sweetener. Others, such as lactose, are not rapidly soluble. Still others yield turbid solutions or solutions of too high a viscosity. There is, therefore, a great problem in extending a small quantity of a nonnutritive sweetener to the bulk density on a sweetened basis of sugar.

In addition, these extenders all suffer from disadvantages including color, taste, hydroscopic nature, tendency to separate, need for preservatives, and the like. More seriously the solid extenders, i.e., corn syrup solids, glycine, lactose, casein and gum arabic, which are the major ingredients by weight in these and similar preparations, are metabolizable and therefore undesirable for weight watchers, diabetics and the like. On a weight basis these extenders are essentially calorically equivalent to sugar. Significantly, all the extended products are dispensed on a volume basis rather than on a weight basis. None of these are intended for use on a weight basis and none can be used instead of sugar in cakes, candy or other high-sugar foods. In effect, none of these extended products contributes anything except sweetness.

According to *R.M.L. Paterson and M.J. Skrypa; U.S. Patent 3,656,973; April 18, 1972; assigned to Allied Chemical Corporation* a glucoside having the structural formula:

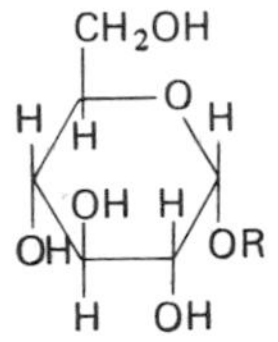

CH_2OH
H H O H
OH H
OH OR
H OH

where R is a radical selected from the group consisting of a lower alkyl of 1 to 4 carbon atoms, and hydroxy-substituted lower alkyl, when used in combination with sweetening agents acts as a functional sugar substitute. Examples of suitable glucosides include methyl glucoside, ethyl glucoside and hydroxyethyl glucoside. The glucoside can occur in either an alpha or beta configuration and the term "glucoside" is meant to include the various stereo and optical isomers and mixtures, e.g., α-dextrorotatory form.

It has been found that the glucoside-sweetener mixture has a unique combination of desirable properties for use in cakes, candies, jams, preserves, frostings, frozen desserts and other foods. The glucosides are white, crystalline solids, having no objectionable taste or odor with a high degree of solubility in water and are nontoxic and nonmetabolizable. With respect to the use of the glucoside-sweetener mixtures in food compositions, they may serve to replace in whole or in part the sugar normally present in these food compositions.

Suitable sweetening agents include glycyrrhizin and noncaloric synthetic sweetening agents, such as perillartine, flavanone glucoside derivatives, saccharin and/or their mixtures. Other potentiating agents, such as maltol and ethyl maltol may be combined with the glucoside-sweetener mixture. If the sweetener is in solid form, the glucoside and sweetener may be intimately mixed in any desired ratio to form a free-flowing mixture. Alternatively, the sweetener may be used in the form of a liquid or solution in which case the sweetener is combined with the glucoside by any well-known spraying technique so that the glucoside sweetener mixture is in the form of a free-flowing mixture.

As mentioned above, the amount of sweetening agent used may vary depending on the degree of sweetness desired and the food product to which the nonnutritive, functional sugar substitute is to be added. However, sweeteners generally are used in an amount ranging about 0.15 to 7.5% by weight based on the weight of the glucoside used. The synthetic sweetening agent may be added either as the free acid, or as the alkali salt, such as the calcium, sodium, potassium or ammonium salts.

Sweetening agents such as sorbitol, glycerol, dextrose, mannitol, maltol and gum arabic may be used in addition to or in lieu of synthetic sweetening agents, except that these are at least partially metabolizable and add calories to the ultimate food composition.

The glucosides are white, crystalline, highly soluble and odorless materials which are nontoxic, noncaloric, nondeliquescent and have good hydration qualities with a mildly sweet taste. Thus, when combined with small amounts of a sweetening agent, there results a combination which is an all-purpose, nonnutritive sugar substitute that can be used in all food areas with little or no recipe change. The use of the glucoside-sweetener mixture as a nonnutritive, functional sugar substitute in the preparation of cakes is of particular advantage since the glucoside-sweetener mixture can replace all or part of the sugar in a typical layer cake with slight modifications in the formula and little or no loss in bulk. If sugar is omitted from a layer cake recipe and replaced with many of the so-called sugar substitutes sold, it is not possible to produce a conventional flour layer cake. The product will not rise, it has a tough gelatinous texture, and is quite unpalatable.

In addition, the glucoside-sweetener mixture may be used in essentially all types of bakery products including the several varieties of bread, cake, cookies, crackers, biscuits, pies, doughnuts, pastries and the like, which normally contain sugar. Furthermore, the other high calorie components of bakery products, such as flour, shortening agents and milk solids may be replaced in whole or part with the use in the bakery product of a nonnutritive, water-insoluble, hydrophilic, cellulosic material to produce a reduced calorie bakery food composition characterized by a reduction of at least 50% in its caloric content in comparison to an equivalent bakery food and has an eating quality of at least about that equal to an equivalent bakery food.

In addition, the nonnutritive functional glucoside-sweetener mixture may be incorporated directly into the food composition in an amount which is essentially the same, on a weight basis, as the sugar present in the corresponding conventional food composition.

Furthermore, because of the high solubility of the glucosides in aqueous solutions, these solutions have extremely high osmotic pressures, quite similar to sugar syrups which prevent the growth of molds and fungi and can be effectively employed in preserved fruit jams and jellies. Thus, unlike other sugarless preserves, no chemical preservatives or fungicides are needed with products prepared using a glucoside.

Moreover, drastic reformulation is not necessary. For instance, in the preparation of low-calorie jams or jellies, the use of low methoxy pectin and a calcium salt is generally required which hampers the reproduction of the texture and mouthfeel of conventional preserves. However, with the use of a glucoside-sweetener syrup, neither a chemical preservative nor a thickening agent (to give body to the juice liquor) is required.

Generally a glucoside-sweetener mixture can be used as a nonnutritive functional sugar substitute and has applicability in any food, food composition, food ingredient, or food product, whether comprising a single ingredient or a mixture of two or more ingredients, whether liquid-containing or solid, whether mainly carbohydrate, fat, protein, or any mixture, whether edible per se, or requiring preliminary conventional steps, such as cooking, mixing, cooling, mechanical treatment, and the like. In other words, the glucoside-sweetener mixture may be used in any food as a nonnutritive functional sugar substitute where sugar and/or synthetic sweetener are conventionally used.

Example 1: Preparation of Glucoside Containing Synthetic Sweetener – Powdered methyl glucoside is mixed in a P-K twin shell blender with a saturated aqueous solution containing calcium saccharin, blended till homogeneous, dried, crushed and screened.

Example 2: A dietetic layer cake was prepared having the following composition given in parts by weight.

Flour	100
Methyl glucoside sweetener	130
Shortening	55
Egg powder	15
Milk	120
Cream of tartar (potassium bitartrate)	4
Sodium bicarbonate	2
Salt	1.5
Vanilla extract	1
Water	40

The shortening and methyl glucoside sweetener were creamed together in the bowl of a Sunbeam Mixmaster mixer at high speed. The egg powder was mixed with the water at high speed, then added to the shortening-sugar substitute mixture, and the system was beaten until it attained a uniform lemon color, about 5 to 10 minutes at high speed. The other dry ingredients were blended together. Increments of the dry blend and of milk were alternately added to the egg-glucoside-shortening mix with the mixer operated at low speed.

When the mixture appeared reasonably homogeneous the batter was poured into round 8 inch diameter. Teflon coated cake pans and baked in an oven at 350°F for 25 minutes. The product was similar in height, texture and appearance to a normal yellow layer cake. The calorie content of the glucoside cake batter was 238 calories/100 grams in comparison to 334 calories per 100 grams for a normal sugar cake.

Example 3: A low calorie layer cake was prepared having the following composition.

Ingredients	Parts by Weight
Avicel*	120
Methyl glucoside sweetener	135
Egg powder	40
Water	70
Milk	165

(continued)

Ingredients	Parts by Weight
Baking powder	5
Salt	1.5
Vanilla extract	1

*A commercially available cellulose derivative comprising crystalline aggregates prepared by acid hydrolysis of cellulose and having a particle size range of approximately 1 to 300 microns.

The egg and water were mixed as in Example 2, and the sugar substitute added. The procedure of Example 2 was then followed. The product was a yellow layer cake similar to that of Example 1. However, the calorie content of the cake batter was only 65 calories per 100 grams compared to 334 calories per 100 grams for the normal flour-sugar layer cake. Similar results are obtained using methyl hydrocellulose having a degree of substitution (DS) of 0.75 in place of Avicel. This material is water-insoluble but hydrophilic.

Example 4: A low calorie fudge candy was prepared having the following composition.

Ingredients	Parts by Weight
Unsweetened chocolate	15
Milk	35
Methyl glucoside sweetener	100
Salt	0.5
Butter	7
Vanilla extract	0.3

The chocolate and milk were heated on a hot plate and stirred till smooth. Then the sugar substitute and salt were added and stirring continued till the mixture was smooth and began to boil. The composition was allowed to boil slowly until the temperature rose to about 236°F, then it was removed from the hot plate.

Butter and vanilla were then added and the system allowed to cool to 110°F. The product was beaten until fairly thick and poured into a greased pan to set and cool. The finished candy was similar to normal sugar fudge but had only 95 calories per 100 grams in comparison to 345 calories per 100 grams of the normal sugar fudge.

Example 5: Low calorie candy bars were prepared having the composition shown below.

Ingredients	Parts by Weight
Methyl glucoside sweetener	100
Tartaric acid	2
Water	20
Flavor and coloring optional, to taste as desired	

The ingredients were heated with gentle stirring, until a clear melt was obtained and continued to 356°F. This system was removed from the hot plate and allowed to cool in a cold greased tin. The product was a hard, clear, brittle, glasslike material which makes an excellent base for a variety of hard candy products.

This material can be colored and flavored as desired, cast into small pieces like sour fruit balls (or boiled sweets), or into plaques with sticks for lollipops or suckers. By cooking to a lower temperature with the addition of shortening various softer toffee-type confections can be prepared, as will be obvious to those skilled in the art. The calorie content of the base material was only 1.5 calories per 100 grams compared to over 350 calories per 100 grams for a similar sugar product.

Example 6: A reduced calorie cake frosting was prepared having the following composition.

Ingredients	Parts by Weight
Butter	57
Methyl glucoside sweetener	200
Vanilla extract	2
Salt	Trace
Water	42
Lemon flavoring	Trace

The butter was creamed and the salt and sugar substitute were added gradually. Vanilla and sufficient water were added to give a good consistency. The frosting was a typical creamed cake frosting but contained only 134 calories per 100 grams in comparison to 402 calories per 100 grams for a normal creamed sugar frosting, a reduction of 34% of the original calories.

If desired, chocolate frosting could be prepared by adding unsweetened chocolate, or using chocolate to replace part of the sugar substitute. A methyl glucoside-glycyrrhizin sweetener is particularly efficacious with chocolate flavorings.

Example 7: A low calorie creamed cake frosting having the following composition was prepared.

Ingredients	Parts by Weight
Avicel	40
Butter or oil	10
Methyl glucoside sweetener	60
Carboxymethylcellulose (Hercules 9MXSP)*	3
Vanilla extract	2
Salt	Trace
Water	100

*DS 0.8 to 0.95

The calorie content of this product is only 41.5 calories per 100 grams. There is no deterioration of the frosting surface upon exposure to air. In comparison, the usual humectants, glycerol, sugar, corn syrup, sorbitol, and the like, while effective, contribute calories, e.g., a similar composition using sugar of glycerol as a humectant had 105 calories per 100 grams.

Methyl glucoside, however, when used as a bulking agent in conjunction with synthetic sweeteners as a sugar substitute, is noncaloric and fully as effective as sugar in reducing the vapor pressure of water in the frosting.

Bulking Agent for Colored Mixes

Particle appearance of beverage mixes containing artificial sweeteners is the object of the process by *L.D. Pischke; U.S. Patent 3,615,672; October 26, 1971; assigned to General Foods Corporation.* The work has been aimed at achieving larger, more uniform particles by means of agglomeration. While agglomeration has been reasonably successful as a technique for obtaining a particle which initially has a desirable appearance, there is a tendency for the agglomerates to break up and appear dusty in the jar, especially as the product is used by the consumer and the volume of product in the jar is diminished.

The work aimed at achieving uniformity of appearance has been primarily directed toward achieving a uniform distribution of the coloring agent throughout the mix in order to avoid having a speckled product. One method of achieving this result in a beverage mix containing large percentages of sugar has been to plate the sugar crystals with a solution of the coloring agent under carefully controlled conditions, such that only the surface of the crystals softens and the coloring agent is adsorbed onto the crystals without causing the crystals to lose their shape. While such a technique is successful, it needs additional processing steps which are costly, time-consuming and require rigid control to avoid melting the sugar.

When attempting to achieve a uniform color in an artificially sweetened beverage mix the difficulties are increased. The techniques of plating sugar crystals which have been successful commercially, have been found unsatisfactory for use with artificially sweetened bulking agents. The wetting and blending of the dried particles significantly changes the density of the bulking agent. The increase in density caused by the wetting makes a previously low-density bulking agent unsatisfactory for use in a beverage mix.

The key to the improved process is the incorporation into a low-density, artificially sweetened bulking agent, of an appropriate coloring agent. The bulking agent is prepared from a mixture of a filler, an artificial sweetener, coloring agents and water. The dried bulking agent consists of uniformly colored and sized particles, and the density of the particles is controlled. A high volumetric ratio of bulking agent to mix imparts uniformity of appearance to the entire mix.

Because the weight ratio of bulking agent to mix is low, it is possible to prepare a mix in which other desirable ingredients such as opacity agents and organoleptic enhancing gum systems are incorporated, without impairing the uniform appearance of the mix.

The product of this process minimizes the processing steps in that it is unnecessary to color the particles of the bulking agent after the bulking agent is prepared. Also, it makes possible the preparation of a low-density, artificially sweetened beverage mix, as the low-density bulking agent is not subjected to a wetting action after drying. As indicated, such wetting action tends to drastically increase the density of the bulking agent.

The filler constitutes the major portion of the bulking agent, generally at least 90% by weight. In order to perform satisfactorily in the beverage mix the filler must be water-soluble. It has been found that the use of dextrin as a filler is quite satisfactory. The most convenient form of dextrin is the class commonly known as corn syrup solids. When manufacturing a sugar substitute it is desirable to form a clear solution which will yield transparent dry particles that have sufficient reflecting surfaces to give the appearance of a crystalline product. Thus in manufacturing a sugar substitute it has been found that the DE (dextrose equivalent) of the dextrin will critically affect the desired properties and a DE of 15 to 20 is optimum. As the product is a colored bulking agent, the clarity of solution is not as critical and it has been determined that a DE of greater than 5 will produce a product with the desired solubility. The upper limit on DE will tend to be governed by the method used to dry the bulking agent solution. Thus, if drum drying is used, the upper limit will be about 25 in order to prevent the formation of a product which resembles cotton more than it does sugar crystals.

The sweetener is added to the bulking agent solution prior to drying in order to insure uniform distribution of the sweetener throughout the beverage mix. The amount of sweetener used in the bulking agent will depend upon many factors such as the actual sweetener used, the percent of bulking agent in the final beverage mix and the taste characteristics being developed for a particular beverage mix. The exact level of sweetener to be used will be readily determined for a particular beverage, and will generally range from about 0.2 to 10.0% of the bulking agent.

By adding the coloring agent prior to drying, all of the particles of the bulking agent are completely and uniformly colored and as the bulking agent comprises about 80 to 95% of the volume of the final beverage mix, a uniformly colored beverage mix is attained. All coloring agents which are normally used in beverage mixes can be used in the product. The exact amount of coloring agent used will vary, depending on the agent itself, the clarity of the bulking agent solution prior to drying and the intensity of color desired in the finished product. Generally, the coloring agent should be present at a level of 0.02 to 0.20% by weight of the dry bulking agent.

Water is an essential ingredient of the bulking agent solution. The percentage of water will be varied in order to obtain a dry product with a desired density. Depending upon the particular beverage mix to be prepared, the density of the dry bulking agent will be varied generally in a range of from 0.05 to 0.30 g/cc. The amount of water used to obtain a dry bulking agent of a specific density will vary greatly depending upon the drying process used. Thus in Example 1 it will be seen that the bulking agent solution is about 52% water when preparing a bulking agent with a density of about 0.10 g/cc by spray drying. In Example 2 it will be seen that a bulking agent with about the same density is achieved when a solution with about 74% water is drum dried.

Example 1: A colored bulking agent is prepared using 97.20 parts corn syrup solids, 0.035 FD&C Yellow 05, 0.075 FD&C Yellow 06 and sodium saccharin adjusted for taste (all parts by weight). About 1.1 parts of water at 140°F were added per part of these blended ingredients and the final solution was then treated and spray dried. The particles had a color which was suited for use in an orange flavored beverage mix.

Example 2: A colored bulking agent is prepared by drum drying. 75.0 grams dextrin (National Starch crystal gum containing trace amounts of dextrose, maltose and triose), 250 ml water, 0.025 grams FD&C Yellow 05, 0.052 grams

FD&C Yellow 06 and sodium saccharin adjusted for taste. These ingredients are blended until a homogeneous solution is obtained. The solution is then dried on a drum dryer with the rolls in a pinched position. The drums are heated at a steam pressure of 40 to 60 psig and are rotated at a drum speed of 1 rpm. A colored, crystalline product is obtained.

Example 3: A water-soluble, artificially sweetened, dry beverage mix having an orange flavor and a bulk density of about 0.26 g/cc is prepared using bulking agent prepared in Example 1 or 2 or other appropriate means and having a bulk density of about 0.10 g/cc by blending the following ingredients:

	% by Weight
Bulking agent	65.00
Citric acid	13.00
Opacity agent (HVO 110)	5.00
Sodium carboxymethyl cellulose (low viscosity)	4.66
Trisodium citrate	1.89
Orange oil (Perma Stable orange flavor)	2.21
Fivefold orange oil (Perma Stable orange flavor)	1.99
Fries & Fries artificial flavor 011736 (at 2.0% fix)	2.91
Fries & Fries orange flavor 011169	0.57
Firmenich imitation orange flavor 059427/AP	1.00
Vitamin C	1.62
Vitamin A	0.15

The ingredients are dry blended for 5 to 10 minutes to obtain a uniform mix. The resulting mix contains 65% by weight and 86% by volume of the filler. The mix has a uniform appearance of an orange-colored beverage and is readily soluble in water. It is obvious that other flavored beverage mixes, e.g., orange-pineapple or pineapple-grapefruit, can be made by changing the coloring agent in the bulking agent and the flavoring ingredients.

Cellulose Bulking Agent

O.A. Battista; U.S. Patent 3,023,104; February 27, 1962; assigned to American Viscose Corporation proposes cellulose crystallite aggregates as a bulking agent in foods to reduce caloric content.

Many foods have been reduced in calories by substituting artificial sweeteners for sucrose, or by reducing fat content. Saccharin, methyl- and carboxymethyl cellulose and seaweed derivatives are some of the nonnutritive materials added to foods, and in each case these products are added at very low levels, since the limiting factor of undesirable flavors is inherent in each of them.

Conventional fibrous cellulose has been used as a bulking agent, but the great defect of this material is its objectionable texture; thus, when mixed with a food or food ingredient and the mixture tasted, it is noticeable per se to the taste, is not smooth, has a fibrous mouth feel when chewed, gives the impression of the presence of an additional insoluble or residual substance, and tends to accumulate in the mouth. The unsatisfactory texture of the fibrous material cannot be remedied no matter how small the fibers are cut. Soluble cellulose derivatives have also been used as bulking agents but tend to form unpalatable gummy masses in the mouth.

Dietetic compositions of this process incorporate cellulose crystallite aggregates, which as described below are a product of the acid hydrolysis of cellulose. The aggregates per se, it may be recorded, are bland in taste and odor, white in color, have a physical appearance resembling starch, and are edible but not nutritious. They are entirely free of the objectionable textural defects of agents like conventional fibrous cellulose, which adversely change the taste and mouth feel of the food with which they are mixed; rather, the aggregates are particularly characterized by having a smooth pleasant mouth feel, by becoming an indistinguishable part of the food with which they are mixed and by simulating its eating quality.

The cellulose crystallite aggregates help to provide low caloric food products while yet enabling the products to meet recognized standards of acceptability, particularly in respect to the essential property of mouth feel or eating quality. In other words, the aggregates afford a reduction in calorie content with no sacrifice of the familiar qualities of time-tested foods.

Beyond this, and coincidently, therewith, the aggregates are capable of improving many foods, and the processing thereof, in a number of other respects. Thus, they permit desirable variations to be made in the physical characteristics of a food, such as its eating quality, appearance, cohesiveness, and texture, including its tactual, visual and taste textures. For example, liquids and pastes may be formed into a crumb, and oily materials into a plastic mass like butter. A less desirable characteristic of a food may be reduced, as in the case of candy of pronounced sweetness, where the presence of the aggregates tends to lessen the sweet taste. In other cases, improvements and advantages are realized in the processing of food material, as by increasing the yield, or simplifying the handling of oily or sticky materials.

As indicated, the cellulose crystallite aggregates are products obtained by the controlled acid hydrolysis of cellulose, there being formed an acid-soluble portion and an acid-insoluble portion. The latter comprises a crystalline residue or remainder which is washed and recovered, being referred to as cellulose crystallite aggregates, or as level-off DP cellulose.

In the acid hydrolysis, the acid dissolves amorphous portions of the original cellulose chains, the undissolved portions being in a particulate, nonfibrous or crystalline form as a result of the disruption of the continuity of the fine structures between crystalline and amorphous regions of the original cellulose. Although hydrolysis may be effected by various specific methods, including the use of various acids, a direct method which is free of secondary reactions comprises the treatment of the original cellulosic material with 2.5 normal hydrochloric acid solution for 15 minutes at boiling temperature.

Another suitable method comprises treating the cellulosic material with 0.5% hydrochloric acid solution (0.14 normal) at 250°F for 1 hour. The cellulose undergoing such treatment reaches, within the time period noted, a substantially constant molecular weight, or in other words, the number of repeating units or monomers, sometimes designated anhydroglucose units, which make up the cellulosic material, becomes relatively constant, from which it is apparent that the degree of polymerization of the material has leveled off, therefore the name level-off DP cellulose (degree of polymerization).

In other words, if the hydrolysis reaction were continued beyond the period noted, the DP would change very little if at all. In all cases, the level-off DP value reflects the fact that destruction of the fibrous structure has occurred as a result of the substantially complete removal of the amorphous regions of the original cellulose.

The term "crystallite" as used herein, means a cluster of longitudinally disposed, closely packed cellulose chains or molecules, and "aggregates" means clusters of crystallites. The aggregates may also be said to comprise straight rigid, relatively nontwistable groups of linear chains. As indicated by x-ray diffraction tests, the crystallite aggregates have a sharp diffraction pattern indicative of a substantially crystalline structure. Although the crystallite chains are of very uniform lengths, particularly by comparison with the original cellulose chains, strictly speaking they do exhibit some variation, and for this reason it is preferred to speak of average length, or of average level-off DP values.

The hydrolysis methods noted are particularly characterized in that in each crystallite aggregate resulting from the hydrolysis, no constituent chain is connected to a chain in a neighboring aggregate; rather all the chains in an aggregate are separate from and free of those in neighboring aggregates.

The cellulose crystallite aggregates, or level-off DP cellulose, is characterized by having a preferred average level-off DP of 125 to 375 anhydroglucose units. Ideally, within this range all of the material should have the same DP, or chain length, but as this is difficult if not impossible to achieve, it is preferred that at least 85% of the material have an actual DP not less than 50 and not more than 550. More preferably, at least 90% of the material should have an actual DP within the range of 75 to 500, and it is still more preferred if at least 95% of the material has an actual DP in the range of 75 to 450.

It may thus be apparent that the chain length of the level-off DP cellulose, or cellulose crystallite aggregates, is very uniform, a consequence of the hydrolysis, wherein the longer chains of the original cellulose were converted to shorter chains and the very short chains were dissolved away. In short, the hydrolysis effected a homogenization of the chain length distribution.

Associated with the foregoing DP properties of the crystallite aggregates is the fact that their chemical purity is very high, the material comprising at least 95%, preferably at least 97 or 99%, polyglucose, or anhydroglucose units, based on chromatographic analysis. In terms of ash content, the aggregates contain less than 100 ppm, although ash may range from 10 to 400 or 500 to 600 ppm. By comparison, conventional fibrous cellulose may have 1,000 to 4,000 ppm of ash.

In connection with the purity of the aggregates, it may be explained that the inorganic contaminants in the original cellulose, which are concentrated in the amorphous regions, are dissolved away by the hydrolyzing acid, and the noncellulose components of the original material are so effectively destroyed that their concentration is reduced to a very low level. Of interest is the fact that the chains produced by the hydrolysis each have on one end a potential aldehyde group, such group being in the 1-carbon position of an end anhydroglucose unit and requiring the assistance of the ring oxygen atom, which is ortho to it, to realize its aldehydic potential. The group has the reducing properties of an aldehyde group. On their other ends the chains have only hydroxy as functional groups.

The source material for the crystallite aggregates may be one or more natural fibers such as ramie, cotton, purified cotton, also bleached sulfite pulp, bleached sulfate wood pulp, etc. Particularly suitable are sulfite pulp which has an average level-off DP of 200 to 300, at least 90% of which has a DP in the range of 75 to 550; and also sulfate pulp which has an average level-off DP of 125 to 175, at least 90% of which is in the range of 50 to 350.

Other suitable cellulose crystallite aggregates may have lower average level-off DP values, in the range of 60 to 125, or even 15 to 60. Aggregates from both of these ranges have the chemical purity and other characteristics of the aggregates from the first noted DP range. Crystallite aggregates in the 60 to 125 average level-off DP range are obtainable from the acid hydrolysis of alkali-swollen natural forms of cellulose, of which a preferred source is cellulose that has been mercerized by treatment with 18% caustic soda solution at 20°C for 2 hours. Aggregates in the 15 to 60 average level-off DP range are suitably prepared from regenerated forms of cellulose, including tire and textile yarns, other regenerated cellulose fibers and cellophane. In general, the cellulosic source material has a DP greater than the level-off DP.

As obtained from the acid hydrolysis and water washing steps, the aggregates in the overall average level-off DP range of 15 to 375 are in a loosely aggregated state, particularly in the larger sizes, from 40 to 250 or 300 microns, and are characterized by the presence of many cracks in their surfaces, including similar surface irregularities or phenomena-like pores, depressions, voids, fissures and notches. Because of such irregularities, the apparent or bulk density of the aggregates is much less than their absolute density.

Furthermore, the cracks and other irregularities persist despite the application of high compressive forces on the aggregates. Thus, when they are compressed at 5,000 psi, they exhibit an apparent density of 1.26; at 10,000 psi the apparent density rises to 1.32; at 15,000 psi it is 1.34; and at 25,000 and 37,000 psi it is 1.38 and 1.38 respectively.

On the other hand, the absolute density of a unit crystal or crystallite is 1.55 to 1.57, from which it is apparent that the aggregates occlude considerable quantities of air in the surface cracks, voids, fissures, etc. The apparent densities of the dried disintegrated aggregates, at the compressive forces noted, are somewhat higher than the foregoing values. Of interest in this connection is the fact that the aggregates, dried or never dried, retain their pressed form after compression; in other words, three-dimensional structures of any desired shape may be formed by compressing the aggregate particles.

Either before or after mechanical disintegration the aggregates may be dried. Where the disintegration is performed in the presence of an aqueous medium, drying is carried out after the disintegration step. Drying may be done in any suitable vacuum or in air room temperature or higher, going up preferably to 60° to 80°C, although the temperature may be up to 100° or 105°C or higher.

Another procedure is to displace the water in the wet aggregates, preferably using a low boiling, water-miscible organic compound such as a low molecular weight aliphatic alcohol like methanol, ethanol, propanol, isopropanol, etc., and then to evaporate off the compound. The resulting dried aggregates tend to be more reactive and to form stable dispersions and gels more readily. Spray drying either in air or in a vacuum is also satisfactory.

Spray drying and also freeze drying and drum drying, are particularly effective to dry the aggregates after the disintegration step. Freeze drying in particular favors the development of a very porous, reactive material which is characterized by the presence of a multiplicity of pores or depressions of extremely small size; such material readily forms stable dispersions and gels.

Mechanical disintegration of the aggregates, as referred to above, may be carried out in several ways, as by subjecting them to attrition in a mill, or to a high speed cutting action, or to the action of high pressures on the order of at least 5,000 or 10,000 psi.

The disintegration of the aggregates is carried out in the presence of a liquid medium, although where high pressure alone is used, such medium, although desirable, is not necessary. Water is a preferred medium, but other preferably edible liquids are suitable, including sugar solutions, polyols, of which glycerol is an example, alcohols, particularly ethanol, and the like. Whatever method is used, the disintegration is carried out to such an extent that the resulting disintegrated aggregates are characterized by forming a stable suspension in the aqueous medium in which they are being attrited, or in which they may be subsequently dispersed.

By a stable suspension is meant one from which the aggregates will not settle out but will remain suspended indefinitely, even for periods measured in terms of weeks or months. The disintegrated aggregates are further characterized by the fact that such suspension forms an extremely adherent film when deposited on a glass panel or sheet or other suitable surface. At lower concentrations of aggregates, the suspension is a dispersion, while at higher concentrations it is a gel.

The preferred disintegration method is to attrite the aggregates by means of a high speed cutting action in the presence of an aqueous medium. The aggregates may be in a dry or never-dried state prior to attrition, although some water should be present during the cutting or shearing of the particles. If they are initially in the never-dried or wet state, that is, as received from the water washing step, they have a moisture content of at least 40% by weight, and it is possible to attrite them without further addition of water, although water may be added if desired. In any event, it is preferred that the water content of the mixture undergoing attrition should be at least 10 to 15 to 20% by weight. The aggregates content of the mixture to be attrited is preferably at least 3% by weight, and desirably is higher as the efficiency of the cutting action increases with the aggregates content.

Suitable consistencies are those of mixtures containing up to about 35% by weight of aggregates and the balance water; such mixtures lend themselves well to good attrition and are convenient to handle both during and after the disintegration; they also have the advantage of directly producing a gel. At consistencies above 35%, i.e., from 35 to 70%, attrition produces a material which, in the lower end of this range, resembles mashed potatoes of relatively soft or mushy appearance, and as the concentration increases, the material acquires a progressively firmer and drier appearance and consistency; above 50% the material tends to become crumbly.

Although the attrited products of consistences above 35% are not gels, they have the distinctive property of forming indefinitely stable, smooth gels of varying thickness and striking appearance upon the addition of water and stirring manually, as with a spoon, for a few minutes.

At about 80% consistency, attrition results in a damp but free-flowing material comprising discrete grains or granules and clumps of grains; the moisture content is apparent to the touch rather than the eye; and the material forms a gel upon being manually stirred or beaten in water. At 80 to 90% consistency, the product of attrition is a crumbly, free-flowing, grainy, dry-appearing material that does not have a damp feel and which requires energetic beating in the presence of water to form a gel.

It may be useful to review briefly the characteristics of the dispersions and gels. They comprise attrited products of an attritable mixture having a solids content of at least 3% by weight during the attrition step. Necessarily, the resulting attrited product will also have at least 3% solids, although, as indicated, some useful materials are obtainable by diluting such attrited product to a consistency of 2% solids, or even 1%, to form a stable dispersion.

In the next place, at least 1% by weight of solids in the product of attrition have a particle size of up to 1 micron. In the third place, the attrited product, in dispersion or gel form, forms substantially adherent films, preferably substantially continuous and self-supporting films, when applied to a suitable surface. Finally, the attrited product is, or forms, a stable and homogeneous colloidal dispersion or gel, the term "homogeneous" referring to the uniform visual aperance of the dispersion or gel.

The preferred dispersions and gels are those that are stable for at least a month, and another preferred group comprises those stable for at least a week. Dispersions and gels that are stable for at least a day, or even an hour, are also useful for some purposes, as where they are to be used almost immediately. The more stable dispersions and gels have the advantage of being storable for a considerable period of time.

Generally, the concentration of aggregates in the aqueous dispersions is at least 4 or 5% by weight. It is possible to make stable homogeneous dispersions having a solids content of up to 6 to 8% by weight, although more usually the solids range from 3 to 6%. In the case of gels, the aggregates content varies from 3 or 4% to about 30 or 35% by weight, the upper concentration being limited only by the capacity of the gel to be handled or worked.

The concentration both of dispersions and gels may be varied not only by varying the consistency of the attritable mixture but also by adding water to the dispersion or gel, and, less preferably, by evaporating water therefrom. Usually the gels are thixotropic when they contain 8 to 10% by weight, or more, of the aggregates. As may be apparent, the more concentrated dispersions may have a solids content which overlaps that of the less concentrated gels.

The fact that gels are obtainable at concentrations as low as 3 or 4% solids is explainable by the presence of considerable amounts of aggregates of a particle size of substantially 1 micron and less, it having been found that gel formation is favored as the concentration of these fine particles increases. In fact, at concentrations as low as about 3% solids, gels are obtainable which are thixotropic provided the aggregates are substantially all of 1 micron size and less.

Following the mechanical disintegration of the aggregates, the resulting product, whether a dispersion or gel, may be taken and used as such; or it may be dewatered and dried; or it may be desirable to fractionate it into fractions having a more uniform particle size and size distribution. If the product is a mixture containing 35 to 90% solids, it may be stirred in water to form a gel, and the latter is handled as indicated below. The dried products are also redispersible in aqueous media with the help of agitation, such as provided by a Waring Blender, to form dispersions and gels.

In respect of the drying of the gels, it should be observed, first of all, that the preferred gels are those obtained by attriting the never-dried hydrolysis product; these gels have very desirable qualities in respect to smoothness, mouth feel, firmness, storage characteristics, etc. They may be dried to any practical moisture content, in which state they are redispersible in water, by the aid of a suitable attrition step, to form a gel, and this latter gel may again be dried if desired and again redispersed to form a subsequent gel. Gels are also obtainable by attriting the dried hydrolysis product, and these gels may be dried and attrited to again form gels.

For producing the dried products, a number of drying procedures are available, and while redispersible materials result from each procedure, some procedures are more advantageous than others, as indicated. For example, freeze drying, spray drying, drum drying and drying by solvent displacement each produce a material which has an appreciably lower bulk density than conventionally oven dried materials, with freeze drying producing the lowest bulk density by far, 9.8 pcf as against 14.1 pcf for oven dried aggregates; each such procedure produces a material which is more easily redispersible in water, by the aid of an attrition step, to form a stable suspension than air or oven dried materials; and each procedure yields a more reactive product than air dried or oven dried products. Freeze dried, spray dried, drum dried and solvent displacement dried materials are noticeably softer to the touch than products of the other drying steps; and freeze drying also produces a more porous product. With regard to the mouth feel of the various materials, those made by freeze drying, spray drying and drum drying are superior.

Fractionation of the attrited products may be accomplished by means of such separation procedures as mechanical sifting, settling in water, or centrifuging, a number of useful fractions being obtainable, including fractions having a particle size of up to 0.2, 1, 2, 5 or 10 microns. Still another desirable fraction is one whose dimensions are all below 100 microns, or below 40 or 50 microns; a fraction in the range of 40 to 250 or 300 microns is of special interest because of the finding that particles in this size range, particularly those having one or two dimensions of up to 250 or 300 microns, tend to have cracks, fissures, notches, voids, depressions, pores and the like in their surfaces.

Preferably, each dimension of the particles should be within the size range noted for each fraction; however, particles having two dimensions within the size range are quite useful, as are particles having but one dimension within the size range although they are less preferred.

Cellulose crystallite aggregates are particularly applicable to bakery products, including the several varieties of bread, cake, cookies, crackers, biscuits, pies and doughnuts; snack items like pretzels and potato chips; other pastries and specialties; also prepared mixes for making any of these products; and in cereals, flours and macaroni products.

Baked goods mainly comprise starch, and this is particularly true of cookies and crackers. The starch content is so predominant that heretofore it has been very difficult, if not out of the question, to reduce the calorie value of these products to any appreciable extent without impairing their taste; but by virtue of the bland quality of the crystallite aggregates and their smooth mouth feel it is possible to reduce substantially the calorie value of many baked products without restricting their appeal.

The aggregates may replace considerable amounts of flour in these bakery products. They may also replace, in whole or part, bread additives like dextrins and starch. In products made of finely ground flour, i.e., cakes and cookies, the aggregates serve to retain moisture and other liquids, in effect increasing the liquid carrying capacity of such flours, thus avoiding staling of the products for longer periods, a result of particular advantage for high-sugar-containing cakes. The use of the aggregates also enables normally liquid ingredients to be handled in a dry free-flowing form; i.e., the liquid ingredients may first be mixed with the aggregates to form a substantially dry granular mixture and the latter can then be added to the recipe mix. In effect, the liquids are dried, although actually they are sorbed, that is, both adsorbed and absorbed, by the aggregates. Such liquid ingredients may include shortening materials, water, milk, syrups and emulsifiers; and other suitable ingredients are lard, butter and eggs.

The aggregates are of value in foods like dressings and spreads which are prepared by the aid of edible fats and oils, including shortening agents, butter, margarine, cooking and salad oils, plastic fats, lard, etc. Besides reducing the calorie value of the product and assisting in holding the ingredients together, particularly the liquids, the aggregates provide a desirable thickening effect.

The preparation of meat and meat products, including sausage, sausage products, meat loaves, etc., comprises another appropriate area for the utilization of the crystallite aggregates. In the manufacture of sausage products and meat loaves the aggregates help to simplify the processing steps by making the meat emulsions and meat mixes easier to handle, an advantage brought about by the sorptive capacity of the aggregates for the liquid-containing meat mixtures. The consistency and stability of a meat emulsion, for example, may be regulated in this way, and thus the cohesiveness of the final product may be varied. Coincident with the drying or sorption effect is the improved retention of flavors, water and fats. As a consequence of fat and water retention, there will be less free fat in the product, and less shrinkage.

Texture, juiciness and other organoleptic characteristics may be improved by varying the firmness, juiciness, chewiness, etc., of a sausage or meat loaf product.

Conventional binders like cereals, starches and flours, which are all nutritious, may be replaced in whole or part by the crystallite aggregates, as may other frequently used meat loaf ingredients like macaroni, cheese, potatoes, etc.

Better distribution of seasoning materials may be secured by first mixing them with the aggregates and then adding the mixture to the meat emulsions or mixes. The crystallite aggregates favor the formation of paste and paste-like products since the aggregates are quite amenable to being spread, especially in the presence of liquids. The aggregates may replace at least partly, the use of ice to obtain a proper consistency and stability in meat emulsions; thus the aggregates may make it possible to achieve a desirable consistency by the use of refrigeration. An advantage of economic importance is the increased yield of the meat emulsion or meat mix obtainable when the aggregates are used.

In one instance, hamburgers were prepared containing 5, 10, 15 and 20% of the aggregates, the balance being ground beef meat as bought in the store, and it was found that all of the patties so prepared, by comparison with controls, had the familiar hamburger flavor and were quite palatable. With respect to mouth feel, none gave any indication of the presence of the aggregates. Of additional interest were the findings that the aggregates-containing patties underwent less shrinkage than the controls, had much less tendency to burn, and had an interesting variation in respect of their chewiness.

Another application of the process is in dairy product foods such as cheese and in foods using milk and cream. Naturally soft and semisoft nongratable cheeses may, by mixing the same with the aggregates, be transformed into a dry granular free-flowing form, making it more convenient to handle such cheeses when they are used as ingredients in a recipe.

Milk and cream may be used more conveniently when mixed with the aggregates. For example, cream as obtained from milk, either with or without the preliminary removal of water from the cream, may be mixed with the aggregates to form a dry free-flowing mixture for use in coffee, and if desired, the mixture may be packaged in inexpensive, single use, disposable paper containers, although it may also be stored in bulk and dispensed in any desired way, that is, by machine or by hand. As a result, there is provided a low calorie cream-containing mix whose use with coffee and other substances simulates natural cream.

The aggregates are also suitable for use in confections, including candy, chewing gum, baker's confections, etc. Particularly in candy, the aggregates have a tendency to decrease or cut the sweet taste to an extent, especially in candy of pronounced sweetness, without resorting to synthetic or substitute sweetening agents. Edible food dyes may be incorporated in the candy in a convenient manner by first sorbing a dye on and in the aggregates and then mixing in the latter with the other candy ingredients. In chewing gum the aggregates are of value for introducing flavors. Thus, a flavor-containing substance may be formed into a granular free-flowing mix with the aggregates, and the resulting mixture may be incorporated with the chicle or other chewable plastic gum base. Blooming of chocolate and other foods comprising the release of oil, may be lessened through use of the aggregates.

Example 1: Honey-flavored doughnuts were prepared having the composition shown on the following page. The crystallite aggregates were prepared by hydrolyzing Ketchikan wood pulp with a 0.5% HCl solution for 60 minutes at 250°F. The aggregates were washed with water and then dried in an oven at 65°F for 24 hours under a vacuum of 29 inches of water. They had an average level-off DP of 220, a moisture content of 2.5% by weight, and a particle size in the range of 1 to 300 microns. The dried aggregates were first mixed with the honey by placing the latter in the bowl of a Model N-50 variable speed Hobart mixer equipped with a paddle or beater attachment and the mixer set in operation at a low speed corresponding to about 61 rpm. The beater comprised a stem having a plurality of spaced arms extending angularly downward from opposite sides, and a substantially V-shaped member connected to the outer ends of the arms; in effect, the paddle resembled an inverted tree.

Ingredient	Grams/Batch	Percent
Eggs, fresh	49.7	7.87
Sugar	31.0	4.91
Crystallite aggregates	79.5	12.60
Honey	105.5	16.72
Milk	185.0	29.31
Shortening	15.0	2.38
Baking powder	10.5	1.66
Salt	1.1	0.17
Cinnamon	0.2	0.03
Nutmeg	0.2	0.03
Baking soda	1.5	0.24
Flour	152.0	24.08
	631.2	100.00

The beater rotated and revolved in the bowl. In other words, it had a compound action: it revolved inside the bowl, and it rotated on its axis, the direction of rotating being opposite the direction of its movement around the bowl. The addition of the finely divided aggregates was done over a period of 5 to 10 minutes, the mixture being converted to a freely flowable dry mass which was removed from the Hobart.

The eggs were then beaten in the Hobart, and to them were added, during agitation, the sugar and the aggregates-honey dry mass. The milk and the melted shortening (hydrogenated vegetable oil) were next added to the mix. Then the baking powder (monocalcium phosphate), salt, cinnamon, nutmeg, and baking soda (sodium bicarbonate) were combined and added with agitation. Finally, the flour was added and mixing continued to form a smooth dough. The latter was chilled in a refrigerator, the doughnuts then formed therefrom and deep fat fried at about 370°F.

They appeared to be normal in every observable respect, including their eating quality, by comparison with control doughnuts made at the same time wherein the aggregates were omitted. A second batch was made in which the aggregates content was only 5%, and these were also indistinguishable from the conventional doughnuts. As a result of the frying process, the doughnuts containing 5% aggregates showed a weight gain of 9.68%, while those containing 12.6% aggregates gained 2.45%. Satisfactory doughnuts were also made containing up to 20% by weight of the aggregates.

Example 2: A peanut butter cookie dry mix was prepared from the following ingredients:

Ingredient	Grams/Batch	Percent
Crystallite aggregates	128.0	13.92
Peanut butter	296.4	32.24
Butter	48.6	5.29
Sugar, brown	125.0	13.60
Sugar, granulated	126.5	13.76
Salt	3.1	0.34
Baking soda	1.8	0.19
Flour, all purpose	190.0	20.66
	919.4	100.00

The crystallite aggregates were prepared as in Example 1.

The peanut butter and the butter were first mixed with the dried aggregates in the Hobart mixer to form a free-flowing dry mass, after which all of the remain- ingredients were added with mixing. Then there were added to the mix in the Hobart, with continued agitation, ½ teaspoon vanilla, 1 egg and 185 ml of water. The resulting heavy dough was rolled into small balls, flattened with a fork on a greased cookie sheet, and baked for 15 minutes in an oven preheated to 375°F. The flavor and eating quality of the resulting cookies were particularly excellent, and in other respects they were equal to the conventional product.

Polyose Bulk Extender

A sugar substitute product is prepared with saccharin using a polyose as a bulk extender. This is described by *P. Jucaitis and I.D. Bliudzius; U.S. Patent 2,876,105; March 3, 1959; assigned to E.I. du Pont de Nemours and Company.*

A polyose is a glucose polymer derived from starch by depolymerization followed by heat polymerization. The polyoses have a considerably different susceptibility to amylolytic enzymes than the original starch or its conventional degradation products. The products on test contain groups which act like very small amounts of reducing sugar, not in excess of 5 to 7% in Polyose A. The reducing sugar is not construed as due to the presence of free glucose.

A preferred polyose for this purpose is Polyose A which can be described as a glucose polymer derived from starch by depolymerization followed by heat polymerization to such an extent that at 67% solids it gives a viscosity (Brookfield) of 20 to 100 poises at 70°F. More viscous polyoses can be used. Polyose B requires only 60% and Polyose C requires only 50%, and Polyose D requires only 40% of solids to give the viscosity described. Polyose B is as good as A and can be used in the same amounts.

The weight of a polyose in this composition can range from the balance of the composition after considering the sweetener content to some smaller figure. Not less than about 20% by weight of polyose would ordinarily be used. It is preferred that the polyose be used in an amount of at least 40% by weight of the composition.

The compositions can also contain up to 30% by weight of gum arabic. The gum arabic acts to make the polyose somewhat heavier so that it does not float upon the surface of a liquid. The gum arabic and the polyose both become more soluble in the presence of each other.

The structure of particles of a sweetening composition containing a polyose is strengthened by the inclusion of a water-soluble cellulose derivative such as hydroxyethyl cellulose. Hydroxyethyl cellulose can be included in amounts up to about 20% by weight.

Instead of the gum arabic there can be used equivalent amounts of other gums such as cold-water-soluble carrageen, guar gum, gum acacia, gum tragacanths, gum ghatti, hydrolyzed collagen, and degraded gelatin. Some of the last named have nutritive value and too much should not be used. Similarly, there can be used gelatin itself or soy bean protein. As agents for strengthening the particles of the composition there can also be used polyvinyl pyrrolidone, cold-water-soluble polyvinyl alcohol, and dextran.

The polyose has a tendency to float on the surface of a liquid and a surface active agent can be included in the composition to minimize this tendency. There can be used lecithin, and polyoxyethylenes sorbitan monooleate. Thus, there can be used Tween 80 which is a polyoxyethylene sorbitan monooleate with 20 mols of ethylene oxide for each of sorbitan or Tween 81 which has 4 mols of ethylene oxide. There can similarly be used Tween 60 which is similar to Tween 80 but is the monostearate.

Inorganic materials can be added such as sodium or potassium hexametaphosphate or tripolyphosphate. These have a tendency to prevent the flocculation of calcium and to broaden the range of materials which can be used with calcium. Traces of sodium chloride or other materials used for saltiness can be added to enhance the flavor.

Example: A sweetening composition having a volume comparable to that of sugar on a sweetness basis was made with the following components in parts by weight as shown.

Saccharin	0.4
Polyose A	80
Gum arabic	60
Hydroxyethyl cellulose	20
Water	100

The dry ingredients were first mixed and added to a Waring Blender. The water was added at 70°C and mixed for 5 minutes in the Waring Blender at high speed. The mixture formed a foam which was spread on heavy duty aluminum foil and dried for 2 hours at 55°C. After drying, the material was stripped from the aluminum. The product had a bulk density of 0.24 to 0.25 g/ml. It was readily soluble in hot water. The granules did not float in the liquid and there was no objectionable cloudiness. Instead of drying as above the mixtures can be spray dried or drum dried at atmospheric pressure or in a vacuum. The solution can be whipped or not.

Arabinogalactan Bulking Agent

There is a considerable loss of bulk when artificial sweeteners replace carbohydrates in a food formulation. These factors require the use of various hydrophilic colloids together with the sweeteners, such as carboxymethylcellulose, alginates, gelatin or carrageenan.

Although the compositions containing hydrophilic colloids in combination with an artificial sweetener overcome some of the problems of using the artificial sweetener alone, they still lack many of the physical and organoleptic properties of the traditional or accustomed sweetening agents such as sucrose.

Illustratively, many of the marine and vegetable gums reduce the incidence of sandiness in ice cream. Gelatin often is smooth and clings to the roof of the mouth. Carrageenan is extremely smooth, airy, and melts in the mouth. Alginates are pasty, rough, almost gritty. Other differences that exist include those relating to solubilities of the gel and various organoleptic properties. Carboxymethyl cellulose in quantities approaching that of sugar produces an excessively viscous product whereas smaller quantities of the cellulose lack the bulking properties of sugar.

Arabinogalactan is used by *G.L. Stanko; U.S. Patent 3,294,544; December 27, 1966; assigned to Richardson-Merrell Inc.* to improve the bulk characteristics of artificial sweeteners. Arabinogalactan has much the same bodying, bulking and other physical and organoleptic properties of sugar (sucrose) except for sweetness which can be supplied by an artificial sweetener. Arabinogalactan is substantially nondigestible and therefore does not contribute calories, does not produce much of an increase in viscosity of liquids, and does not possess many of the shortcomings of other materials used as bodying or bulking agents with artificial sweeteners.

Arabinogalactan is a natural gum that occurs in the genus Larix. Western larch contains large amounts of arabinogalactan, ranging from 8 to 25% on a dry wood weight basis. It appears to be a highly branched polymer of arabinose and galactose in the ratio of 1:6 respectively. Its average molecular weight is from 70,000 to 95,000. The particle size of the arabinogalactan for use in this process is not critical, provided that it can be easily admixed in the food product. Thus, it can be powdered or in small lumps, or in solution with water, preferably together with the artificial sweetener.

Illustrative of artificial sweetening agents which can be used in this process, there can be mentioned sulfimides, e.g., saccharin; ureas, e.g., p-ethoxyphenylurea; m-nitroanilines, e.g., 2-propoxy-5-nitroaniline; oximes, e.g., perillaldehyde oxime; amines, e.g., 2-hexyl-2-chloromalondiamide; hydrazines, e.g., succinic acid dihydrazides; imino nitriles, e.g., β-(p-tolyl)-β-imino-β-propionitrile; aromatic keto carboxylic acids, e.g., 2-(p-methoxybenzoyl)benzoic acid; triazine derivatives, e.g., glucin; and benzimidazole derivatives, e.g., 2-benzimidazole-propionic acid. The preferred artificial sweeteners are the water-soluble sodium saccharin, calcium saccharin and ammonium saccharin.

The proportions of arabinogalactan and artificial sweetener to each other and in the food product will vary dependent on the particular food, the artificial sweetener and the desired physical and organoleptic properties of the final composition.

As a sugar substitute, the arabinogalactan preferably replaced sugar (sucrose) on the same weight basis, e.g., 1 pound of sugar is replaced with the same weight of arabinogalactan. A quantity of sweetener sufficient to give an equivalent sweetness to the arabinogalactan on a weight basis as the same quantity of sugar is preferably admixed with the arabinogalactan.

Foods containing the bulking and sweetener composition are beverages such as carbonated beverages and canned fruit juices; baked goods such as cakes and cookies; icings; candy; frozen desserts such as ice cream and sherbets; jams and jellies; canned fruit; puddings and custards; pie fillings; syrups and sauces; liqueurs and sweetened wines; and salad dressings. Additionally, syrup bases for orally administered pharmaceutical products can be prepared with the bulking and sweetener composition. These foods and pharmaceutical base syrups can be prepared by simply following the normal preparation of such products but simply substituting arabinogalactan for sugar and also adding an appropriate quantity of artificial sweetener.

Example 1: A low calorie maple syrup can be formulated by mixing the following ingredients with a sufficient quantity of caramel for color and maple flavor to the desired taste.

Arabinogalactan	500 g
Saccharin, sodium (to match sucrose)	
Sorbitol solution	50.0 ml
Benzoic acid	1.0 g
Water, purified, q.s. to make 1,000 ml	

Example 2: A sugar substitute can be prepared by mixing and granulating the following ingredients.

Arabinogalactan	900 g
Saccharin, sodium (to match sucrose)	

The product of this example approximates the sweetness, in bulk, of natural sugar without the caloric content. This product can be used in dietetic ice cream, low calorie ice cream toppings, puddings, pies, icings, cakes, liquors, etc. The above product fairly replaces the body, viscosity and sweetness of about 900 grams of sugar (sucrose).

Example 3: Chocolate candy can be prepared with the below listed ingredients, together with the use of the following procedure.

Powdered arabinogalactan	2 cups
Saccharin, sodium (to match sucrose)	
Corn syrup	2 tsp
Milk	1 cup
Butter	2 tbs
Vanilla	1 tsp
Chocolate	2 oz
Salt	½ tsp

Heat the arabinogalactan, syrup, milk, chocolate and salt over low heat, with stirring to about 230°F. Add butter and vanilla. Cool and press into desired shapes.

Example 4: Uncooked fudge can be prepared with the ingredients listed below.

Chocolate	4 oz
Powdered arabinogalactan	1 lb
Sodium saccharin (to match sucrose)	
Egg, beaten slightly	1
Condensed milk	1/4 cup
Vanilla	1 tsp
Butter	1/2 cup

Melt chocolate and butter in top of double boiler. Mix egg with the arabinogalactan and saccharin; add milk, stir in the chocolate-butter mixture, add vanilla. Turn into pan; chill.

A sherbet can be prepared by intimately admixing the following ingredients and cooling the mixture to about 28°F until it has solidified.

Powdered arabinogalactan	11.0 lb
Artificial sweetener	
Corn syrup solids	110 lb
Ice cream mix (12% fat, 11% MSNF and 15% arabinogalactan)	17.5 lb
Stabilizer	0.4 lb
Fruit puree	15.0 lb
Water and 10.75 oz of 50% citric solution and color	46.1 lb

Example 5: Ingredients for white butter icing are as follows.

Butter or other shortening, soft	1/3 cup
Powdered arabinogalactan	3 cups
Sodium saccharin (to match sucrose)	
Vanilla	1.5 tsp
Cream or rich milk	3 tbs

The shortening, arabinogalactan and saccharin are blended and the vanilla and cream are stirred into the blend until smooth.

Example 6: Orange jelly can be prepared with the below listed ingredients and by following the below described procedure.

Gelatin	2 tbs
Cold water	1 cup
Boiling water	1 cup
Powdered arabinogalactan containing 2% by weight of artificial sweeteners	1 cup
Orange juice	1.5 cups
Lemon juice	3 tbs

Soak gelatin in cold water 5 minutes; dissolve in boiling water. Add arabinogalactan and sweeteners. Stir well; add orange and lemon juice. Pour into wet mold. Chill until firm.

Starch and Arabinogalactan

In another process involving arabinogalactan described by *A.L. La Via, R.L. O'Laughlin and R.W. Walton; U.S. Patent 3,704,138; November 28, 1972; assigned to E.R. Squibb & Sons, Inc.* a composition was developed containing an artificial sweetener which has a granular appearance similar to cane sugar.

The product obtained according to the process has a bulk density of 0.06 to 0.2 g/cc, and, on a volume basis has a caloric value of 5 to 25% of household granulated sugar.

As sugar has a caloric content of 18 calories per teaspoon, the product has a caloric value of from 0.9 to 4.5 calories per teaspoon, preferably from 0.9 to 3 calories per teaspoon. The product may be used on an equal volume basis for sweetening coffee, tea and cereals.

The artificial sweeteners used in the compositions include saccharin (2,3-dihydro-3-oxobenzisosulfonazole), pharmacologically acceptable salts of saccharin and mixtures thereof. The pharmacologically acceptable salts of saccharin can be used and are preferable since they are more readily soluble. Suitable salts include the sodium, potassium, calcium, ammonium and magnesium salts.

The artificial sweetening agents are used at a level sufficient to bring the sweetening power of the composition up to the sweetening power of the identical volume of household granulated sugar. This level may vary from 0.5 to 2%. Thus, for a product intended as a direct volume-to-volume substitute for granulated sugar, although the proportion of artificial sweetener to the bulk extenders may vary on a weight basis as the bulk density of the product varies, the product will always contain a proportion of artificial sweetener so that a given volume of the product provides the same sweetening power as an equal volume of granulated sugar.

The bulk extenders used in the preparation of the compositions are specialized dextrin-maltose syrups derived from acid or enzymatic hydrolysis of cornstarch. These liquors are especially formulated to yield relatively low dextrose contents as compared to the common variety of corn syrups. Such liquors are relatively low in natural sweetness and have a Dextrose Equivalent (percent reducing sugar calculated as dextrose) of about 13 to 28%. It is also contemplated to use mixtures of such hydrolysates.

It was found that hydrolyzed corn starches having a DE below about 10% have the disadvantage of poor solubility characteristics tending to become gummy and sticky instead of entering into solution. Further, they possess various off-taste characteristics (such as flavor which tends to be starchy in character) which render their use in a bulk sugar substitute undesirable. Hydrolysates having a DE above 28% are, on the other hand, too hygroscopic to be useful in such compositions.

The third essential ingredient of the compositions is arabinogalactan. Arabinogalactan is a water-soluble polysaccharide of natural origin derived from western larch trees. It is a complex highly branched polymer of arabinose and galactose in a ratio of 1:6 and has a molecular weight in the range of 72,000 to 92,000.

It may be used in the range of from 4 to 10%. The amount of arabinogalactan is always less than that of the hydrolyzed conrstarch but greater than that of the artificial sweetener.

The compositions are formulated to contain a major amount (over 75%) of hydrolyzed cornstarch having a DE of from 13 to 28%, from about 3 to 15% arabinogalactan and from 0.5 to 2% of artificial sweetening agent. They have a bulk density of from 0.06 to 0.2 g/cc and a caloric content of from 0.9 to 4.5 calories per teaspoon. Most preferably they have an arabinogalactan content of from 4 to 10%, a bulk density of from 0.06 to 0.15 g/cc and a caloric content of from 1.2 to 3 calories per teaspoon.

The products are made in a manner which provides a low bulk density, substantially below those heretofore available, and at the same time provide a product which will not separate or segregate during handling and packaging. The artificial sweetening agent remains intimately dispersed among the particles of the bulk extenders thus insuring that all portions of the product are substantially uniform in composition even after packaging, handling and long periods of storage. In addition, the products are highly nonhygroscopic, and thus do not require the addition of flow controllers or other such additives, nor do they require the use of moistureproof packaging as is the case with prior art materials.

The products are prepared by dissolving the artificial sweetening agent, the cornstarch hydrolysate, and the arabinogalactan in a quantity of water, introducing finely divided inert gas, such as air or nitrogen, into the solution, and drying under vacuum the aerated solution to provide material having a particle size range similar to that of granulated sugar. In the vacuum drying process, the final bulk density of the product is controlled by the solids content of the solution treated, the amount of inert gas introduced into the mixture, by the conditions of drying, and the extent of comminution of the dried product.

In order to produce a product having the desired bulk density, it is necessary to introduce inert gas to a level such that a foam density of 0.6 to 0.8 is produced. Increasing the amount of inert gas introduced into the mixture decreases the density of the product obtained and, conversely, decreasing the amount of inert gas introduced increases the bulk density of the product. The gas may be introduced into the liquid by any means to be suitable for this purpose, such as beater bars, pumps, whippers, gas injectors or dispargers, high speed shredder plates and the like.

The product may be dried under vacuum by any known method, such as on trays, belts, drums, etc. The conditions of drying may be varied depending on the bulk density desired. It is normally desirable to use as high a vacuum as possible. A pressure of less than 2 inches Hg absolute has been found to be particularly advantageous. Any temperature may be used, including room temperature. It is undesirable to use a temperature above 95°C, as charring of material begins to occur at this temperature. A temperature of 90°C has been found to be particularly advantageous, resulting in rapid setting up of the foam structure, and a moisture content in the final product of less than 2%.

Any convenient means may be used to comminute the dried foam to the desired size range. A gentle mill in conjunction with a particle classifying device is quite suitable. Milling of the dried foam with an oscillating granulator has been found to be quite advantageous. A solution of starch hydrolysate, arabinogalactan and artificial sweeteners can be subjected to a pressure of from 2,500 to 4,000 psi. The pressurized solution is then aerated with an inert gas such as air, nitrogen, carbon dioxide, and the aerated solution subsequently spray dried to produce a free-flowing solid product having the appearance of granulated sugar.

The solids content of the feed solution may range from 50 to 80 weight percent. In general, other variables remaining constant, a greater solids content in the feed solution results in a denser product. In order to achieve a bulk density of 0.12 g/cc, a density particularly preferred in the process, the use of a feed solution having a solids content of about 65 weight percent has been found to be particularly advantageous.

The bulk density of the final product of the spray drying process is also dependent upon the amount of aeration gas used. In general, it has been found that the addition of 1 to 2 scfm of aeration gas for each 10 gallons per hour of liquid fed to the spray dryer results in a satisfactory product bulk density. The aeration gas is introduced into the pressurized liquid stream and thoroughly mixed into the liquid just prior to the actual spraying operation.

The exact pressure and amount of aeration gas chosen, of course, is dependent upon the particular character and design of spray drying apparatus chosen. Such features as diameter of the atomizing nozzle, velocity of air flow, and the like, will dictate specific operating conditions for any particular spray drying unit. The use of a single nozzle having an orifice diameter of 0.016 to 0.021 inch is preferred. When such a nozzle is used, it is particularly preferred to employ a pressure of about 3,000 psi.

The solution temperature may be virtually any temperature at which the solution possesses a workable viscosity. Usually such a temperature will range from 50° to 180°F, although to obtain improved solution viscosities it is preferable to use temperature of between 150° and 180°F. These temperatures have the added advantage of providing pasteurization which may be important if the solution must be held for any substantial period of time before drying.

The spray drying apparatus and its operating conditions are based upon a concurrent flow of liquid and heated air in the conventional manner. The apparatus is so operated that the product, as removed from the base of the chamber, has a moisture content of below 3% by weight. The fines produced are recovered in a second tower and continuously returned through a duct system and introduced into the liquid spray system. The air temperature in the drying chamber is in general controlled by controlling the exhaust air temperature to between 200 and 250°F, although the specific drying temperature used will vary dependent upon the particular equipment used.

The product produced by the above spray drying technique is a granular free-flowing solid having a moisture content of between 1 and 3% by weight and a caloric content of less than 3 calories per 5 cc (the volume of 1 level teaspoonful). The material has a combination of shapes, including, but not necessarily limited thereto, partial spheres, hollow irregular shapes modified by penetrant and nonpenetrant hollowed tubes, and hollow tubes closed at the ends.

Example 1: 1.19 kg of magnesium oxide is dispersed in water heated to 70°C. To this dispersion is added 10.90 kg of saccharin with stirring until a clear solution results. 52.50 kg of arabinogalactan is then added to the solution and stirred until homogeneous. 881.13 kg of cornstarch derived, liquid malto-dextrin (77.5% solids) having a DE of 27, heated to 70°C is then added to the mixture and stirred until homogeneous to give 1,000 kg of a solution containing 75% solids.

This solution is then cooled to room temperature, and air, in the form of finely divided bubbles is introduced by means of beating the solution in a Hobart mixer for 5 minutes to form a foam having an apparent density of 0.786 g/cc. This foam is spread in a tray, placed in a preheated oven and dried under a vacuum of 1 to 1.5 inches Hg absolute at a temperature of about 90°C for 4 hours to produce a cellular "loaf" which is then particulated and sieved through a 14 mesh screen to give a product having a white, granulated appearance, a bulk density of about 0.096 g/cc, and a caloric value of about 1.9 calories per teaspoon, and which has approximately the same sweetening power, on a volume basis, as ordinary granulated sugar.

Example 2: The procedure of Example 1 is followed except that in place of the magnesium oxide and saccharin there is employed 11.25 kg of sodium saccharin. An equivalent product is obtained.

Example 3: A solution is formed as set forth in Example 1. After cooling to room temperature this solution is beaten for 8 minutes in a Hobart mixer to form a foam having an apparent density of 0.696 g/cc. The foam is spread on a tray and placed in a cool oven under a vacuum of 1 to 1.5 inches Hg absolute. The oven is then brought to a temperature of 90°C after 1 hour and thereafter maintained at a temperature between 89° and 90°C for 3 hours. The dried product is then particulated and sieved through a 14 mesh screen to give a product having a white granulated appearance and a bulk density of 0.118 g/cc, and a caloric value about one-eighth that of ordinary granulated sugar.

Coated Sugars

Crystals of sugar (e.g., sucrose, glucose or fructose) are coated with a synthetic sweetening agent by *L.L.F. Deadman; British Patent 977,482; December 9, 1964; assigned to Ashe Chemical Limited, England* to give increased sweetening power while retaining an appearance similar to that of the original sugar. Crystals of a sugar are treated with a synthetic sweetening agent of sweetening power greater than the sugar in the presence of a solvent either for the sugar or for the synthetic sweetening agent or for both, the amount of solvent being in any case insufficient to dissolve all the sugar crystals, and the resultant mixture is dried to give a nonsticky mass.

By using a liquid medium that has some solvent action either for the sugar or the sweetening agent or for both, it is ensured that, when the liquid medium is removed leaving only the solid constituents, the sweetening agent is strongly adherent to the surface of the sugar crystals. The materials may be mixed dry and a small quantity of solvent added to the mixture, but it is preferred to treat the sugar crystals with a solution of the synthetic sweetening agent, which may also contain more of the dissolved sugar.

The removal of the liquid medium may be effected, for example, by centrifuging, by heating to cause it to evaporate, by allowing it to evaporate at room temperature, by blowing a current of cold air through the mass, or, preferably, by blowing a current of hot air through the treated sugar crystals using a fluidized bed technique.

Example 1: Sugar (sucrose) crystals (99 parts) are treated with a solution of sodium saccharin (1 part) in water (3 parts). The sodium saccharin solution is spread onto the sugar crystals and the greater part of the water allowed to evaporate. There results a product containing substantially 99% by weight sucrose and 1% sodium saccharin in which the latter adheres firmly to the surface of the sugar crystals. The final product has approximately four times the sweetening power of ordinary sugar.

Example 2: Sugar (sucrose) crystals (50 parts) are treated with 2 parts of the following solution.

Ingredients	Percent by Weight
Water	25.6
Sodium hydroxide	4.4
Saccharin	20.0
Sucrose	50.0
	100.0

The sticky mass is dried by passing a current of hot air through it until dry, using the fluidized bed technique. The resulting product contains about 1% by weight sodium saccharin and 99% sucrose.

Saccharin-Sugar Product

One of the primary difficulties for persons advised to reduce their calorie intake, whether for medical reasons or simply for reasons of diet, is the necessity of suddenly reducing their intake of sweets. Generally, they are told to stop taking sugar because of the high caloric value and to substitute artificial sweetener such as saccharin. For a very few persons this might be satisfactory, but for the great majority of people this is both difficult to do and undesirable.

The difficulty and undesirability of depending on the artificial sweeteners is due mainly to the following: people taking the artificial sweeteners actually miss the caloric effect of the sugar; and such people also miss the syrupy effect on the tongue and taste of the sugar. This latter "natural" effect of the sugar may be divided into two parts: (a) the effect on the buccal membranes in the mouth; and (b) the syrupiness of the sugar which gives the only true natural feeling of sweetness.

It may therefore be seen that the elimination of natural sugar from the diet and the substitution of the artificial sweeteners therefore, which are taken in thin watery solutions of tiny, and in fact minute amounts, is unsatisfactory from both the physiological and psychological standpoints.

E.A. Ferguson, Jr.; U.S. Patent 2,761,783; September 4, 1956 found a way to provide sweetening compositions which have little caloric value and which satisfy the craving for sweets much better than do the artificial sweeteners, themselves, in fact to an extent which is equivalent to natural sugar, while supplying much fewer calories than natural sugar.

The term "artificial sweetener" is meant to refer to all those artificial sweetening agents of little or no caloric value which are utilized as a substitute for sugar, i.e., saccharin and salts thereof such as sodium saccharin.

It was found that a small amount of the natural sugar component such as sucrose, is all that is necessary to supply the physiological and psychological need of the individual, provided the difference in degree of sweetness between the natural sugar and the usual amount of natural sugar is made up for by an artificial sweetener. Although the full amount of the natural sweetener will supply the physiological and psychological need of the individual for sweetness, this is undesirable because of the high caloric value, which in fact is the reason for the need for artificial sweeteners. Similarly, although the artificial sweeteners alone can supply all the sweetness necessary without supplying calories, as explained above, these artificial sweeteners do not satisfy the craving for sweets.

However, the combination of the artificial sweetener with the natural sugar, particularly in the ratios which will be indicated, supplies both the physiological and psychological sweetness without supplying a large amount of calories. Moreover, this combination also has the advantage of avoiding the bitter metallic aftertaste often found by people taking the artificial sweeteners.

Apparently, only a small amount of natural sugar, such as cane sugar, is necessary to provide the syrupy effect on the tongue and buccal membranes so that a person has the physiological lift due to the syrupy effect and part of the psychological lift due to the full-bodied caloric taste of the natural sugar.

It has been found that the amount of natural sugar which a person ordinarily uses may be reduced to one-quarter by replacing the sweetness of the other three-quarters by an artificial sweetener. For example, one who normally uses two teaspoons of natural sugar in a cup of coffee can achieve the same sweetening effect, not only as to sweetness per se, but also with respect to the physiological and psychological effect of sweetness, utilizing only one-half teaspoon of sugar and making up the balance of sweetness per se with an artificial sweetener, i.e., three-eighths grain of saccharin.

The lower limit of one-quarter the normal amount of natural sugar utilized by the person, with the remainder of the sweetness made up by an artificial sweetener, has been found necessary since below this amount of sugar the physiological and psychological advantages of utilizing sugar in the combination is lost. Although from the psychological and physiological viewpoint there is no upper limit to the amount of natural sugar to be utilized in the composition, as a practical matter, the upper limit is one-half the normal amount of sugar since the purpose of the composition is to achieve natural sweetness with lowered caloric intake and supplying more than half the number of calories would be impractical in its effect.

The amount of artificial sweetener necessary to make up the difference in sweetness between the amount of sugar utilized and the normal amount of sugar depends upon the particular artificial sweetener. The saccharins are about 200 times as sweet as sugar. It is simple to determine from this what amount of the artificial sweetener is necessary.

Organoleptic tests show that within the limits given above, the following favorable results are achieved. Although the caloric intake is cut from one-half to one-quarter of the normal caloric intake for the purpose of sweetening, individuals utilizing the special compositions could not distinguish between the compositions and natural sugar. Thus a sufficient sweetening effect was obtained from both the physiological and psychological standpoint to achieve natural sweetening while lowering the caloric intake at least 50%.

It is of course inconvenient to utilize the compositions by mixing the natural sugar with the artificial sweetener just prior to use. Also, the ordinary mixing of the ingredients by the manufacturer and sale of the mixed composition has the disadvantages that the artificial sweetener is a light powder in comparison to the heavy granules of sugar so that a sifting and floating occurs during handling of the package resulting in improper proportioning of the components.

With the object of overcoming this difficulty, a process of producing a sweetening composition, comprising the steps of mixing an artificial sweetener with an approximately equal amount of natural sugar and a minor proportion of at least one hydrophilic colloid, wetting the formed mixture with an aqueous alcohol solution, drying the wetted mixture, grinding the dried mixture, and mixing the ground mixture with an amount of natural sugar such that the formed composition is between two to four times as sweet as natural sugar.

The usual pharmaceutical practice of mixing a light substance with a heavy substance by dissolving both substances in a solvent such as water and recrystallizing, the well-known granulation process, is highly expensive and could not be utilized from the point of view of economy. Furthermore, this pharmaceutical granulation process is usually utilized in the preparation of tablets and it is not suitable for the manufacture of a product which is intended to be kept in powdery or granular state.

It was found that the above disadvantages in the manufacture of the compositions can be overcome by the following process. The total amount of artificial sweetener is mixed with an equal amount of granular natural sugar. To this mixture is added a small amount of a hydrophilic colloid such as a colloid gum of the type of acacia, gum arabic, karaya gum, etc. or carboxymethylcellulose, preferably about 0.8 gram per pound of artificial sweetener. Then, per pound of artificial sweetener there is added 100 cubic centimeters of an aqueous alcohol solution, preferably 90% alcohol plus 10% water, so as to just wet the particles of the mixture of artificial sweetener.

Upon drying, the product cakes. The caked product is then ground by gentle means to the size of granular sugar, taking care not to powder the product. This granular product may then be mixed with the remaining amount of natural sugar and is found to mix readily and to remain as a homogeneous mixture even after standing for long periods of time.

Example: 0.35 pound of sodium saccharin is mixed with 100 pounds of granulated cane sugar. 2.1 grams of guar gum is added to the mixture. 175 cubic centimeters of 90% alcohol is then admixed with the mixture until all the particles are wetted therewith. The resulting mixture is then allowed to dry whereupon it cakes. The caked mass is then ground gently until the particles are about the size of granulated cane sugar whereupon it is mixed with 98.25 pounds of cane sugar until homogeneously mixed. The resulting product is approximately twice as sweet as ordinary sugar.

Organoleptic tests showed that it was impossible to tell the difference in taste or effect between natural sugar and this composition although the composition has one-half the calories of natural sugar per unit of sweetness.

Starch Hydrolysates

A sugar substitute is prepared by acid hydrolysis of starch and artificial sweetener is added for equivalent sweetness to saccharose in the process of *E. Conrad and G. Frostell; British Patent 1,169,538; November 5, 1969; assigned to Lyckeby Starkelseforadling AB, Sweden.*

Examples of starches which may be used in the hydrolysis include cornstarch and potato starch. The hydrolysis may be carried out under acid, weakly hydrolyzing conditions, for example, at a pH of 2 to 4 and may be conducted at an elevated temperature, for example, 150°C. A weak organic acid or an inorganic acid may be used to provide the desired pH. The hydrolysis temperature, pH and pressure are chosen so as to give the desired partial hydrolysis.

The hydrolysis is carried out by use of a weakly active enzyme, the hydrolysis being performed under the optimum conditions for the particular enzyme used. The hydrolysis is stopped when the product has the desired dextrose equivalent value. When the hydrolysis is an acid hydrolysis carried out at an elevated temperature, it may be halted by neutralization of the acid and/or by decreasing the temperature while if the hydrolysis is enzymatic it can be stopped by increasing the temperature so that the enzyme is destroyed.

If the hydrolysis product is to be hydrogenated directly the hydrolysis is halted before the hydrolysis product has a dextrose equivalent value of 15%. However, the hydrolysis product may be subjected to a fermentation step whereby mono- and disaccharides are removed, the product remaining after the fermentation consisting of a mixture of dextrins and poly- and oligosaccharides having a dextrose equivalent value of at least 1.2% but less than 15% and the product hydrogenated.

In this case the dextrose equivalent value of the hydrolysis product may exceed 15% as the fermentation step reduces the DE to below 15%. Critical control of the hydrolysis conditions will often be necessary in order to ensure that only the desired degree of hydrolysis occurs.

In order to determine the conditions necessary in any particular process it will generally be necessary to conduct a few preliminary tests. For example, if in one such test it is found that the hydrolysis conditions used result in the hydrolysis going too far then clearly the hydrolysis conditions are too strong. Another test then has to be conducted using milder hydrolysis conditions or shorter hydrolysis times to see if the desired degree of hydrolysis then occurs. The hydrolysis product, or the product remaining after the fermentation, if a fermentation step is used, is hydrogenated in a known manner, the conditions being such that all reducing groups present in the mixture undergoing hydrogenation are hydrogenated substantially without any reduction of the number of monosaccharide units per molecule. Thus, the hydrogenation step results in substantially no hydrolysis of the hydrolysis product or of the product remaining after the fermentation step, if a fermentation step is used. The hydrogenation can conveniently be carried out at an elevated temperature and using, for example, Raney nickel as the catalyst.

The hydrogenation serves to provide a product which is considerably less fermentable than the material which is hydrogenated because any reducing groups present in this material are hydrogenated to hydroxy groups.

The sugar substitute obtained by the hydrogenation has a lower degree of sweetness than saccharose, which is due to the fact that mono- or disaccharides have only been formed to a minor extent in the hydrolysis step or have been removed in the optional fermentation step. For this reason, if desired, the product obtained by the hydrogenation step can be mixed with an artificial sweetening agent such as saccharin, preferably in an amount which gives a degree of sweetness corresponding to that of saccharose.

The result is that the mixture obtained can directly replace a corresponding amount of saccharose in the preparation of food products of many kinds including candies and sweets, etc. The product of the process can be prepared in the form of a viscous syrup, that is, in solution, but it is possible to subject the solution to a drying step in order to obtain the product in the form of a white powder which is hygroscopic only to a very minor extent. The drying may be performed in any suitable manner, such as spray drying, roller drying or freeze drying.

Example 1: Acid Hydrolysis – A suspension of potato starch containing 45% of dry solids (2,000 kg of potato flour containing 1,800 kg of dry starch suspended in 2,000 liters of water) was admixed with 5 liters of 37% technical hydrochloric acid. The pH of the mixture was 2.2.

The hydrolysis was performed in thirty minutes at 130°C and at a gauge pressure of 3 atmospheres in order to transform the starch to a dextrinous decomposition product. The amount of reducing substances, calculated as dextrose, was 12%. The batch was neutralized with 20 liters of 15% sodium carbonate solution to a pH of 6.5.

The solution was decolorized by the use of active carbon and solid substances were removed by centrifugation. Then Raney nickel catalyst was added to the hydrolysis product in an amount of 2%, that is, 36 kg. The hydrolysis product was then hydrogenated at a hydrogen pressure of 75 atmospheres and at a temperature of 145°C.

The hydrogen consumption was 30 cubic meters hydrogen at NTP. The hydrogenation was performed conventionally and was continued until an equilibrium was reached, that is until all reducing substances were hydrogenated. The catalyst was then removed by separation. The solution was passed through a cationic and an anionic ion exchange resin for removal of metal and acid ions. The solution was then concentrated by vacuum evaporation and spray dried to yield a nonhygroscopic powder.

Example 2: Enzymatic Hydrolysis – The 45% starch suspension according to Example 1 was admixed with 3 kg of enzyme (bacterial alpha-amylase). The pH of the solution was 6.2. By the addition of sodium carbonate solution the pH was adjusted to 7.5. The suspension was heated for 15 minutes to a temperature of 90°C. Then the enzyme was inactivated by heating for 5 minutes at a temperature of 110°C.

The end product was a solution containing mainly dextrineous decomposition products. The content of reducing substances, calculated as dextrose, was 4.6%. The solution was decolorized and separated and hydrogenated as described in Example 1. In the hydrogenation step the hydrogen consumption was 26 cubic meters at NTP. When it is desired to obtain a product which can be used as a direct substitute for saccharose the concentrated solution is mixed with saccharin.

Example 3: Intermediary Fermentation of Mono- and Disaccharides – In order to obtain the lowest possible amount of mono- and disaccharides in the final product the following intermediary step was used in the process according to Example 1. By increasing the hydrolysis time in Example 1 from 30 to 60 minutes a reducing (dextrose equivalent) value of 27% was obtained. The solution was neutralized by the use of 18 liters of a 15% sodium carbonate solution to give a pH of 5.0. The temperature was adjusted to 30°C and 10 kg of brewers' yeast (*Saccharomyces cerevisiae*) were added.

The fermentation was carried out conventionally and was finished after 48 hours when the formation of carbon dioxide was ended. The amount of reducing sugars had then been decreased to give a dextrose equivalent value of about 1.2%. The solution was decolorized by the use of active carbon and filtered. The clear solution obtained was hydrogenated as previously described. The amount of hydrogen used was 12 cubic meters at NTP. The hydrogenated solution was separated from the catalyst, filtered and treated with ion exchange resins. In the subsequent evaporation step, the alcohol formed during fermentation was removed.

MALTOL SWEETNESS POTENTIATOR

E.F. Bouchard, C.P. Hetzel and R.D. Olsen; U.S. Patent 3,409,441; November 5, 1968; assigned to Chas. Pfizer & Co., Inc. gives a description of the use of maltol as a potentiator for sweetness without itself contributing flavor. Maltol, also known as 2-methyl-3-hydroxy-gamma-pyrone, has been enjoying increased use in enhancing the flavor and aroma of foods. However, maltol has not been known to increase the apparent sweetness of sugar. It has been found that maltol has a powerful lifting effect on the sweetness of sugar and, as a result, it is possible to replace part of the sugar in sweetening compositions with maltol.

Maltol has been found to actually potentiate the sweetness of sugar without adding a taste of its own and without merely replacing one taste with another. Indeed, maltol, by itself, does not possess a sweet taste. This enhancement of sweetness is so pronounced that as much as 15 parts by weight of sugar in a composition containing 100 parts of sugar can be replaced with only from about 5 to 75 ppm of maltol. Since the price of sugar represents a significant portion of the total manufacturing cost of many foods, such as, for example, baked goods, candies, carbonated beverages and fruit drinks, the reduced sugar levels in the varied formulations of this process allow highly significant economic savings to be obtained. For example, consideration might be given to the typical savings obtained on using 10% less sugar in a lemonade formulation. Ordinarily, 800 pounds of sucrose is used to make 1,000 gallons of lemonade. Eleven cents a pound, for example, as the price for sucrose, which tends to fluctuate from time to time, represents a value of $88 for the sugar. Decreasing the amount of sucrose by 10%, to 720 pounds, results in a new cost of sugar in 1,000 gal of lemonade of $79.20, but the lemonade is less sweet, flatter and more acid to the taste.

It is found that the addition of 58 grams of maltol to 1,000 gallons of lemonade containing the decreased amount of sugar causes such an enhancement in sweetness that the lower sugar-containing lemonade is identical with the original, 100% sugar-containing formulation. At a maltol market price of $12 a pound, the amount of maltol added represents $1.50 per 1,000 gallons of lemonade. Thus, $8.80 saved on lowering the sugar content has been achieved at a net maltol cost of $1.50 and a total net savings of $7.30 per 1,000 gallons of lemonade is realized.

Maltol is freely available commercially. It can be prepared, for example, by the combination of fermentation and chemical synthesis processes disclosed in U.S. Patent 3,130,204. The process involves the oxidation of kojic acid, which is obtained by fermentation, to comenic acid, the decarboxylation to pyromeconic acid, treatment with formaldehyde to form 2-hydroxymethyl pyromeconic acid, and reduction to form maltol, 2-methyl-pyromeconic acid.

The term "sugar" used in this description refers to carbohydrates having a sweet taste and the general formulas, $C_nH_{2n}O_n$, $C_nH_{2n+2}O_n$ or $C_nH_{2n-2}O_{n-1}$. Among the sugars whose sweetening power is enhanced by the addition of maltol are, for example, fructose, invert sugar, sucrose, glucose, xylose, maltose, rhamnose, galactose, raffinose, lactose, mannitol, sorbitol, xylitol and arabitol.

An important commercial use of the process would be illustrated by the preparation of carbonated beverages and fruit-type beverages. These are ordinarily prepared by adding citric acid and sugar (usually enough to provide from 9 to 13% of the total) to the acid-containing fruit juice or flavors.

Maltol is a crystalline substance and can be used as is or in solution in a solvent such as water. The flavoring compositions can be mixtures of solid sugars and solid maltol or solutions of maltol. It is not necessary to premix both ingredients since the addition of sugar may precede or follow the separate addition of maltol.

It will be recognized that, since the compositions contain up to 90 parts sugar and maltol in an amount to provide from 5 to 75 parts by weight based on the food, if desired, only 5% or more, or even less, of the sugar may be replaced in any given formulation. The amount of maltol to be used will, of course, depend on the amount of sugar to be replaced but will fall within the stated range. Since maltol itself does not taste sweet, a total replacement of sugar cannot be made but rather a sparing technique where partial replacement of sugar, with attending economic advantage, is attained.

Example 1: Fresh lemonade and limeade containing 9.0% sugar and 15 ppm maltol were prepared. Control lemonades and limeades containing 9.45 and 10.35% sugar but no maltol were also prepared. When these samples were given to a trained and experienced taste panel, the panel members matched the control and test samples as being equal in sweetness.

Example 2: Three lemonades were prepared with the following compositions, respectively.

	1	2	3
Fresh lemon juice, ml.	35	35	35
50% sugar syrup (w./v.), ml.	50	45	45
Water, ml.	175	177	177
1% maltol solution, ml.	-	-	0.4

Each of the three solutions is presented to seven tasters; seven out of seven tasters judged 1 and 3 most alike; 2 is described as less sweet, more acid or flatter. Thus, it is found that 10% or 10 parts per hundred parts, of the sugar in the lemonade can be replaced with 15 ppm of maltol based on the beverage, while the acid level is maintained at about 8.4 parts of citric acid per 100 parts of sugar originally taken, and there is obtained lemonade with sweetness and acid taste equivalent to the original sample. Substantially the same results are obtained when 10 parts per hundred of the sugar are replaced with 25 ppm of maltol based on the beverage.

Example 3: An acidulated, sweetened mixed fruit-type punch drink is prepared which contains 100 grams of sugar per liter of drink. A second drink is prepared containing 95 grams of sugar per liter. A third drink is prepared containing 95 grams of sugar and 0.015 gram of maltol per liter. The third drink, which contains 15 ppm of maltol, is fully as sweet and acceptable as the first drink; the second drink is less sweet than the first and third. Thus, 100 parts of sugar have been replaced with a composition comprising 95 parts of sugar and $\frac{1}{67}$ part of maltol, or it can also be said that 5,000 parts of sugar have been replaced with 15 parts of maltol and 1 part of maltol has replaced 333 parts of sugar.

Example 4: A mayonnaise-type salad dressing is made, which contains 1% of sugar and 1% of acid as acetic. This is used as a control in an evaluation of the sweetness and acidity of a number of dressings of the same formulation wherein up to 10% of sugar has been replaced and to which maltol has been added in an amount to provide from 5 to 75 parts by weight based on the dressing. It is found that the dressings containing less sugar and also containing maltol are fully equivalent to the control dressing.

A particularly efficacious combination comprises 0.95% sugar and maltol in an amount to provide 15 ppm based on the dressing. The amount of acid in the dressing is varied from 0.5 to 100 parts based on the sugar. It is found that 100 parts of the sugar can be substituted with up to 90 parts sugar and from 5 to 75 ppm by weight of maltol based on the dressing and the same sweetness and acidity as the control dressing is obtained.

Example 5: A cherry flavored beverage was prepared by adding a cherry flavor extract in equal amounts to the following formulations.

	Sample A	Sample B	Sample C
Sucrose, g.	13.2	11.9	11.9
Fumaric acid, g.	0.15	0.15	0.15
Maltol, ppm	0	0	50

Sample C was as sweet as Sample A and tasted sweeter than Sample B. When 75 ppm of maltol are added to Sample B, containing 10% less sugar than Sample A, it is found to taste sweeter than Sample A and sweeter than Sample C.

Example 6: Maltol is added to a number of sugars and sugar syrups: sucrose, brown sugar, maple syrup, corn syrup, fructose, invert sugar, glycose, xylose, maltose, rhamnose, galactose, raffinose and lactose. Each of the sweeteners has been dissolved in water in amounts corresponding to 0.33 and 0.66%, respectively. The amounts of maltol added are 75 ppm, and 250 ppm based on total solutions. A portion of each of the sugar solutions is reversed for use as a control.

Sweetness of the solutions is tested by presenting each to a taste panel, the members of which are requested to state whether there is a positive sweetness, borderline sweetness, or no sweetness present. Maltol is found to definitely enhance the sweetness of the sugars. Maximum enhancement is obtained at the 75 ppm level. At the 250 ppm level, the taste of the maltol is apparent. When this experiment is repeated using 100 ppm of maltol based on total solution, the taste of maltol is apparent.

AEROSOL SWEETENER

A noncaloric sweetener is dispensed from an aerosol-type container as described by *N. Grober; British Patent 1,091,370; November 15, 1967.* When the concentrated sweetener is prepared in liquid form, the adaptability of the chemical sweetener is broadened. The liquid concentrated sweetener easily dilutes and mixes in cold as well as in hot media. However, the practical problem of measurement of small amounts of concentrated liquid and the use of such small amounts over an area larger than the physical measured size of the liquid, creates measurement problems and therefore overuse problems. One drop or 0.05 cc of the liquid concentrate or some multiple thereof is generally the unit of liquid to be measured and used.

Various commercial suppliers have adjusted sweetness concentration formulas to varying concentrations by the addition of water and/or other bulking agents to arrive at a practical measurable amount of a liquid sweetener concentrate. Too much water cannot be added, as the required volume would affect the usefulness of the sweetening product. If excessive liquid is added to the product to be sweetened the appearance and taste of that product would be affected.

The spreading of concentrated liquids to dry or semidry objects such as dry cereals or fruits is not practical. The application of one or several drops to a general area the size of a grapefruit, or a bowl of cereal or similar areas to be sweetened, results in pockets of concentration of sweetness. The sweetening effect is lost unless the specific area containing the drop of concentrate is tasted, and if that specific area is tasted there is an overconcentration of sweetness at that spot and a lack of sweetness in areas not physically controlled. There is an alternative in that the drop measure can be further diluted before use.

There is not in use a reliable method of directly applying a concentrated liquid sweetener to an area. A drop measurement when normally made, except under laboratory conditions may vary from 0.025 cc or less to over 0.2 cc. The variations in the size of the drops result from the type of container from which the drop is dispensed, the aperture through which the liquid is forced to produce the drop, the pressure applied to the container to expel the liquid, the viscosity of the liquid being squeezed out and/or the amount of liquid in the container at the time it is being dispensed. The variables are so many that standardization in dispensing the small amount of liquid has not been an accomplished fact.

The dispensing of the composition may be in the form of a spray, a mist or a film issued through a valve on the aerosol container. Containers of this type are well known and may be formed of metal, glass, plastic coated glass or the like. A metering or measuring valve can be used.

It is important that the propellant used be substantially inert so as to have little or no effect on the odor, flavor, appearance or texture of the dispensed product. Also, since pressurized packaging is involved, care must be taken to prevent hazards during packaging, storage and use of the dispenser.

The propellants may include as constituents one or more of the known commercially available propellants. These include perfluorocyclobutane which is known commercially as Freon C-318. Other propellants which may be utilized include Genetron, Isotron and Ucon. Carbon dioxide and nitrous oxide gases are also used to pressurize a container. Certain other gases such as propane are also used to assist or build up the pressures within the aerosol container.

The propellant may be a liquefied, dissolved and/or compressed gas which propellant, within the confines of the container, combines in varying degrees with the product and creates or exerts a pressure within the confines of the container which forcibly expels the product through a restricted orifice upon the release of the pressure by a suitable valve. The gas both carries and/or propels the product.

As the product is expelled or propelled through the opening, the liquid is broken up into finer individual droplets. The size of these droplets can be predetermined by the parameters of the restricted orifice through which the product is to be forced, by the valve selected and by the amount of the pressure exerted on the product in expelling or propelling the product from the container. The droplet of product may be further broken up if the droplet contains any of the propellant which may vaporize at specific temperatures, this vaporization causing an explosive force within the droplet and a further breaking up of that droplet.

The quantity of propellant used will vary and will be determined by the pressure desired which should be adequate to propel or expel the composition through the orifice in the pattern desired. Generally, it is preferred to use propellants which at 70°F will exert vapor pressure in the range of from 10 to 100 psig. Pressure variations within this range can be attained by blending or combining different proportions of propellants. The liquefied gaseous propellant recommended is a specific product which has a vapor pressure at 70°F of 25.4 psig.

This propellant is available under the trade name Freon C-318. A composition is prepared with approximately 4.5% Freon C-318 by weight and 95.5% of concentrate solution. These percentages however, may vary substantially and the concentrate solution may be in the range from 75 to 99.5%.

It is desirable to use chemical sweeteners which will provide the desired concentration of sweetness in its most effective form for use in a spray dispenser for direct application to the product to be sweetened. A low viscosity solution such as one of soluble saccharin salts is especially desirable for dispensing in a spray pattern, however, higher viscosity solutions which result from different combinations of sweetening agents also may be used.

The method of preparing these liquid compositions of chemical sweeteners involves incorporating in a water solution, a soluble saccharin salt and preservatives such as benzoic acid and methyl para-hydroxybenzoate. While it is intended to use a water carrier for these chemical sweeteners, there is no intent to be limited thereto as ethyl alcohol, propylene glycol or olive oil among others may be used especially if dulcin or mannitol are used.

The proportions by weight of the saccharin, the sorbitol, the dulcin and/or the mannitol to be used is a function of the measurement device to be incorporated in the valve or spray unit so that the sweetening power of the dispensed sweetener coupled with the volume dispensed by the spray will, taken together, give a unit of sweetness such as the equivalent to a teaspoon of sugar. In the absence of an automatic control measurement device, the unit of measure may be a factor of elapsed time in applying the spray.

There may be also added small amounts of maltol in order to mask the possibility of metallic aftertaste and to enhance the flavor of the product to be sweetened thereby further masking the aftertaste sensations by increasing the intensity of the basic taste. The maltol should be thoroughly and carefully mixed in the sweetener solution, especially since only very small amounts of this ingredient are used.

In order to retard rapid surface evaporation of the concentrated sweetener solution after it has dispersed or been dispensed in droplet form, it being understood that the total surface of these sprayed droplets in a finely atomized spray is enormously large, and in order to prevent rapid evaporation which may also cause particles from the sweetener concentrate to show and clog in and around the orifice and valve area which will affect the proper functioning of the spray apparatus, required amounts of humectants such as propylene glycol are added to the sweetening solution.

This increases the ability of the carrier for the chemical sweetener to retard evaporation, aids in clearing the orifice of crystals, aids in preventing the rapid formation of crystals, and aids in retention of moisture in the droplets thereby increasing the ability of the droplet, by retention of the moisture, to further spread a film of sweetener to its immediately surrounding area on application of the spray.

Example 1:

	Percent by Weight
Sodium or calcium saccharin	20.50
Benzoic acid	0.10
Methyl para-hydroxybenzoate	0.05
Water	79.35
	100.00

Example 2:

Sodium or calcium saccharin	20.50
Maltol	0.0015
Benzoic acid	0.1000
Methyl para-hydroxybenzoate	0.0500
Water	79.3485
	100.0000

To prepare a sweetening formulation for a 25 gallon batch, for example, the following procedure would be followed. Warm 10 gallons of deionized water to 160° to 180°F and add 3.5 oz of benzoic acid and 1.75 oz of methyl para-hydroxybenzoate. After these preservatives are dissolved, add 10 gal more of deionized water. Cool the solution to room temperature. Then add the required amounts of sodium or calcium saccharin. At this point, desired amounts of maltol and/or one of the humectants may be added and thoroughly mixed into solution. Add deionized water to bring the volume to 25 gal and filter the solution.

The amount of sweetening concentrate to be dispensed at one time is to be controlled and measured in units of 0.05 cc and generally in the range from 0.025 to 0.5 cc. Such measurements can be accomplished by the utilization of aerosol dispensing valves such as those of the type conventionally used. The controlled amount to be dispensed as a spray is to be a multiple of the sweetening power of 1 teaspoon of sugar.

The concentrated sweetener is now to be combined with the selected propellant in the following proportion in order to form a self-propellant concentrated chemical sweetener: Freon C-318, 4.5%; concentrated chemical sweetener, 95.5% by weight in the preferred embodiment. The resultant composition is packaged in an aerosol container and the resulting product can then be dispensed from the aerosol container through manual release of the valve.

STABILIZED SACCHARIN DRY EMULSION

In order to minimize the aftertaste of the soluble salts of saccharin, *A.R. Globus; U.S. Patent 3,730,736; May 1, 1973; assigned to Farah Manufacturing Co., Inc. and Guardian Chemical Corporation* developed a formula containing:

(1) An emulsifying agent
(2) Saccharin formed in the presence of the emulsifying agent by reaction of the water-soluble saccharin salt and a nontoxic acid
(3) A stabilizer for the emulsion
(4) A food grade carrier for the emulsion

The first component is a nonionic surface active agent, that is, complex esters or ester ethers conventionally known as Span type materials and Tween type materials. The Span type materials are partial esters of the common fatty acids (lauric, palmitic, stearic and oleic) and hexitol anhydrides (hexitans and hexides) derived from sorbitol. The Tween type materials are derived from the Span products by adding polyoxyethylene chains to the nonesterified hydroxyls.

The second component constitutes the active or sweetening ingredient and is prepared in situ by reacting a metal salt of saccharin such as a sodium, potassium or calcium salt of saccharin with at least the stoichiometrically necessary amount and preferably an excess of a fairly strong nontoxic acid such as tartaric acid, citric acid, malic acid, gluconic acid, etc.

It is also possible to use as the acid component a compound which gives rise to the acid in the presence of water as for instance glucono-delta-lactone which forms gluconic acid in the presence of water. The preferred acid is citric acid.

The saccharin salt, for instance, saccharin sodium, is reacted with the acid, for instance, citric acid by introducing both of these compounds into the first component, there being formed a thick syrup containing the acid salt, that is, sodium citrate and free saccharin in a highly emulsified state. This thick syrup is colorless and highly water-soluble.

The third component is a food grade polyphosphate, as, for example, potassium polyphosphate and serves as a stabilizer for the emulsion.

The fourth component is a carrier such as glycine, alanine, aspartic acid, lycine and the like or a mixture thereof. There can also be used as carrier, glucono-delta-lactone or sorbitol. It is necessary only that the amino acid or other compound selected for use as carrier be a solid, freely soluble in water and nontoxic in the amounts involved.

In general, the first component is prepared first by mixing a small amount of water with relatively large amounts of glycerin and the surface active agent. The second component, consisting of a dry mixture of soluble saccharin and food grade solid acid, is then added to the first component or vice versa. A thick syrup is formed containing in a highly emulsified state the products of the reaction and namely saccharin and the acid salt. The polyphosphate is then incorporated into the syrup.

The introduction of the solid polyphosphate may bring about a gelling and/or hardening of the emulsion. If this takes place, it is necessary that the gell be broken up before the carrier is added. This may even require that the hardened gell be subjected to grinding.

The fourth and final component, the carrier, is then added. The important feature here is that sufficient mixing should be utilized by working, grinding, homogenizing, or otherwise to secure a complete and thorough dispersion of the semi-finished product formed from the first three components throughout the carrier and thereafter the formation of a unitary material inseparable into its components without further treatment. The final product is required to be a dry powder. If the composition obtained following incorporation of the carrier is not entirely dry, the working, that is, grinding can be carried out under warm air.

The special sweetener is a white powder similar in appearance to flour and/or confectionary sugar. In some instances, it may be semicrystalline in nature and resemble crystalline sugar more closely than powdered sugar. It is not particularly hygroscopic and is highly soluble in water.

Example:

	Amount (percent)
Sodium saccharin	2.0
Citric acid	5.0
Water	1.0
Glycerin	2.0
Glycerol monostearate	2.0
Potassium polyphosphate	8.0
Glycine or sorbitol	Balance

The water, glycerin and glycerol monostearate were mixed together. The sodium saccharin and citric acid were introduced into the water-glycerin-glycerol monostearate mixture and allowed to react in the cold. The potassium polyphosphate was introduced into the resultant thick syrup. The syrup thereupon underwent gelling and hardening. The hardened material was broken up by grinding and in the ground form added to the glycine. This mixture was homogenized by further grinding, the grinding being continued until a unitary powder form material was obtained.

The product of the example was tested for aftertaste by being offered to a large number of subjects in a form of a sweetening agent in coffee, soft drinks, jams, jellies, gelatins and ice cream. In complete contrast to the same products in which sodium saccharin had been used as the sweetening agent, in no instance was any aftertaste reported for the products containing the sweetening composition of the process.

The ammonium salt of glycyrrhizic acid was substituted for the sodium saccharin and substantially the same results were obtained. The sweetening compositions are seven times as sweet as commercial sugar (sucrose) and eight times as sweet as fructose. They contain 2.4 calories per gram. These calories are not, however, derived from carbohydrate.

SACCHARIN COMPLEX

A. Zaffaroni; U.S. Patent 3,876,816; April 8, 1975; assigned to Dynapol discusses the approach of producing a type of nonnutritive sweetener by covalently bonding a biologically active sweetening group to a controlling molecule that resists active and passive transport in vivo. The complex is represented by the following general formula: $(AM\sim)_n C$ wherein Am is an active sweetening group $\sim$ is a covalent bond, C is a polymeric molecule, and n is at least one.

The representation Am as the sweetening moiety or group generally includes any chemical group that is capable of producing a sweet effect in animals including humans and can be covalently bonded to a controlling group, C, while simultaneously retaining its ability to produce a sweet effect. The sweetening group can be bonded directly or through other functionally equivalent covalent bonding moieties attached to the sweetening group that are nonessential for a sweet effect. The sweetening group can be of naturally occurring or synthetic origin and it can have a nutritive or nonnutritive value, which latter properties are not available because the sweetening group is covalently bonded to the controlling molecule. Generally, the sweetening group can be any group that causes sweetness or intensifies sweetness in an in vivo environment.

Example 1: The covalent bonding of a sweetening group to a polymeric material is illustrated as follows. First, 1-chloro-2,4-dinitrobenzene is reacted with 2-hydroxyethyl ether in the presence of an alkali, NaOH, to yield the corresponding product 1-(2'-hydroxyethoxy)-2,4-dinitrobenzene. This latter compound is then reduced with sodium disulfide to form the ortho- and para-amino isomers which are separated on a silica gel column. Next, 1-(2'-hydroxyethoxy)-2-amino-4-nitrobenzene is acetylated with one equivalent of acetyl chloride at room temperature to yield the corresponding 2-(2'-hydroxyethoxy)-5-nitroacetanilide, which is reacted with epichlorohydrin under mild alkali conditions (OH^-) to give 2-[2'-(1,2-epoxy-3-propoxy)ethoxy] -5-nitroacetanilide.

This compound is then optionally reacted with polyvinyl alcohol in the presence of a strong base and an organic solvent such as dimethylsulfoxide or in an aqueous medium to yield the nonnutritive sweetener after hydrolysis under acid conditions of the amide function to the free amine.

Example 2: Following the procedure as set forth in Example 1, a sweetener is prepared as follows. First, 0.31 mol of 2-amino-4-nitrophenol is added to 200 milliliters of acetic anhydride in 700 ml of acetic acid and the reactants refluxed overnight at room temperature to give 2-acetamido-4-nitrophenol. Next, 0.209 mol of the nitrophenol in 4.25 liters of 1-propanol is gently mixed with 250 ml 1 N NaOH and the reaction carried out with stirring at 50° to 60°C for 10 min to form the sodium salt of 2-acetamido-4-nitrophenol.

Next, 0.048 mol of the just prepared and recovered 2-acetamido-4-nitrophenol is mixed with 0.06 mol of 3-bromo-1-propanol in 100 ml of n-propanol and the reactants refluxed for 24 hours under normal atmospheric conditions, to form 2-(3'-hydroxypropoxy)-5-nitroacetanilide, which is then reacted at 0°C for 20 hours in pyridine with 0.09 mol of p-toluenesulfonyl chloride to form the 3'-tosylate ester. The ester (0.015 mol) is added to a mixture of 0.015 mol of dimsyl sodium and 0.05 mol of polyvinyl alcohol and 100 ml of dimethylsulfoxide and the reaction carried out at room temperature for 24 hours. Then

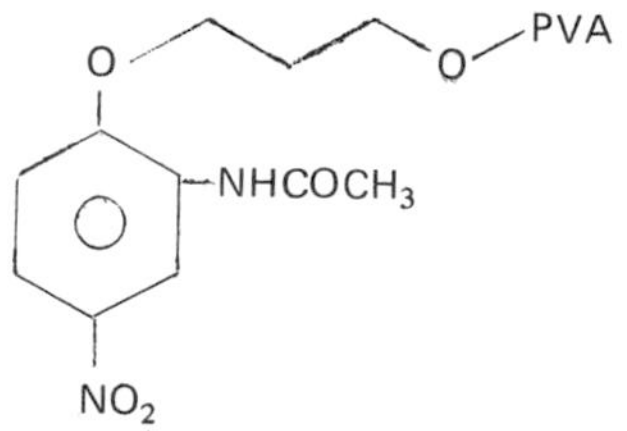

O O PVA

$NHCOCH_3$

NO_2

is recovered and hydrolyzed with 250 milliliters of 2 N HCl under refluxing conditions for 1.5 hours. This gives the sweetener of the formula shown on the following page, after neutralization with base, wherein PVA is the polymer polyvinyl alcohol. The acid addition salts of the NH_2 functionality can also be readily prepared according to this procedure.

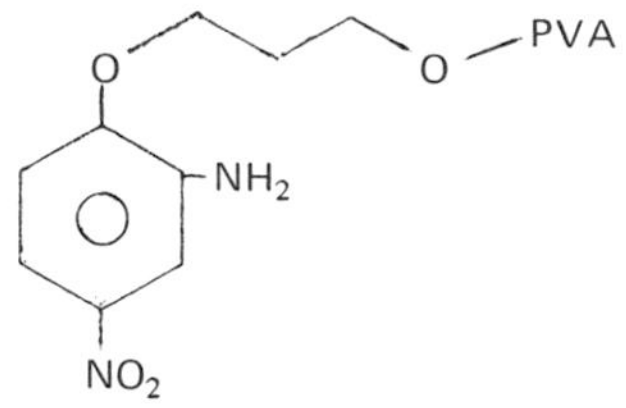

The nonnutritive sweeteners are distinct from naturally occurring sugars since they are not converted to carbon dioxide, water and energy by the body while retaining their sweetening properties. Additionally, they do not have an insulin requirement, and because of the chemistry they are usually nonhygroscopic and do not caramelize as naturally occurring sugars do during processings of foods.

Moreover, since the active sweetening group, Am, is essentially permanently joined through a covalent bond to a controller that transports it and allows the active agent to give a sweet result, while preventing its absorption and assimilation, any adverse effects inherent in the active sweetening group Am are retained by the compound $(Am)_n C$ for eventual elimination from the gastrointestinal tract by the host.

AMINO ACID INTENSIFIER

J.F. Lontz and R.T. D'Alonzo; U.S. Patent 3,833,745; September 3, 1974 gives the following general background on sweetness mechanisms and special sweetener formulations using amino acids as sweetness intensifiers. The concept of sweetness, in one sense of the term, is self-evident but in the physiological or sensory respect the quality of sweetness involves varied and complex factors many of which have yet to be understood. Thus, often initial sweetness, especially with artificial sweeteners when overdone or overindulged in, can develop in the taste buds of the host a markedly varying quality of taste or can impart to taste buds or palate an opposite sensation through a gradation of decreased sweetness leading ultimately to a bitter taste.

The mechanism of this change, sometimes alluded to in a psychophysical scale or spectrum of taste, is not clearly understood, particularly with the ingestion of benzenoid or aromatic ring derived sweeteners, notably that of saccharin and dulcin. It can be conjectured that the range from sweetness to bitterness is a latent effect especially after some time has elapsed after the ingestion of the benzenoid artificial sweeteners as a delayed secretionary response, quite like many undiscerned counteractive mechanisms that nature is wont to provide.

The degree to which such counterresponses, that is, from sweetness to bitterness, are stimulated can be complex and of differing order with different individuals and with different sweetening additives. Thus, cyclohexylamine cyclamate, while devoid of the aromatic ring structure and having a much lower sweetening action than saccharin, appears to impose less of the bitter aftertaste or sensation compared to the benzenoid saccharin or dulcin type. In short, to gain a desirable or favorable taste by the use of an artificial or synthetic sweetener there is imposed the disadvantage of an annoying and unfavorable aftersensation.

Thus, the sensation of palatability has far reaching implications involving, or derived from salivary physiology with its complex secretionary responses on the one hand and the digestive state on the other. Taste is the least informative of the senses when graded on a psychophysical scale or spectrum, albeit it is usually categorized into an empirical quality spectrum ranging from sweet to sour, bitter and salty.

In order to understand some of the secretionary aspects of taste, numerous studies have been conducted in two general directions. The first of these relates to obtaining a quantitative or quantitative-like rating in a conventional panel test of participating, individual tasters who rate sweetness by ranking a concentration range, thereby providing a sort of a psychophysical basis involving only the initial sensation to the taste buds.

The other, more basic type of studies relates to the biochemical changes in the composition and activity of human saliva in the regeneration of enzymic activity and transient changes in the salivary composition. Both approaches obviously have to be applied along with the conventional assessment of the changes in the biochemical constituents of blood and other circulatory transport systems, all related to the total surveillance of the merits of a candidate sweetening composition or structure.

However, what has been particularly lacking in the psychophysical or sensory quantification of a sweetener is the rating of the aftertaste that in the panel tests described in the ensuing discussion and examples has been found to acquire equal stance or importance with sweetness scale. Moreover, the aftertaste which ultimately dictates the individual's preference for a sweetener involves not only the intensity of the initial sweetness but also its persistence.

These two related yet distinctly different qualities have been utilized as the inseparable bases for uncovering or finding and selecting preferred sweetening ingredients and optimal formulation comprising the following four components, namely: (1) a base of the simplest amino acid, glycine; (2) a benzenoid substituted amino acid, typtophan, as synergist; (3) an amido substituted amino acid, asparagine, serving as intensifier; and (4) a polyhydric alcohol, sorbitol, serving as a carrier.

Firstly, the base sweetener in the formulation is a member of the amino acid metabolites which in themselves are endowed with varying degrees of sweetness and divergent aftereffects on the taste buds. A specific combination of these amino acids has been discovered to have surprisingly synergistic effect on sweetness with regard to initial intensity and persistence. Thus, the simplest amino acid, glycine, on a sweetness scale of 100 for sucrose, has a relative rating of 65 while one of its analogs, DL-alanine, has a sweetness rating of 107. The corresponding sweetness rating for saccharin is 30,600, just to illustrate the markedly lower order of sweetness of the amino acids.

Initial subjective taste screening tests by a panel on an extensive series of D- and L- forms of amino acids also revealed a variety of sweetness quality, intensity, and aftertaste among which D-tryptophan proved most amenable for combinative formulations and most synergistic in providing the panel with the most favored or acceptable sweetening taste. D-tryptophan in combination with the least expensive, elementally simple amino acid, glycine, which is available as a large scale production item, provides a balance of initial intensity and neutral aftertaste unmatched by any of the combinations screened for acceptable taste. From taste panel ratings, it was established that the combination of D-tryptophan as a synergist with the base glycine provided a bland sweetness rating of the combinations of these two components that far exceeded the expected, arithmetic sum of the proportionated ingredients.

In order to match the initial taste intensity so characteristically associated with saccharin and cyclamates to which most people are accustomed, it was found that the required sharp taste could be provided by a small amount of L-asparagine. L-asparagine is referred to as an intensifier. This important function derives from the evaluation by the tasting panel noting that with only the first two ingredients there was lack of bite or tang which could be made up, or imparted with the L-asparagine component.

Finally, to gain the quality of dispersion and uniform dissolution of the composite formulations as dispensed into prepared brews of coffee or tea or other beverage preparations, hot or cold, it was found that the D-tryptophan tended to aggregate into a floating flaky residue by some undiscerned phenomenon. It was found that this slightly hydrophobic character could be overcome in any one of several ways. One of these is to disperse the D-tryptophan crystalline form in the presence of a water-avid carrier with fairly high hydrophilic yet powdery characteristics, such as found with polyhydric alcohols, mannitol and sorbitol.

Another method is by the selective grinding of the four functional components in a controlled sequence. Still another method is by converting a portion of the D-tryptophan to the corresponding sodium salt. These features may not be needed however when the sweetening formulation is applied to prepared concentrates including such nondietary items as mouth washes, oral antiseptics, toothpastes, etc.

Example 1: A series of formulations based on the above functional features was prepared with the combination of the four components such as are summarized in Table 1 describing the responses of a five-membered panel of tasters who tested the formulations in the form of dry, sugary powders.

TABLE 1: PANEL TEST RATING OF FORMULATIONS

	Ingredients	Parts by Weight Formula A	Formula B	Formula C
Function				
Base	Glycine	50	50	50
Synergist	D-tryptophan	40	40	40
Intensifier	L-asparagine	0	1	2
Carrier	D-sorbitol	10	9	8
		Panel Rating (composite)		
Initial sensation		Sweet but bland	Sweet	Sweet
Aftertaste		None*	None*	None*

*Compared to saccharin.

The panel members were young women between the ages of 16 and 19 with a strong preference for beverages other than milk. Each participant was asked to rate the initial taste for sweetness and for aftertaste fifteen to twenty minutes afterwards, the latter done on separate days to eliminate overlap of the effects of each of the formulations tested.

Without the presence of the D-tryptophan in a control formulation for the above, the panel concensus accorded a rating of a mild bland sweetness. Furthermore, without the intensifier, L-asparagine, the mixture imparted a transient sweet taste lasting only a few seconds. Table 1 illustrates only one of an extensive series of dry powder testings of the four component formulations with the preferred range of the base amino acid, glycine, being from 30 to 50 parts with 60 to 40 parts of D-tryptophan, the remaining 10 parts being apportioned between the intensifier L-asparagine and the hydrophilic carrier D-sorbitol.

Example 2: A brew of Ceylon tea of moderate strength was prepared with a series of dilutions using sweetener Formula B described in Example 1 for tasting by a five-membered panel of young college women selected for their preference for artificially sweetened beverages. The tasters were asked by means of blind test to rate the presence or absence of sweetness with results summarized in Table 2.

TABLE 2: PANEL RATING OF INITIAL TASTE IN HOT (WARM) TEA

	Members Acknowledging Sweetness at .... Dilutions 1:x				
	500	1,000	2,000	4,000	8,000
Formula B	5	5	3	3	0
Saccharin (tested separately for control)	5	5	5	5	5

While the saccharin clearly induced a sweetness response through the above dilutions greater than that of Formula B, the consensus of the panel was that the latter gave no aftertaste reaction at the lower dilutions which was pronounced with the saccharin.

A similar test of brewed tea using only D-tryptophan at the above dilutions evinced a sweetness response only at the 1:500 dilution level, clearly indicating a synergistic effect with the glycine and possibly with the presence of L-asparagine used as an intensifier. The test requires a waiting period, with tests preferably one day apart in order to avoid lingering sensitivity complicated by incidental ingestions by the panel members of various carbonated beverages.

DRINKS, JELLIES, FRUITS AND CHEWING GUM

DRINK PRODUCTS

Dry Cola Beverage Mix

Commercial dry carbonated beverage powders are based on a sodium bicarbonate-citric acid mixture, with the carbon dioxide being released by the interaction of the citric acid and the sodium bicarbonate. Unfortunately, the sodium bicarbonate-citric acid system adversely affects cola flavorings. As a result, it was not possible to produce a carbonated cola-flavored beverage from a dry powder composition that possessed a satisfactory cola flavor. Attempts have been made to solve the problem by modifying the cola flavor. However, this has not proved to be satisfactory.

In the process of *R.L. Swaine and D.W. Beusch; U.S. Patent 3,510,311; May 5, 1970* the sodium bicarbonate is used in the range of from 30 to 38 weight percent, and preferably about 36 weight percent. The citric acid should be present as anhydrous citric acid in the range of 39 to 48 weight percent, and preferably about 46 weight percent. If the low concentration range of the citric acid is used, such as about 39 weight percent, then the citric acid should be reinforced with up to about 4 weight percent of sodium citrate, so that the combined weight percent of citric acid and sodium citrate totals about 43 weight percent. The weight percentage of sodium saccharin is variable and is adjusted for sweetness and flavor balance.

Disodium inosinate is used as a flavor potentiator. Thus, its presence insures a greater flavor effect from the cola flavoring than can be achieved in its absence, although at the level used it contributes no flavor of its own. The disodium inosinate should be present in the mixture in concentration of from 0.0056 weight percent to 0.112 weight percent, with an optimum concentration of about 0.01848 weight percent.

The disodium guanylate is also a flavor potentiator. It should be present in the amount of from 0.0044 weight percent to 0.088 weight percent, and preferably about 0.01452 weight percent. The presence of the disodium inosinate and the disodium guanylate gives the user the feeling of a viscous product in the mouth without physically affecting the product's viscosity. In addition, it is desirable to have from 1.4 to 2.0 weight percent of powdered tribasic calcium phosphate present to keep the mixture free-flowing and prevent caking. It has also been found that up to 15 weight percent of D-mannitol may be added. The mannitol will serve to smooth out the flavor of the composition.

All of the components are dissolved in isopropanol, and then the mixture is granulated while driving off the isopropanol. The mixture should be dried until all moisture is removed, at a temperature of the order of 140°F in a hot air oven. The use of the technique of dissolving all of the components in isopropanol has the advantage of preventing layering, which may be present if spray-dry flavors are mechanically mixed. A very satisfactory cola-flavored carbonated beverage is obtained when 3 grams of the composition are mixed with 6 ounces of water.

Composition

Component	Weight Percent
Sodium bicarbonate, USP	30.79
Anhydrous citric acid	39.19
Sodium citrate	4.2
D-mannitol	14.0
Sodium saccharin (suitable amount)	
Vanilla powder	0.56
Caramel color	2.8
Disodium inosinate	0.0784
Disodium guanylate	0.0616
Tribasic calcium phosphate	1.4
Cola flavor	0.16
Dry cinnamon	0.56

Basic Effervescent Process

In the production of effervescent compositions in the form of free-flowing mixtures or in the form of mixtures suitable for the production of compressed tablets, the basic effervescent mixtures, when employed for medicinal or internal application, involve the use of mixtures of alkali metal carbonates or bicarbonates, such as the sodium, potassium or calcium compounds (singly or in combination) together with a nontoxic dry organic acid such as tartaric, citric or malic acid, or with a mixture of these acids.

To these basic effervescent mixtures various therapeutic and flavoring ingredients or additional effervescent ingredients may be added, depending upon the specific type of product desired. When the effervescent mixture is added to water, the acid or acids present react with the carbonate or bicarbonate or mixture of these present in the mixture, and the release of the gaseous carbon dioxide formed as a product of this reaction produces the desired effervescence.

Effervescent compositions for applications other than internal medicinal use can be prepared and may be employed as dental cleansers, as bath salts and for other analogous uses where not only the effervescent ingredients mentioned above may be used, but others as well. To form tablets from an effervescent mixture such as that described above, the mixture must usually be in granular form in order that it will be sufficiently free-flowing to be easily fed to the die cavity of the tableting machine for shaping the tablets. Powders do not flow freely and these powders must be granulated in some convenient fashion. The usual methods of preparing such free-flowing granulations include the heat fusion method, the use of steam or water injection, or the use of a double granulation method.

The heat fusion method consists of mixing the particular alkali metal carbonate or bicarbonate or mixture with the desired organic acid or combination of acids. The mixture should include monohydrated citric acid in an amount of from about 8 to 30% of the total acid present, placing the mixture in a suitable container and heating until the water of crystallization in the monohydrated citric acid present is released. This treatment causes a partial reaction and results in the formation of a plastic mass which, when broken down into coarse granules and then screened, dried and lubricated, can be compressed into tablets.

The steam or water injection method of granulation is similar to the heat fusion method except that monohydrated citric acid does not have to be used since the required moisture is added either in the form of steam or as water, which is sprayed or injected into the mixture as it is agitated. Heat is not essential in this method of forming the plastic mass which, after being broken up into coarse granules, is subsequently screened and dried. When used for the production of tablets, the granulated mixture is usually lubricated and the lubricated granulation compressed into tablets.

The double granulation method consists of preparing granules of the alkali metal carbonate or bicarbonate by moistening the latter with a solution of a binding material such as sugar, acacia, gelatin or lactose, screening the moist mixture to form granules and then drying the moist granules. A separate granulation is made in a like manner of the acid components. The medicinal or flavoring ingredients are then incorporated into either granulation or divided between the granulations and the two granulations are then mixed in the proper formula proportions. For tablet formation by compression, the granulation mixture is lubricated prior to compression.

When the granulations are then tableted, the tablets formed are not always sufficiently hard for handling in subsequent packaging operations and, in general, are easily chipped, flaked apart or cracked. When greater pressure is used in compression, other physical defects, such as capping, may occur. Not only are chipped, flaked or cracked tablets unmarketable, but when the breakage occurs during the packaging operations, the tablet feeding and packaging mechanism is easily clogged and the frequent jamming greatly impairs the efficiency of the packaging operation.

Formulations and processes are given by *R. Millard and C.A. Balmert; U.S. Patent 2,985,562; May 23, 1961; assigned to Warner-Lambert Pharmaceutical Co.* for effervescent free-flowing granules and also for tablets.

Free-flowing granulated effervescent compositions containing an alkali metal carbonate or alkali metal bicarbonate or a mixture and an organic acid, such as tartaric acid, citric acid or malic acid or mixtures of these acids, may be obtained by the usual granulating procedures if a small amount of a monocarboxylic amino acid is incorporated. Preferably, citric acid is used for at least part of the organic acid in the composition. All of the citric acid may be in the anhydrous form. By this process, not only are highly satisfactory free-flowing granulated mixtures formed, but these granulations may be used very satisfactorily in the production of tablets by compression molding procedures. Not only are the tablets formed from the granules so obtained of a satisfactory degree of hardness for further processing but, on subjecting these tablets to a heat treatment following the compression molding operation, the degree of hardness of the molded tablets is substantially increased without loss of solution speed and the difficulty of chipping, capping or splitting of the tablets is substantially eliminated.

In forming free-flowing effervescent mixtures, 0.5 to 3.0% by weight of amino acid is used in the composition. Such amino acids as glycine or alanine may, for example, be incorporated in the mixture. The amino acid should be a monocarboxylic acid. The composition to which the amino acid is added should be of relatively fine particle size and should not include any appreciable proportion of particles over 40 mesh in size.

The mixture thus formed is then heated to a temperature between 60° and 120°C and held at this temperature for from about 15 to about 35 minutes. The mixture becomes slightly plastic under these conditions. Following this heat treatment, the mixture is cooled and then screened. The free-flowing screened mixture thus obtained forms the basic mixture to which any desired medicinal agent, flavoring agent, color or tableting lubricant may be added. For the production of free-flowing granules which are to be used without tableting, and granules of 5 to 8 mesh size, the acid should include about 3% by weight of monohydrated citric acid and the mixture should be reduced to at least 100 mesh particle size before heating.

Where the effervescent composition is formed of a mixture of an alkali metal bicarbonate, such as sodium bicarbonate and citric acid, the precise stoichiometric ratio of citric acid to sodium bicarbonate in the basic effervescent formulation is 192 parts by weight of anhydrous citric acid to 252 parts by weight of sodium bicarbonate, with three mols of sodium bicarbonate being necessary to neutralize each mol of citric acid.

In the case of tartaric acid, only two mols of sodium bicarbonate are necessary to neutralize each molecule of tartaric acid. When a somewhat tart taste is desired in the product, a higher level of the acid is used. An excess of up to 35% by weight of that theoretically required may be used in such instances. Aside from this, the use of an excess of acid is not necessary and entirely satisfactory formulations are obtained where the exact stoichiometric ratio is used. An excess of the alkali metal bicarbonate may also be desired where it is advantageous to maintain the formulation on the alkaline side, for example, where aspirin is used and solubilization dictates that alkaline conditions be used. Heating the tablets to a temperature of 70° to 90°C for about one to two hours is usually sufficient to impart the desired degree of hardness to the shaped tablets.

Example 1: The following dry powdered ingredients are thoroughly mixed in the proportions given below, the particle size being from 40 to 200 mesh:

	Percent
Citric acid anhydrous	63.73
Sodium bicarbonate	34.71
Glycine	1.56

The mixture obtained is spread in broad shallow trays and heated for 28 minutes in an oven maintained at 115°C. The trays are then removed and the contents allowed to cool to room temperature. The fused mass is then removed from the trays and passed through an oscillating or rotary granulator equipped with 10 times 10 mesh stainless steel screen. The granulated mass is then dried for 90 minutes in an oven maintained at 70°C. From the above basic mixture is prepared the following blend:

	Parts
Basic mixture	88.69
Saccharin (suitable amount)	
Orange flavor	2.71
Orange color	0.26
Vitamin C	0.74

The mixture is thoroughly blended and compressed into tablets by a rotary tableting machine. The tablets produced are ¾ of an inch in diameter, 5/16 of an inch in thickness, weigh 2.25 grams and have a hardness of 3 to 6 on the Strong Cobb hardness scale. After heating at 70°C for 90 minutes the tablets reach a hardness of 20 to 30 on this scale.

Example 2: The following dry powdered ingredients are thoroughly mixed in the proportions given below, the particle size of the ingredients being from 40 to 200 mesh:

	Percent
Citric acid anhydrous	63.55
Sodium bicarbonate	34.60
dl-Alanine	1.85

After thorough blending in a ribbon mixer, 2.0% or 2 g of water per 100 g of powder mix is added to the blender by means of a fine spray while the mix is being agitated. After the water has been added the blender is permitted to operate for an additional three to five minute period. The wet mass is then removed and put through an oscillating or rotary granulator equipped with a 10 x 10 mesh stainless steel screen. The granulated mass is then dried in an oven maintained at 70°C for 90 minutes. In this procedure identical quantities of the granulated mixture are used as in Example 1, with the same quantities of basic mix. The tablets obtained exhibit the same increase in hardness on the same heat treatment.

Example 3: The following dry powdered ingredients are thoroughly mixed in the proportions given below, the particle size of the ingredients being from 40 to 200 mesh:

	Grams
Citric acid anhydrous	768.3
Sodium bicarbonate	1,008.3
Saccharin (suitable amount)	
Glycine	135.0

After thorough mixing, the blend is sprayed with 20 cc of water while the mixture is still being agitated. Following a post-mix period of 3 to 5 minutes the material is passed through a 10 x 10 mesh stainless steel screen. The granulated mass is then dried for 90 minutes at 70°C. To the dried granulation is added the following:

	Grams
Acetyl p-aminophenol	135
Methapyrilene	9.3
Ascorbic acid	13.5
Red color	0.9

After thorough blending of the above mixture the material is fed to a rotary tablet press. The tablets produced are 1 inch in diameter, 5/16 of an inch thick, and weigh 3.25 grams. The average hardness of the tablets thus produced is 8 to 10 Strong Cobb units. The tablets are then heated to 90°C for 90 minutes. After the heating period the tablets have a hardness of from 20 to 30 units on the Strong Cobb scale.

Example 4: The following dry ingredients are thoroughly mixed in the proportions given below, the particle size of the ingredients being from 20 to 200 mesh:

	Parts by Weight
Citric acid anhydrous	20.0
Sodium bicarbonate	10.5
Glycine	0.5

The mixture obtained is spread in broad shallow trays and heated to a temperature of 80°C for 35 minutes. The mixture is then cooled and put through a 10 mesh screen on an oscillating granulator. The free-flowing composition obtained comprises the basic mixture and can be used in the preparation of tablets by blending the following:

	Parts by Weight
Basic mixture	31.0000
Saccharin (suitable amount)	
Dry flavor	1.0000
Color	0.0625

The whole is thoroughly mixed after being lubricated with 0.5% light mineral oil or with cottonseed oil. The lubricated mixture is then fed to a tablet machine, preferably a rotary type, equipped with 27/32 dies and punches and 35-gram tablets, 3/16 inch in thickness are punched out. The tablets formed have a hardness of from 4 to 6 on the Strong Cobb scale. After heating the tablets to 90°C for 1 hour the tablets reach a hardness of from 12 to 15 on this scale. The tablets taken directly from the tablet machine dissolve in water at 50°F in about 60 seconds and, after the heat treatment, dissolve in 70 seconds.

Example 5: The mixture shown on the following page is blended. After being mixed thoroughly the mixture is fed to a rotary tablet machine equipped with one-inch dies and punches, and 50-grain tablets approximately 3/16 inches thick are punched. The tablets obtained have a hardness of from 5 to 8 on the Strong Cobb scale. After heating the tablets to 90°C for 1 hour the tablets reach a hardness of from 15 to 20 on this scale. The untreated tablets dissolve in water at 50°F in 50 seconds, and after the heat treatment are found to dissolve in 60 seconds.

	Parts by Weight
Basic mixture (from Example 4)	31.000
Sodium bicarbonate	16.500
Acetanilide	2.5

Example 6: The following dry powdered ingredients are thoroughly mixed in the proportions given below, the particle size of the ingredients being from 40 to 200 mesh:

	Parts by Weight
Glycine	1.00
Tartaric acid	5.00
Malic acid	5.00
Citric acid anhydrous	11.56
Sodium bicarbonate	16.44
Sodium carbonate anhydrous	2.00
Potassium bicarbonate	4.00
Sodium bromide	5.00

The mixture obtained is spread in broad shallow trays and heated to a temperature of 110°C for 35 minutes. The mixture is then cooled and put through an 8 mesh screen on an oscillating granulator. A free-flowing granular composition is obtained which can be submitted to a heat treatment at 120°F for 1 hour for further stabilization.

If the granulation is to be punched into tablets, a 10 mesh screen is used on the oscillating granulator and the free-flowing composition suitably lubricated and fed to a tablet machine, preferably of the rotary type, and provided with suitable dies and punches. When 50-grain tablets are prepared in this way with 1-inch dies, the tablets are approximately 3/16-inch thick, with a hardness of from 7 to 10 on the Strong Cobb scale. After heating to 70°C for 1 hour the tablets reach a hardness of from 15 to 25 on this scale. The tablets from the tablet machine dissolve readily in 50°F water in about 50 seconds and, after the heat treatment, dissolve in 60 to 70 seconds.

Modified Bicarbonate Method

In another process a double granulation method is used by *B. White; U.S. Patent 3,105,792; October 1, 1963; assigned to Warner-Lambert Pharmaceutical Co.* The process also requires conversion of the alkali bicarbonate to the corresponding carbonate by controlled heat treatment.

During the heat treatment the bicarbonate may be exposed to temperatures up to 200°C until the desired degree of conversion is effected. Usually, the desired degree of stability is attained if about 2 to about 10% by weight of the bicarbonate is converted in this fashion to the carbonate. The reaction which takes place during the heat treatment consists of two mols of the bicarbonate reacting to form one mol of the corresponding carbonate, together with one mol each of water and carbon dioxide.

The actual degree of conversion is readily determined by acid titration of the modified bicarbonate with sulfuric acid employing methyl orange as the indicator. Sufficient movement and air circulation should be provided during the heat treatment so that all of the surfaces of the alkali metal bicarbonate particles being heat treated will be uniformly exposed and at the same time the carbon dioxide and water vapor formed as reaction products will be removed. Preferably the heating should be carried out at a temperature of 100°C or higher to ensure the vaporization and removal of the water formed during the reaction.

It is believed that the stabilization of the effervescent compositions formed with the modified bicarbonate results from the chemical change on the surface of the bicarbonate particles produced by the heat treatment. By providing sufficient circulation, the outer surfaces of the alkali metal bicarbonate particles are apparently converted to the less active carbonate and the presence of the latter on the particle surfaces serves in part as a barrier to hinder any reaction with the organic acid in the mixture prior to the time the effervescent composition is added to water and dissolved in use.

The improved stability characteristics of the effervescent mixtures prepared with stabilized alkali metal bicarbonate are retained when these mixtures are subsequently tableted. In addition, it has also been observed that tablets prepared from these mixtures are self-hardening and, upon being allowed to stand for about a week after tableting, will go from a hardness of 3 to 4 on the Strong Cobb scale (see U.S. Patent 2,645,936) to a hardness of 12 to 14, yet without any loss in solution time when added to water.

Accordingly, no subsequent heat treatment or curing of the tablets is necessary to improve their hardness characteristics. The supplementary hardening agents heretofore used as an aid in producing tablets which may be packaged successfully without undesirable chipping or cracking may be eliminated, although they are still desirable for the flavor note they add. This is particularly applicable to the use of glycine.

In the case of flavored tablets, the elimination of any heat-curing step substantially improves flavor retention since the more volatile of the flavor constituents present have no opportunity to escape as is the case where the tablets are exposed to elevated temperature when heat cured.

One very substantial advantage achieved by the process is the ability of the modified bicarbonate to yield an aspirin-containing effervescent composition which is quite stable in either granular or tablet form. Exposure of these aspirin-containing effervescent compositions to elevated temperatures of 45°C produces little evidence of decomposition as measured by the degree of carbon dioxide formation. Since this thermal stability test is normally conducted with the composition in sealed packages, any evidence of thermal decomposition is immediately apparent due to the puffing of the foiled package, which is produced by any carbon dioxide formed.

Example 1: Fifty pounds of granular sodium bicarbonate are modified by heating in trays in a draft oven set at a temperature of 100°C. The total heating time used is 45 minutes, the heating being divided into three 15-minute periods, with the granular sodium bicarbonate being mixed thoroughly between each heating period.

The modified material obtained is then stored immediately in air-tight containers to avoid any moisture absorption. When titrated with 1 N H_2SO_4 employing methyl orange as the indicator to determine the extent of the chemical change, an average of about 7 to 9% by weight of the sodium bicarbonate is found to have been converted to sodium carbonate. The degree of conversion in the case of sodium bicarbonate is readily determined by the following equation:

$$\text{Weight of sample in grams} = 84(E - X) + 53X$$

E = equivalents of total acid required, X = equivalents of acid required for Na_2CO_3

An equivalent degree of conversion is obtained in the case of potassium bicarbonate by heating in the same manner, but at a temperature of 125°C.

Exposure to a temperature of 50° to 60°C may require 48 hours or more for any significant degree of conversion to be effected, whereas exposure to temperatures of 200°C or more with suitable air circulation and agitation will cause a noticeable reaction in but minutes. The reaction conditions used are merely adapted to the result desired so that if, for example, the conversion to sodium carbonate of a specific percent by weight of sodium bicarbonate is desired, an almost infinite number of combinations of time and temperature are available.

The higher the temperature the more rapid the conversion. The ultimate result achieved with respect to the degree of conversion is, however, readily determined by means of the acid titration described. For effervescent compositions containing aspirin, a bicarbonate in which 7 to 9% by weight has been converted to the carbonate is preferable in order to produce a satisfactory thermally stable effervescent composition.

While stable effervescent compositions are also obtained with a sodium bicarbonate, for example, where as much as 20% by weight has been converted to the carbonate, the compositions thus formed are found to have a taste which is somewhat bitter and, therefore, objectionable. Aside from this subjective factor the compositions and tablets made therefrom are quite satisfactory as far as their thermal stability is concerned.

Example 2: A flavored beverage tablet is formed of the following composition:

Ingredients	Parts by Weight
Citric acid, anhydrous	18.26
$NaHCO_3$, modified (7 to 9% conv.)	12.74
Glycine	0.5
Ascorbic acid	0.25
Saccharin sodium (suitable amount)	
Color	0.125
Sealva grape imitation flavor	0.750

The glycine and citric acid are first mixed and to this mixture is added 0.1% by weight (on the total ingredients) of propylene glycol. The remainder of the ingredients are added, mixed well and 0.5% by weight of heavy mineral oil then added as a tableting lubricant. After thorough mixing, the composition is compressed into 36-grain tablets and the latter packaged by sealing in aluminum foil. The packaged tablets thus formed are found to be stable for from 5 to 10 days at a temperature of 70°C without puffing, whereas tablets formed in the same fashion without employing the modified bicarbonate are stable only for from 4 to 7 hours before puffing due to the formation of CO_2 within the sealed foil. A less drastic stability test and one more likely to be similar to conditions encountered in commercial operations consists of exposing the foil-packaged tablets to a temperature of 45°C. Under these conditions the usual tablets commonly exhibit noticeable puffing in about two weeks, whereas the tablets prepared in accordance with this process show no sign of puffing even after exposure to 45°C for three months.

Example 3: A flavored beverage tablet is formed of the following composition in the manner described in Example 2, with propylene glycol and mineral oil being included.

Ingredients	Parts by Weight
Citric acid, anhydrous	18.42
$NaHCO_3$, modified	13.05
Glycine	0.5
Ascorbic acid	0.25
Saccharin sodium (suitable amount)	
Color	0.44
Flavor (lemon-lime)	0.094

The composition is compressed into tablets weighing 34 grains each. The thermal stability of the tablets obtained is similar to that of the tablets of Example 2.

Gum Coated Bicarbonate

According to *P.F. Smith and L.D. King; U.S. Patent 3,082,091; March 19, 1963*, the use of gum protected bicarbonate has special advantages in the preparation of effervescing compositions used for making soft drinks. It was found that the combination of the gum coated bicarbonate, particularly when a vegetable gum of the hemicellulose is used as the coating, with citric or tartaric acid and formulated with a sweetening agent such as saccharin and the usual powdered flavors and dyes, produces effervescing soft drinks of excellent properties.

The bicarbonate is mixed with a gum mucilage and the mixture is agitated to form a homogeneous slurry. The proportions may vary over a considerable range; for example, a proportion of 25 parts of mucilage to 100 parts of bicarbonate has been found specially suitable with guar gum. The slurry is dried sufficiently to permit screening the mass through a fine mesh screen. The drying process then may be completed, preferably while avoiding a temperature exceeding about 95°F, since at higher temperatures the decomposition of bicarbonate to carbonate begins to accelerate.

The dried and coated particles are then mixed in the desired proportions with the other ingredients of the composition. In preparing medicinal preparations, it may be desirable to have either an excess of bicarbonate or an excess of the fruit acid, depending upon whether an alkaline or acid solution is desired. For preparation of a soft drink effervescing composition, however, it is important to have a ratio of citric acid to bicarbonate providing a pH in aqueous solution in the range of about 3 to 4.5. A proportion of sweetener is added to give the desired degree of sweetness. A small amount of a glutamate may be added to enhance sweetness.

Buffering agents to stabilize pH at the desired level may be used. The mono- and disodium citrates, for example, have value. Also, a small amount of sodium benzoate may be added as a preservative and moisture repellent. Tricalcium phosphate in small quantities may be used to prevent the citric acid from caking during mixing and storage. It may be necessary to add a small amount of a compatible potable nonionic detergent such as polyoxyethylene sorbitan monooleate or other polyglycol esters of fatty acids.

The detergent is added in small concentrations of about 1% by weight and provides clarity and sparkle to the soft drink product. The use of sorbitol (from equal proportions to 2 or 3 parts to 1) in combination with the sucaryl enhances the flavor of the finished drink. The sorbitol appears to supply "body," making the drink less watery to the taste. It also helps to mask the taste of saline by-products and promotes the sweetening effect of saccharin.

In the experiments, a mucilage of the gum was first prepared and treated so as to insure complete hydration. Sodium bicarbonate then was placed in a mixing vessel and mucilage was added in increments, with thorough mixing. The mixture was partially dried by applying a current of warm air while continuing agitation until the material started to "ball." The mass then was forced through a 20-mesh wire screen and was placed in a constant temperature, circulating air oven to complete the drying. The drying temperature was maintained at 95°F maximum temperature, and the average drying time in the oven was three hours. A series of runs was made with the following gums:

Gelatin, USP–10% mucilage

Pharmagel A (pork skin gelatin having an isoelectric point about pH 8 and exhibiting maximum solubility in acid solution)–10% mucilage

Pharmagel B (bone gelatin exhibiting maximum solubility in alkaline solution)–10% mucilage

Methylcellulose (Methocel-Dow) 100 cp grade–5% mucilage

Guar gum (Jaguar A20A and Jaguar A20D-Hall)–1% mucilage

In these experiments, it was found that the guar coated samples gave less foaming, less floating of solids on top of the foam and less foam stabilization than those coated with the other gums. The gelatins appeared somewhat less desirable from the standpoint of foam stabilization and showed a relatively slow rate of hydration. The methylcellulose also tended to stabilize the foam to a greater extent and hydrated less rapidly than guar.

Also, the celluloses were found to be somewhat less readily compatible with other ingredients for use in aqueous solution. The guar coated bicarbonates were outstanding in properties, and analysis showed that there was substantially no conversion of sodium bicarbonate to sodium carbonate during the coating process. After drying, the coated product was found, by analysis, to contain less moisture than the original bicarbonate.

In another series of experiments the above procedure was followed using samples of guar mucilage both freshly prepared without heating and after heating in a water bath for half an hour to insure complete hydration. The mucilage concentration was 1%. It was found that the heated mucilage was superior in producing better dispersion of the gum. Although the heating process increased the viscosity of the mucilage, this did not appear to be disadvantageous. In other experiments, the ratio of mucilage to bicarbonate was varied from 1:3 to 1:5, and a ratio of about 1:3 was found to be most favorable. Also, the drying time was increased from one hour to four hours at 90°F and then to one and a half hours at 105°F, with only negligible conversion of bicarbonate to carbonate.

Example 1: This example illustrates the preparation of a lime-lemon effervescent powder. First, the sodium bicarbonate was coated and colored according to the following formula: sodium bicarbonate, USP, 250.0 grams; and guar mucilage (1%, Jaguar A20A), 83.3 cubic centimeters (containing 0.260 gram FD&C Yellow 5 and 0.017 gram FD&C Blue 1). The resulting stabilized and colored bicarbonate was then mixed thoroughly with the remaining ingredients, after all had been reduced to 30-mesh particle size, in the following proportions:

	Parts
Sodium bicarbonate (coated and colored)	241.1
Citric acid, anhydrous	418.9
Lime-lemon flavor (Sealva)	15.0
Saccharin (suitable amount)	
Tricalcium phosphate, NF	1.1

Example 2: This example illustrates a formula using a mixture of sodium cyclamate and saccharin:

	Parts
Sodium bicarbonate (coated and colored)	409.8
Citric acid (anhydrous, fine grain)	837.8
Calcium phosphate (tribasic)	2.2
Lime-flavor powder (Polak Flav-O-Lok)	4.68
Lemon-flavor powder (Polak Flav-O-Lok)	4.68
Saccharin sodium (suitable amount)	

In the above example, the bicarbonate was prepared by slurrying 500 parts of USP sodium bicarbonate with 133.32 parts of a 1.25% mucilage of guar gum (Jaguar A20A), 0.52 part of FD&C Yellow 5 dye and 0.034 part of FD&C Blue 1 dye and, thereafter, drying and granulating.

Example 3: This example illustrates the preparation of a root beer effervescing composition. First, sodium bicarbonate was coated and colored (as described above) using the following materials:

	Parts
Sodium bicarbonate, USP	250.0
Guar mucilage (1% A20A)	83.3
Containing–	
FD&C Red 4	1.040
FD&C Red 2	0.520
FD&C Yellow 5	2.080
FD&C Blue 1	0.135

The resulting bicarbonate was mixed thoroughly with the remaining ingredients, after reduction of all particles to 30-mesh size, to obtain the following composition.

	Parts
Sodium bicarbonate (coated and colored)	244.4
Citric acid, anhydrous	418.9
Tricalcium phosphate, NF	1.1
Root beer flavor (Sealva)	20.0
Saccharin (suitable amount)	

About 7.0 cc (7.2 g), which is equivalent to one rounded teaspoonful of the above formulation prepares one drink of eight fluid ounces.

In the above examples, it was found that the ingredients should be divided so the particles are all finer than 30-mesh to provide rapid solution. On the other hand, it was found preferable to have the particles coarser than 60-mesh because excessive foaming tends to result if the particles are too fine and there is a greater tendency to pick up moisture.

The gum coated bicarbonate has a wide range of applications not limited by the above examples. Where ordinarily bicarbonate is thermally unstable and resists tableting so that it is difficult to stabilize by mechanical means, the gum coated preparations show excellent storage stability. For example, the stability of the powdered products as prepared in the examples was checked by storage for nine weeks in containers maintained at 68°F at varying relative humidity. The product was stable and unspoiled at relative humidities up to 52%. After twelve weeks in an open beaker in an air-conditioned office at 77°F, no evidence of deterioration could be detected.

The product is free-flowing and lends itself to packing by automatic machinery. Advantageously, the packing conditions should be maintained below about 50% relative humidity and 77°F.

A particular advantage of the gum coated bicarbonate compositions is that they may be mixed with materials such as vitamin C (ascorbic acid) which are normally inactivated by alkalies.

Hydrophilic Gum Additive

To improve the body and acceptability of carbonated nonalcoholic beverages or soft drinks formulated with artificial sweeteners, *H.R. Schuppner, Jr.; U.S. Patent 3,413,125; November 26, 1968; assigned to Kelco Company* uses a Xanthomonas hydrophilic colloid.

An additive to a beverage must be readily and completely soluble and remain so under the acid environmental conditions. Soft drinks are often highly acid, especially in the concentrate syrup form in which they may be stored, the syrups being substantially more acid than the diluted drink when consumed. This problem of highly acidic concentrates is of concern since bottlers and fountains customarily purchase syrup concentrates and store them for extended periods prior to mixing them in the finished drink. Hence, the stability of the concentrate is extremely important.

The Xanthomonas hydrophilic colloid is an additive which is highly stable despite customary acidity of such drink concentrates. A related problem is the cold-water solubility of bodying additives since it is customary to prepare beverage concentrates, and especially soft drink syrups, by what is known as a "cold process." The concentrate ingredients are mixed at or about room temperature, no heat being supplied to facilitate the dissolving of the ingredients or the destruction of microorganisms. Thus, a beverage additive must be soluble in cold water for desirable versatility in use.

The beverage and liquid-food trade is becoming increasingly concerned with dietetic beverages. In such dietetic drinks, and more particularly those of the low-calorie type, a dietetic sweetener is customarily substituted for the usual soft drink syrup, including one or more of the following: a sugar, an invert sugar and dextrose at about 11 to 13% concentration by weight of the diluted drink. While providing a low-calorie drink, such artificial sweeteners are devoid of the mouth-feel characteristics usually supplied by sugar.

Xanthomonas hydrophilic colloid is a colloid produced by the bacterium *Xanthomonas campestris.* This colloidal material is a polymer containing mannose, glucose, potassium gluconate and acetyl radicals. In such a colloid, the potassium portion can be replaced by several other cations without substantial change in the property of the material.

This colloid, which is a high molecular weight, exocellular material, may be prepared by whole culture fermentation of a medium containing 2 to 5% commercial glucose, organic nitrogen source, dipotassium hydrogen phosphate and appropriate trace elements. The incubation time is approximately 96 hours at 28°C, aerobic conditions. In preparing the colloid, it is convenient to use corn steep liquor or distillers' dry solubles as an organic nitrogen source. It is expedient to grow the culture in two intermediate stages prior to the final inoculation in order to encourage vigorous growth of the bacteria. These stages may be carried out in media having a pH of about 7.

In a first stage, a transfer from an agar slant to a dilute glucose broth may be made and the bacteria cultured for 24 hours under vigorous agitation and aeration at a temperature of about 30°C. The culture so produced may then be used to inoculate a higher glucose (3%) content broth of larger volume in a second intermediate stage. In this stage, the reaction may be permitted to continue for 24 hours under the same conditions as the first stage.

The culture so acclimated for use with glucose by the aforementioned first and second stages is then added to the glucose medium. In the method of preparing a *Xanthomonas campestris* hydrophilic colloid, a loopful of organism from the agar slant is adequate for the first stage comprising 200 milliliters of the glucose medium.

In the second stage, the material resulting from the first stage may be used together with nine times its volume of a 3% glucose medium. In the final stage, the material produced in the second stage may be mixed with 19 times its volume of the final medium. A good final medium may contain 3% glucose, 0.5% distillers' dry solubles, 0.5% dipotassium phosphate, 0.1% magnesium sulfate having 7 mols of water of crystallization and water. The reaction in the final stage may be satisfactorily carried out for 96 hours at 30°C with vigorous agitation and aeration. The resulting *Xanthomonas campestris* colloidal material can be recovered by precipitation in methanol of the clarified mixture from the fermentation. This resulting material may be further characterized as a hydrophilic colloid produced by the bacterium species *Xanthomonas campestris.* Alternative Xanthomonas colloidal material was prepared by repeating the above preparation procedure with other Xanthomonas bacteria; namely, *Xanthomonas incanae, Xanthomonas carotae, Xanthomonas begoniae, Xanthomonas phaseoli* and *Xanthomonas malvacearum.*

However, these alternative Xanthomonas hydrophilic colloid materials are not full equivalents; for example, they must be substituted in different proportions from that of the *Xanthomonas campestris.* Accordingly, equivalent Xanthomonas colloids found satisfactory for substitution for the campestris variety in the examples to be described hereinafter are listed below. Listed also are the relative proportions generally substituted for one-part *Xanthomonas campestris.*

Xanthomonas Bacteria	Substituent Colloid Quantity (parts)
Xanthomonas campestris	1
Xanthomonas malvacearum	1.25
Xanthomonas carotae	1.25
Xanthomonas begoniae	
Strain 3	1.65
Strain 9	1.1
Xanthomonas incanae	1.5
Xanthomonas phaseoli	1.1

In order to produce drinks of unusually good stability and clarity, it is desirable to treat the Xanthomonas hydrophilic colloid by the following method: A 1½% by weight solution of a Xanthomonas hydrophilic colloid is heated to an elevated temperature of about 170°F for a period of 20 minutes, cooled to 70°F and filtered through a commercial filter using a filter aid (Dicalite Speed Flow). The filtrate was then reduced to a pH of 6.5 by the addition of hydrochloric acid, and the Xanthomonas hydrophilic colloid product recovered by precipitation with isopropyl alcohol. The resulting precipitate was then dried.

Example 1: A Sugar-Free Lemon-Lime Soft Drink –

	Parts
Carbonated water	98
Lemon-lime oil flavoring	0.1
Citric acid	0.1
Sodium citrate	0.05
Saccharin (suitable amount)	
Clarified *Xanthomonas campestris* hydrophilic colloid	0.05

This beverage has about ⅓ calorie per fluid ounce and good mouth-feel.

Example 2: A Dietetic Carbonated Cherry Drink –

	Parts
Carbonated water	97
Saccharin (suitable amount)	
Citric acid	0.10
Sodium citrate	0.05
Ascorbic acid	0.02

(continued)

	Parts
Clarified *Xanthomonas campestris* hydrophilic colloid	0.05
Cherry flavoring to taste	
Artificial coloring for color	

This beverage has only about ½ calorie per fluid ounce and yet has good body, despite the absence of sugar. The same formulation can be modified to give differently flavored drinks by altering the coloring appropriately and substituting for the cherry flavoring the following:

Root beer and herb flavoring for root beer
Essential oils of orange, lemon and lime for orange
Oil of sassafras and ascorbic acid for sarsaparilla
Caffein, phosphoric acid and cola flavor for cola
Ginger root extractive, essential oils, caramel color and ascorbic acid for ginger ale
Vanilla and creme flavoring, phosphoric acid, caramel coloring and 0.05% benzoate of soda for creme soda
Quinine, oils of lemon and orange for quinine water

Example 3: Dietetic Maple Syrup –

	Parts
Water	95
Sorbitol	1.0
Saccharin (suitable amount)	
Xanthomonas campestris hydrophilic colloid	0.3
Benzoate of soda (preservative)	0.05
Citric acid	0.20
Caramel coloring	0.20
Salt	1.0
Imitation maple flavoring	*

* Balance to taste

The colloid gives this syrup good mouth-feel, while lacking high-calorie content of the heavy sugar syrups, having only about one calorie per ounce.

Example 4: Dietetic Artificial Sweetener –

	Parts
Calcium saccharin soluble USP (suitable amount)	
Benzoic acid preservative	0.1
Methyl paraben	0.05
Xanthomonas campestris hydrophilic colloid	1.0
Propylene glycol USP	5.0
Water	85
Artificial flavoring	*

* Balance to taste

In order to further evaluate the effectiveness of the method for preparing non-alcoholic beverages of improved mouth-feel, a beverage syrup was prepared as follows:

	Parts
Saccharin (suitable amount)	
Sodium citrate	0.365
Citric acid	1.27
Additive	1.1
Water	95.805

For preparation of the final drink for taste evaluation 13.7 grams of the above syrup was diluted with 86.3 grams of water to give a concentration of 0.15% of the additive. Samples of such a beverage were prepared using as the additive gum arabic, carboxymethyl cellulose, crystalline sorbitol, low methoxy pectin, and a *Xanthomonas campestris* hydrophilic colloid. The beverages so prepared with the different additives were then submitted to a taste panel of six members. Each of the six members of the panel considered a beverage sample having the *Xanthomonas campestris* hydrophilic colloid additive included to have the best or most natural mouth-feel of the various additives tested.

The carbonated beverages, such as those referred to above, are prepared as is customary, namely, by preparing a syrup from sugars or artificial sweeteners and water; adding acid and flavoring; blending this mixture and transferring a measured amount to a bottle or other container; filling the container with a purified, carbonated water; capping, labeling and shipping (the sweetened syrup or drink concentrate being sometimes shipped by itself, of course). The syrup is usually prepared by a manufacturer and shipped to various bottlers and carbonated water added to the container at the bottling plant or at the dispensing point, such as a soda fountain or vending machine.

In addition to the sweet syrup and carbonated water, such soft drinks often include one or more of the following: beverage acids, coloring, flavoring and preservative materials. For a dietetic liquid, the level of fats and proteins will usually be insignificant and carbohydrates no more than a few percent (for example, about 0.1 to about 9%). For low-sodium or salt-free foods, sodium will be held to from about 1 to 10 milligrams per 100 grams of food material.

Flavoring materials used in making beverages, and especially those of the carbonated type, generally take the form of alcoholic extracts for the oily types of flavoring which cannot be carried in water alone, aqueous solutions for water-soluble ingredients, etc.

In the latter, a preservative such as benzoate of soda is often added. Essences of the natural or synthetic type are commonly used. Typical "beverage-flavoring" ingredients used are: caffein for cola type drinks, ginger and citrus oils for ginger ale, artificial fruit flavoring with or without a fruit extract for imitation fruit drinks, vanilla, vanillin or bourbonal for cream-soda flavor; at least one from: oil of wintergreen, oil of sweet birch and methyl salicylate for root beer; the same for birch beer, with methyl salicylate predominating; and the same also for sarsaparilla, including additionally oil of sassafras. One of the commonly used beverage acids is a 50% citric acid solution, the citric acid adapting itself well to nearly all light or fruity flavors. Phosphoric acid is widely used in cola drinks and the heavier leaf, root, nut or herbal flavors, while tartaric acid is used in grape flavors. In lesser amounts citric, adipic, fumaric, succinic, malic and lactic acids are also used (the malic acid for apple flavors).

Glycerol Sweetener

In certain pathological conditions, particularly some forms of diabetes, there is a need to provide the patient with a diet low in sugar. While this can be done by using artificial sweetening agents, such as saccharin, this course of action has the disadvantage that the artificial sweetener has no energy value in itself, so that it is necessary to provide the patient with other sugar-free energy-giving foods. Moreover, the patient is likely to retain his liking for sweet things and it is, consequently, psychologically desirable to be able to provide him with energy-giving foods which taste sweet.

J.H. Briggs, J.S. Pryor and G.R. Fryers; British Patent 1,279,392; June 28, 1972 found that glycerol can be administered to patients as a sweet-tasting source of energy. The use of glycerol is particularly advantageous in the case of diabetic patients, since glycerol is metabolized independently of insulin. It is readily absorbed and rapid in action. It counteracts ketosis caused by insufficient insulin in the blood, and will also correct hypoglycemia of various origins; for example, neonatal hypoglycemia, that associated with prediabetes, that associated with cyclical vomiting in childhood, and idiopathic hypoglycemia.

Glycerol causes a sensation of satiety and is thus useful in calorie-controlled diets for obese persons. Furthermore, as glycerol is a normal constituent of fats and is liberated in the digestive tract, it is a normal metabolite the administration of which can be expected to give rise to no difficulties. The products provide sugar-free drinkable compositions comprising water, 10 to 40% by weight of glycerol and a flavoring agent, but no pharmacologically or therapeutically active compound (other than the glycerol itself).

Conventional flavoring agents used in drinks for human consumption may be used. For example, a mixture of citric acid and lemon essence can be used to give a lemon-flavored drink, and carbon dioxide may be injected under pressure in the normal way to give an effervescent drink. The proportion of glycerol can be varied within the stated limits, depending upon the degree of sweetness required of the drink.

Besides the water, glycerol and flavor already referred to, the compositions may also contain other ingredients conventionally included in drinkable compositions for human use (for example permitted colorings), but, as already stated, they do not contain any sugar or other pharmacologically or therapeutically active compound.

Example 1:

Glycerol (99%)	300 ml
Water	700 ml
Citric acid	7.5 g
Lemon Essence, qs	

Example 2:

Glycerol (99%)	1 fluid oz
Water	2 fluid oz
Orange Essence, qs	

Both these compositions may be diluted to taste. In the treatment of diabetes mellitus, especially of the type called "brittle" characterized by frequent spontaneous major fluctuations in blood sugar level, the patient may be given 150 g a day of glycerol in three equal doses in the form of an oral composition, as described above. The glycerol is antiketogenic and provides energy for intracellular metabolism. Its entry into cells does not depend on insulin.

All types of hypoglycemia may be treated by oral administration of glycerol, for example, in dosage of 50 grams as a 25% aqueous solution. In ketotic states, the same dosages may be used as in the treatment of hypoglycemic states. Glycerol is particularly efficacious in this context because it rapidly supplies intracellular energy.

ARTIFICIALLY SWEETENED JELLIES

Carrageenan-Pectin Base

Artificially sweetened jellies are desirable food products for those persons with dietary restrictions who must limit their normal intake of sugar, or for those persons who desire a low-calorie diet. However, it has been found difficult to provide suitable gelling properties for jelly which has been artificially sweetened. Conventionally, jellies containing sucrose have been made with pectin as the gelling agent. However, pectin requires a combination of sugar and acid to fully utilize its gelling properties and obtain a product with good appearance and texture.

It has been found that those pectins having a low degree of methylation (DM) in combination with calcium ions are effective in jelly compositions containing artificial sweetening materials. While such low methoxyl pectins provide a suitably gelled or thickened artificially sweetened jelly, the jelly product tends to have a very cloudy or opaque appearance that is quite different from the clear, glossy appearance associated with conventional jellies made with sugar and pectin having a high DM. In addition, in using low methoxy pectin, boiling is required to deaerate the product and prevent fast gelling, and the low methoxy pectin tends to burn onto the jacket of the kettle in which the jelly mixture is boiled. Consequently, a high degree of agitation is required during processing of artificially sweetened jelly containing low methoxy pectin to minimize this burn-on.

An improved gelling system for dietetic jellies is given by *L.J. Horn; U.S. Patent 3,563,769; February 16, 1971; assigned to Kraftco Corporation.* The gelling system is a polysaccharide gum extracted from seaweed in combination with pectin. In a process for manufacturing artificially sweetened jelly with the gelling agents, it is not necessary to boil the jelly ingredients, and artificially sweetened jelly can be produced without burning of the jelly product onto the sides of a process kettle.

The seaweed extract used in the polysaccharide gum combination is selected from those seaweed extracts which provide clear, firm water gels with a mouth-feel similar to that of conventional jellies containing sugar and gelled with pectin containing a high degree of methylation. Carrageenan and eucheuman extracts are preferred. Particularly preferred are those carrageenan extracts derived from *Chondrus crispus* and *Gigartina stellata.* A eucheuman extract derived from *Euchema spinosum* may also be used. Carrageenan from *Chondrus crispus,* commonly referred to as Irish moss, is preferred because of its ready availability and its low cost.

It is preferred that the seaweed extract be recovered from an extraction process by precipitation with a soluble alcohol or ketone, such as isopropanol, ethanol, methanol or acetone. This provides an extremely pure polysaccharide extract which provides maximum clarity in the finished jelly product.

The carrageenan may be further treated by ion exchange to increase the cation content. It is well known that the gelling strength of carrageenan is related to the level of certain cations, such as potassium, sodium or calcium, associated with carrageenan. However, since cations may be supplied by other jelly ingredients, such treatment is not essential. When grape jelly is produced by the described method, it is best to provide at least part of the total level of cations by maintaining the potassium ion level of the grape concentrate in the preferred range of 1,200 to 3,500 ppm of potassium.

For other types of jelly, such as apple, cations may be added to provide the desired level of cations of from about 1,200 to about 3,500 ppm. The carrageenan is used at levels of from about 1 to about 3% by weight based on the finished jelly product. A level of about 1.5% by weight is preferred.

Pectin is primarily used in the gelling agent combination of this process to limit the breakdown of the carrageenan in the finished jelly product. Carrageenan is readily depolymerized and loses its gelling power when the pH of the gelled system is below about 4.0, as occurs in most jelly products. The pectin serves to stabilize the carrageenan and prevent the depolymerization and subsequent loss of gel structure through syneresis.

The pectin has a degree of methylation of at least about 45%, that is, it is not a low-methoxy pectin. It is, of course, well-known that pectins containing a high degree of methylation (above about 45%) do not form gels in an artificially sweetened system. Such high-methoxy pectin would require sugar and acid to provide a gel structure. The high-methoxy pectin is used at a level of from about 10 to about 50% by weight based on the levels of carrageenan. At pectin levels below about 10% of the level of carrageenan, the stabilizing effect is insufficient and the carrageenan may be depolymerized after addition to the jelly mixture.

The use of the gum combination as a gelling agent eliminates the problems of burn-on encountered when low-methoxy pectin is used as a gelling agent for artificially sweetened jellies. This elimination of burn-on permits the use of processing temperatures below boiling, since boiling is not required for agitation of the mixture during processing to prevent burn-on. Such low-temperature processing further enhances the flavor of the finished jelly product in that volatile flavor and aroma components are less likely to be driven off. Since the carrageenan is particularly susceptible to depolymerization in a heated acid environment, it is preferred to maintain the carrageenan as a separate solution prior to addition of the carrageenan into the mixture of jelly ingredients. It is preferred to make such addition of carrageenan at as late a point in the process as is practical. A preferred method is to introduce the carrageenan as a solution into a flowing stream of jelly ingredients just prior to packaging of the jelly ingredients in suitable containers. The carrageenan, however, is not affected by relatively high temperatures under basic conditions, and it is preferred to maintain solutions of carrageenan at a temperature of at least about 160°F to prevent a weak gel from forming during storage of the carrageenan in solution.

Example 1: Apple jelly was prepared using the gum combination by the following process: A carrageenan solution was prepared by slowly pouring dry carrageenan. After thoroughly mixing, the solution was held at 170°F to prevent a weak gel from forming. The carrageenan used was extracted from Irish moss and was recovered from the extraction process by alcohol precipitation and subsequent drying.

Three-hundred and forty-two pounds of water were then added to a jacketed kettle equipped with an agitator and steam was introduced into the jacket; 214.2 pounds of 47° Brix apple concentrate and 62.5 pounds of a solution containing apple pectin at a level of 4% by weight were then added and the mixture was heated to 170°F.

The apple pectin had a degree of methylation of 65%. 10.8 ounces of sodium saccharin was then dissolved in 40 pounds of water and added to the jacketed kettle. 1.25 pounds of potassium citrate and 0.75 pound of potassium sorbate were then dissolved in 15 pounds of water and the mixture was added to the jacketed kettle. 2.44 pounds of potassium chloride was dissolved in 20 pounds of water and the mixture was added to the jacketed kettle. One pound of anhydrous citric acid diluted with an equal amount of water and 5.0 pounds of 150-fold apple essence were then added to the jacketed kettle.

The above mixture was then stirred and the temperature was brought back to 170°F. To the mixture was then added 290 pounds of the carrageenan solution which had been previously prepared. The pH was then checked and adjusted to 3.60. The percent solids, as measured by °Brix, was then adjusted to 12.2° by the addition of water. The apple jelly mixture was then heated to 185°F and was filled into jars at a temperature of 160°F or above. The jars were capped and were cooled at a rate sufficient to prevent trapping air bubbles in the jelly.

An artificially sweetened apple jelly product was produced by the above procedure which had extremely good clarity when compared to apple jelly produced in a conventional manner with sugar and high-methoxy pectin. The process used to manufacture the artificially sweetened apple jelly eliminated the problem of burn-on on the sides of the mixing kettle usually encountered when producing artificially sweetened jelly.

Example 2: Artificially sweetened grape jelly was produced with the gelling agent by the following process: A carrageenan solution was prepared in accordance with the procedure of Example 1. 77.5 pounds of water and 369.8 pounds of 29° Brix grape juice concentrate were added to a jacketed kettle equipped with an agitator and steam was supplied to the jacket. 62.5 pounds of a solution containing apple pectin at a level of 4% by weight (65% degree of methylation) were then added to the kettle and the mixture was heated to a temperature of 170°F. 10.8 ounces of sodium saccharin was dissolved in 40 pounds of water and the mixture was added to the kettle. 1.25 pounds of potassium citrate and 0.75 pound of potassium sorbate were dissolved in 15 pounds of water, and the mixture was added to the kettle. 20 pounds of a 25 weight percent sodium hexametaphosphate solution were added to the kettle and the temperature was equalized at 170°F. 33 pounds of 25-fold grape essence were then added and the pH was adjusted to 3.50.

The solids content, as measured by °Brix, was adjusted to 13.7°F by the addition of water and 375 pounds of the previously prepared carrageenan solution was added. The mixture was heated to a temperature of 185°F and was packaged into jars. The jars were capped and cooled at a rate sufficient to prevent trapping air bubbles in the jelly.

Polyose Base

A nonnutritive pectin-gelling agent is used to prepare low-calorie and dietetic gelled products in the process of *I.D. Bliudzius, P. Jucaitis and N.P. Rockwell; U.S. Patent 2,876,101; March 3, 1959; assigned to E.I. du Pont de Nemours and Company.*

The gelling agent is polyose, a glucose polymer derived from starch by depolymerization followed by heat polymerization. The polyoses have a considerably different susceptibility to amylolytic enzymes than the original starch or its conventional degradation products. The products on test contain groups which act very much like small amounts of reducing sugar, not in excess of about 5 to 7% in Polyose A. The reducing sugar is not construed as due to the presence of free glucose.

A preferred polyose for this purpose is Polyose A which can be described as a glucose polymer derived from starch by depolymerization followed by heat polymerization to such an extent that at 67% solids it gives a viscosity (Brookfield) of 20 to 100 poises at 70°F. More viscous polyoses can be used. Polyose C requires 50%, Polyose B 60% and Polyose D requires only 40% of solids to give the viscosity described.

The amount of polyose required to form a gel is not strictly related to the amount of sugar normally used in the conventional products, but will depend to a greater extent upon the amount of pectin and acidity present, the processing conditions used, and other gelling agents used besides pectin. The amount of polyose used is further dependent upon the particular grade and concentration of polyose used. Broadly, the amount of polyose that can be used in these products will vary from about 25 to 60% by weight. The exact relationship between the quantity of pectin and the quantity of polyose used in a pectin gel will vary with the particular type of gelled food product being made.

Ordinary commercial pectins form gels by dehydration and electrical neutralization of the colloidally dispersed and highly hydrated pectin agglomerates or micellar aggregates. The pH should be below 3.5, and the concentration of pectin above 0.3% by weight for forming pectin gels. The rate of gel formation depends upon a number of factors, such as sugar and pectin concentration, pH, type of pectin and the temperature.

For fixed amounts of the same pectin, the lower the pH and the temperature and the higher the polyose concentration, the faster the rate of gel. In preparing fruit jellies, it is best to use slow-setting pectins so that, on a commercial scale, capping, labelling, casing and stacking operations can be done before gelation takes place. Pectins which cause rapid gel formation are of particular value in the preparation of jams because thickening and gelation should occur before the solid food rises in the container.

Various other materials can be added to a pectin-containing foodstuff to assist in forming a gel structure and to improve flavoring. For example, fruit flavors, spices, salts and the like can be added to enhance palatability characteristics. To improve gel structure, glycerol, sorbitol, low-methoxyl pectin, starch, pregelatinized starch, gelatin, corn syrup, carboxymethylcellulose, syrups, alcohols, Irish moss and the like can be used. But, as most of these substances have nutritive content, too much of them should not be used.

Example 1: Low-Calorie Jellies – Pectin jellies are made from the juices of fruit containing pectin by boiling out and straining the fruit to separate the juices. The fruit juices used can be separated from the fruit pulp by any standard means such as a rack and cloth, press or the like. Any one of the three general methods of clarifying the juices (centrifuging, filtering and enzymatic) can be used. Fruit juice jellies have a tendency to increase in firmness during the first week after manufacture. The firmness of a fruit juice or jam can be determined by the use of the Exchange Ridgelimeter. A jelly food composition containing polyose as a pectin-gelling agent is prepared using the following proportions:

Fruit juice (current)	72 lb
Polyose C	80 lb
Mint flavor	2 oz
Green color	4 oz
Artificial sweetener (suitable amount)	

Boil until the mixture will form a suitable gel. With the proper amount of pectin and acidity present in the fruit juice, a jelly containing about 60% of polyose will result. The time factor of preparing jelly is most important and should not be underrated. The jelly mixture should be concentrated to its finish point as rapidly as possible in order to avoid flavor losses and changes in color and pectin content. Similarly, filling and sealing should be accomplished without time delay.

In order to obtain good, firm jelly, it is desirable to prepare small quantity test batches using varying amounts of polyose in order to determine the optimum amount of polyose needed to obtain a good pectin gel. A standard formula once derived can be used for the remainder of the juice in any one total batch. Usually, each test batch consists of about one pound of juice, for which the amount of a polyose needed on gelling varies from about one-quarter to one pound.

Example 2: Low-Calorie Jellies – This example demonstrates the production of a gelled dietetic food composition containing smaller amounts of polyose than used in the previous example, but the amount of pectin and acidity required in this case is greater. The same general techniques are used in the preparation as disclosed above.

If sufficient pectin and acidity is not present in the fruit juice, it is necessary to add ingredients to make the mix shown in this example:

Fruit juice	50 lb
Polyose D	15 lb
Powdered pectin	6 oz
Powdered citric acid	2 oz
Artificial sweetener (suitable amount)	

Upon boiling and gelling, the resulting product will contain about 25% by weight polyose.

Example 3: Low-Calorie Jams and Preserves – Preserves are jams of semisolid or viscous foods made by boiling the mixture of fruit and sugar until such mix becomes thick or syrupy. At this point the sugar or sugar substitute (in this case a polyose) is added. Then, the preserves are concentrated to the finishing point by continued cooking in an open kettle, although a vacuum evaporator can be used. The following proportions by weight of ingredients are here used for making a preserve:

Peeled, cored, trimmed peaches	8 lb
Polyose B	1.6 lb
Powdered pectin	0.15 lb
Water	0.80 gal
Citric acid	0.08 lb
Saccharin	0.008 lb

When the finish point is reached, the product was handpacked into containers which were promptly sealed. The resulting product contains 10% by weight of polyose.

Example 4: Low-Calorie Marmalades – Marmalades are usually citrus-fruit jellies in which sliced or chopped portions of fruit skins are embedded or suspended. Prior to the addition of the skins to fruit pulp and sugar the skins must be removed from the fruit and preferably softened by cooking. Prior to being softened, the skins are usually chopped or shredded. The fruit pulp used may be strained or put through a cyclone-finisher to make a more uniform textured product.

After skins, pulp, sugar or sugar substitute (in this case a polyose) are mixed, the product is then handled the same as a jam or preserve. A marmalade is prepared using polyose and pectin by employing the following proportions by weight:

Oranges (pulp and skin)	33 lb
Polyose A	67 lb
Powdered pectin	4 oz
Citric acid	2 oz
Artificial sweetener (suitable amount)	

The resulting product will contain about 70% polyose.

Example 5: Low-Calorie Butters – Butters are semisolid pastes made by cooking a strained fruit pulp or a mixture of fruit pulps that have been concentrated.

The fruit is cooked with a small amount of water until it becomes a pulpy mass. Then it can be strained or run through a cyclone-finisher to break the pulp into finely divided particles of uniform texture. The strained pulp is then mixed with the sugar or sugar substitute (in this case a polyose). Occasionally boiled cider or fresh cider boiled four parts to one is added and the mixture is concentrated to the desired finishing point.

If desired, spices can be added about ten minutes before cooking is completed. The product is then handled the same as a jam or preserve. A butter is prepared using a polyose and pectin by employing the following proportions by weight:

Fresh apples	6.5 lb
Polyose	2.0 lb
Boiled cider	0.08 gal
Cinnamon	0.02 lb
Artificial sweetener (suitable amount)	

FRUIT TREATMENTS

Frozen Fruit–Low Density Syrup

In the conventional preparation of frozen fruit (peaches, for example), the following procedure is used: Fresh peaches are subjected to preliminary operations which include washing, peeling, removing of pits and cutting into slices. The peach slices are then placed in cartons, covered with a syrup containing about 50% sucrose and a small proportion of ascorbic acid. The cartons are then sealed, frozen and maintained in frozen storage until ready for consumption.

The product is not stable and upon storage the fruit slices turn brown. The rate of browning is greatly accelerated by increase in temperature. It would be expected that the ascorbic acid in the syrup would prevent this browning because of its well-known antioxidant properties. The ascorbic acid cannot protect the fruit because at least part of the fruit is simply not in contact with the ascorbic acid-containing syrup. The point is that the syrup, because of its high sucrose content, has a density much greater than that of the fruit pieces. As a result, some of the fruit pieces float above the level of syrup. The proportion of these floating pieces to the total amount of fruit depends, of course, on the free space at the top of the carton. A certain proportion of free space must be left in the carton to allow for expansion of the product when it is frozen.

As the products are handled in distribution channels, warehouses, retail freezer cabinets, home freezers, etc., the cartons will be oriented in different positions. This allows different pieces of fruit to rise above the syrup, particularly if there occur even temporary increases in temperature sufficient to cause thawing of the syrup. Under these conditions, a large proportion of the fruit pieces eventually becomes brown, so that the product has greatly reduced flavor and appearance.

D.G. Guadagni; U.S. Patent 3,025,169; March 13, 1962; assigned to the U.S. Secretary of Agriculture proposes that the fruit be packed with a protective solution which has a density less than that of the fruit. The protective solution used to cover the fruit generally contains water, ascorbic acid and a sweetening agent. The water is required to furnish enough bulk to permit covering of the fruit and to carry the other ingredients. The ascorbic acid acts as an antioxidant to prevent the fruit from browning. The proportion of ascorbic acid in the solution is generally in the range from about 0.04 to about 0.1%. A larger proportion of ascorbic acid may be used if desired but is not necessary for preserving the color of the fruit. The sweetening agent is provided to give the solution the degree of sweetness as may be desired. Thus various compounds and proportions may be used depending on the taste desired.

The critical factor of the protective solution is that its density be lower than that of the fruit. This desideratum cannot be attained where a sugar is used at conventional sweetness levels. Such degree of sweetness requires a syrup of about 50% sucrose and even greater concentration of sugars other than sucrose is used. These syrups have densities far above that of the fruit.

Accordingly, it is preferred to use the noncaloric, ingestible sweeteners which will provide solutions of the proper degree of sweetness yet of density less than that of the fruit. Among the compounds which fulfill these requirements are saccharin, and saccharin sodium. These compounds exhibit such a high degree of sweetness that dilute solutions having substantially the same density as water itself will provide the proper degree of sweetness to the packed fruit. As noted, the concentration of the sweetening agent in the protective solution may be varied according to the taste desired in the packed fruit.

However, in no case should so much of the sweetener be used as to increase the density of the solution above that of the fruit. The possibility of this occurring is out of the question in practical consideration because such a high concentration of sweetener would render the product so exceedingly sweet as to be unpalatable.

The protective solution, in addition to the water, ascorbic acid and noncaloric ingestible sweetener, may contain any desired food ingredient or mixture of food ingredients. Examples of food ingredients which may be incorporated in the protective solution are given below.

Flavoring Substances: Sucrose, dextrose, fructose, maltose, invert sugar, corn syrup, molasses, maple sugar or syrup, etc. Particularly desirable are fruit juices, concentrated fruit juices, and especially full-flavored concentrates, that is, those in which the volatile essences usually lost in evaporative processes are recovered by condensation or other techniques and returned to the concentrate.

Fruit-juice concentrates are desirable in that they not only add their sweetening effect but also the characteristic fruity taste so that the flavor of the product is intensified. Other flavoring agents which may be used are spices, salt, citric acid, alcoholic extracts of vanilla, lemon, mint, etc.; synthetic flavorings such as vanillin and methyl anthranilate; natural fruit essences such as those recovered by condensation from the vapors evolved in the evaporation of fruit juices, etc.

Nutritive Substances: Vitamins, vitamin precursors, mineral salts, proteins, protein hydrolysates, solubilized starch, etc.

Coloring Materials: Dyes suitable for food use, juices or extractions from highly pigmented fruits or vegetables, for instance, Concord grape juice.

Preserving Agents: Browning inhibitors such as sodium sulfite, sodium bisulfite, lemon juice, etc. Agents for preventing or inhibiting microbial spoilage such as sodium benzoate, sodium para-hydroxybenzoate, antibiotics, etc. Agents for firming the texture of the fruit, such as calcium chloride, pectin, low-methoxyl pectins, methyl cellulose, etc. Agents for thickening the liquid surrounding the fruit, such as gelatin, tragacanth, soluble starch, pectins, algins, etc.

It is evident from the above that the food ingredients to be added to the preservative solution may be chosen as desired for any particular function. In any case, the total amount of material added to the solution should be limited so that the density of the solution remains below that of the fruit to be preserved.

In one modification of the process, the fruit is deaerated prior to being covered with the preservative liquid. By removal of air from the fruit tissues, the density of the fruit is increased so that one is assured that the fruit will remain submerged in the preservative liquid. Although the deaeration may be carried out in various ways, it is preferred to use a vacuum technique whereby removal of air may be accomplished rapidly.

To this end, the fruit pieces are placed in a bath of an edible liquid in a closed vessel which is subjected to a vacuum to remove most of the air from the fruit tissue. Then the vacuum is broken, whereupon the edible liquid rapidly fills the tissue interstices formerly occupied by air. The edible liquid used in this procedure may be plain water or an aqueous solution containing sweetening agents, preservative agents, etc. Usually the same preservative liquid as used in packing the fruit is used as the edible liquid in the vacuum deaeration steps.

Example 1: Fresh Elberta peaches were steam-peeled, sliced into twelfths and packed into one-pound waxed paper tubs provided with slip-on covers. (A) In one set of samples, the peach slices were covered with 50% sucrose syrup containing sufficient ascorbic acid to provide 200 mg per pound of finished pack. (B) In another set of samples, the peach slices were covered with a water solution containing sufficient saccharin to match the sweetness of 50% sucrose syrup and sufficient ascorbic acid to provide 200 mg per pound of finished pack. It was observed that the peach slices floated in the sucrose syrup, whereas, in the saccharin solution, the peach slices submerged.

The packed tubs were sealed by placing on the covers. Various lots of the peaches were stored at 0°, 10° and 30°F. At periodic intervals, the packs were opened and examined for color changes by reflectance measurements. The following data were obtained:

		Proportion of Original Color Lost by Fruit, %	
Temperature of Storage °F	**Time of Storage**	**Sucrose Pack (A)**	**Saccharin Pack (B)**
0	1 year	12	0
10	1 month	18	0
20	2 weeks	12	0
30	2 days	30	9

Example 2: Fresh apples were peeled, cored and cut into slices. The apple slices were placed in a vessel and covered with an aqueous solution containing saccharin

as in Example 1, and 0.1% of ascorbic acid. The vessel was closed and a vacuum applied. The vacuum was maintained until the ebullition caused by removal of air from the fruit ceased. The vacuum was then broken and the fruit pieces allowed to stand in the solution for a few minutes. The fruit pieces were then removed, packed into containers and covered with an aqueous solution of the same composition as used in the vacuum impregnation. The containers were sealed and frozen.

Conditioning Prior to Freezing

Another process by *D.G. Guadagni; U.S. Patent 2,788,281; April 9, 1957; assigned to the U.S. Secretary of Agriculture* comprises a conditioning procedure for treating frozen fruit and vegetables for better flavor and texture quality.

Briefly described, this is accomplished by first preparing the frozen fruit in the conventional manner, that is the fruit is packed into a container together with dry sugar or syrup. The sugared fruit is then frozen in the usual way. After the product has been frozen it is removed from the freezer and subjected to what is termed a conditioning operation. This operation involves holding the product in a thawed or unfrozen state, preferably at a temperature just above its freezing point, for a substantial period of time.

After the conditioning operation the product may be consumed or otherwise utilized without further treatment. If the product is to be stored, it is refrozen and maintained in frozen storage until used.

During the conditioning operation the fruit is in contact with the syrup surrounding it and sugar will diffuse from the syrup into the fruit tissue. This conditioning is so effective that the sugar content of the fruit will become equal to the sugar content of the syrup, whereas originally the fruit had a lower sugar concentration than the surrounding syrup. As the conditioning proceeds, sugar diffuses into the fruit tissue, with the result that the sugar content of the fruit increases and the sugar content of the syrup decreases. Finally an equilibrium point is reached at which the sugar contents of the tissue and the syrup are equal.

The reason for the criticality of freezing prior to conditioning is not understood and cannot be scientifically explained. Regardless of any theoretical considerations, the process results in absorption of large amounts of sugar, whereas in the known method only minor amounts of sugar are absorbed.

In applying the process to raw solid fruit, the procedure involves conditioning the fruit while it is in intimate contact with an aqueous solution containing at least 20% sugar. The conditioning is continued until the fruit and the solution each contain substantially the same concentration of sugar, within the range of about from 20 to 30%, whereby the fruit and the liquid have substantially the same degree of sweetness. The resulting mixture of fruit and liquid is then frozen and maintained in frozen condition until consumed.

Fresh produce such as fruit, vegetables, meat, etc. is first subjected to the usual preparatory steps (washing, peeling, pitting, slicing, etc.). The preparative steps to be used will, of course, depend on the nature of the produce and the type of product desired. For example, fruit such as apples and pears are usually washed, peeled, cored and sliced. Peaches and apricots are washed, peeled, pitted and sliced. Small fruit such as berries, grapes, etc. are merely washed. Vegetables such as beans, squash, broccoli, spinach, are washed and cut into convenient pieces. Root vegetables such as carrots, beets, potatoes, etc., are washed, peeled and cut into pieces.

Since the impregnation of the solid food with the food ingredient involves a diffusion process, the size of the food pieces will have an effect on the distribution of the food ingredients within the solid food. Thus, when the food is reduced to small pieces, the path of diffusion is decreased with the result that the impregnation will be more uniform throughout the internal structure of the food. Thus, to promote uniform impregnation in a reasonable time, it is preferred that, where necessary, the food be reduced to slices, dice, slabs, or other pieces in which the smallest dimension is not over about 1.5 inches. However, it may be noted that where the conditioning time is extended, uniform impregnation will be obtained regardless of the size of the food pieces.

Where necessary the food may be subjected to such treatments as: blanching in steam or hot water to inactivate enzymes; treatment with agents such as sulfur dioxide, sodium sulfite, sodium bisulfite or ascorbic acid to prevent browning; complete or partial cooking to tenderize the food, etc.

The solid food units, that is whole small fruit or pieces of fruit, vegetables or meats, are placed in a suitable container such as a can, carton, etc. Over the contents of the container is then poured a quantity of an edible liquid containing the selected food ingredient, for example, sugar dissolved in water. In many cases the food ingredient may be added in a dry state. In such case, a mixing of the dry ingredient with the food will develop a solution of the ingredient in juice issuing from the solid food. Such technique is convenient with fruits and similarly juicy foods.

Sweetening Agents: Sucrose, dextrose, fructose, maltose, invert sugar, corn syrup, molasses, maple sugar or syrup, etc. For the so-called "dietetic" foods, saccharin may be used. For the impregnation of fruits, particularly desirable sweetening agents are the concentrated fruit juices and especially full-flavored concentrates, that is, those in which the volatile essences usually lost in evaporative techniques are recovered by condensation or other techniques and returned to the concentrate. Fruit-juice concentrates are desirable in that they not only add their sweetening effect, but also the characteristic fruity taste so that the flavor of the treated fruit is intensified.

To promote the diffusion of the food ingredient into the solid food units during the conditioning operation, the concentration of the food ingredient in the edible liquid should be greater than the concentration of the same food ingredient within the solid food. For example, if the solid food to be treated contains 5% sugar, then the concentration of sugar in the edible liquid should be higher than 5% thus to provide a driving potential for the diffusion process.

The freezing may be accomplished in any of the devices available for freezing purposes, for example, a plate freezer or tunnel freezer. The frozen product directly after freezing or storage at freezing temperatures is then subjected to the conditioning step. A simple way of doing this involves placing the containers in a room maintained at a temperature just above the freezing point of the food, such conditions being maintained for a substantial period to allow the food ingredient to diffuse into the solid food units.

The temperature of conditioning to be used with any particular food will depend largely on the type of food and particularly on its content of soluble solids. Thus, foods low in soluble solids, such as cabbage, lettuce, broccoli, etc., have freezing points at or near 32°F and with such foods the conditioning temperature may be just above 32°F, for instance, 32.5°F. Other foods which are higher in solids content, such as cherries, prunes, apricots, peaches, apples, grapes, etc., have freezing points below 32°F and, in such cases, the conditioning may be carried out at temperatures of 32°F or below depending on the freezing point of the product in question.

In general, conditioning temperatures from about 25° to 35°F are preferred, depending on the nature of the food being treated. Temperatures higher than 35°F may be used, but to avoid danger of microbial spoilage during the conditioning operation, it is preferred to use a temperature just above the freezing point of the food, that is, about 0.5 to 5°F above the freezing point.

Since the penetration of the food ingredient into the solid food involves diffusion, sufficient time must be allowed for the penetration to take place. Obviously the time for penetration will depend on the degree of penetration desired and on the nature of the food, that is, its porosity or denseness, the size of the food units, etc. To secure maximum penetration, the conditioning is usually continued for a period from several hours to several days.

After the conditioning operation is completed, the food may be treated in several ways. Thus, for example, the impregnated food with the accompanying edible liquid may be frozen and maintained in frozen storage. If desired, the solid food units may be separated from residual edible liquid and the solid units frozen and maintained in frozen storage until needed for consumption. In the alternative, the separated units may be subjected to dehydration to reduce their moisture level to a low level so the product will be self-preserving. The impregnated food units, with or without the residual edible liquid, may also be preserved by canning. If the impregnated food is not to be preserved, it may be consumed directly or used directly in the preparation of such products as pies, stews, creamed products and so forth.

Example 1: A lot of fresh strawberries were washed then sliced into pieces about ⅜ inch thick. The slices were then mixed with sucrose in the proportion of 4 lb fruit to 1 lb sucrose. During this mixing, the sucrose was dissolved by the strawberry juice so that the final mixture consisted of strawberry slices surrounded by syrup containing water, sucrose, and fruit extractives. The resulting mixture was placed in cartons, the cartons being then sealed and placed in a plate freezer at minus 20°F.

After the product was frozen, it was removed from the freezer and placed in a refrigerator maintained at 30°F. Packages of the product were opened at various times and the fruit and syrup separated by draining through a sieve.

The soluble solids content (largely sucrose) in the fruit and in the syrup was then determined by the use of a refractometer. The results obtained are set forth on the following page.

Period of Storage at 30°F, days	Soluble Solids Content, Expressed as Percent of Sucrose	
	Fruit	Syrup
0	15	35
1	20.6	27
3	22.5	25
6	23.5	24

Example 2: Red sour cherries were washed and pitted then placed in cans. The fruit in the cans was covered with a 60° Brix solution of sucrose in water. The cans were sealed then frozen at minus 10°F. The cans of frozen product were then placed in a refrigerator maintained at 30°F and samples were withdrawn at intervals and tested as described in Example 1. The following results were obtained:

Product	Period of Storage at 30°F, days	Soluble Solids Content, Expressed as Percent of Sucrose	
		Fruit	Syrup
A	0	16	45
B	1	22.8	26
C	3	26	27.6

In addition to the above tests, the products conditioned at 30°F (B and C) and the product which had not been conditioned (A) were tasted. It was found that products B and C were sweet and there was essentially no difference in sweetness between the fruit and the syrup. In the case of product A, the fruit was very tart, whereas the syrup was very sweet.

Fruit Sections–Canned Product

It has been found that sections of citrus may be canned for extended shelf life while retaining their natural flavor, color and texture by immersing the sections in a sweetening solution within a container and sealing the container while maintaining a steam atmosphere in the head space in a manner such that a 10 to 15 inch (and preferably an approximate 12 inch) vacuum will be attained after sealing the container. Then the container is conveyed through a heating bath to attain a temperature at an internal point spaced from the bottom center of the container ranging from 165° to 215°F (and preferably 165°F) within 20 to 60 minutes and a low pressure of 15 to 25 psi. Then the contents is cooled to a temperature of about 50°F and then stored at a reduced temperature.

The quick heating and cooling under low pressure uniquely accomplishes sterilization while avoiding loss of flavor, color and texture; and the meat substantially resembles the natural meat one encounters from fresh fruit.

The process is described by *M. Verlin; U.S. Patent 3,592,664; July 13, 1971* in the following examples:

Example 1: Oranges – Oranges are carefully selected from groves known to produce high-quality fruit. The fruit is picked and hauled without bruising and then aged in bins until it will peel properly. The oranges then are carefully culled in final grading on the way to the peeling room. However, before peeling, and depending on the condition of the oranges, they are heated to above 125°F at a point ¼ inch under the peel so the peel may be easily removed without damage to the segments. The peeled fruit is then placed in wire baskets and submerged in, or flooded over with, lye solution (1 to 3% by weight) and at a temperature of approximately 190°F to remove the albedo.

The oranges are then spray-rinsed of the lye solution and conveyed through a chilled water bath where they are chilled to about 60°F. The oranges are then sectionized and the sections placed in glass jars to a point approximately ¼ inch below the jar lip, together with artificially sweetened water with a stabilizer (propylene glycol alginate, guar gum or the like) and with or without preservatives (usually sodium benzoate) as may be desired.

The open jars are then passed under a device designated to displace a specific amount of the liquid portion of the pack to provide a predetermined head space, which is 15 to 20 milliliters in the case of pint (1 pound) jars and 10 milliliters for quart (2 pound) jars. While maintaining a steam atmosphere in the head space, the jars are sealed with plastic-gasketed metal lug caps, in such manner as to produce an approximately 12 inch vacuum. The capped jars are conveyed through a hot water cooking bath for 29 minutes, the water temperature being maintained at 190°F for pint jars and at 200°F for quart jars.

Under these cooking conditions, a pressure of about 20 to 25 psi will be attained in the jars of each size and a temperature of 160° to 165°F is reached at an internal point in the product about two inches from the bottom center of the jar.

Subsequently, the jars are discharged from the cooker to a conveyor, which delivers them to the entrance to a chiller after a travel time of five minutes. The temperature of the product in the jars at a point two inches above bottom center should remain approximately the same from discharge at the cooker to the entrance to the chiller. The jars are then slowly conveyed through the chiller under a series of water-fogging nozzles, starting at ambient temperature, and they are gradually subjected to water sprays at lower temperatures so that they emerge from the chiller with a product temperature of approximately 50°F.

The jars are then cased and trucked directly to a chill room or to ambient temperature storage. After several months storage, the orange sections will be found to have a "natural" taste and texture and color. For product canned without calcium lactate as preservative, the product has an even closer natural taste.

Example 2: Grapefruit – In similar fashion to Example 1, grapefruit are carefully selected from groves, hauled without bruising and then aged in bins until proper peeling is possible.

After culling, and depending on the condition of the grapefruit, they are heated to above 125°F at a point ¼ inch under the peel so that the peel may be easily removed without damage to the segments. The peeled fruit is then placed in wire baskets and submerged in or flooded over with lye solution (1 to 3% by weight) and at a temperature of approximately 170° to 190°F to remove the albedo. The grapefruit are then spray rinsed of the lye solution and conveyed through a chilled water bath, where they are chilled to about 60°F. Sections of the fruit are placed in glass jars to a point approximately ¼ inch below the jar lip, together with artificially sweetened water with a stabilizer and with or without preservatives, as desired (see Example 1 for specific compounds). A predetermined head space of 15 to 20 milliliters for pint (1 pound) jars and 10 milliliters for quart (2 pound) jars is provided. Then, while maintaining a steam atmosphere in the head space, the jars are sealed in such manner as to produce an approximately 12 inch vacuum. The capped jars are conveyed through a hot-water cooking bath for 30 minutes, the water temperature being maintained at 190°F for pint jars and at 200°F for quart jars. Under these cooking conditions, a pressure of about 25 psi will be attained in the jars of each size and a temperature of 160° to 165°F is reached at an internal point in the product about two inches from the bottom center of the jar.

Subsequently, the jars are slowly conveyed through a chiller under a series of water-fogging nozzles, starting at ambient temperatures. They are gradually subjected to water sprays at lower temperatures so that they emerge from the chiller with a product temperature of approximately 50°F after ten minutes from hot to chilled condition. The jars are then cased and trucked directly to a chill room or to ambient temperature storage.

FREEZE DRIED FRUIT

Impregnation with Sweetener

With the advent of freeze drying it has been possible to prepare freeze dried fruits by various methods which result in dried fruits which can be rehydrated in milk in from 30 to 60 seconds. These fruits are particularly advantageous for use with a dried breakfast cereal, as the rate of rehydration in milk is sufficiently rapid that the fruit is rehydrated before the cereal becomes undesirably soft and soggy. The result of blending a dry cereal and rapidly rehydratable fruit is a breakfast food which, after addition of milk, consists of cereal with the desirable characteristics of crispness and rehydrated fruit which has the desirable texture of fresh fruit.

It has been found that consumer acceptance of the rehydrated fruits is greatly enhanced if the fruit is sweetened, as is consumer acceptance of commercially available canned or frozen fruits. It is usual for canned or frozen strawberries, for example, to contain from about 20 to 30% added sugar by weight.

However, attempts to sweeten freeze dried fruits prior to or after the freeze drying process by dipping the foodstuff in a sugar solution, spraying a sugar solution on the foodstuff or dusting the sugar on the foodstuff have been notably unsuccessful, the sweetened product having an undesirable appearance. Also, in many instances the known methods of adding sugar result in serious processing difficulties.

When fresh fruit, for example strawberries, are dipped into a concentrated sugar solution (having a sugar concentration of about 40%) they tend to coalesce into large lumps during the freezing and freeze drying operations due to the inherent stickiness of the sugar solutions. In addition, the procedure of dipping results in a coating of sugar on the surface of the fruit, which gives the fruit a glazed, foamy, unnatural appearance. Dipping in dilute sugar solutions (having a sugar concentration of about 20%) is unsatisfactory in that it is very difficult to impart sufficient sweetness to the product.

Also the large amount of sugar solution which must be added to the fruit to impart the desired sweetness results in an undesirable increase in moisture content of the fruit, which prolongs the freeze drying operation. Attempts to introduce the sugar to the fruit after freezing is unsatisfactory. Spraying a sugar solution onto the fruit after freezing in sufficient quantity to impart the desired sweetness causes a thawing of the fruit due to a lowering of the freezing point. The partially thawed strawberries tend to lump and coalesce as they are being refrozen and additional processing steps and product appearance are then subject to the same disadvantages associated with dipping or spraying before freezing.

Dipping or spraying the freeze dried fruits with a sugar solution is unsatisfactory in that the product then tends to absorb a significant amount of moisture and an additional drying step is necessary. Also, the surface becomes coated with a film that destroys the natural appearance of the product.

W.L. Vollink, R.K. Scharschmidt and R.E. Kenyon; U.S. Patent 3,511,668; May 12, 1970; assigned to General Foods Corporation found that the artificial sweetener is applied in such a manner as to penetrate the outer surface of the foodstuff to be freeze dried. A critical feature of this product is that it is rapidly rehydratable, for example, within 30 to 60 seconds in milk. Rapid rehydration is necessary in order to permit the hydrating fluid to carry the sweetener further into the freeze dried product before it can be absorbed by the hydrating fluid.

While the exact mechanism is not fully understood, it is theorized that because the food product absorbs the hydrating fluid quickly, a mass-flow situation is set up whereby the artificial sweetener is carried into the food product, similar to a board being carried along in a fast current. The mass-flow is such an overriding force that the artificial sweetener does not have a chance to diffuse out into the hydrating medium while the food product is being rehydrated. It is probable that given sufficient time after rehydration the artificial sweetener would be absorbed into the remainder of the hydrating medium in contact with the food product.

Therefore, the product is particularly useful as a food product which will normally be consumed in a period of sufficiently short duration so that the artificial sweetener will not migrate to the unabsorbed fluid before the product is completely consumed. In this regard, the freeze dried fruits are particularly adaptable for use with dried cereals as a ready-to-eat breakfast product. Normally after milk is added to a breakfast cereal, the product is consumed within about five minutes, and in this period of time no noticeable sweetness is imparted to the milk.

The necessary characteristic of rapid rehydration may be imparted to the food product by any known process which results in a product having the desired characteristic. One means of producing such a product is as follows: Freshly picked foodstuffs are frozen before they have had time to deteriorate in quality. The freezing step is started by placing the foodstuff in trays in a freezing room. The air in the freezing room may be static or force draft, as long as the temperature is maintained sufficiently low to cause the food to cool to a temperature of about 30° to 32°F. At this point the driving force between the cooling air and the fruit should be sufficiently small to allow the fruit to slowly freeze over a period of several hours.

When the bulk of the moisture in the food has been frozen, the temperature curve of the food will start to dip from a plateau and descend toward the temperature of the cooling air. At this point the air temperature can be lowered and the food again allowed to exhibit a plateau at some other freezing temperature such as 20° to 25°F.

This process of freezing the fruit slowly should be continued until all of the water in the food product has been completely frozen. The freezing is carried on over a period of several hours so that the water in the food is allowed to crystallize and the crystals are allowed to grow into large crystals sufficient to cause the cellulose structure in the food product to be ruptured. After all of the water has been frozen, the temperature of the foodstuff is then lowered sufficiently to insure that the product will be completely frozen and remain in a completely frozen state when it is transferred to a vacuum freeze drying chamber.

In the vacuum freeze drying chamber the foodstuff is then dehydrated by sublimation to a moisture content of about 2 to 3%. The dried product may then be removed from the chamber and packaged in water-vapor-resistant packages, or may be stored in a low-humidity area for further processing. A foodstuff, and particularly a fruit, which is subjected to the foregoing freezing and freeze drying treatment will be found to have the desired property of rapid rehydration in milk.

The freezing technique may be modified in that after the initial freezing, the fruit may be warmed to a temperature of about 10° to 25°F so that it may be comminuted into pieces of a desirable size while still sufficiently frozen to retain its natural juices but not sufficiently brittle to shatter upon comminution. The pieces of frozen fruit would then be chilled down to a temperature sufficiently low to insure that they remain frozen while being put into the freeze dryer and the freeze drying process is completed as before. The freezing time may be varied depending upon the fruit which is being frozen and the size of the desired pieces.

In any event, the freezing variables will be controlled such that the final freeze dried fruit, if it is to be used with a cereal for a breakfast food, will have a rehydration rate in milk of from 30 to 60 seconds.

A foodstuff prepared by these techniques will have the desired property of rapid rehydration. It remains to be shown how such a product may be prepared with a sufficient amount of an artificial sweetener to impart the desired sweetness to the food product, and how the artificial sweetener may be added in such a manner that it will not be absorbed by the hydrating medium when the freeze dried foodstuff is being rehydrated.

It has been found that the techniques of dipping and spraying an artificial sweetener onto the fruit can be used. These techniques are satisfactory if used prior to freezing or after freezing but prior to freeze drying. When applying the artificial sweetener to the foodstuff, it is found that the artificial sweetener impregnates the surface of the foodstuff and that after freeze drying there is no visible film of sweetener remaining on the surface. The dry product has a desirable, natural appearance. It has been found that addition of the artificial sweetener to the food after it has been freeze dried is not desirable. If the artificial sweetener has been added by dusting, the sweetener tends to cling to and coat the outer surface of the product resulting in an unnatural appearance. Also, the dusted-on sweetener tends to flake off on additional handling. If the artificial sweetener is added by spraying after the freeze drying, it is found that an undesirable surface film remains which causes a white, spotty appearance.

Therefore, it is considered essential that the artificial sweetener be added prior to freeze drying. Furthermore, when the sweetener is added after freezing but prior to freeze drying, it is desirable to warm the product to a temperature of about 10° to 25°F (and most preferably 21° to 24°F) so that there is some softening of the ice on the outer surface of the product, thus permitting the sweetener to penetrate the surface. After adding the sweetener, the product is again reduced in temperature prior to transferring it to the freeze dryer.

The quantity of artificial sweetening agent used will be determined by taste, enough being present to give a taste equivalent to that of natural sweetening agents. The procedure outlined in the foregoing paragraphs is particularly applicable to fruits in general and particularly to strawberries, peaches, bananas, blueberries, raspberries, blackberries, pineapples, apples and cranberries.

However, the process is not limited to fruits, but is equally applicable to other foodstuffs, for example, tomatoes, peas and other vegetables. In the case of strawberries, peaches and bananas, the frozen fruit is sliced to from about one-quarter to three-eighths inch sections after the frozen fruit has been rewarmed to a semifrozen state (10° to 25°F). In the case of blueberries, the whole berries are frozen, warmed to a temperature of about 10° to 25°F and then pricked to develop holes in the skin and cellulose structure of the berry. The pricking operation enables the blueberries to be freeze dried in a much shorter time and prevents shriveling, without destroying the whole-berry appearance, texture and structure.

Example 1: Ten pounds of freshly picked whole strawberries whose stems had been removed were washed in water and graded for uniformity. Strawberries having a particle size of about three-quarters to one inch in diameter were then arranged in monolayers on freezing trays and the freezing trays were placed in a large freezing room having an ambient temperature of 0°F. The strawberries took about one-half hour to cool from room temperature to about 28°F and about six hours for substantially all of the water to be frozen at 28°F. The strawberries were then allowed to cool to 0°F.

A cooling curve in which temperature was plotted against time showed a temperature profile wherein an initially relatively steep slope down to the freezing point of the moisture in the fruit was followed by a flat line or plateau during the actual change from a liquid moisture state to a frozen moisture state, the flat line again sloping rapidly when substantially all of the water was frozen and the product temperature lowered to 0°F. The strawberries remained in the plateau for six hours and in this time developed a growth of ice crystals sufficiently large to at least partially rupture the cellular walls of the strawberries.

The frozen strawberries were then stored at 0°F to protect the strawberries against enzymatic or bacterial degradation, arranged in monolayer fashion on solid aluminum trays and freeze dried in ten-pound charges in a freeze dryer under a vacuum of 100 microns Hg, a platen temperature of 100°F and a condenser temperature of minus 40°F for about 20 hours until a terminal moisture content of 1.5% was attained. The resulting product has the appearance and taste of natural unsweetened strawberries.

Example 2: Eighty grams of saccharin and 150 pounds of water were mixed in a 50-gallon metal container until the saccharin was completely dissolved. The temperature of the solution was maintained between 50° and 60°F. This solution was approximately equal in sweetening power to a 30% aqueous sugar solution. Ten pounds of washed, whole, fresh strawberries, whose stems had been removed, were placed into a wire mesh stainless steel basket. The basket was then submerged in the saccharin solution for 60 seconds, during which time the strawberries were gently agitated to assure uniform contact with the solution.

After this 60-second immersion, the berries were allowed to drain for 60 seconds and then were placed on freezer trays. The fruit was cooled, frozen and freeze dried to 2.0% moisture according to the procedure of the previous example. The resulting product had the appearance of fresh strawberries and a taste equivalent to that of strawberries sweetened by the addition of natural sweeteners and had an acceptable level of sweetness as determined by consumer tests, whereas this level of sweetness could not be obtained with any amount of sugar which could be used in the process. The degree of preference for the artificially sweetened strawberries was two to one over unsweetened strawberries.

The freeze dried artificially sweetened strawberries (moisture content 2.0%) were then combined with corn flakes (moisture content 2%) at a level of about 7 to 10% by weight strawberries and 90 to 93% corn flakes. The blending operation was conducted in a packing room having a relative humidity of 30% in a period of less than 10 to 15 minutes, thereby limiting moisture pickup of the strawberries to less than 1%.

The strawberry-corn flake cereal was then packaged in a water-resistant, wax-laminated foil liner, which was placed inside a chip-board shell and enclosed with a wax-laminated foil over-wrap. The packaged product having a terminal moisture content of less than 2.5% for the strawberries and about 2% for the corn flakes was stored in this form at 70°F and 50% relative humidity for 3 to 6 months without any degradation in product quality.

At the end of this period, the strawberries were found to reconstitute in milk or cream in about 60 seconds to a flavor, texture, and appearance equivalent to that of fresh strawberries sweetened by the addition of natural sweeteners. The breakfast cereal could then be eaten within 1 to 5 minutes with the berries being fully reconstituted, but not mushy and the cereal still in a crisp form. The berries were found to have a lasting sweetness similar to that obtained by the use of natural sweeteners, instead of the fleeting bloom of sweetness ordinarily experienced with artificial sweeteners, and had the desired blend of sweetness and tartness similar to sugar-sweetened strawberries. The milk or cream used as rehydrating liquid did not get noticeably sweet.

Example 3: The procedure of Example 1 was followed with the exception that the frozen strawberries were kept in the freezing plateau for only two hours at a temperature of 15°F. The strawberries were then frozen to below 0°F, stored and then warmed or tempered to about 23°F in preparation for the cutting operation. Tempering was desirable to render the berries less brittle. The berries were then cut into halves and placed on a screen container. Then 10 milliliters of a solution prepared as described in Example 2 were sprayed on the surface of the frozen strawberries using a fine pressure spray at 50 psig. The strawberries were agitated during spraying to assure uniform coverage. After spraying, the strawberries were refrozen to below 0°F and then freeze dried in about 18 hours to a terminal moisture content of less than 1.5%, according to the procedure of Example 1.

Flavor Enhancement

The steps for flavor enhancement of freeze dried fruit are given by *D.J. Ewalt and R.E. Kenyon; U.S. Patent 3,501,319; March 17, 1970* as follows: (1) freeze piece of fruit to a temperature above the eutectic point; (2) cause at least a portion of the water-soluble pigments located at the surface of the piece to liquefy; (3) apply a solution of water-soluble, flavor-enhancing material to the surface of the piece; and (4) cause the solution to intermix with the liquefied pigments to an extent such that the piece can then be recooled and freeze dried without formation of a grossly visible white crystalline substance on the surface of the piece.

The process is carried out by lowering the freezing point of an aqueous solution containing the enhancing agent (for example, artificial sweeteners), as well as the liquefiable constituents at the surface of the fruit piece, so as to promote intermixing of the plant pigments and the flavor-enhancing agents while the interior of the fruit remains in a substantially solid condition.

A preferred practice is to employ in the flavoring solution a freezing point depressant, such as an edible mono- or polyhydric alcohol like ethyl alcohol or propylene glycol, or an oligosaccharide such as invert syrup. In the case of fruits, the flavor-enhancing agent is an artificial sweetener; thus, an aqueous solution of the artificial sweetener has its freezing point sufficiently reduced so that it will cause the aqueous ingredients at the surface of the fruit to melt and cause the artificial sweetener solute to intermingle with the thus liquefied constituents on the surface of a fruit piece.

Alternatively, a frozen plant tissue piece adjusted to a temperature only slightly below the ice crystallization point in the fruit may have a flavor-enhancing aqueous solution, such as one containing artificial sweetener, applied to the surface of the piece; the surface of the piece with the flavoring solution will assume a temperature where its surface liquids, particularly the coloring matter, will intermingle with the enhancing material sufficiently until such time as the piece may be frozen and freeze dried without the added flavorants being grossly visible.

This practice is less preferred because of the prolonged holding period required to cause the flavor-enhancing solution to sufficiently intermix with the liquefied surface portion of the plant tissue, and the companion problem of controlling temperature of the piece during such period so as to avoid undue liquefaction on the surface of the piece which can result in surface deformation incident to such thawing and consequent clumping or clustering of the fruit.

Among the plant tissue that may be treated in the process are fruits such as strawberries, peaches, blueberries, cherries, raspberries and similar fruits. Other fruits which can be similarly enhanced by the addition of a flavorant can also be treated. The benefits of the process are most applicable to treatment of those tart fruits which are sweetened before eating and which call for addition of a sweetening agent to enhance the natural flavor of the fruit. Generally fruit pieces will be frozen by reduction to a temperature of 0°F or below upon harvesting, whereafter the fruit will be stored at that temperature until further processing is performed. The stored piece will be sliced, and to achieve this result, the whole or subdivided piece will be tempered by allowing it to elevate to an average product temperature below 28°F (generally in the neighborhood of about 10° to 25°), whereby the piece can be sliced, punctured or otherwise subdivided preparatory to flavoring.

In the case of blueberries and other fruits having an integument which may impede rehydration, the fruit will be punctured after such tempering. Other fruits, such as strawberries, will be similarly tempered to the average temperature whereby the strawberry can be sliced without shattering prior to flavoring.

The fruit piece is tempered to a temperature just below the melting point of the liquefiable constituents at the surface so that a portion of the fruit will become liquefied by subsequent addition of the solution containing a freezing-point depressant and will, thereby, release mutual solvents for the flavor-enhancing solutes and natural plant pigments so that these fruit coloring agents are in a condition to intermix with the flavor-enhancing additives.

However, it is not critical or essential that the fruit be so tempered before application of the additive solution, since such liquefaction of natural pigments at the surface can be caused to take place during, and subsequent to, application of a flavor-enhancing solution containing a freezing-point depressant. Typically the fruit is warmed or tempered to a temperature in the neighborhood of 10° to 25°F preparatory to addition of the flavor-enhancing solution.

This may be achieved by depositing the substantially frozen fruit on a continuous belt traveling within a tunnel freezer, wherein air at a temperature of 25° to 30°F is passed through the fruit on the belt as it travels through the tunnel. The resident time of the fruit required to achieve such tempering is dependent upon such factors as the rate of travel and the bed depth, but usually is for a period of 10 to 15 minutes.

On the other hand, the fruit may be tempered under stationary conditions wherein the fruit is held in an atmosphere of air at a temperature of 25° to 30°F for a period in the neighborhood of 24 hours and air is passed over and around the fruit at the rate of 20 to 25 feet per minute. By so preparing the fruit pieces, they are in a condition whereby a minimum of surface treatment, both from the standpoints of time and amounts of flavor-enhancing solution, are required to cause surface liquefaction. Also, eventual clumping or clustering of the refrozen product following the application of the flavor-enhancing agent is avoided.

Refreezing is accomplished simply by relowering the temperature of the piece to below 5°F if the product is to be freeze dried directly thereafter, or to a temperature 0°F or below if the product is to be held for any significant period of time preparatory to freeze drying.

The flavor-enhancing solution containing saccharin in aqueous solution is preferably sprayed onto the tempered fruit piece, the solution being previously formulated with a sufficient concentration of flavorants to assure the desired degree of sweetness or other enhancement concomitant with natural flavor level of the fruit, while not calling for an undue amount of aqueous carrier such as can occasion excessive liquid water application to the fruit surface. Similarly, the duration of the application of the enhancing solution should be limited to that required to simply adequately cover the fruit surfaces with the flavor-enhancing solution, uniform and complete coverage being generally preferred. However, the flavoring enhancer may be applied to the fruit piece by a brief dip during which period the elapsed time of contact between the enhancing solution and the fruit piece is controlled to a period of less than ten seconds.

The flavor-enhancing solution is formulated to contain material such as an alcohol like propylene glycol or ethyl alcohol, or a low-molecular weight saccharide such as invert sugar. The term "alcohol" refers to an edible alcohol with one or more hydroxyl groups. Although it is intended to embrace the most typical such alcohol, namely glycerol or ethyl alcohol, it also applies to a variety of other water-miscible sugar alcohols. Preferably the alcohol is of a low-molecular weight so as to offer a substantial effect in decreasing the freezing point of the flavor-enhancing solution and the liquefiable materials on the surface of the frozen fruit piece treated therewith.

"Oligosaccharides," however, may also be so used as freezing-point depressants. Typical of this group are common commercially available reducing hexoses or the di-, tri- and other lower saccharides. The term "oligosaccharides" is intended to embrace not only sugars like total invert syrup, dextrose and maltose, but also the broad class of reducing saccharides. The oligosaccharides, however, should be only those which do not crystallize readily.

The freezing-point depressants should be compatible flavorwise with the flavor-enhancing additive as well as the natural flavor of the fruit itself, and should also serve to sufficiently lower the freezing point of the aqueous solution containing the enhancers for example, saccharin as well as the natural liquefiable constituents on the surface of the fruit itself, to assure the desired surface intermixing of the additive with the natural fruit pigments. Usually a reduction in the freezing point of the flavor-enhancing solution of 4° to 22°F (and most typically 6° to 12°F) will suffice to provide the desired effects, the freezing point reduction being dependent to some extent upon the nature of the fruit.

In the case of fruits like strawberries and peaches which have a water crystallization zone in the neighborhood of 28°F, that is, the solidification temperature of water contained in the fruit, it is best to use a flavor-enhancing solution which has its freezing point reduced to about 26°F or below. On the other hand, fruits such as bananas and like fruits having higher water-soluble solids therein will have a free-water crystallization zone much lower than 28°F, depending upon the relative concentration of such soluble solids in the fruit, but usually will call for a use of a flavor-enhancing solution whose freezing point is reduced well below 26°F.

Generally, the level and character of the freezing-point depressant should only be such as to lower the freezing point of the liquefiable constituents at the surface of the fruit and the flavor-enhancing additive in solution to assure that amount of intermixing of the enhancer with the surface liquid required, while avoiding excessive thawing of the fruit piece and consequent softening and liquefaction during subsequent handling in the course of refreezing. The preferred freezing-point depressant of use is invert sugar. Generally, the flavor-enhancing solution may contain anywhere from 7.5 to 35% of such an oligosaccharide. In the case of alcohols such as ethyl alcohol, on the other hand, a concentration of 2.5 to 20% in flavor-enhancing solution may be used; and in the case of propylene glycol specifically, another preferred additive, a concentration of 5 to 15% by weight of the flavor-enhancing solution can be used.

The concentration of such freezing-point depressants in solution will be dependent to some extent also upon the procedure used to apply the enhancer, as well as the character of the plant material being treated. But, typically, it will be found that the solution may be applied at a ratio of 1.5 to 4.5 pounds per 100 pounds of frozen fruit and plant material. A flavor-enhancing solution containing an artificial sweetener at such a concentration can be applied to any number of fruits which have been tempered to a product temperature ranging from 18° to 22°F.

Example 1: A five-gallon batch of sweetening solution was prepared by mixing together the following ingredients: sodium saccharin (suitable solution); 2.0 lb of propylene glycol; 36.85 lb of hot water (130°F).

Whole frozen fruit (for example, strawberries) at 0° to -5°F was tempered in a tunnel freezer for 12.5 to 15 minutes on a belt using an air temperature of 22° to 27°F, the fruit being metered onto the belt continuously at a bed depth such as assured that the fruit emerged from the belt after the elapsed period of 12.5 to 15 minutes with an internal pulp temperature of 20° to 22°F.

The tempered fruit (typically strawberries, but commonly any one of a variety of popular fruits such as peaches and pears) have the surface portions substantially solid. The sweetening or flavor-enhancing solution was sprayed onto the product, preferably after it had been directed into rotating slicers or halved. In the case of strawberries, the sliced fruit was passed into a three-foot diameter reel revolving at 12 rpm, nozzles being positioned within the reel so that at least the final 25% of the reel length was available to tumble the sweetened fruit and adequately coat all of the product with the sweetening solution sprayed thereon. The sweetening solution was sprayed on at the rate of 37.5 lb per 2,500 lb of frozen fruit.

By virtue of the reduced freezing point of the sweetening solution, the solution penetrates the fruit and intermixes with fruit pigments prior to refreezing, thereby assuring that a sufficient length of time is provided for the sweetener solution to harden on the surface and further assuring that increased sweetener penetration of the fruit surface is obtained. In this way, a high-level of addition of sweetening solution may be used without encountering a crystallizable residue of sweetener on the surface of the fruit at any such high level of addition.

Immediately after addition of the sweetening solution, the fruit was recooled to a temperature from 0° to -10°F by passing the fruit through a refreeze tunnel using an air temperature of -20° to -35°F and a residence time of 5 minutes. The refrozen fruit was then freeze dried in a conventional manner. The freeze dried fruit did not have a white residue of sweetener on the surface and had an appearance which was substantially unchanged as compared to a freeze dried fruit having no sweetener added. By virtue of the generally solid condition of the interior of the fruit, despite the addition of the sweetening solution to the surface and the minimal surface softening that occurs incident to surface liquefaction, the fruit is in a substantially discrete condition and the pieces are not deformed during their handling in the refreezing tunnel. As a consequence the fruit has an appearance, both in the dried and the reconstituted form, that is acceptable in that it is free of surface deformation and is not clustered. Although this specific example has been described by reference to a product, such as strawberries, which is destined to be freeze dried, the product may also be reduced to a stable moisture content by having at least a major part of the moisture content of the fruit sublimed, the remainder of the drying operation being carried out by vacuum air drying which may precede or follow sublimation and be carried out under conditions where the fruit is allowed to warm to temperatures above 0°F and the moisture is removed by evaporation rather than sublimation.

In all such applications, application of the sweetening or flavor-enhancing solution containing the freezing-point depressant promotes the adequate penetration of sweetening agent to the surface such that the dried product is substantially free of any visible crystallizing sweetening agent such as the saccharins specified.

Example 2: A five-gallon batch of sweetening or enhancing solution was prepared by mixing the ingredients listed as follows: sodium saccharin (suitable solution); 2.1 lb of 190 proof ethyl alcohol; 36.7 lb of hot water (130°F).

The sweetening solution was applied as outlined for the procedure set forth using propylene glycol in Example 1. The resulting freeze dried product had an excellent appearance, being quite similar to the unsweetened product in that it was free of any "frost" or other visible manifestation of a sweetening additive on the surface. Generally in following the procedures of Examples 1 and 2 it is preferred to use alcohol concentrations in the neighborhood of 5 to 10% of the sweetening solution, freedom from any grossly visible white crystalline residue being increased as the alcohol level is increased.

Rapid Sweetening Method

Rapid and uniform sweetening of freeze dried fruit pieces is claimed in a process by *N.A. Lemaire and R.D. Peterson; U.S. Patent 3,356,512; December 5, 1967; assigned to Kellogg Company.* A uniform level of sweetness, for example, equivalent to 40% by weight of sucrose, can be obtained to raise the level of the fruit to a uniform and desired degree of sweetness regardless of the climatic, geographical or seasonal conditions of growth and harvest of the fruit.

The integral pieces of freeze dried fruit do not have a "snowy" or glazed appearance as a result of surface deposition of sweetener, the sweetener moreover being impregnated within pores of the dried fruit so that it will not be readily washed off upon rehydration with, for example, cream or milk on consumption.

Sucrose has, of course, been used as a sweetening agent for fruits and upon long standing in a syrup whole or sliced fruit can become impregnated with the sucrose. However, use of sucrose as a sweetening agent has several technological disadvantages when used to sweeten fruits for freeze drying. Sucrose, when used prior to freeze drying, lowers the normal freezing point of the fruit, thus causing a loss in drying efficiency since lower vacuum chamber pressures are required for sublimation. Moreover, sucrose may yield an unnatural appearing fruit since surface glazing may occur when the sweetener is applied both prior to and after freeze drying. Artificial sweeteners will also yield a glazed fruit upon freeze drying if applied to the fresh or frozen fruit pieces being treated without long periods of storage, since rapid impregnation or penetration of integral pieces of fruit is negligible. Further attempts to bring freeze dried tart fruit up to a desirable level, such as to 40% sucrose by weight, require use of high amounts of water such that the freeze dried fruit will have its texture and rehydration characteristics irreversibly ruined. At best this would require refreezing and freeze drying a second time, an uneconomical procedure, which still would not bring the product back to its original freeze dried condition and the rehydration time of the final sweetened product would be substantially increased.

Artificial sweetening agents applied to freeze dried fruits do not result in glazing as does sucrose, since the artificial sweetener is able to penetrate the fruit due to the inherent porosity of freeze dried fruit. Sweetener, when applied to freeze dried fruit where penetration has occurred, does not dissipate but is carried into the fruit by rehydration medium, further resulting in a uniformly sweetened fruit product.

The limit of permissible rehydration of freeze dried fruit with aqueous sweetener has been found to be approximately 6%, thus generally prohibiting the use of aqueous solutions of sucrose when it is desired to sweeten the freeze dried fruit to a level of 40% by weight of sucrose equivalent of sweetness. However, with artificial sweetening agents, such as saccharin sodium, which have sweetening power many times greater than that of sucrose, they can be applied in adequate amount to bring the fruit up to the desired level of sweetening without incorporation of more than 6% by weight of moisture. This small amount of moisture can be substantially removed by heat to the original moisture content of the freeze dried fruit (that is, below about 2.0% by weight) without detriment to the rehydration properties of the freeze dried fruit and without necessity for re-freeze drying.

For example, 10 grams of artificial sweetener can be dissolved in 100 milliliters of water at 140°F. Peaches sweetened with this solution at a level of 40% sucrose equivalent require the addition of less than 3.0 grams of water per 100 grams of freeze dried peaches. This amount can be added in such a manner, or removed by heat such that virtually none remains behind in the peach, and the quality of the freeze dried product is not materially altered. Despite the fact that at 140°F the solubility of sucrose is considerably higher than that for the artificial sweeteners, its relatively low sweetness requires that for the equivalent amount of sweetness, 7.7 times as much water must be used as solvent. Thus, the more than 23% of water which would be added to the freeze dried product to achieve a 40% sweetness level would irreversibly ruin the texture of the freeze dried peach and consequently the rehydration characteristics.

In a typical example, fresh sliced peaches having a moisture content of about 88% by weight were spread out and disposed on shelves in a freeze drying chamber having a shelf temperature of about 250°F for about 2 to 3 hours, followed by gradual reduction to a temperature on the order of about 120°F for a total drying cycle of about 8 to 10 hours. The surface temperature of the product was not permitted to rise above 110°F and the pressure in the vacuum chamber was approximately 200 to 600 microns mercury. The ice that sublimed off was condensed in an adjacent chamber, wherein the condenser temperature was -30° to -40°F.

The porous freeze dried fruit is then sweetened by spraying a hot concentrated aqueous solution of artificial sweetener over the fruit in an amount to bring the fruit up to the desired sweetness level and which will not increase the moisture content beyond approximately 6% by weight, followed by heat drying to remove excess moisture and to bring it back to substantially the original freeze dried moisture content.

The freeze dried fruit pieces are first subjected to heating as by forcing warm air at approximately 10% relative humidity through a bed of the freeze dried pieces until the fruit temperature has reached approximately 130° to 150°F. At this time there is a marked and unexpected decrease in friability of the fruit product without increasing the moisture content.

In another procedure, the aqueous solution of sweetener is sprayed onto the freeze dried fruit at a temperature above about 212°F (such as, for example, 220°F) to result in a "flashing" on and into the fruit of the sweetener, with substantial evaporation of the moisture so that no further heating or drying may be necessary, although such further drying may be used if found necessary to reduce the moisture content to below about 2% by weight.

Referring to the drawing, Figure 8.1, **10** indicates pieces of raw freeze dried fruit which are spread out onto the endless conveyor belt **11** and carried through a warming chamber **12** where the dried fruit is preheated to a temperature of 130° to about 150°F to decrease the friability of the particles and facilitate the subsequent handling.

FIGURE 8.1: RAPID SWEETENING METHOD

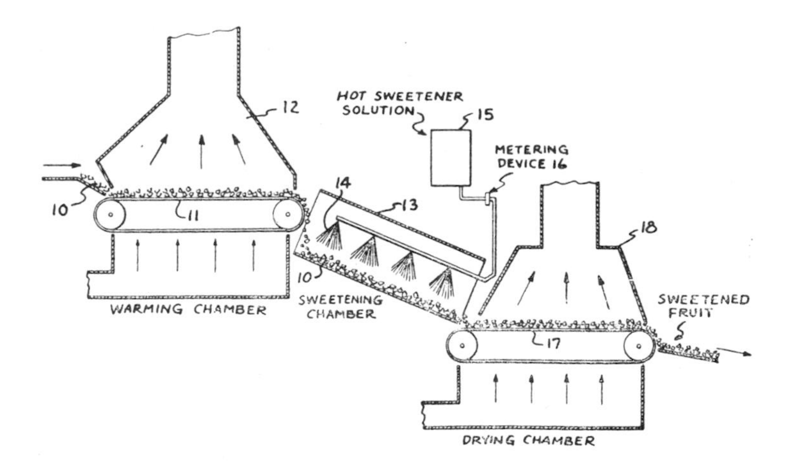

Source: U.S. Patent 3,356,512

The particles are then passed from the warming chamber **12** onto the endwise open inclined rotating sweetening chamber **13** where they are subjected to a spray **14** of aqueous sweetener solution delivered from the chamber **15** in controlled amount by means of the metering device **16**. The sweetener solution in chamber **15** is suitably at a temperature of about 220°F so that as the solution is sprayed onto the particles **10** passing through the sweetening chamber **13** they are simultaneously impregnated with the sweetening agent and the moisture content flashed off so that as they are delivered from the exit end of the sweetening chamber onto the conveyor belt **17** the particles are substantially dry and may not require any further drying. However, if further drying is required, this is accomplished in the drying chamber **18** through which conveyor belt **17** passes and the particles therein dried by means of hot air at a temperature of from about 125° to about 175°F and discharged at a moisture content of less than 2% and preferably less than 1.5%.

HONEY-MALT FLAVORING

A honey-malt flavor material is produced in the process of *W.A. Mitchell and H.D. Stahl; U.S. Patent 3,622,349; November 23, 1971; assigned to General Foods Corporation* by the heat treatment of low jelly grade pectin and saccharin.

The process generally involves drying an aqueous solution of low jelly grade pectin mixed with artificial sweetener. The artificial sweetener may include sodium and calcium salts uf saccharin. The product is a light tan colored powder with a bulk density of about 0.10 to 0.15 g/ml. The powder is stable and remains pourable indefinitely under usual room humidity conditions. The product has excellent cold and hot water solubility and exhibits reflecting surfaces similar to the crystal faces of brown sugar.

Although pan and other conventional methods of drying are suitable, atmospheric drum drying of the aqueous mixture of low jelly grade pectin and artificial sweetener is a convenient method for attaining the desired time-temperature relationship for producing the chemical changes of the low jelly grade pectin which are apparently similar to the caramelization of dextrose and which yield the unique honey-malt flavor. Dehydrating the mixture by drum drying forms light tan-colored film which is removed from the drum dryer rolls as a fine attractive appearing powder with a moisture content of about 3 to 4%.

The chemical reactions which produce the distinctive honey-malt flavor are not well understood. It has been established, however, that the presence of the artificial sweetener in combination with the low grade pectin is not necessary to yield the unique flavored product of this process. A solution of low jelly grade pectin alone, when dehydrated under specified conditions, yields a material with a desirable flavor and aroma but which, because of the absence of the artificial sweetener, has a less sweet malt flavor characteristic.

The chemical reactions brought on by the elevated temperature conditions are, evidently, a temperature rearrangement of the pectin composition, much like the caramelization of dextrose upon heating. Increasing the ratio of artificial sweetener to low jelly grade pectin in the mixture will, upon proper heat treatment, create a product with the unique honey-malt flavor but with a greater impact of sweetness of honey factor in the taste of the product. Conversely, heat treatment of low jelly grade pectin alone yields a product characterized by having a less sweet malt flavor. Pectins originating from any natural source (such as citrus fruits, beets, apples, etc.), when properly modified, are suitable for purposes of this process. It is important that the pectin be chemically altered to what is termed a low jelly grade material, that is, pectin having a minimal sugar solution gelling characteristic. Conventional methods for decreasing the jelly grade of the pectin, such as treatment with acid, alkali or enzyme (all of which are well-known and practiced in the respective art) yield a low jelly grade pectin material suitable for use in this process.

The upper limit of the jelly grade of the pectin to be used is considered to be about 5 to 10 grade. A more objective and critical measure of the degree and extent of polymerization of the desirable pectin material can be obtained by the ferricyanide reducing method of analysis. By this method, pectin having a grade of 100 gelling units has a ferricyanide number range of about one to five, whereas galacturonic acid (pectin hydrolyzed to its monomer) has a ferricyanide number of 198 to 200. The pectin material best suited for this process has a ferricyanide number of 80, with a preferred range being 60 to 100.

While low jelly grade pectin obtained by hydrolysis of natural pectin via an acidic or a basic treatment is suitable, it is preferred to use low jelly grade pectin which has been hydrolyzed by treating natural pectin with enzyme. It has been found that enzymatically degraded pectin, when combined with the artificial sweetener and properly heat treated and dehydrated, yields a product with a more intensive and better balance honey-malt flavor than low jelly grade pectins which have been produced from natural pectins via an acidic or basic hydrolysis treatment.

Any of the pectinases are satisfactory for the purpose of catalyzing the hydrolysis of natural pectin to provide the low jelly grade pectin material. The reason for the improvement in flavor of the finished low-calorie sweetening agent with pectin which has been hydrolyzed with pectinase versus acidic or basic hydrolyzed pectin as one of the essential ingredients is not clearly understood.

It is believed, however, that the pectinase enzyme, being a proteinaceous material and thus having amino groups, may enter into the reaction to a limited degree during the heating and dehydration of the process. This is brought about by the fact that the pectinase is present as residual material in relatively small amounts and is intimately mixed with the hydrolyzed pectin material in the solution to be drum dried to produce the product of the process.

The low jelly grade (5 to 10) pectin has excellent cold-water solubility. A 20% by weight solution, which is the upper limit for the preferred pectin concentration range of the aqueous solution to be heated and dehydrated, has a room temperature viscosity of approximately 10 centipoises.

The degree of sweetness of the finished product can be controlled by proper adjustment of the ratio of soluble pectin solids to soluble artificial sweetener solids in the mix used as feed to the drum dryer. The lower the ratio of pectin to artificial sweetener solids, the higher the sweetness or honey factor of the drum dried product. The acceptable range of ratios of pectin to artificial sweetener is from two parts pectin to one part artificial sweetener, to 20 parts of pectin to one part artificial sweetener, and the preferred range of pectin to artificial sweetener ratio is from 5:1 to 15:1.

The total solids content of the aqueous solution to be drum dried is not critical. It is preferred to use solutions ranging in concentration from about 5 to about 20%. Higher concentrations, having higher viscosities, are more difficult to transfer to the rolls of the drum dryer and, in general, are less amenable to the production of a satisfactory product. Solutions with solids content appreciably below 5% yield a satisfactory product but are necessarily too dilute and require the evaporation of excessive water in the drum drying step of the process.

Since the development of the unique honey-malt flavor is the result of a chemical reaction which is influenced by both time and temperature, careful adjustment of drying condition is essential. Proper control of drum dryer roll speed and roll surface temperature is necessary to obtain the proper heat treatment conditions for a particular solids concentration feed material.

In general, for a constant and particular temperature of the drum dryer rolls, the higher the solids concentration of the feed solution, the faster the drum speed (shorter residence time of the film of the surface of the drum) because of the smaller amount of water in the solution to be evaporated. This relationship holds true as long as the residence time on the heated rolls is sufficient to permit the development of the desired degree of intensity of honey-malt flavor.

It was determined that, at the preferred atmospheric drum dryer roll surface temperature range of 100° to 150°C developed by using a drum internal steam pressure of 50 to 70 psig, a film residence time on the rolls of 20 to 30 seconds is necessary for the proper development of satisfactory flavor notes in the dehydrated product. Higher drum roll temperatures are balanced with faster drum rotational speeds to obtain the desired honey-malt flavor product. However, it has been found that drum roll temperatures exceeding 170°C yield a product having a burnt or charred flavor characteristic, even though the film residual time is held to a minimum.

Example 1: The preparation of the low jelly grade pectin entailed mixing 360 grams of high ester, high jelly grade (approximately 250) pectin with 0.48 gram of pectinase, 300 milliliters of water and sufficient hydrochloric acid to adjust the mixture to 4.0 pH. The mixture was then held for 65 hours in an air oven at 55°C.

At the end of the holding period the viscosity of the 14.5% solids solution was 3.7 centipoises at 24°C, and the solution was then partially neutralized to a pH of 5.7 by adding approximately 15 milliliters of 5 N NaOH, after which the neutralized solution was diluted to 4.5% solids concentration. To a 250 milliliter aliquot of the above solution was added sufficient sodium saccharin to sweeten.

This aqueous mixture was then drum dried under atmospheric pressure conditions to a light tan colored product, using a drum drier in which the roll surfaces were heated to approximately 125°C by using an internal steam pressure of 60 psig within the rolls. The roll speeds were adjusted to 2 rpm to provide a film residence time of one-half minute. The film doctored from the rolls readily into a finely divided powder having a pleasant sweet honey-malt flavor. The product had a bulk density of 0.12 g/ml and was nonhygroscopic.

Example 2: A second 250 milliliter aliquot of the diluted solution of Example 1 without the addition of artificial sweetener was drum dried in a manner similar to that described in Example 1. This material also was easily removed from the drum dryer rolls as a finely divided powder having a pleasant malt flavor but lacked the sweetness of the product of Example 1.

Example 3: Hydrolyzed pectin was prepared as follows: 400 grams of a high ester, high jelly grade (approximately 200 jelly grade) pectin was mixed with 2,000 milliliters of water, and the acidity of the mixture adjusted to pH 4 with 0.5 N HCl, followed by the addition of 3.2 grams of Pectinol 59-L. The mixture was held for 66 hours at 50°C, after which it was filtered and the filtrate partially neutralized to a pH of 5.0 by adding a dilute solution of $NaHCO_3$. The solids content of the filtered, partially neutralized solution was 12.6%.

To 1,700 milliliters of the above solution was added sufficient sodium saccharin for desired sweetness. This aqueous mixture was then drum dried under atmospheric pressure conditions to a light tan product using a double drum dryer in which the rolls were heated with steam applied internally to the rolls at 70 psig. The drums were driven at a rotation speed of 2 rpm to provide a film residence time of one-half minute. The film doctored readily from the rolls into a fluffy (0.13 g/ml) finely divided powder having a sweet honey-malt flavor without imparting any bitter aftertaste.

Example 4: The product of Example 3 was added to a commercial nondairy dry whipped topping mixture at a level of 3.76 grams to 61.5 grams of topping mixture. The prepared topping made from this mixture was compared with a topping prepared from the nondairy topping mixture without the addition of the pectin sweetener product of Example 3, all other conditions being the same.

The topping prepared from the mixture containing the pectin sweetener had a more creamy and whipped-creamlike texture and color than that prepared from the control. It also had a superior flavor in that the pectin sweetener added some "buttery-malt" notes and seemed to mask the undesirable "caseinate" taste associated with the control.

Example 5: The product of Example 3 was added to a syrup (50% industrial sucrose, 50% corn syrup) to the extent of 2% by weight. The flavor enhanced syrup had a pleasant honey flavor with some maple notes and had an attractive amber-honey color.

SUGARLESS CHEWING GUM

Xylitol Formulation

Sugarless chewing gums are compounded with xylitol in the process of *J.E. Hammond and T.K. Streckfus; U.S. Patent 3,899,593; August 12, 1975; assigned to General Foods Corporation.*

Xylitol is a pentahydric alcohol which is prepared most commonly by the hydrolysis of xylan (a common constituent of wood, corncobs and oilseed hulls) to form xylose, followed by the reduction of xylose to xylitol by hydrogenation under pressure in the presence of a nickel catalyst. Xylitol appears as a crystalline compound which possesses a sweetness level of about 90% that of sucrose, and xylitol is metabolized in the body to glycogen by way of the pentose-phosphate pathway and is thus safely consumed by diabetics. Xylitol is readily soluble in water, possesses a relatively large negative heat of solution and is capable when present in sufficient quantities of producing a cooling effect in the mouth.

It is essential that the xylitol be present in the formulation at a level which will lend sufficient sweetness to the composition and which will also yield the desirable cooling effect. This will normally require the presence of xylitol in the chewing gum composition in the amount of at least 50% by weight. The gum bases include natural gums, synthetic resins, waxes, fillers and softeners. The commercially available chicle gum bases have proven quite satisfactory for use, as have the synthetic gum bases now also commercially available.

The glycerol component is used in order to soften the gum base to the extent that the base will be capable of binding the crystalline material. The final texture and chewing characteristics of the gum can be controlled by varying the amount of glycerol in the composition.

Example: A sugarless chewing gum consisting of 71% xylitol, 24% gum base, 4% glycerol and 1% flavor, is prepared by heating the gum base to 150°F in a mixer, adding the glycerol to the base and mixing for about five minutes, adding the xylitol to the mixer and continuing the mixing for another five minutes. The mixture is then cooled to 40°F and the flavor is added and mixed. The resulting composition is then formed into sheets, cut into sticks and wrapped.

Miraculin Coating

T.R. Johnson, Jr.; U.S. Patent 3,681,087; August 1, 1972; assigned to Meditron, Inc. found that miraculin incorporated in or coated on a chewing gum composition has the effect of preserving the refreshing taste of chewing gum for long periods. Furthermore, the addition of miraculin to chewing gum compositions substantially reduces, or eliminates, the amount of sugar or artificial sweeteners now used to render the gum pleasant-tasting for long periods. Apparently the miraculin in the gum composition not only enhances the taste of the gum components and/or flavoring, but itself effects a pleasant taste that is maintained for an hour or more.

The miraculin either can be mixed with the chewing gum base or can be coated around a core comprising the chewing gum base. It is best that the miraculin be incorporated in a coating rather than the core since less miraculin is required to obtain the desired taste improvement and prolongation. The coating can be formed by any convenient method, such as by dusting or by forming a glazed coating. Suitable gum bases include unmodified potianac gum, Gulta Katian, chicle, rubber latex base and jelutong base or mixtures of these bases modified with paraffin waxes.

The miraculin is used in amounts of about 0.1 to 50 milligrams per unit dosage form, such as a gum stick or gum ball. Additional miraculin can be used, but no substantial advantage is obtained. Generally, a mint or fruit flavoring is added to the gum base and sugar can be added if desired.

In another aspect of this process, the gum composition can include either a nontoxic alkaline material or acid material. Alkaline materials neutralize the mouth acids and increase the effectiveness of the miraculin in that degradation normally caused by mouth acids is reduced or eliminated. This reduces the amount of miraculin needed to obtain the desired results. Suitable nontoxic alkaline materials include magnesium carbonate, sodium bicarbonate, aluminum trisilicate, aluminum hydroxide complexes such as aluminum hydroxide-magnesium carbonate codried gels, aluminum hydroxide in amounts between 10 and 500 milligrams. The alkaline material is added to, or forms the coating, so that it neutralizes the mouth acids prior to substantial contact of miraculin with the mouth acids. Sugar can also be added to the alkaline material to improve its taste.

Acids incorporated in the gum composition enhance the taste in that they taste sweet after contact of the taste receptors with miraculin. Thus, when acids are used, the miraculin is incorporated in an outer layer separate from the acid so that the miraculin contacts the tongue prior to the acid. The miraculin contacting the tongue prior to the acid thereby reduces miraculin degradation while obtaining the desired taste-modifying effect.

Suitable nontoxic acids include carboxylic acids such as citric, malic, ascorbic, formic, acetic, tartaric or inorganic acids in lower concentrations or mixtures thereof. The acids are added in amounts to obtain an effective molarity of between 0.001 M and 0.1 M.

Example: A lemon flavored gum is prepared from the following formulation:

	Grams
Chicle	130.00
Paraffin wax	37.30
Tolu balsam	6.20
Peruvian balsam	3.10
Glucose	150.00
Citric acid	55.00
Corn syrup	50.00
Salt	5.00
Water	170.00
Yellow artificial color (Bates)	1.00
Artificial lemon flavor (IFF)	10.00

The chicle is soaked in water and mixed hot with the paraffin and balsam. The remaining ingredients are kneaded into the chicle until a homogeneous mixture is obtained. The gum then is molded into sticks. The sticks are dusted with stable miraculin powder so that each stick is coated with about 20 to 30 milligrams miraculin.

Dipeptide Sweetener

Chewing gum having a longer sweetness and flavor impact may be prepared from compositions comprising a chewable gum base, flavor, sweetener in an amount sufficient to impart normally accepted sweetness to the gum, and an amount of L-aspartyl-L-phenylalanine methyl ester (APM) effective to produce a longer-lasting sweetness and flavor in the gum. The chewing gum compositions may be either sugar-containing or sugarless.

Further, it has been found that a sugarless chewing gum having longer-lasting sweetness and flavor while retaining a substantial initial release of sweetness may be prepared by modifying a portion of the L-aspartyl-L-phenylalanine methyl ester (APM) present in the gum composition, the modification being directed at increasing the solubility rate of the APM. Chewing gums so prepared have been found to retain discernible levels of sweetness and flavor after as much as 30 minutes chewing time.

The modification of APM to impart initial acceptable sweetness to chewing gum and long-lasting character is described by *B.J. Bahoshy, R.E. Klose and H.A. Nordstrom; U.S. Patent 3,943,258; March 9, 1976; assigned to General Foods Corporation.*

The sweetener, or sweeteners, is added to the chewing gum composition in amounts sufficient to impart normally accepted sweetness to the chewing gum. The loss of flavor and sweetness impact in chewing gums so prepared, whether they be sugar-containing or sugarless, generally occurs after about four to five minutes of chewing. Increasing the level of sweetener employed has the disadvantage of providing excessive sweetness during the initial portion of the chewing period. In sugarless gums, such sweetener increases are further hampered by the undesirable bitter after-taste imparted to the gum by excessive amounts of saccharin.

It has been found that the addition of L-aspartyl-L-phenylalanine methyl ester (APM) to the chewing gum composition in effective amounts extends the time period over which both flavor and sweetness are discernible during chewing. The amount of APM necessary to achieve this flavor and sweetness extension may vary according to the gum base, sweeteners and flavors used. Generally, however, it has been found that the addition of APM at levels as low as about 0.1% by weight of the final gum composition is effective. The upper limit at which APM may be added is subject solely to considerations of the undesirability of imparting excessive initial sweetness to the gum product, since APM is itself a sweetening agent.

While this consideration depends greatly upon individual preference, it has been found generally that levels of greater than about 1.5% APM result in a chewing gum of excessively sweet taste throughout the initial portion of the chewing period. The use of APM within the above-mentioned ranges in either a sugar-containing or a sugarless gum results in a product whose flavor and sweetness is sustained at a discernible level for a period far in excess of that achieved in conventional chewing gums.

Chewing gums prepared from compositions of this process have been found to exhibit sustained flavor and sweetness release for anywhere from ten to thirty minutes of chewing time. Even with such significant increases, it is found that the chewing gums do not exhibit excessive sweetness during the initial chewing period. These results are found both in sugar-containing and sugarless gums.

When APM itself is used as the synthetic sweetener in chewing gum, it has been found that the amount needed to impart normally accepted sweetness to the chewing gum is about 0.2% to about 0.25% by weight of the final gum composition. Sugarless gum compositions employing APM as the synthetic sweetener display a significant flavor and sweetness extension over commercially available gums when the total amount of APM used is 0.3% by weight of the final composition or greater. The rapid dissolution of the sugar portion of a sugar-containing chewing gum effects a relatively strong, desirable initial perception of sweetness of flavor. This initial sweetness perception is generally unable to be matched by sugarless chewing gums due either to the insolubility of the synthetic sweetener used or a limitation on the level of synthetic sweetener used, for example, the bitter after-taste of saccharin. It is possible to still further heighten this initial sweetness perception by modifying a portion of the "sweetness-extending" amount of APM. The modification is directed at increasing the solubility of the APM and may be accomplished by treating the APM with organic or inorganic acids to convert it to its salts, co-grinding with acid, finely grinding the APM, or other such methods aimed at increasing solubility.

Example 1: Sugar-containing gum sticks were prepared from the following formulations:

Ingredient	Percent Composition A	Percent Composition B
Chewing gum base	23.993	23.993
Sugar, 6X	59.737	59.387
Corn syrup 46° Brix	14.245	14.245
Peppermint oil	1.276	1.276
Glycerol	0.749	0.749
APM	-	0.350

The gum base was softened for ten minutes at 150°F in a one-gallon Sigma mixer. The glycerol was added and mixed for ten minutes and the corn syrup was then added and mixing continued for ten minutes. For Sample A, the sugar was added and blended for ten minutes, while in Sample B the APM was blended with the sugar and added to the mixer as above.

The heating medium was turned off, the peppermint oil was added and mixing continued for two minutes. Mixing was then stopped, the gum was removed from the mixer, rolled into sheets, tempered and scored into sticks. On the basis of panel result, Sample A was found to have reached the point where flavor and sweetness are barely detectable after five minutes of chewing, while Sample B was found not to have reached the barely detectable level even after over ten minutes of chewing.

Example 2: Gum sticks were prepared according to Example 1 from the following ingredients:

Ingredient	Percent Composition A	Percent Composition B
Chewing gum base	23.993	23.993
Sugar	59.736	58.736
Corn syrup	14.246	14.246
Peppermint oil	1.276	1.276
Glycerol	0.749	0.749
APM	-	1.000

Results of the panel testing indicated that Sample B with 1% APM added had a flavor and sweetness extension of 30 minutes over Sample A and exhibited no undesirable after-tastes.

Example 3: Sugarless chewing gums were prepared according to Example 1 from the ingredients listed on the following page:

Ingredients	. . Percent Composition. . . A	B
Sorbitol	51.5	51.3
Gum base	34.9	34.9
Mannitol	8.1	8.1
Glycerol	2.8	2.8
Water	1.4	1.4
Flavor	1.1	1.1
Saccharin	0.2	-
APM	-	0.4

Sample B was found to have a better overall sweetness impact than Sample A and a flavor and sweetness extension over Sample A of greater than six minutes.

Example 4: Sugarless gum sticks are prepared as in Example 1 from the following formulation:

Ingredient	Percent Composition
Sorbitol	51.3
Gum base	34.9
Mannitol	8.1
Glycerol	2.8
Water	1.4
Flavor	1.1
Saccharin	-
APM	0.3
Modified APM	0.1

The modified APM is prepared by co-drying a solution of three parts citric acid and one part APM and then finely grinding the dried material.

Gum Acacia for Improved Texture

Sugarless confectionary compositions, and in particular chewing gum, have increased in importance over recent years primarily because of efforts by the dental profession to discourage use of sugar-containing confectionaries. Consumption of sugar-containing confectionaries has been discouraged primarily because of reported studies which allege, or conclude, that such confectionaries constitute a chief cause of dental caries.

Although sugarless chewing gums heretofore produced have met with acclaim by the dental profession, they have not met with great public acceptance because of poor taste and texture characteristics upon consumption. Such sugarless chewing gums have consisted primarily of gum base synthetic sweeteners and suitable flavoring ingredients.

The synthetic sweeteners employed in sugarless chewing gums have been used to replace corn syrup and sugar customarily used in conventional sugar-containing chewing gums. Typically, the synthetic sweeteners include sorbitol or mannitol alone, although various combinations of these ingredients have been tried with very little commercial acceptability. Although these synthetic sweeteners have the desirable properties of being water-soluble while having an acceptable taste or sweetness, they usually fail to impart a desirable texture to chewing gum compositions which have public acceptance.

In attempting to provide a suitable sugarless chewing gum for acceptance by consumers, it was found that corn syrup which constitutes an ingredient of sugar-containing chewing gums has the unique property of binding the sugar to the gum base; and, thus, there results a sugar-containing chewing gum having a desirable texture. From this discovery, it became apparent that a suitable ingredient was necessary to substitute for corn syrup and all its properties without contributing sugar properties for use in preparing a sugarless chewing gum having the desirable characteristics of acceptable taste and texture requirements.

In attempting to develop a suitable corn syrup substitute for sugarless chewing gums, various agents such as starch derivatives and dextrins were examined as possible substitutes for corn syrup of sugar-containing chewing gums, but these agents were all found undesirable since they readily break down in the digestive tract of the human body and form undesirable sugars. Gelatin and various combinations of gelatin were also examined and were found to yield poor chewing gum products, both as to taste and as to texture characteristics. Solutions of gum tragacanth, locust bean gum and agar were also examined and found to be unsatisfactory for inclusion as a corn syrup substitute for use in sugarless chewing gum compositions.

The addition of gum acacia solution to the chewing gum base was found by *A.G. Bilotti; U.S. Patent 3,352,689; November 14, 1967; assigned to Warner-Lambert Pharmaceutical Company* to improve the taste and texture qualities of sugarless chewing gum. Gum acacia solution or aqueous suspension, as the system may appear, may be prepared by slowly dissolving gum acacia powder in water heated to about boiling at standard conditions. Powdered gum acacia is desirably used since it may be expeditiously dissolved in water. Other forms, such as granular gum acacia, may also be used as desired. The following formulations generally represent the combination of ingredients useful for preparing the sugarless chewing gums.

Ingredients	Parts by Weight	
	Broad Range	Preferred Range
Gum base	20 to 50	20 to 30
Gum acacia solution	3 to 15	5 to 10
Sorbitol	20 to 50	30 to 40
Gum acacia powder	0 to 10	2 to 4
Mannitol	20 to 50	20 to 30
Flavor and artificial sweeteners	0.1 to 5	0.5 to 2.5

It is found desirable to add the gum base, gum acacia solution and artificial sweeteners to a preheated kettle wherein the ingredients are blended by mixing blades. Desirably, the temperature of the mixture is maintained below 70°C, although higher temperatures may be tolerated.

After the ingredients are thoroughly blended, sorbitol is added and, after an additional period of mixing the batch, the gum acacia powder, when used, along with mannitol is added. The batch is blended to a substantially uniform mixture and flavoring is added as desired. The chewing gum is maintained at a temperature of 35° to 50°C during further processing, such as rolling and kneading, to the finished form which is then wrapped and prepared for shipment. A specifically preferred composition is one containing 28 parts by weight of gum base, 6 parts by weight of gum acacia solution (45 to 50%), 36 parts by weight of sorbitol, 27 parts by weight of mannitol, 3 parts by weight of gum acacia powder and 1 part by weight flavoring oils and artificial sweeteners. The finally prepared composition is found to have a moisture content of 3 to 5% by weight and typically 3.5 to 4.5% by weight dry basis.

The gum base which is used generally includes natural gums, synthetic resins, waxes, fillers, and softeners. Gum acacia powder is one having a particle size such as less than 100 mesh (U.S. Sieve). Larger particles of gum acacia powder may be used, but the product is found to have a less desirable granular or sand-like texture.

Flavoring oils, which may be used to impart suitable flavor to the gum base, include all natural, essential and synthetic flavoring oils acceptable by the Food and Drug Administration. When preparing the chewing gum portion of the composition, the gum base containing various ingredients, such as natural gums, synthetic resins, waxes and fillers is added to a mixing kettle having suitable agitators for blending additives. Thereafter, the additional ingredients are added to the gum base under constant agitation conditions.

Example: A gum acacia solution is prepared in a steam-jacketed kettle, preheated to 190° to 200°F. To 52 parts by weight of heated water is slowly added 48 parts by weight of gum acacia powder under constant agitation while cooking at slightly less than a temperature of 200°F and until a homogeneous solution results. Upon completion, the homogeneous solution is strained through a 30 mesh screen to remove any foreign matter which may remain undissolved therein.

Thirty-two parts by weight of chewing gum base is heated in a mixing kettle to a temperature of 150°F. Thereafter, 8 parts by weight of the gum acacia solution previously prepared is blended with the chewing gum base along with 3 parts by weight of gum acacia powder to fortify the gum acacia solution, 28 parts by weight of mannitol and 30 parts by weight of sorbitol. About 1 part by weight of equivalent amounts of flavoring and saccharin is added and the mixture is blended to form a homogeneous chewing gum composition. The gum is removed from the heated mixing kettle and rolled into slabs of a convenient size. The slabs are thereafter cut into sticks, which are aged for a brief period such as one day and wrapped for shipment. Upon consumption, the sugarless chewing gum so formed is found to have acceptable taste and texture.

Although the process chewing gum is desirably formed into sticks, any other suitable form may also be used. In addition to the ingredients used in formation of a chewing gum base, additional active ingredients, such as phosphates, chlorophyllins, vitamins, enzymes, antacids, fluoride, etc., in solid or liquid form may also be included in the composition. The amount of these ingredients used in the chewing gum composition is from 1 to 15% by weight, based on the weight of the chewing gum product. The solid active ingredients are generally used in the form of solid particles of 100 mesh (U.S. Sieve) or less and preferably of a mesh size of about 200.

Other confection items to which the process may find application include bubble gum and candy-coated gum. The use of any of the approved FD&C dyes or lakes may also be employed in the sugarless chewing gum to provide suitable coloring.

Releasable Phosphate

One approach to the problem of inhibiting dental caries has been to incorporate various chemically active ingredients into vehicles which will come into contact with the teeth, such as dental preparations, lozenges or chewing gums, in order to counteract the effects of any sugar products taken into the oral cavity. However, for sugarless gum formulations, no methods exist describing the incorporation of an additive such as dicalcium phosphate dihydrate into the gum formulation so as to yield substantial release of the additive into the oral cavity upon chewing of the gum, while still maintaining the flavor consistency, and storage stability necessary for consumer acceptance of the gum product.

U.S. Patent 3,352,689 gives the formulation of a sugarless gum prepared from gum base, gum acacia in water, gum acacia powder, sorbitol, mannitol, sweeteners and flavoring agents which may contain additional active ingredients such as phosphates. However, no statement is made concerning the form in which these active ingredients must be or the manner for incorporating these active ingredients into the sugarless gum formulation so as to insure the release of effective amounts of the active ingredients into the oral cavity.

In another process, *A.G. Bilotti; U.S. Patent 3,655,866; April 11, 1972; assigned to Warner-Lambert Company* describes a composition containing freely releasable dicalcium phosphate dihydrate in sufficient quantities to insure that therapeutically beneficial amounts of this additive will be brought into contact with the dental enamel upon chewing of the gum.

Dicalcium phosphate dihydrate can be incorporated into a sugarless gum formulation containing gum base, sorbitol solution, sorbitol, mannitol, artificial sweeteners and flavoring agents so that it is freely released upon chewing and brought into contact with the teeth by coating or agglomerating dicalcium phosphate dihydrate powder with a water-soluble polyol or combination of polyols in granular form, prior to incorporation into the sugarless gum formulation.

Suitable water-soluble polyols include mannitol, sorbitol, xylitol and arabitol or combinations. Among these, mannitol is preferred. Water-soluble polyol or a combination of water-soluble polyols is mixed with one part by weight of dicalcium phosphate dihydrate powder and water is added in small portions until the mixture has a bread or dough-like consistency. The polyol and water present form a solution which surrounds or coats the smaller particles of the insoluble dicalcium phosphate dihydrate powder; this material is spread on trays and dried, forming an agglomerated product.

Techniques used in forming the agglomerated product are commonly called "granulation" in tableting operations. Subsequent milling or pulverizing of the agglomerated product produces particles of a size such that no more than 25% by weight of the particles remains on a 100 mesh screen. It is preferred that the bulk of the particles be between 100 and 270 mesh in size.

The polyol agglomerated product may also be prepared by a spray drying technique. In this procedure, the water-soluble polyol is dissolved in an excess of water. One part of dicalcium phosphate dihydrate is suspended in the solution and the resultant suspension is sprayed into a large cone and dehydrated by a current of warmed, dry air circulating in the cone. The dry powder consisting of dicalcium phosphate dihydrate agglomerated with the polyol or combination of polyols is removed from the bottom of the cone. A variety of commercial spray dryers may be used for this process.

A particularly preferred product contains 28% by weight gum base, 9% by weight Sorbo 70% solution, 31% by weight sorbitol, 28% by weight mannitol agglomerated dicalcium phosphate dihydrate containing 2.7 parts by weight of mannitol per each part by weight of dicalcium phosphate dihydrate, a further amount of 3% by weight mannitol, 1% by weight artificial sweetener and flavoring agent, based on the weight of the entire formulation.

In formulating the sugarless gum containing the freely releaseable dicalcium phosphate dihydrate, the following sequence of addition and blending of ingredients is used to insure optimum quality in the final product. The polyol-agglomerated dicalcium phosphate dihydrate is prepared separately, according to one of the previously described methods.

The gum base, sorbitol solution and artificial sweetening agent are added to a gum mixing kettle. These ingredients are mixed until a homogeneous mass is obtained. The sorbitol (crystalline) is added to the kettle and all ingredients are further mixed. The polyol-agglomerated dicalcium phosphate dihydrate is added and all ingredients are again homogeneously blended. At this point, the mannitol (granular) and flavoring agent are added and all ingredients are mixed until a uniform mixture results. The total mixing time for all mixes is 15 to 17 minutes. After the mixing is completed the bulk formulation is ready for further processing into the desired gum slabs.

Example 1: The preparation of mannitol-agglomerated dicalcium phosphate dihydrate is as follows: 73 grams of mannitol granules are dissolved in 500 milliliters of water and 27 grams of dicalcium phosphate dihydrate powder is dispersed in the resultant solution. The suspension is sprayed into the cone of a small spray dryer and dehydrated by a current of warmed, dry air circulating in the cone. The dry powder consisting of dicalcium phosphate dihydrate and mannitol is removed from the bottom of the cone. The agglomerated product is screened to remove any large flakes. The finished product is collected and checked for particle size. No more than 25% of the material should remain on a 100 mesh screen.

Example 2: The preparation of the sugarless gum containing mannitol-agglomerated dicalcium phosphate dihydrate is as follows: 28 grams of gum base, 9 grams Sorbo 70% solution and 0.26 gram of a 37% sodium saccharin solution are added to a gum mixing kettle. The agitators are started and the ingredients are mixed in reverse for two minutes and then mixed with a forward motion for two minutes. Thirty-one grams of sorbitol are added and all ingredients are mixed with a forward motion for two minutes. Twenty-eight grams of the mannitol-agglomerated dicalcium phosphate (prepared according to the procedure in Example 1 above) are added and all ingredients are forward-mixed for another three minutes.

The final ingredients, 3 grams of mannitol and 1.0 gram of natural and/or synthetic flavor are added and all ingredients are forward-mixed for approximately three and one-half minutes, then reverse-mixed for approximately one-half minute. The mixing is stopped and the sides of the kettle are scraped down. The agitators are again started and the mixture is again forward-mixed for approximately three minutes. The gum product is then unloaded from the kettle, rolled, scored, conditioned and wrapped.

To illustrate the advantage of the improved product, the following tests were run: A sugarless gum product containing mannitol-agglomerated dicalcium phosphate dihydrate was made, following the procedure of Examples 1 and 2. A second sugarless gum product was made following the procedure described in Example 2 except that an equivalent amount of unagglomerated dicalcium phosphate dihydrate and mannitol were added in place of the mannitol-agglomerated ingredient.

Samples of each of the above chewing gum products were analyzed to determine the amount of dicalcium phosphate dihydrate actually present in a representative sample. To determine release, each product was chewed for 30 minutes; each sample was then analyzed to determine the amount of dicalcium phosphate dihydrate remaining in the gum bolus.

Percentages of dicalcium phosphate dihydrate released into the oral cavity by chewing were calculated from these figures. The chewing was performed by a panel of four subjects. In every case, the same subject chewed both types of gum, so that a true comparison of the releases from both types of gum could be obtained. The following results were obtained, in percent by weight.

Average Dicalcium Phosphate Dihydrate Released

	Subjects			
DCP	1	2	3	4
Unagglomerated	22.68	37.01	31.54	24.30
Mannitol-agglomerated	47.02	56.27	49.34	41.09

These results indicate that the release of mannitol-agglomerated dicalcium phosphate dihydrate from the sugarless gum product of this process is from 1.52 to 2.1 times greater that the release of the unagglomerated dicalcium phosphate dihydrate. Dicalcium phosphate dihydrate is known to inhibit the formation of caries, particularly when brought into contact with the teeth using a vehicle such as a sugar-containing chewing gum. However, the use of a sugarless gum vehicle for the same purpose offers the advantage of not supplying sugar to the oral cavity at the same time the caries-inhibiting agent is released.

By use of this particular gum formulation, release of incorporated dicalcium phosphate dihydrate is improved to the extent that effective quantities of dicalcium phosphate dihydrate are released into the oral cavity. Thus, the product has a three-fold advantage: since it is a sugarless gum formulation, there is no danger of subjecting teeth to a sugar environment; additionally, it is a pleasurable substitute for the more deleterious forms of confectionaries; and still further, an effective quantity of dicalcium phosphate dihydrate is released into the oral cavity to combat the untoward effects of other sugar-containing foods normally part of the diet.

SUGARLESS CANDY

In recent years the confectionary industry has found it desirable for a variety of reasons to produce a confection called "sugarless hard candy" which resembles ordinary hard candy but is usually prepared from sorbitol, mannitol and other materials, including coloring and flavor, in the place of the sucrose, corn sugar, coloring and flavor normally used in the manufacture of hard candy. It is well-known, however, that sugarless confections of the kind are extremely difficult to prepare owing to the disadvantages presented by the inherent tendency of the sugarless piece to remain soft or tacky instead of solidifying into a hard piece as desired.

Hard confections made from sorbitol solution, crystalline sorbitol and crystalline mannitol usually include a gum, such as gum arabic, acacia or tragacanth, to promote the "setting-up" or solidification of the confection. Notwithstanding its inclusion for this purpose, however, the presence of the gum necessitates relatively low cooking temperatures and a relatively high moisture content in the solution from which the confection is prepared in order to keep the gum suspended.

Because of these factors, the crystallization time for the confection is excessively long, frequently necessitating a delay of as much as several days in the processing time. Moreover, when the confection does crystallize, the clarity of the piece is destroyed owing to the presence of the colloidally dispersed gum and to the surface crystallization of the sorbitol and mannitol. The desirable optimum, therefore, is a sugarless confection which "sets-up" in a relatively short time into a hard solid piece and one which is also substantially clear or transparent.

A procedure is used whereby an anhydrous melt of ingredients containing sorbitol and mannitol is seeded in the process of *J.W. Du Ross; U.S. Patent 3,438,787; April 15, 1969; assigned to Atlas Chemical Industries, Incorporated.* The sorbitol and mannitol may be initially in the crystalline forms or in aqueous solution where practicable. Sorbitol and mannitol may be contained in the confectionary composition in a ratio of one to the other within the approximate ranges indicated below in the table:

	Parts by Weight
Sorbitol	75 to 100
Mannitol	0 to 25

The composition produced by the method containing crystalline sorbitol, or an intimate blend of crystalline sorbitol and mannitol, in a ratio within the approximate limits set forth in the table above, to which suitable flavors and colors may be added as desired. The composition, however, is substantially anhydrous in that the moisture content ordinarily may not exceed a maximum of 0.5% by weight.

Crystalline sorbitol or a combination of crystalline sorbitol and crystalline mannitol are dissolved in an amount of water not more than sufficient to effect solvation of the crystals. Ordinarily the ratio of solids to liquid will be approximately 2.33 parts by weight of solids to 1 part by weight of water. The amount of water required for complete solution of a blend of crystalline sorbitol and mannitol will vary, however, upon the weight ratio of sorbitol to mannitol selected. The solution is then rapidly heated to the boiling temperature of the solution and cooking is continued with continual reduction of the moisture content to the temperature at which the moisture content of the solution does not exceed 0.5% by weight. Usually the cooking temperatures are maintained between a minimum of 385°F and a maximum of 390°F. Vacuum cooking may be used to reduce charring effects. When the moisture content of the solution has been reduced to 0.5% by weight, the solution is cooled with moderate agitation to a temperature suitable for seeding the solution with crystals to effect the crystallization; the desired temperature, accordingly, lies within the range of 150° to 200°F. Since effective seeding of the solution with appropriate crystals is critical to the crystallization process, that is, to the solidification of the sorbitol or sorbitol/mannitol blend, seeding of the solution at temperatures substantially higher than 200°F will be less effective because the higher temperatures will cause at least some of the seeding crystals to melt, while seeding at temperatures substantially lower than 150°F will be less effective because the increased viscosity of the solution at such low temperatures will tend to prevent thorough dispersion of the seeding crystals throughout the relatively viscous mass.

When the temperature of the solution has been reduced to an appropriate seeding temperature, the solution is seeded with a finely ground sorbitol/mannitol glass, or other appropriate seed, with constant agitation to assure that the seeding crystals are thoroughly dispersed throughout the solution. Ordinarily about 0.50% by weight of the composition to 5% by weight may be added in the form of seeding crystals. If seeding crystals in an amount of less than 0.50% by weight of the composition are added, the crystallization of the solution into a glass-like solid is likely to be disadvantageously slow; if more than 5% by weight of seeding crystals are added, the resulting viscous mixture is likely to be difficult to handle.

In those cases where subsequent handling of the crystallizing solution is not important, even greater amounts of seeding crystals may be used. The speed with which solidification of the melt is accomplished depends upon the quantity of seeding material used, the degree of agitation, the particular blend of sorbitol and mannitol which may be used and the character of the seeding crystals themselves.

When the seeding material has been added, the solution is then placed in an environment maintained at a constant temperature within the range of 80° to 90°F, preferably from 83° to 87°F, and at a relative humidity not more than 50%, preferably between 30 and 40%, until the solution crystallizes. The time required for crystallization of the solution may vary from as little as a few minutes to as much as several hours, depending upon the conditions set forth above.

Preferably the seeding crystals used to effect solidification of the sorbitol solution or sorbitol/mannitol solution are of the same character as that of the solution with which they are used, that is, a sorbitol solution is seeded with crystals of finely ground sorbitol and a solution which is a blend of sorbitol and mannitol is seeded with finely ground crystals the composition of which contains sorbitol and mannitol in approximately the same proportion as that of the solution being seeded.

This process is not limited, however, to the use of seeding crystals which are substantially identical in their composition to that of the solution being seeded; on the contrary, other seeding crystals which contain no sorbitol or mannitol and are compositionally foreign in their character to that of the solution to be solidified may be used. Examples of other types of seeding crystals which may be used are finely ground crystals of dextrose, sucrose and other crystalline carbohydrates. The crystals should be sufficiently finely ground to pass through a

140 mesh screen. Usually, after the seeding operation but before solidification of the melt into a hard piece has taken place, it will be desirable to pour the seeded melt into a suitable mold to shape the final piece. Decantation, of course, must be performed at a temperature at or above the flow point of the viscous mass. In practice, it has been found that the melt is desirably cooled to 150°F prior to the decantation thereof, but ordinarily should not be cooled below 140°F for this purpose.

Example 1: Eighty grams of sorbitol and 20 grams of mannitol were dry mixed and added to 50 grams of water in a glass beaker. The solution was boiled for 20 minutes until the temperature thereof reached 390°F. The solution was then cooled by stirring to 197°F at which point 1.3 grams of seeding crystals made from a finely ground sorbitol/mannitol glass having the proportion of 95 parts by weight of sorbitol to 5 parts by weight of mannitol were added and dispersed throughout the solution by agitation. The solution was poured into molds and then placed in a cabinet having controlled temperature and humidity and there maintained in an atmosphere in which the ambient temperature was 85°F and the relative humidity 40% for a period of one hour. After one hour, the piece was totally solidified.

Frequently, it is desirable to produce a confectionary composition which is substantially clear, that is, unclouded, owing to the relative attractiveness of confection of such character. A hard sugarless confectionary composition which is substantially clear may be prepared by combining sorbitol and mannitol in the composition in a ratio of one to the other within the approximate ranges indicated below in the table:

	Parts by Weight
Sorbitol	89.0 to 92.5
Mannitol	7.5 to 11.0

Example 2: 92.15 grams of sorbitol and 7.85 grams of mannitol were dry mixed and added to 50 grams of water in a glass beaker. The solution was boiled for 20 minutes until the temperature thereof reached 390°F. The solution was then cooled by stirring to 197°F at which point 1.3 grams of seeding crystals made from a finely ground sorbitol-mannitol glass having the proportion of 95 parts by weight of sorbitol to 5 parts by weight of mannitol were added thereto and dispersed throughout the solution by agitation. The solution was poured into molds and then placed in a cabinet having controlled temperature and humidity and there maintained in an atmosphere in which the ambient temperature was 85°F and relative humidity 40% for a period of one hour. After one hour, the piece was totally solidified and very clear.

XYLITOL DENTIFRICE FORMULATION

In the process of *J.B. Barth; U.S. Patent 3,932,604; January 13, 1976; assigned to Colgate-Palmolive Company* xylitol functions as a humectant-sweetener in dentifrice products. The essential ingredients of the improved dentifrice are an abrasive, water, detergent and xylitol. If desired, the dentifrice may also contain additional ingredients, such as gums, fluorine-containing compounds, flavors, etc. Suitable abrasives include, for example, dicalcium phosphate, tricalcium phosphate, insoluble sodium metaphosphate, aluminum hydroxide, magnesium carbonate, calcium carbonate, calcium pyrophosphate, calcium sulfate, silica, sodium aluminum silicate, polymethacrylate, bentonite, etc., or mixtures of these materials. The detergent may be an organic anionic, nonionic, ampholytic or cationic surface-active agent, preferably one which imparts detersive and foaming properties.

Suitable detergents are water-soluble salts of higher fatty acid monosulfates, such as the sodium salts of the mono-sulfated monoglyceride of hydrogenated coconut oil fatty acids, higher alkyl sulfates, such as sodium dodecyl benzene sulfonate, higher alkyl sulfoacetates, higher fatty acid esters of 1,2-dihydroxy propane sulfonates, and the substantially saturated higher aliphatic acyl amides of lower aliphatic amine carboxylic acid compounds. Examples of the last-mentioned amides are N-lauroyl sarcosine, and the sodium, potassium and ethanolamine salts of N-lauroyl, N-myristoyl or N-palmitoyl sarcosinates, all substantially free of soap or similar higher fatty acid material.

Suitable gums or gelling agents include the natural and synthetic gums and gum-like materials, such as Irish moss, gum tragacanth, sodium carboxymethylcellulose (CMC), polyvinylpyrrolidone, starch and inorganic thickeners such as "Cab-O-Sil" (fumed silicon dioxide), "Laponite" (hydrous magnesium silicate clay), Syloid 244 (aerogel silica), and other organic and inorganic materials.

The fluorine-containing compound may be sodium fluoride, stannous fluoride, potassium fluoride, potassium stannous fluoride (SnF_2, KF), sodium hexafluorostannate, stannous chlorofluoride, sodium fluorozirconate, and sodium monofluorophosphate.

The xylitol content of the dentifrice is from about 10% to about 25%. To avoid a bitter taste (or aftertaste) the amount of water present should be no more than that required to give at least a 10 or 15% solution of xylitol. The amount of standard humectant used will depend on the xylitol content. If a dentifrice contains at least about 15 to 20% xylitol it is not necessary to use an additional humectant because orifice plugging will not occur when the toothpaste tube is left uncapped overnight. Preferably about 0.2% of a standard sweetening agent will be used. The exact amount of this ingredient will depend mainly on the taste desired. The amounts of the other ingredients may be varied as desired. The amounts of ingredients in the following dentifrice formulation are by weight.

	Percent
Xylitol	5.00
Carboxymethylcellulose	1.10
Glycerine	9.99
Sorbitol	11.90
Sodium benzoate	0.50
Water	24.00
Na_2PO_3F	0.76
TiO_2	0.40
Insoluble sodium metaphosphate	36.85
Hydrated alumina	5.00
Anhydrous dicalcium phosphate	1.00
Sodium lauryl sarcosinate	2.00
Flavor	1.50

When the teeth are brushed with this dental cream a pleasant taste develops and remains in the oral cavity during brushing and for a short time thereafter.

DESSERTS AND BAKED GOODS

FROZEN DESSERTS

Polyose as Freezing Point Depressant

In conventional frozen desserts sugar, besides contributing sweetness, also helps in lowering the freezing point of the product. Sugar affects the freezing point not only by contributing to total product solids, particularly in low-fat-content frozen desserts, but also by affecting product texture and freezing and melting characteristics. Fat in frozen desserts not only increases viscosity of the mix by contributing to total product solids but also gives good body and texture characteristics by influencing, for example, product hardness.

In general, insofar as physical effects are concerned, the lower the fat content, the greater the effect of sugar upon the product. The relationship between sugar content and fat content in frozen desserts is illustrated in the following table, which shows the average composition of, respectively, ice cream, ice milks, sherbet, and water ice, in terms of sugar, fat, nonfat milk solids, and total solids. The table demonstrates that when fat content is diminished, as usually occurs in the preparation of dietetic frozen desserts, the sugar content becomes an increased percentage of total solids.

	. . . Ice Cream . . .			. . . .Ice Milk. . . .			. . . Sherbets. . .			. . . Water Ices . . .		
Composition	**Per-cent**	**Cal/ Pt***	**Percent Cal**	**Per-cent**	**Cal/ Pt***	**Percent Cal**	**Per-cent**	**Cal/ Pt****	**Percent Cal**	**Per-cent**	**Cal/ Pt*****	**Percent Cal**
Fat	12	308	53	4	102	25	2	66	12	0	0	0
MSNF†	10	103	18	12	123	30	3	39	7	0	0	0
Sugar	15	171	29	16	182	45	32	464	81	35	570	100
Total solids	37	582	100	32	407	100	37	569	100	35	570	100

*At 80% overrun.
**At 40% overrun.
***At 20% overrun.
†Milk-solids, nonfat.

Hence, the lower the fat content, the greater the effect of sugar upon the physical properties of the products usually containing it. To prepare a dietetic low-calorie or low-carbohydrate frozen dessert, then, a nonnutritive substance or substances must be used in place of sugar which closely resemble sugar in capacity to affect product freezing and melting characteristics.

I.D. Bliudzius, N.P. Rockwell and P. Jucaitis; U.S. Patent 2,876,104; March 3, 1959; assigned to E.I. du Pont de Nemours and Company use polyose as a nonnutritive freezing point depressant having a volume, consistency and appearance comparable to food compositions containing usual nutritive freezing point depressants, such as sugar. These nonnutritive freezing point depressants are used like sugar but usually in smaller quantities.

To this nonnutritive frozen food one may add a noncaloric sweetener to obtain a sweetness level which one is accustomed to obtain with sugar. Thus, low calorie and carbohydrate frozen desserts can be made which are easily and pleasantly usable as conventional sugar-containing foodstuffs. The frozen desserts which can be prepared using a nonnutritive polyose as a freezing point depressant include ice cream, ice milk, sherbet and water ice.

In such foodstuffs a low fat content is usually used and such fat as remains can be furnished by conventional sources, such as creams (including sweet, frozen or plastic), unsalted butter or butter oil. The nonfat milk solids, or serum solids, can be furnished by using whole milk, skim milk, dried whole or skim milk or concentrated whole or skim milk. Contributing to the nonfat milk solids present, of course, are those serum solids in any cream used.

The polyose is used in place of the bulk sugar normally added to a mix because the resulting frozen product has substantially all of the palatability qualities of the conventional nutritive counterpart. Thus, instead of cane, beet, corn, invert, or perhaps malt sugar, a polyose is added in an amount sufficient to give the product the desired physical characteristics. Occasionally other substances as described more fully below are added to enhance palatability characteristics.

Due to the wide range of frozen-dessert-type products that can be produced and the various combinations of ingredients used therein, the exact amount of polyose that can be used in these products varies considerably. The amount used is dependent upon the product desired and other constituents present in the frozen dessert formula involved.

Generally, however, the amount of polyose that can be used in a frozen-dessert-type composition varies from 5 to 30% by weight. The result is a dietetic foodstuff which has a low carbohydrate or low calorie content but which has normally high nutritive value as regards protein, mineral and vitamin content. The product is particularly useful for persons who must watch their caloric and carbohydrate intake.

In addition to polyose, varying amounts of other ingredients can be used in the frozen food compositions, many of them in ways well understood. For example, to freeze ices and ice creams successfully at home refrigerator temperature conditions it is sometimes advisable to add a thickener such as one of the following: gelatin, flour, cornstarch, eggs, gum acacia, methylcellulose, and other natural and synthetic gums. Some of these substances are nutritive and too much should not be used. Other substances can also be added to help prevent an icy consistency and promote a palatable blend, such as corn syrup, honey, inorganic salts of various kinds, hydrolyzed protein, and glycerin. While these substances are also valuable as freezing point depressants some are nutritive and too much of them should not be used. Stabilizers and emulsifiers can also be used, such as gelatin, pectin, carboxymethylcellulose, sodium carboxymethylcellulose, sodium alginate, gum tragacanth, karaya gum, starch, locust bean gum, and Irish moss.

In the nonlimitative examples that follow, the various ingredients used in preparing the dessert mix are usually weighed and mixed while cold. The mix is then pasteurized by heating it in a container or vat to a temperature of from 150° to 165°F, holding at this pasteurization temperature for about 30 minutes. The mix can then be homogenized at pasteurization temperature at a pressure of 2,000 to 3,500 psi. Homogenizing prevents the churning in the freezer of any fat present and thus improves the whipping properties of the mix and the texture of the finished product. The mix is then frozen in the usual manner.

Example 1: The fat and the nonfat milk solids are supplied by concentrated whole milk and skim milk in the following low calorie ice cream formula.

	Parts	Grams (pint, 522 g)
Fat	12	62.64
Nonfat milk solids	10	52.20
Polyose	15	78.30
Sodium carboxymethylcellulose	0.3	1.57
Saccharin	*	-
Water	62.3	325.21
Vanilla	0.2	1.04

*Amount for sweetness.

Example 2: Condensed skim milk supplies the fat and the nonfat milk solids, together with concentrated whole milk in the following low calorie ice milk formula.

	Parts	Grams (pint, 522 g)
Fat	6	31.32
Nonfat milk solids	12	62.64
Polyose	5	26.10
Sorbitol	5	26.10
Glycerin	2	10.44
Karaya gum	0.128	0.67
Sodium alginate	0.128	0.67
Saccharin	*	-
Water	65.68	342.85
Chocolate flavoring	4.0	20.88

*Amount for sweetness.

Example 3: Source of fat for this low calorie ice milk mix is cream testing 30% butterfat and the balance of the nonfat milk solids is from condensed skim milk.

	Parts	Grams (pint, 522 g)
Fat	4	20.88
Nonfat milk solids	13	67.86
Polyose	16	83.52
Stabilizer	0.26	1.36
Emulsifier	0.10	0.52
Saccharin	*	-
Water	66.06	344.83
Vanilla	0.25	1.31

*Amount for sweetness.

Example 4: The fat and part of the nonfat milk solids are supplied by fresh homogenized whole milk testing 3.2% butterfat in the low calorie sherbet formula below.

	Parts	Grams (pint, 522 g)
Fat	1.5	7.83
Nonfat milk solids	3.0	15.66
Polyose	34.0	177.48
Gelatin	0.5	2.61
Saccharin	*	-
Water	58.37	305.69
Orange juice (conc. 3:1)	2.58	13.47

*Amount for sweetness.

Example 5: A low calorie water ice formula is given below.

	Parts	Grams (pint, 522 g)
Polyose	25.0	130.5
Glucose	7.0	36.54
Gum tragacanth	0.4	2.09
Saccharin	*	-
Water	64.52	336.79
Orange juice	2.58	13.47

*Amount for sweetness.

Low Calorie Dessert

T.P. Finucane and P.J. Capasso; U.S. Patent 3,702,768; November 14, 1972; assigned to General Foods Corporation prepare a low calorie frozen dessert by the use of a combination of liquid vegetable oil and hydrogenated vegetable fat emulsified in a matrix phase comprising water, protein and sugar.

Also the product contains a critical amount of a high capacity water binding material as well as effective amounts of emulsifying and flavoring agents. The product is able to function as low calorie desserts primarily due to the ability of the composition to achieve an overrun about twice as large as can be obtained with conventional ice cream. In general, the product contains an edible composition, including a liquid vegetable oil-vegetable fat mixture emulsified in a water-protein-sugar solution, which composition is whipped to a high overrun (at least

200) and frozen to produce a low calorie dessert. The composition also includes as an essential ingredient a critical amount of a high capacity water binding material. The composition on a percent weight basis is as follows.

	Percent
Protein solids	2-7
Water binding material	0.7-2
Sugar	10-15
Water	60-70
Hydrogenated vegetable fat	10-15
Liquid vegetable oil	2-5
Emulsifier	2-6
Color-flavor agents	0.05-0.5

The protein solids include any of the whippable proteins such as sodium caseinate, casein, lactalbumin, whey solids, soy protein, hydrolyzed soy protein and other hydrolyzed cereal proteins. These whippable protein materials are capable of being aerated to produce a high overrun. The high capacity water binding material such as a gelatin of high Bloom, is capable of binding water in the frozen dessert composition in such a manner that the consumer will be unable to detect, by taste, the large percentage of water present in the frozen dessert.

The gelatin can be used in amounts of from 0.9 to 1.5% by weight of the composition. Generally speaking, the higher the Bloom of the gelatin the less is the amount required. Care must be taken to avoid the incorporation of too much gelatin since the texture of the final product will be adversely affected. This results from the fact that virtually all the water will be bound with the gelatin and there will be insufficient free water remaining. This free water is a necessary component of the final composition since it is only this water which will produce the desired melting sensation in the mouth of the consumer when the frozen composition is eaten.

The sugar in the composition is sucrose; however, other sweetening agents such as corn syrup solids, dextrose, fructose, or mixtures of any of these can also be used. It has been found that the ice-cream-like taste and consistency can be produced by a particular combination of 3 to 5 parts by weight of fat consisting of hydrogenated vegetable fats and 1 part by weight of liquid vegetable oil. The fats are those which melt easily in the mouth and which soften and melt in the neighborhood of 25°F.

The vegetable oils used both in the production of the fat component and as the liquid oil component can include cottonseed oil, soybean oil, corn oil, palm oil, coconut oil, peanut oil and the like. The preferred fats are the hydrogenated mixtures of coconut and palm oils available commercially as Wecotop A. The preferred liquid oil is cottonseed oil.

The emulsifiers can be any of those commonly used in the food industry or mixtures of any of these such as egg yolk, various monoglycerides such as glycerol monostearate, etc., and other natural or synthetic products which are capable of forming stable oil-in-water emulsions. Particularly good results in obtaining a high overrun have been obtained by using as the emulsifier a principal amount (85 to 95%) of propylene glycol monostearate together with a lesser amount (5 to 15%) of lecithin. It is believed that the lecithin acts to modify and extend the whippable protein in such a manner that the protein is capable of whipping to an increased overrun. The formulations are whipped to overrun in excess of 200. This high value is not able to be attained with conventional ice cream which normally has an overrun in the range of from 80 to 100. The ability to attain the high overrun is thought to be due to the presence and amount of the high capacity water binding material in the formulations.

While it is recognized that conventional ice cream may contain some gelatin-like material, the gelatin is present to help stabilize the butterfat emulsion and to regulate the size of ice crystals in the ice cream to a small size and is present in amounts of only about 0.3% by weight.

It is estimated that a 45% reduction in calories is able to be achieved by the product as compared with conventional ice cream. The process involves heating the water component to boiling and then dissolving and mixing the gelatin in the hot water. The whippable protein and sugar components should be preblended in dry form prior to their addition to the water so that, upon blending, a thoroughly homogeneous foamed mixture is produced.

The selected fats, oils and emulsifiers should be intimately combined before forming the final emulsion. This is accomplished by melting the fats together with the oil and emulsifiers, pouring the liquid fat-oil-emulsifier mixture into the foamed and heated protein-sugar-water mixture and homogenizing the combined mixture to form an emulsion. This emulsion is then chilled and whipped at a high speed until stiff peaks form. The mixture is finally frozen at -35°F and then stored at 25°F.

The color-flavor agents may be added at virtually any point in the mixing process; however, preferably they are added to the protein-sugar-water mixture just prior to the blending step. The formulations prior to whipping and freezing could be reduced to a dry mix, which, when reconstituted with water and whipped, could be frozen into a dessert similar to that described above. The dry mix product would have a composition on a weight basis as follows.

	Percent
Protein solids	6 - 20
Gelatin	2 - 6
Sugar	30 - 45
Hydrogenated vegetable fat	30 - 45
Liquid vegetable oil	6 - 15
Emulsifier	6 - 18
Color-flavor agents	0.2 - 1.5

Example:

	Grams	Weight Percent
Gelatin (235 Bloom)	4.00	0.95
Sodium caseinate	10.00	2.38
Sugar (sucrose)	51.00	12.11
Water	275.00	65.35
Hydrogenated vegetable fat (Wecotop A)	56.00	13.31
Propylene glycol monostearate	13.00	3.09
Lecithin	1.30	0.31
Liquid vegetable oil (cottonseed)	10.00	2.38
Color-flavor agents	0.52	0.12

The water is heated to boiling and the gelatin is dissolved therein. A preblended mixture of the sodium caseinate and sugar is added to the water gelatin solution and the mixture together with the color-flavor agents, is blended, in a Waring Blender, for about 2 minutes, in order to produce a foamed blend. The hydrogenated vegetable fat is melted in a double boiler, and mixed with the liquid vegetable oil and emulsifiers. This mixture is then poured into the (still warm) foamed protein-sugar-water mixture and homogenized by two passes through a Manton Gaulin Homogenizer at 3,000 psig.

The resulting emulsion is poured into a bowl and chilled in an ice bath to 59°F. The chilled emulsion is whipped at a high speed until stiff peaks form, frozen at -35°F, and finally stored at 25°F. The frozen whipped emulsion of this example whips to an overrun of 217 and possesses a caloric content, based on a standard one-sixth quart serving, of 111. An equal amount of a commercial medium fat content ice cream possesses a calorie count of 197. The product has the physical appearance of ice cream and the product possesses a taste and mouthfeel closely resembling that of conventional ice cream.

Dry Ice Cream Mix

A.B. Avedikian and S.Z. Avedikian; U.S. Patent 3,355,300; November 28, 1967 describes the problems of formulating a low calorie frozen dessert such as ice cream. Powdered fat compositions comprising fats and emulsifying agents, proteinaceous materials and other compositions capable of being whipped into toppings and cake frostings are well-known.

To attain suitable overrun and suitable properties, these fat compositions are usually prepared by drying emulsions of the fat and the emulsifying agent containing the various stabilizing and encapsulating solids. However, the frozen end-products of such powdered fat compositions, after whipping, have the appearance and texture of frozen whipped cream and not the body, creaminess and melting properties of ice cream.

A free-flowing ice cream mix with approximately one-half the calories of regular ice cream uses ingredients in about the following proportions, namely, 4 kg of sugar, 4 kg of corn hydrolysate (corn sweetener), 150 grams of food grades of Methocel or carboxymethylcellulose, 496 grams of a vegetable oil, a blend of vegetable oils, a fat or a blend of such fats and oils stabilized, 303 grams of an emulsifying agent composition containing the following individual substances, 27 grams of a polyoxyethylene sorbitan monostearate, 83 grams of a polyoxyethylene sorbitan tristearate, 28 grams of a polyoxyethylene sorbitan monooleate and 165 grams of a propylene glycol fatty acid derivative of which 45% is the monostearate derivative and 55% is the palmitate derivative, 7.8 kg of nonfat milk solids, instantly soluble, and 100 grams of the desired flavoring agent, such as a composition containing vanilla or its equivalent.

When about 170 grams of the above-indicated mixture is uniformly dispersed in 270 grams of water and agitated at high speed, for example, the speed at which cream is usually whipped, by means of an egg beater type of mixer, a smooth, creamy product is obtained. The finished volume is about one quart and shows an overrun of 100%, the initial volume being about 16 fluid ounces. This creamy product is frozen in a metal ice tray or other suitable container by placing it in the freezer compartment of a refrigerator.

The frozen product is a smooth, creamy ice cream which is ready to serve in two to four hours depending on the temperature of the freezer. When this operation is carried out at a speed of 500 rpm, the speed at which cream would usually be whipped, the desired creamy consistency and overrun of 100% is obtained in about 10 minutes. At lower speeds it takes longer to achieve the proper dispersion of the ingredients and the desired overrun.

The following procedure is followed in preparing the dry ice cream mix. The oils and the emulsifying agents are heated until they are homogeneous or blend thoroughly into a uniform mixture (hereinafter referred to as fats) and are sufficiently fluid to pour. A temperature of 100°F has been found suitable.

The sugars, corn hydrolysates and one-half of the Methocel or the carboxymethylcellulose are blended into a uniform, dry mixture. This mixture (hereinafter referred to as sugars) is added to the fats while the fats are still warm and, hence, liquid enough to coat all particles of the sugars when the mass is subjected to continuous attrition. This is similar to the creaming of sugar with shortening or butter as practiced in cooking, and the operation is continued until the composition becomes uniform in consistency and appearance with each discrete particle properly coated.

This well-mixed composition of fats and sugars is then mixed with the nonfat milk solids, flavoring agent and the other half of the Methocel or the carboxymethylcellulose. These may be previously mixed together or they may be added individually. It has been found that when this mixing is carried out in this manner, so as to obtain a thoroughly uniform end product, this end product is a dry, free-flowing, essentially powdery composition which has good keeping qualities and is stable in storage. The sugars used in the amount of 8.15 kg (18.6%) in the preferred embodiment of the process comprise the following:

(a) Sucrose, e.g., cane sugar, 4 kg (9.13%), functions as the primary sweetening agent. It was found that a portion of this can be substituted with artificial, nonnutritive sweetening agents such as saccharin. The ratio is 1 part of saccharin to about 300 parts of of the sucrose. As indicated in Example 1, the sucrose may also be completely substituted by saccharin.

(b) Corn hydrolysate, 4 kg (9.13%), functions as a secondary sweetening agent, and also acts as a bodying agent and assists in the freezing and thawing processes of ice cream. The corn hydrolysate is in the form of a dry powder and is made from corn syrup by evaporation of the water. The corn hydrolysate is a mixture of polysaccharides. It may contain a small percentage of unhydrolyzed cornstarch.

(c) Methocel, 150 grams (0.34%), functions as a protective agent and a coupling agent which makes possible the uniform dispersion of the fats into the aqueous phase when the dry ice cream mix is beaten with cold water. It helps to produce the creamy texture and to maintain this texture all through the freezing operation by coupling with the Methocel which is added as part of the nonfat milk solids. It protects the colloidal structure.

Instead of using Methocel, sodium carboxymethylcellulose (CMC of commerce) can be used or natural gums, such as guar, sodium alginate, carrageenan (Irish moss extract). The proportions of the natural gums may be varied to give the desired consistency to the ice cream upon whipping, and to impart to the final product more organoleptically acceptable properties. By varying the proportions of these ingredients, different appearance of ice cream structure, gel-like properties, more rapid or slower melting characteristics have been obtained.

The fats used in the amount of 799 grams (1.82%), comprise 496 grams (1.13%) of a blend of vegetable oils, preferably bland and neutral in taste and flavor. It is best to use oils (fatty acid glycerides) which have a high percentage of polyunsaturates and are vegetable in origin, although glycerides of animal origin and low in polyunsaturates or even of wholly saturated fatty acids have also functioned well. Vegetable oils which have iodine numbers in the range of 109 to 145 are most suitable.

The emulsifying agents used in the amount of 303 grams (0.69%) comprise complex mixtures of polyoxyethylene ethers of mixed partial fatty acid esters of sorbitol anhydrides, for example, polyoxyethylene 20 sorbitan monostearate, tristearate, and monooleate. These are supplemented with the monoester of propylene glycol in which the fatty acid could be any fatty acid but preferably is palmitic or stearic or any combination. The ratio of palmitic to stearic acid should be 55 to 45, for example, as in a product known as PGMS-45S (Drew Chemical Corp.).

The nonfat milk solids used in the amount of 7.8 kg (17.82%) comprise the dried product obtained preferably from spray-drying of skimmed milk produced by any of the well-known methods. Such a product is commonly known as instantly soluble and usually occurs in well-known bead form. The nonfat milk solids constitute the major portion of dry ice cream mix and are also the main source of protein in the ice cream mix composition.

The following are specific examples of dry ice cream mix compositions, each of which, when dispersed in cold water, whipped and frozen at a temperature of 20°F produces an ice cream that is smooth and is organoleptically acceptable to the consumer as this popular dessert.

Example 1:

	.... Parts by Weight	
Ingredients	**Dry Mix**	**Final Ice Cream Product**
Nonfat milk solids	54.74	18.22
Saccharin	*	-
Corn hydrolysate	28.90	9.11
Methocel (high viscosity)	1.95	0.62
Soybean oil	5.30	1.67
Polysorbate 60	0.48	0.15
Polysorbate 65	0.72	0.22
Polysorbate 80	0.96	0.30
PGMS-45S	1.93	0.61
Flavoring (vanilla type)	0.51	0.22
Cold water	-	68.40

*Amount for sweetness.

Example 2:

Ingredients	Parts by Weight Dry Mix	Final Ice Cream Product
Nonfat milk solids	45.15	16.62
Sucrose (cane sugar)	4.52	1.66
Saccharin	*	-
Corn hydrolysate	38.50	14.18
CMC (high viscosity)	1.74	0.64
Soybean oil	4.52	1.66
Polysorbate 60	0.45	0.17
Polysorbate 65	0.45	0.17
Polysorbate 80	0.23	0.09
PGMS-45S	2.70	1.00
Flavoring (vanilla type)	0.62	0.23
Cold water	-	63.17

*Amount for sweetness.

ORGANIC ACID SALT ADDITIVE FOR MILK BASE PRODUCTS

In the process of *British Patent 1,275,347; May 24, 1972; assigned to Pfizer Incorporated* an aftertaste-masking organic acid salt is used for milk or milk product sweetened with nonnutritive sweetening agents. These products are characterized by a lack of bitter and metallic flavors and are much more acceptable than the same products sweetened with the nonnutritive sweetening agents to which the organic acid salt has not been added.

The process provides a method for sweetening a milk or milk product based food which contain a nonnutritive sweetening agent at a level equivalent in sweetness to form 5 to 25% w/w sugar, that is, sucrose, in the final food composition, and a sodium or potassium salt of citric, adipic, gluconic, fumaric, tartaric, succinic, malic or acetic acid, the salt being added at a level of from 0.6 to 1.5% w/w based on the final food composition.

Example 1: Low calorie frozen desserts essentially free of aftertaste are formulated as shown below where parts are by weight.

	Parts by Weight
Butterfat	4.0
Nonfat milk solids	23.0
Sorbitol	5.0
Calcium saccharin	0.083
Calcium carrageenan	0.5
Chocolate flavor	0.2
Permitted color	to suit
Water	65.0
Sodium citrate	0.6

Example 2: Diet liquid breakfast drinks are formulated as follows.

	Parts by Weight
Nonfat dry milk	3.4
Sodium caseinate	1.6

(continued)

	Parts by Weight
Corn syrup solids	1.0
Carrageenan	0.2
Sodium saccharin	0.083
Vanilla flavor	0.2
Permitted color	to suit
Vitamins and minerals	*
Milk	92.0
Sodium citrate	0.6

*To support label claims.

The above drinks are pleasant tasting and free of aftertaste. The use of sodium adipate, sodium fumarate or sodium succinate in place of sodium citrate, or sodium gluconate also produces equivalent products, as does the corresponding potassium salts of all the acids indicated.

Example 3: A pleasant-tasting pie filling is formulated as follows.

	Parts by Weight
Cornstarch	3.6
Nonfat dry milk	1.2
Sodium saccharin	0.083
Calcium carrageenan	0.06
Lemon flavor	0.1
Permitted color	to suit
Sodium citrate	1.5
Milk	93.0

FLUFFY FROSTING

Fluffy frostings conventionally consist predominantly of sugar and water and also must contain a foaming/aerating agent. As prepared in the home, fluffy frostings almost always utilize egg white as the foaming/aerating agent. Dry mixes which can be whipped into a fluffy frosting upon the addition of water are commercially available. These products generally utilize egg albumen or a similar protein source as the foaming/aerating agent and they also contain a stabilizer for the foam which is usually a gum-type material such as algin, gelatin, or a cellulose derivative such as carboxymethylcellulose.

A primary disadvantage possessed by conventional fluffy frostings is their instability over a period of time. For example, fluffy frostings, although highly aerated and of desirable eating quality when fresh, generally lose air and/or liquid and become rubbery or marshmallow-like upon storage or while standing on a cake overnight. This disadvantage is seen both in homemade fluffy frostings and in prepared mix products. This instability characteristic also explains why ready-to-spread prepared fluffy frostings have not been made commercially available to any extent.

A further disadvantage possessed by commercially available fluffy frosting mixes is the fact that the protein foaming/aerating agent is generally a very expensive ingredient and is suceptible to microbiological attack. For example, the most commonly used protein foaming/aerating agent, egg albumen, is expensive and is known to be susceptible to microbiological problems.

The characterizing ingredient in the formula and process of *G.F. Brunner, B. Lawrence, N.B. Howard and P. Seiden; U.S. Patent 3,592,663; July 13, 1971; assigned to The Procter & Gamble Company* comprises certain polyglycerol esters of fatty acid. These materials are polyglycerol esters containing from 5 to 12 glycerol units and from 1 to 4 fatty acid groups per molecule.

The polyglycerol essentially is a polymer which is formed by the dehydration of glycerin. For each unit of glycerin that is added to the polymer chain there is an increase of 1 hydroxyl group and, from 1 to 4 of these hydroxyl groups of the polyglycerol molecule form ester links with fatty acids having from 8 to 26 carbon atoms, at least 40% of fatty acids having at least 22 carbon atoms.

As with ordinary glycerol or other polyols, polyglycerols can be esterified by reaction with fatty acids. Esterification can take place at any or all of the hydroxyl groups but generally occurs predominantly at the secondary hydroxyl positions, leaving the terminal hydroxyl group unaffected. Depending upon the reaction conditions and the ratio of fatty acid to polyglycerol, the number of secondary hydroxyl groups which are esterified varies.

By controlling the balance of esterified to unesterified hydroxyl groups, the lipophilic-hydrophilic balance of the polyglycerol ester can be varied. With an increasing number of esterified hydroxyl groups, the polyglycerol esters become progressively less hydrophilic. This lipophilic-hydrophilic balance in the polyglycerol ester is important in the fluffy frosting compositions. It has been found that sufficient lipophilic properties are imparted to the polyglycerol ester by the fatty acid esterification of a single hydroxyl group.

However, to maintain sufficient hydrophilic properties in the molecule, the polyglycerol ester cannot contain more than 4 fatty acid radicals. Preferably, the polyglycerol ester will contain 3 fatty acid radicals. In this same regard, the polyglycerol ester used in the fluffy frosting compositions can contain from 5 to 12 glycerol units, and preferably contains 10 glycerol units.

By far the most important requirement of the polyglycerol ester is the carbon atom chain length of the fatty acid groups. This chain length can range from 8 to 26 carbon atoms, but at least 40% by weight of the fatty acids must contain at least 22 carbon atoms. It has been found that the 40% minimum amount of C_{22} or higher fatty acid provides fluffy frostings with outstanding stability properties.

The fatty acid groups can be derived from suitable naturally occurring or synthetic fatty acids and can be saturated or unsaturated, but are preferably substantially saturated. Examples of these fatty acids are caprylic, capric, lauric, myristic, palmitic, stearic, oleic, linoleic, arachidic, behenic, and lignoceric. The latter mentioned fatty acids contain at least 22 carbon atoms.

Behenic (C_{22}) is the preferred fatty acid of the polyglycerol esters in this fluffy frosting. While it is customary to esterify polyglycerol with a single type of fatty acid, polyglycerol which has been esterified with a mixture of fatty acids can be used in the fluffy frosting compositions. The polyglycerol ester of mixed fatty acids can comprise mixed fatty acid radicals on each individual polyglycerol molecule (so that all the polyglycerol molecules are substantially the same) or can comprise the same fatty acid radical on each individual polyglycerol ester molecule (so that all the individual polyglycerol molecules are not the same).

It should also be understood that in actual practice polyglycerol esters usually contain a mixture of molecules that average the specified number of glycerol units and fatty acid ester groups per glycerol unit; individual molecules within the mixture can vary from the average. For example, decaglycerol tribehenate contains an average of 10 glycerol units per molecule and an average of 3 fatty acid ester (behenoyl) groups per molecule while some individual molecules could contain 8, 9, 11 or 12 glycerol units and 2 or 4 fatty acid ester groups.

Fatty acids per se or naturally occurring fats and oils can serve as the source for the fatty acid component of the polyglycerol esters. For example, rapeseed oil provides a good source for C_{22} fatty acid. The C_{16} to C_{18} fatty acid can be provided by tallow, soybean oil or cottonseed oil. The shorter chain fatty acids can be provided by coconut, palm kernel, or babassu oils.

When using naturally occurring fats and oils as the fatty acid source, it is preferred that they be substantially completely hydrogenated, for example, to an I.V. of less than 10. The polyglycerol esters can be prepared by conventional direct or interesterification techniques. More specifically, decaglycerol tribehenate, a preferred polyglycerol ester for use in this formula can be prepared by the following procedure:

Equipment	50 lb reaction vessel
Ingredients:	
Polyglycerol (decaglycerol)	60.29%
Behenic acid (practical grade)	39.41%
85% phosphoric acid	0.30%
Conditions:	
Reaction time	2 hours
Reaction temperature	450°F
Reaction pressure	⅓ atmosphere
Atmosphere	Nitrogen (sparge)
Agitation	Mechanical and N_2 sparge

At the end of two hours at reaction temperature, the charge is cooled to 250°F and withdrawn from the reaction vessel. After 16 hours the product separates into two distinct layers: (1) solid, fat-like, top layer (about 50%) of total (polyglycerol ester) and (2) viscous liquid bottom layer (about 50%) of total (unreacted polyglycerol). The top layer is separated and purified to obtain decaglycerol tribehenate. Fluffy frosting based on decaglycerol tribehenate which can be prepared according to the above procedure is illustrated in the example.

Decaglycerol triester of certain mixed fatty acids, another preferred polyglycerol ester for use herein can be prepared by the following procedure. One mol decaglycerol is reacted with 0.5 mol behenic acid (80% C_{22}), 0.1 mol palmitic acid (95% C_{16}), 0.07 mol myristic acid (95% C_{14}), and 0.1 mol lauric acid (95% C_{12}). The fatty acid and decaglycerol are reacted under reduced pressure (⅓ atm) at 410°F for 60 minutes. The reaction mixture separates into two liquid layers where the top layer is the decaglycerol fatty acid ester with a fatty acid mol ratio of about 2.8. The top layer is separated and purified to obtain the decaglycerol triester. The fatty acid radicals are randomly distributed on the decaglycerol molecules:

	Weight Percent
Behenic (C_{22})	58
Arachidic (C_{20})	10
Stearic (C_{18})	6
Palmitic (C_{16})	11
Myristic (C_{14})	7
Lauric (C_{12})	7
Capric (C_{10})	1

Fluffy frosting based on decaglycerol mixed fatty acid esters can be prepared according to the above procedure. This process provides fluffy frosting compositions based on the above-described polyglycerol esters. As with conventional fluffy frostings, water is a component of the composition when it exists in final form ready for using and eating. Thus, the polyglycerol ester can be added to water (preferably with sufficient heating to at least partially dissolve the polyglycerol ester) and the water-polyglycerol ester blend can then be whipped to form an aerated, aqueous fluffy frosting suitable for spreading, filling, and/or eating.

In terms of foaming/aerating properties and stability, the fluffy frosting need only contain the specified polyglycerol ester and water. However, in order to provide a product with desired eating characteristics, additional ingredients can be combined with the polyglycerol ester either before or after the addition of water. Thus, a sweetening agent, for example, sugar or an artificial sweetener, is combined with the polyglycerol ester in the preparation of an aerated, aqueous fluffy frosting.

The sweetening agent included in the fluffy frosting compositions can be any suitable sugar such as sucrose, dextrose, lactose, glucose, galactose, and the like or mixtures. These materials can be used in such conventional forms as cane sugar, beet sugar, corn syrup, brown sugar, maple sugar, maple syrup, honey, molasses, and invert sugar. Sucrose and/or dextrose are preferred sweetening agents.

In place of all or part of the above-described sugars, artificial sweeteners such as saccharin can be used. A minor amount of salt can be added and any conventional flavor material can also be added to the fluffy frosting composition. For example, vanilla, vanillin, chocolate, fruits and fruit extracts, nuts, and the like can be used as desired.

Conventional foaming/aerating agents for fluffy frostings can be added but they are not required and are preferably omitted. These agents most often are protein or protein-containing materials such as whole milk, nonfat milk solids, soy protein, egg white, egg yolk, and egg albumen. Conventional thickening agents for fluffy frostings can also be added to affect eating characteristics such as mouthfeel, but are not required for stabilization. Among these agents are gums such as carrageenan, tragacanth, arabic and ghatti; seaweed colloids such as agar, carrageen and sodium alginate; seed extracts such as locust bean and guar; water-dispersible cellulose derivatives such as sodium carboxymethylcellulose; starch; and gelatin.

Shortening is often an ingredient of so-called creamy-type frostings but is conventionally not present in fluffy frostings. In these fluffy frosting compositions, it is preferable to omit shortening from the formulation. Shortening or other fat, particularly in liquid form, can decrease the foaming/aerating properties of the polyglycerol ester-based fluffy frosting compositions. However, a small amount, for example, less than 5% of the composition, of a plastic fat can be added to affect eating characteristics.

The fluffy frosting compositions can be prepared and made available to consumers in a variety of forms. For example, a liquid frosting composition comprising a suitable polyglycerol ester and sugar can be mixed with water and packaged in a suitable container. The user then merely whips the liquid composition to incorporate air. More preferably, the liquid frosting composition can be whipped to form an aerated, aqueous frosting that is ready to spread and/or eat. Such prepared frostings can be packaged in suitable containers, for example, sterilized airtight can, and then distributed in this form to be used by consumers without further preparation.

Alternatively, a fluffy frosting composition can be placed in a pressure-dispensing container from which it can be removed in aerated form. The propellant for the pressure container can be any conventional nontoxic, odorless, tasteless gas including nitrogen, nitrous oxide, carbon dioxide, dichlorodifluoromethane (Freon), and the like. These containers, conventionally known as aerosol dispensers, can have a dispensing orifice of 0.03" in diameter or less and the frosting is able to pass therethrough and be whipped during such passage.

The fluffy frosting composition based on the above-described polyglycerol ester can be prepared and packaged also in the form of a dry mix. Thus, the consumer adds water to the mix and then whips it into an aqueous, aerated fluffy frosting for use. A dry mix based on the polyglycerol ester per se is not desirable since the specific polyglycerol esters suitable for use exist in the form of hard, brittle waxy solids. However, dry granular ingredients, preferably sugar, can be blended with the polyglycerol ester to provide a dry, prepared mix.

The dry mixes can be prepared by combining the polyglycerol ester with an aliphatic polyhydric alcohol carrier. The lower monohydric aliphatic alcohols of 2 to 4 carbons such as ethanol and butanol can serve as a carrier for the polyglycerol ester in a fluffy frosting dry mix. Aliphatic polyglycerols such as pentaglycerol can also be used. Alkylene glycols of from 2 to 6 carbon atoms such as the propylene glycols and butylene glycols are also suitable as the carrier.

Within the group of aliphatic polyhydric alcohols, those having from 2 to 4 carbon atoms and from 2 to 3 hydroxyl groups are preferred. Glycerol and propylene glycol represent the most preferred carriers with glycerol being the most highly preferred.

The combination of the aliphatic polyhydric alcohol carrier with the polyglycerol ester component allows the preparation of a dry mix by providing for a uniform dispersion of the hard brittle ester. Moreover, the carrier promotes and contributes to the foaming/aerating ability of the polyglycerol ester. When preparing the frosting compositions in the form of a dry mix, it is preferred to follow the following procedure. The polyglycerol ester is first combined with the polyhydric alcohol carrier and the mixture is heated, preferably to above the melting point of the polyglycerol ester, to dissolve the ester in the carrier. The mixture is heated to at least 130°F. The heated mixture is then rapidly chilled, for example, in a scraped wall heat exchanger such as a Votator or in a mixing bowl placed in an ice bath, to form a pasty mass.

Thus, the polyglycerol ester and alcohol carrier are plasticized together. Sugar and additional dry ingredients are then blended into the pasty mass, for example, in a ribbon blender, to form a dry mix. The dry mix can then be impact milled, for example, in an Entoleter, to remove lumps and large particles, resulting in the formation of a granular, free-flowing and uniform dry mix.

When the fluffy frosting compositions are prepared in the form of a dry mix they require only the addition of water and beating in a bowl to yield within a few minutes a highly aerated, aqueous fluffy frosting with a smooth viscous texture. For example, mixing in a conventional household electric mixer for a period of less than 10 minutes is sufficient to whip the composition into an aerated fluffy frosting. The compositions tend to reach minimum density faster, for example, they reach minimum density in less than 3 minutes, than do conventional fluffy frostings when mixed in a household electric mixer. Additional mixing, for example, for 1 to 5 minutes beyond this point is desirable to assure uniformity.

The aerated, aqueous, fluffy frostings have a smooth viscous texture at least comparable to commercial products and good aerated structure. The frostings are more aerated than conventional fluffy frostings as indicated by their density which is generally less than about 0.3 g/cc and preferably less than 0.2 g/cc. Because of the high level of air which they contain, the frostings have a very desirable glossy-type appearance.

The frostings are extremely stable; for example, they can be spread on a cake in peaks and the peaks remain over substantial periods of storage time. Moreover, the icings do not have a tendency to leak, that is, lose liquid upon storage. Further, the frostings do not change in appearance or eating quality upon storage.

Example:

Ingredients	Percent by Weight
1:1 Mixture of decaglycerol tribehenate and decaglycerol tristearate	5.00
Carboxymethylcellulose (CMC)	4.00
Artificial sweetener and flavor	2.00
Water	89.00

One part decaglycerol tribehenate (DGTB) was added to one part decaglycerol tristearate and the mixture heated on a steam bath until the DGTB was dissolved (melting point about 150°F). The heated mixture was rapidly chilled by agitating in a mixing bowl placed in an ice water bath for 10 minutes to form a smooth creamy paste.

Two parts CMC (artificial sweetener mixture) and one part of the paste were blended in a mixing bowl to form a cream-like mass. Additional CMC was blended into the cream until a dry granular mixture was formed. The mixture was then passed through an impact mill (Entoleter) to reduce lumps and large particles. A granular, free-flowing and uniform dry mix was thus obtained.

Frosting Preparation – Aerated, aqueous fluffy frosting was prepared by adding 88 grams of water at 95°F to 3.5 ounces of the above-prepared dry mix, blending at slow speed (about 100 rpm) on a conventional household electric mixer for one-half minute, and mixing at high speed (about 850 rpm) for 5 minutes. The finished aqueous, aerated frosting was fluffy and delectable and had a smooth, viscous texture.

DIETARY DRY CAKE MIX

The following descriptions are given on reduced calorie cake by *S.B. Radlove; U.S. Patent 3,658,553; April 25, 1972.* It is desirable, for reasons of diet, to provide a reduced calorie cake, and such a cake may be obtained by means such as reducing the shortening content. The lower calorie cake is particularly obtained by providing a form which has little or no sugar. Not only are calories reduced with sugar elimination, but this carbohydrate form is eliminated to meet certain health requirements such as those of diabetes mellitus.

It is known that the material sorbitol is not inimical to diabetics, and has inherent sweetness, but of a lower order. It is known that sorbitol has about one-half the sweetness of sucrose, consequently this material is a popular subject of investigation as a possible substitute for sugar in food preparations. Sorbitol has been used to replace sugar in candies, jams, cookies, and sorbitol has been added to freshly prepared cake batters specially mixed for dietetic purposes.

The substitution of sorbitol for sugar in special baking techniques is a procedure which requires some skill and careful attention. It is not convenient nor simple for the average user to acquire such skills or to direct the necessary attention to details required for successfully making such dietary cakes containing sorbitol.

It would be desirable to provide a dry cake mixture which can be simply and conveniently handled by the average user so that a successful and tasty cake may be prepared. It is desirable that a mixture be provided which preferably requires only added water prior to baking. It is likewise desirable that the cake baked from such a simple mixture have desired levels of moisture retention and palatability. The moisture retention is a feature of palatability, and it is also a property which extends the keeping quality of the cake.

It has been found that a dietary dry cake mixture may be provided with a non-shortening portion having a particular sorbitol ratio, and a shortening portion having an emulsifier part and a fat part which may be a liquid vegetable oil or a plastic shortening. The term "sorbitol ratio" is used in a way equivalent to the term "sugar ratio."

For example, a sugar ratio of 100% means a ratio of 1 part sugar to 1 part cake flour, by weight. A sugar ratio of 50% would mean one-half part sugar to one part cake flour by weight, and so on. In the same way, the sorbitol ratio will be defined in percentage terms to represent the ratio of parts by weight of sorbitol to parts by weight of cake flour. It has been found that particular sorbitol ratios are required to obtain the successful prepared dietary dry cake mixes. It has also been found that the shortening portion requires particular minor amounts of a primary emulsifier to be used in combination with the particular sorbitol ratios. A primary emulsifier in the cake baking art means a material which leads to incorporation of air into a cake batter. This is important to attain the desired lightness, volume, and tenderness of the cake.

The specific gravity of the cake batter is one index for determining whether the cake will have the desired high volume, lightness and tenderness. The primary emulsifiers or surfactants may be selected from the many which are suitable for addition to foodstuffs. In particular, successful results are obtained by using from 5 to 12% by weight of the emulsifiers in the shortening portion.

It has also been found that successful cakes are obtained from a prepared cake mixture in which small amounts of the shortening portion are present relative to the nonshortening portion. This is desirable to further reduce the caloric content of the cake. In general, about 4½ parts of the nonshortening portion is combined with 1 part of the shortening portion, or less than 1 part, say about 0.8 part. A minor amount of the shortening portion includes the emulsifier by which is meant one emulsifier or a mixture of emulsifiers. In one example, the emulsifier part of the shortening portion includes an emulsion enhancer which is stearyl monoglyceridyl citrate.

Representative illustrations of the type of primary emulsifiers which can be used in the dry cake mixture include lactic acid esters of monoglyceride and diglyceride, also known as glyceryl-lacto esters of fatty acids or lactated monoglycerides and diglycerides, also referred to collectively by the symbols LMG. Other useful primary emulsifiers are propylene glycol monostearate (PGMS); glyceryl lactostearate (GLS); glyceryl monostearate (GM); stearoyl lactic acid (SLA); stearoyl-lactylic acid (S_2LA); SLA + S_2LA. The stabilizer enhancer, stearoyl monoglyceridyl citrate, will also be referred to by the further symbols SMGC.

Example 1: A white cake mix is made up of a nonshortening portion and a shortening portion. The nonshortening portion is made up of 203 grams of cake flour, 14 grams of starch, 168 grams of sorbitol, 25 grams of skim milk solids, 28 grams of dried egg whites, 7 grams of vanillin on powdered base, 14 grams of salt (1 teaspoon equals 5.3 grams), 11 grams of baking powder and artificial sweetener as needed. The shortening portion is made up of 70 grams of a liquid vegetable oil shortening (corn or cottonseed) and 10 grams of a primary emulsifier (PGMS).

The nonshortening portion and the shortening portion may be mixed and stored until ready for conversion into a batter and baking into a cake. The shortening and nonshortening portion may, alternatively, be maintained separately until mixed just prior to combining with liquid for preparing the batter.

Some stability problems may arise by combining the shortening portion with the nonshortening portion, especially upon prolonged standing. Any threat of instability may be markedly reduced by using a more stable liquid shortening such as Durkex 500, a high stability oil. Plastic shortenings may be used in place of the liquid triglycerides which also extend stability.

Example 2: The nonshortening and shortening portions of Example 1 are mixed, and 8 ounces of cold tap water are added to the mixture. The water and dry portions are then mixed at speed No. 1 on a Sunbeam Mixmaster, with the aid of a spatula to move the dry ingredients into contact with the liquid. Mixing is continued for one minute at speed No. 12, thereupon an additional 4 ounces of cold tap water are added, blended in gently, and mixed for three minutes at speed No. 7, using a spatula to obtain a uniform batter.

The specific gravity of the batter is 0.78 to 0.80. The cakes are baked by depositing 13 ounces of the batter in each of two 8-inch greased pans, and baking at 365°F for 25 to 30 minutes. The remaining batter for the mixture is used to bake cupcakes. The resulting cake has a desirably sweet taste, is tender and light, and has a good texture. The volume of the cake is between 1,150 and 1,120 ml. The lightness and high volume of the cake is related to the desired specific gravity determination of the batter.

Example 3: The nonshortening portion of Example 1 is combined with a shortening portion prepared from the following ingredients: 4 grams of PGMS, 4 grams of LMG, 2 grams of SMGC and 70 grams of hydrogenated plastic shortening. The above shortening portion includes a mixture of primary emulsifiers and the emulsion enhancer SMGC, which makes up 12% of the mixture, the plastic shortening being 88% of the mixture.

The dry cake mixture is combined with water according to the process steps described in Example 2, and a white, smooth batter is obtained having a specific gravity of 0.93. A cake is baked by depositing 14 ounces of batter in each of two 8-inch greased pans. The resulting cake has a volume of 1,120 ml. The cake is tender and acceptable as to flavor.

Five white cake mixes were made by using the ingredients of Example 1, except that different primary emulsifiers were used, each emulsifier present in an amount of 12.5% by weight of the shortening portion. Table 1 lists the different primary emulsifiers, describes some properties of the batter, lists the volume of the prepared cake, and describes some qualities of the prepared cake. The palatability rating is a semiquantitative taste test utilizing numbers from 1 to 10, No. 1 indicating the highest rating and No. 10 the lowest. It is a composite rating of taste, aroma, mouthfeel and ease of swallowing.

Table 2 represents data from six different white cakes prepared from the portions of Example 1, except that different primary emulsifiers are used, and an emulsifier enhancer is added in different amounts. It will be seen from Table 2 that varying percentages of primary emulsifiers and the emulsifier enhancer, within the prescribed ranges, generally lead to acceptable cakes. It can be further seen that specific gravities of the batters fall within the range of 0.6 to 0.8 except for one cake which has a higher specific gravity of 1.02 and a smaller resulting volume.

Table 3 presents the results of three cakes essentially from the mixtures presented in Example 1, except that different amounts of sorbitol are used for each cake; and the shortening portion is cottonseed oil instead of plastic shortening.

It is seen that the palatability rating is high for the sorbitol ratios employed, and that the specific gravities of the batters are between 0.7 and 0.8. The volumes are high for all the cakes and the general cake structure is of good quality.

TABLE 1

		Batter			Cake		
Emulsifier Type	Percent of Shortening	Specific Gravity	Comments	Volume (ml)	Taste	Structure	Palatability Rating 1-10
PGMS P-06 (90% PGMS)	12.5	0.74	Thick, white and smooth	1,150	Slightly dry	Slightly coarse, even, firm	4
Myverol 18-100 (90% GM)	12.5	0.75	Fluid, white and smooth	1,200	Dry	Slightly coarse, even, split top	6
Drewpol 3-1-S (PGS)	12.5	0.96	Thin, white and smooth	*	Dry	Coarse, poor cake	10
Marvic acid (SLA + S_2LA)	12.5	0.64	Very thick, white and smooth	1,155	Dry	Slightly coarse, even, very pale top, firm	6
LMG	12.5	0.81	Thick, white and smooth	1,200	Slightly dry	Slightly coarse, firm, brown top, even	4

*Small sides, small volume.

TABLE 2

			Batter		Cake		
Emulsifier Type	Percent of Shortening	Percent Citrate Ester of Shortening (SMGC)	Specific Gravity	Comments	Volume (ml)	Structure	Palatability Rating 1-10
PGMS P-06	12.5	2.5	0.74	White and smooth	1,155	Tender, good mouthfeel	3
Myverol 18-100	12.5	2.5	1.02	Thin, slightly yellow	1,040	Pale top	5
Marvic acid	10.0	2.5	0.64	Very thick, white and smooth	1,160	Very firm	5
Marvic acid	6.25	6.25	0.66	Thick, white and smooth	1,190	Center fell slightly, firm	5
GLS	8.85	3.75	0.75	Thick, white and smooth	1,180	Tender, pale top	3
LMG PGMS	5.0 5.0	2.5	0.77	Thick, white and smooth	1,170	Tender, good eating	2

TABLE 3

Weight of Sorbitol to 7¼ oz Flour	Specific Gravity of Batter	Comments	Volume (ml)*	Cake Tops (Appearance)	Cake Structure	Palatability Rating 1-10
4 oz (55%)	0.73	Smooth, white and thick	1,100	Pale yellow, not attractive	Sides and shape good, tender but slightly firm, fine and even	2
6 oz (83%)	0.77	Smooth, white and thick	1,140	Medium brown, attractive	Sides and shape excellent, fine and even, tender, excellent mouthfeel	1
8 oz (110%)	0.78	Smooth, white and fluid	1,110	Darker medium brown,	High shrinkage, structure very tender, fine and even, excellent tasting and mouthfeel	1

*Of cake for 14 oz of batter.

DIETETIC FLOUR

Insoluble Protein Approach

As a sweetener, sugar can easily be substituted by noncaloric chemical sweetening agents, such as saccharin. On the other hand, no product is known to be capable of satisfactorily replacing the starch polysaccharide for use in food products which is not converted into glucose or other assimilable sugars during the digestion process.

There are many polysaccharides which do not release assimilable sugars during digestion following ingestion by living animals, including humans. In this group of polysaccharides, for example, belong the polysaccharides composed of nonassimilable sugars, the polysaccharides that are composed of assimilable sugars but which cannot be broken down, and, lastly, the polysaccharides that are composed of nonassimilable sugars and which cannot be broken down. While natural starch is insoluble at low temperature and merely swells by absorbing water at 60° to 70°C, these polysaccharides often dissolve readily in cold water forming gels or highly viscous pastes which are very unpleasant to the taste, and therefore are unsuitable for the manufacture of food products.

It has been found that these polysaccharides which do not yield assimilable sugars during the digestive process, when suitably treated, become insoluble in cold water, and merely swell in warm water by water absorption in the same manner as natural starch and, when mixed with the suitable binding agent, yield a flour which is eminently suitable for production of foodstuffs.

This treatment for polysaccharides was developed by *R.F. Menzi; U.S. Patent 3,097,946; July 16, 1963; assigned to Dr. A. Wander, SA, Switzerland.* The treatment or process involves the preparation of a substantially homogeneous mixture, in the presence of or with the subsequent addition of a liquid to form a paste consistency, of at least one polysaccharide which does not yield assimilable sugar during digestion with at least one portion which becomes insoluble under the action of heat.

The paste thus produced is formed into bodies having a large surface area and then dried and heated to a temperature between 100° to 250°C. The drying and heating is, of course, terminated before any impairment of the edibility properties of the product results. The resultant polysaccharide-protein product is ground and, when the occasion arises, mixed with at least one edible binding agent which holds the polysaccharide-protein in an agglutinated form, such as a protein having the physical properties of gluten, to produce an edible flour.

The polysaccharide which does not release assimilable sugar during digestion is mixed with an aqueous mixture of the protein which becomes insoluble under the action of heat to form a substantially homogeneous mixture having the consistency of a paste.

A quantity of protein which becomes insoluble under the action of heat ranging from 2 to 15% by weight of the total solids is sufficient to achieve the desired results. This low protein requirement is of considerable practical importance because a product having a high protein content would to some extent lose its value as a dietetic flour. The protein which becomes insoluble under the action of heat is used in quantities ranging from 3 to 10% by weight, but preferably in quantities of about 5%. The lowering of the water solubility of the polysaccharides which do not release assimilable sugar is effected by mixing them with a solution of at least one protein which becomes insoluble under the action of heat. The drying and heating of a mixture of this kind makes the protein become insoluble and then brings about the formation of a net of protein insolubilized in situ inside the polysaccharide mass.

Further, the free amine groups of the protein react with the sugar moieties of the polysaccharide in accordance with Maillard's reaction, forming insoluble macromolecules. There are thus obtained, depending upon the quantity and kind of the protein used as well as on the temperature and duration of the heating, products which have limited water solubility at low temperature but which swell in warm water by water absorption.

As polysaccharides, any number of substances of the abovementioned three types can be used. As a polysaccharide which is composed of nonassimilable sugars, polymannan, carubin, guar, agar, alginate, polygalactane or pectin, for example, can be used. Certain soluble cellulose derivatives may be used as nondecomposable polysaccharides composed of assimilable sugars, for example, cellulose, inulin, dextran and chitin. As a polysaccharide which is composed of nonassimilable sugars and is not broken down, there may be used, for example, carboxymethylcellulose, methylcellulose, ethylhydroxyethylcellulose or tragacanth.

Carubin, that is, carob seed flour, is of great practical importance as a source of polysaccharide. It has been found that the most satisfactory results are obtained with this cheap starting material only when it has previously been subjected to weak acid hydrolysis. Hydrolysis can be performed with an aqueous carubin paste or with a suspension of carubin in ethanol, isopropyl alcohol, acetone or any other suitable solvent which can be easily removed. Hydrochloric acid or any other strong acid is added as required to this paste or suspension and the latter heated until the desired degree of hydrolysis is reached, whereupon it is neutralized with a base, such as a sodium hydroxide solution or soda solution, and if necessary the solvent is removed.

As a protein which becomes insoluble under the action of heat, a great many animal or vegetable proteins can be used. Among the former, ovalbumin is very suitable. Among the latter, soybean or ground nut protein can, for example, be employed. As a binding agent cereal glutens, such as corn gluten, or any other protein which has similar physical properties, such as soy protein, can be employed. The binding agent is employed in a proportion in the range of between 10 and 20% by weight of the composition.

Example 1: Ninety-five parts of highly viscous carboxymethylcellulose are mixed by kneading with an aqueous solution containing 395 parts of water and 5 parts of ovalbumin. The paste thus obtained is extruded in "spaghetti" form, and first dried in this form at 40°C for 5 hours and then heated in an oven to 200°C for 8 minutes. The product thus obtained is ground and mixed with 20 parts of dry corn gluten to form a flour. This flour can be used for the production of dietetic food products, such as wafers, cookies, bread, and the like, as with conventional wheat flour or other cereal flours.

Example 2: Ninety-five grams of carubin flour are kneaded to a paste with a mixture of 250 ml of water and 5 ml of 37% hydrochloric acid. The paste is heated to 80°C for 75 minutes and then neutralized to pH 6.5 by adding 2.3 grams of caustic soda in 20 ml of water. Five grams of ground nut protein in 30 ml of water are added directly or after previous drying to the paste, which is then dried and heated to 200°C for 3 minutes. The product thus obtained is further treated as in Example 1.

Example 3: One hundred grams of carubin are suspended in 200 ml of ethanol and 5 ml of 37% hydrochloric acid are added. The suspension is heated, while stirring, to the reflux temperature (80°C) for 2 hours. It is then filtered, the residue is washed three times with 50-ml portions of ethanol and dried at 60°C. The flour thus obtained is further processed in the way described in Example 1, using ovalbumin.

Low Calorie Doughs

The reduction in calorie and/or assimilable carbohydrate content which can be effected by introducing more water into farinaceous products, or by decreasing or eliminating their fat content altogether, is limited. Also, it has not been possible to dilute farinaceous products extensively with nonnutritive solids to reduce their assimilable carbohydrate content without a concomitant loss of the desirable properties characteristic of the products.

Finally, the substitution of proteins for assimilable carbohydrates, for example in products like the commercially available protein breads, does not significantly reduce the caloric content of the products because proteins and assimilable carbohydrates each contain about the same number of calories per gram. This substitution effects only a limited reduction of carbohydrate content before the character of the product changes.

For example, the partial substitution of gluten flour for ordinary flour is not successful in reducing their carbohydrate content by more than from 60 to about 40%, with no significant reduction in calories. With higher gluten contents, rubbery doughs are formed which are "bucky," that is, are difficult to work and develop.

The highly tenacious protein bonding (probably attributable to the presence of sulfhydryl groups and disulfide linkages) observable in such doughs of high gluten content is modified in ordinary farinaceous materials by the presence of the starches (assimilable carbohydrate), naturally found in flours. This permits the development, in ordinary baked bread for example, of a more open crumb, forming a network to which the pleasing texture of the bread is largely attributable.

A system containing nonnutritive filler and binder, and an edible vegetable gum which appears to bind to the gluten protein was developed by *R.L. Singer; U.S. Patent 3,574,634; April 13, 1971.* The presence of the nonnutritive filler and binder makes the otherwise dense, rubbery doughs workable and convertible by heating, for example, boiling, steaming, baking, into palatable products poor in, or free of, assimilable carbohydrate and of reduced caloric content. Because the products do not involve a mere substitution of starch by protein of equal caloric content, they are products significantly reduced in caloric content as well as in carbohydrate content.

The calorie content of a typical synthetic bread is no more than about 1.2 calories per gram as baked. The least palatable commercial diet breads now available contain about 2 calories per gram as baked. Conventional bread contains about 2.6 calories per gram as baked.

The doughs of the synthetic bread should contain "vital" gluten, that is, gluten proteins which have not been denatured by the processes used to separate starch from the gluten. The doughs may be either substantially free of assimilable carbohydrate (starch) or may contain up to at most about 10 to 15% by weight of starch.

Doughs containing vital gluten and a low starch content are prepared from commercially available vital gluten wheat flours. These flours are generally prepared by extracting starch from ordinary flours with water or other solvents and then drying the gluten residues without excessive heating which would tend to denature the gluten protein. A typical commercially available vital gluten wheat flour has the following approximate analysis.

	Percent
Moisture	9.6
Protein	38.7
Ash	0.7
Fat (acid hydrolysis)	4.2
Carbohydrate	46.8

A vital gluten wheat flour of still greater gluten content is commercially available as Vicrum and has the following approximate analysis.

	Percent
Moisture	6
Protein	71
Ash	1.2
Fat (acid hydrolysis)	6.5
Carbohydrate*	15.3

*Defined as composed of N-free extract and fiber.

The material is a light tan free-flowing powder 97% of which passes a standard 60 mesh screen. Although this material, or equivalent gluten flours from other sources, may be used directly in preparing the doughs, additional starch can be removed by washing, leaching, or treatment with diastatic enzymes. The desirability of such additional treatments depends, among other things, on their cost, and must be balanced against the desire or need for a product totally free of assimilable carbohydrate.

If an untreated substantially starch-free gluten flour like the Vicrum product is used, the content of assimilable carbohydrate present in a typical finished baked product is usually less than about 4% by weight after dilution with other ingredients. Lower carbohydrate contents will result in leavened doughs because of carbohydrate removal by fermentation.

Doughs containing vital gluten and less than 10 or 15% by weight of starch can be prepared from ordinary flour by repeatedly leaching such flour with water to remove starch prior to combination with the other ingredients characteristically present in the doughs of the process. Alternatively, the starch content of doughs prepared from ordinary flour or gluten flour having a substantial percentage of residual starch can be reduced by treatment of the dough with diastatic enzymes and by repeated leavening with yeast, thus to convert the starch to sugar and ultimately to alcohol and carbon dioxide. The utilization of leaching, enzymatic, and leavening processes to provide vital gluten doughs of low starch content from conventional flours as starting materials is of little commercial importance because of the time which these processes consume and their cost.

In practice, it is much better to use commercially available vital gluten flours from which starch has been extensively removed. Cellulosic materials are used as the nonnutritive edible filler incorporated into the product. These materials include vegetable matter derived from spinach, squash, cabbage, kale, beets, rhubarb, corncobs, cucumber, melon rind, straw, peanut shells, wood pulp, and the like.

Nonnutritive edible fillers particularly suitable in the process are cellulose crystallite aggregates like those described in U.S. Patent 3,023,104, and wheat middlings (hereinafter referred to as bran), particularly middlings from which assimilable carbohydrate has been removed. An example of such crystallite aggregates is Avicel.

The starch content of the bran fillers may be reduced or removed by simple grinding, which liberates starchy particles adhering to the middlings, followed by sifting. Starch may also be removed by washing the middlings with water or with weak alkali, with or without prior grinding, or by treatment with diastatic enzymes.

A variety of vegetable gums can be used as binders. These material include gum karaya, psyllium husk, tragacanth, guar, pectin, and locust bean gum. Other gums including algin and acacia can be used but are less suitable as tending to inhibit the rise of leavened doughs. In general, the best gums are those that are highly retentive of water, but not necessarily those which absorb the most water.

The presence of cellulose and gums in bakery products such as breads makes the products materially more absorptive of water than conventional breads. This in turn promotes a feeling of satiation caused by increased bulking in the stomach after the ingestion of only modest quantities of the product. This aids the dieter in restricting his food intake. The low calorie doughs fall within the following composition range.

	Parts by Weight
Flour component	50
Inert filler	10-50
Vegetable gum	1-10
Water	50-100

In the doughs having a content of starch less than about 10%, the protein:starch ratio in the flour component should be at least 2:1. These ingredients alone are sufficient for the manufacture of pasta products such as spaghetti and macaroni, or for the manufacture of unleavened baked goods such as cocktail snacks, breakfast foods in flake or particle form, and the like. For the production of leavened products, bakers' yeast and yeast nutrients are added to the dough. In general, an amount of yeast conventional for ordinary farinaceous products is used in the synthetic foodstuffs, that is, the minimum amount sufficient to give the desired degree of rise within an economically feasible time period. There is no theoretical upper limit on the amount of yeast which can be used because it is killed during the baking process. In practice, the amount of yeast is kept to a minimum because of its cost. For a dough composition of the kind specifically set forth above, yeast is generally added in amounts of from 0.5 to 6 parts by weight.

Leavening is achieved by the action of the yeast on residual fermentable carbohydrate in the product, for example, those small quantities which may be present in a dough prepared with low-starch gluten flour. To increase the rate and amount of leavening, fermentable sugar may be formed in the compositions by the action of added diastatic enzymes.

Alternatively, sugar may be added to the compositions to promote leavening. If sugar is added, the minimum amount necessary to produce the desired degree of rise, usually between 0.5 and 1.5 parts by weight, is used in order to minimize the carbohydrate and calorie content of the final product. A high degree of utilization of the sugar can be assured by judicious choice of the relative amounts of yeast and sugar. A dough suitable for the formation of unleavened products is prepared by combining the following ingredients.

	Parts by Weight
Low-starch gluten flour*	50
Cellulose crystallite aggretates**	20-40
Gum***	1.5-5
Water	60-50

*For example, Vicrum having a protein:starch ratio of about 5:1.
**For example, Avicel.
***For example, karaya or pectin.

If desired, the dry ingredients can all be first combined to form a dry mix, adaptable to long storage, to which the water can be added when convenient to form the dough. The same dough can be used for the production of leavened products by the addition of about 3 parts by weight of yeast and minor amounts of yeast nutrients, for example, 0.25 to 0.5% by weight of total solids. About 2 parts by weight of sugar may be added to accelerate leavening, or fermentable sugar may be formed in the dough by diastatic action.

For the production of bread and rolls it is particularly advantageous to replace a portion of the cellulose crystallite aggregates by bran from which substantially all assimilable carbohydrate has been removed. Palatable products simulating whole wheat bran in color, flavor and texture have been obtained using a mixture of equal parts by weight of bran and cellulose crystallites as the nonnutritive filler.

In all of the compositions, additional coloring and flavoring agents may be present in amounts to taste. Such agents include salt, onion, caraway, herbs such as dill and the like, minor amounts of residual yeast and sugar, and other natural artificial food coloring and flavoring agents including synthetic sweeteners and food dyes. Mold retardants such as the propionates added to bread doughs may also be used in amounts of from 0.05 to 0.2% by weight of total solids.

The vegetable protein supplied by gluten can be nutritionally upgraded by adding about 0.25% by weight of the amino acid lysine to the products, the percentage being by weight of the total protein content. Leavened doughs like those described above are suitably prepared by combining the gluten flour and yeast together with a portion of the water to form a dough and permitting the dough to ferment and rise, for example, in a fermentation room, whereby the carbohydrate content is reduced. The raised dough is then mixed with the other solids and remaining water and permitted to rise.

In an alternative preparation of either leavened or unleavened dough, the solid ingredients are first combined and then blended with water which is warmed to about 125° to 150°F. If the mixture is yeast-leavened, the dough is next divided into the shape of the bread, rolls, or other product desired and is proofed and baked. The optional use of a pan cover during baking helps to regulate the formation of the crust. Unleavened dough products are chilled to about 40°F to decrease their stickiness and improve their workability. For the formation of spaghetti or other pasta products, or of breakfast foods, for example, the dough can be pulled into sheets and cut, or can be extruded.

For pasta products, the cut or extruded dough may then be dried and sold in the form conventional for spaghetti, macaroni, noodles, etc. The cut or extruded dough may be sold also without prior drying, for example, frozen in plastic bags or other containers resistant to the transmission of water vapor, and then cooked inside or outside of such containers.

Pasta products of particularly good texture and consistency on chewing are made from doughs containing about 15% by weight of assimilable carbohydrate (starch). These doughs are conveniently prepared from a flour component in which vital gluten flour is combined with a flour of higher starch content, preferably semolina. The nonnutritive edible filler and edible vegetable gum are critically present in these doughs, just as in the doughs of lesser starch content.

These pasta doughs are prepared from a flour component comprising from 15 to 25 parts by weight of a vital gluten flour starch as Vicrum and from 35 to 25 parts by weight of semolina (or a comparable flour). The flour component is combined with nonnutritive edible filler and edible gum in amounts previously disclosed.

The filler is present in an amount of 40 to 50 parts by weight per 50 parts of flour component. If the vital gluten content of the flour component drops below about 15 parts in 50, the pasta products produced fragment badly on boiling. At vital gluten contents greater than about 25 parts in 50, excellent structure is maintained. However, the products are perhaps subjectively less desirable since their texture does not imitate as well to the mouth, teeth, and tongue the texture of ordinary high calorie pasta.

In these pasta doughs, about 3 to 6 parts by weight of edible gum are combined with 50 parts of flour component. Lower gum contents lead to products which fragment on boiling, while higher gum contents tend to impart an undesirable slippery feel. The gum cannot be omitted or masses are obtained which cannot be extruded to form a pasta article. A dry mix for making pasta dough has 20 parts by weight of vital gluten flour such as Vicrum, 30 parts of semolina, 50 parts of cellulosic filler such as Avicel, and 5 parts of edible gum, such as karaya.

The mix contains about 24 to 25% of available carbohydrate. One hundred parts of the mix are combined with 80 parts of water to give a dough having a starch content of only 13 to 14%. The dough is extruded to form pasta.

Since pasta products of this type are more difficult to rehydrate than ordinary products of high starch content, the pasta is not dried after extrusion, but cooked in boiling water. Cooking adds from 100 to 120 parts of water per 180 parts of extruded dough, further reducing the available carbohydrate content to 9 to 10%, corresponding to a calorie content in the cooked product of less than 1 calorie per gram, indeed of only 0.5 calorie per gram. The cooked pasta is conveniently marketed in cans or boilable plastic pouches, alone or with suitable sauces.

Example 1: A low starch, low calorie bread was prepared by suspending 0.3 gram of a heat-stable bacterial diastase (Rhozyme H-39) in 10 ml of water. 1.5 milliliters of this suspension were added to 90 milliliters of water warmed to 60°C and then combined with 100 grams of flour comprising (1) 60 grams of low starch vital gluten flour having a protein content of about 71% by weight and a carbohydrate content of 15% by weight and (2) 40 grams of a vital gluten flour having a protein content of about 39% by weight and a carbohydrate content of 47% by weight. The resulting dough ball was kept at a temperature of 60°C for 15 to 30 minutes.

Meanwhile a crumbly mass was prepared by combining 80 ml of water with 50 grams of cellulose crystallite aggregates, 3 grams of yeast, 0.6 gram of yeast nutrients, 0.15 gram of lysine, 5 grams of gum karaya, 5 grams of caraway seed, 4 grams of salt, 0.15 gram of sodium calcium propionate, and 20 drops of an artificial sweetener solution. This mass was then blended into the dough ball. The dough was proofed for two hours and then oven-baked. The resulting bread had a starch content of 8.2% and contained about 1.2 to 1.3 calories per gram as baked.

Example 2: A bread further reduced in caloric content was prepared as in Example 1 above except that 10 grams of wheat middlings (bran) from which assimilable carbohydrate has been removed were combined into the dough together with the 50 grams of cellulose crystallite aggregates, yeast, gum, etc.

Example 3: A synthetic bread product was prepared by dry blending 40 grams of a commercial gluten flour having a protein content of about 71% and a carbohydrate content of 15%, 10 grams of a commercial gluten flour having a protein content of 39% and a carbohydrate content of 47%, 20 grams of cellulose crystallite aggregates, and 3 grams of guar gum. The mixed ingredients were then added to 80 ml of agitated water containing 2 grams of salt, 5 grams of yeast, and 4 grams of sugar. The dough was kneaded until fully developed, proofed, and then oven-baked at 425°F.

Example 4: One hundred grams of a commercially available gluten flour having a protein content of 71% and a carbohydrate content of 15% were dry blended with 7.5 grams of guar gum, and 75 grams of wheat middlings from which assimilable carbohydrate had been removed by leaching with water. The resultant dry mixture was combined with 180 ml of water containing 4 grams of salt, 5 grams of yeast, and 2 grams of sugar. The dough was kneaded until developed, then proofed and baked.

Example 5: Thirty-five grams of a commercially available gluten flour having a protein content of 71% and a carbohydrate content of 15% and 15 grams of a second commercial gluten flour containing 39% of protein and 47% of carbohydrate were combined with 30 grams of cellulose crystallite aggregates, 2 grams of gum karaya (or pectin), 2 grams of salt, and 80 ml of water.

After thorough mixing, the resultant dough was extruded into preferred shapes, for example, as spaghetti, macaroni, and noodles. Portions of the shaped dough were then sealed into polyethylene bags to each of which an amount of water equal to the water content of the dough therein was added in the form of ice. The bag was then sealed and frozen. The product was easily prepared for eating by placing the sealed bag in boiling water for 10 minutes. All of the water added to the bag was absorbed to produce a cooked pasta product containing 0.5 to 0.7 calorie per gram.

Dietetic Low Protein Bread

Low protein foods and calorie reduction are required for those on special diets. The problems associated with low protein bread is described by *G.H.R. Watson; British Patent 1,242,350; August 11, 1971; assigned to General Mills Incorporated.* Dietetic low protein breads are generally prepared for people observing diets restricted in protein or specific amino acids.

Examples of people who must restrict or modify their protein intake are: those suffering from celiac disease and nontropical sprue, that is, those who must restrict their intake of certain specific grain proteins; individuals suffering from PKU or phenylketonuriac disorders, that is, those who must reduce their intake of the amino acid phenylalanine; and those suffering from uremia or severe kidney disorders, the latter class being unable to dispose of urea and other nitrogenous waste products due to kidney malfunction.

The problem of producing a palatable baking mix for bread acceptable to those suffering from uremia is the most difficult. Those suffering from celiac disease may have milk protein and milk protein aids significantly in bread production. This is not so with uremia sufferers because these patients must severely limit their intake of all protein.

Uremia sufferers and others such as heart patients must restrict their intake of sodium and potassium. Salt is normally added in bread-making processes to control fermentation. Uncontrolled fermentation in a gluten-containing bread is not desirable because the gluten must be aged to fully attain its elastic properties and if it is not it will not function properly as a cell structure elasticizer. Fermentation must therefore be slowed to allow the gluten to attain its desired properties and salt is used to slow this process.

Furthermore, although attempts have been made in the past to produce low protein bread by eliminating the gluten from wheat flour and by eliminating milk or milk protein from the bread mix, loaves made in this manner have proven to be unacceptable. These low protein loaves have a very close-grained structure, lack the typical resiliency found in bread, are crumbly with a cake-like texture and are very dry and highly compact.

A bread mix for a dietetic low protein bread that overcomes these disadvantages contains starch, fat, a sugar and a total amount of up to 10% by weight of the wheat starch, fat and sugar, of at least one structure enhancer selected from pectin, pregelatinized waxy maize starch, pregelatinized tapioca starch, pregelatinized wheat starch, methylcellulose and carboxymethylcellulose, the mix containing not more than 0.3% by weight of wheat gluten.

The bread mix produces a baked loaf similar to bread in every way. The fat is present in the bread mix in the form of shortening. The wheat starch used may be a commercially available wheat starch and is usually made by subjecting wheat flour to a series of water washing steps to separate the starch from the gluten. An example of this type of wheat starch is the product Paygel.

The structure enhancer or structure preservative is desired to perform a function similar to the gluten and other proteins present in conventional bread. Bread normally contains a specific type of wheat gluten noted for its elasticity. It often contains milk protein. The gluten and added milk protein provide elasticity of structure to the cell walls of the foam created when the dough is subjected to a leavening agent.

By providing this strength and resiliency, gas produced during the leavening process is able to expand the cell walls to a point where the bread rises and a soft, springy, relatively open even cell structure characteristic of bread is present on the inside of the loaf. The outside of a normal bread loaf is characterized by a smooth tan or brown color with no apparent fissures on the surface due to uneven cell wall expansion.

Therefore, this smoothness is also enhanced by the presence of protein. Since no gluten is added in the bread mix, yeast growth may be allowed to proceed normally and salt is not needed. The mix may have a sodium content of 38 milligrams per 100 grams of mix, for example, at a moisture content of 8.8%. This is compared to a conventional gluten bread which contains 507 mg Na per 100 grams of bread (at 35.6% moisture). The mix also generally contains 8 mg of K/100 grams as opposed to 105 mg K/100 grams using the moisture figures above. The bread mix may therefore be used for the preparation of a dietetic bread for people on a low salt diet.

The bread has an even cell structure, a smooth brownish crust, a fair degree of resiliency and elasticity, an open crumb structure, and a moist white inner surface thereby producing a loaf highly similar in every respect to normally available gluten-containing wheat bread.

Other related, binding agents such as pregelatinized corn, potato, and arrowroot starch and vegetable gums such as carageenan do not provide the desired results. Some of these agents settle to the bottom of the loaf, others do not retain sufficient moisture or else they produce a loaf which is unappetizing in color, while others do not provide sufficient strength for proper rising of the bread or result in uneven expansion leaving fissures on the crust and/or throughout the interior of the loaf.

The low protein bread has a total protein content of less than 6% of the protein content of an equivalent weight of conventional enriched white bread. Conventional enriched white bread in general contains 8 to 9% of protein. Thus, the low protein bread generally has a total protein content of less than 0.9% and preferably less than 0.8%, the total protein content being less than

0.6%. A total protein content of 0.5% is particularly preferred, this protein level corresponding to 3.8 grams of protein in a 1.75 lb loaf of the low protein bread.

The procedure for preparing the mix, the proofing and the baking conditions were the same for all of the loaves prepared. First, a premix was made consisting of 86.69% of wheat starch (0.3% protein), 7.08% of sugar (granulated), and 6.23% of shortening. To this premix was added the structure enhancer (the amounts of the structure enhancer will be given as percent of the total dry mix excluding yeast). The mix was added to an aqueous yeast solution, the yeast being present at a level of 4% by weight of the starch and the amount of water sufficient to make a dough, here an amount about equal in weight to the mix.

The mix and the yeast solution were then blended till homogeneous at low speeds in a conventional mixer. The bowl was scraped with a rubber blade and blending was then continued for one-half minute at medium speed. The dough was then poured into greased pans and allowed to proof at a relative humidity of 80 to 85% until the pans, which were initially half full, were full to the top with dough. This proofing generally takes about thirty minutes. The dough was then baked in a conventional oven at 400°F for thirty minutes. Of the total protein content, less than one-half was due to the gluten in the wheat starch.

Example: A number of loaves were prepared by the procedure outlined above. Each loaf contained one of the structure enhancers at the levels indicated below.

Structure Enhancer	Percent
Pectin (highly purified, unmodified)	0.5, 1.0
Waxy maize starch (pregelatinized)	2-10*
Tapioca starch (pregelatinized)	2-10*
Wheat starch (pregelatinized)	5
Methylcellulose	0.5, 2
Carboxymethylcellulose	0.25, 2

*In 1% gradations.

When waxy maize starch or pregelatinized tapioca starch was used, it provided the best loaf at a 5% level and outside of the shiny appearance of the inside crumb structure was adjudged identical with the 5% wheat starch sample. Moisture levels were taken after 20 hours of storage at ambient temperature on the 5% tapioca starch loaf and the loaf was found to contain 43.5% moisture. At a level of 10% the loaf developed a more open crumb structure and was noticeably more moist than the loaf containing the 2% level. This was true with all of the loaves as the level of structure enhancer increased.

The major differences noted among the loaves prepared in this example using pectin, carboxymethylcellulose and methylcellulose as structure enhancers were in crumb structure and moisture. Both cellulose derivatives and pectin proved to be useful at lower levels. Within the ranges given for the particular mix formulation the 10% waxy maize starch and tapioca starch loaf showed more of a wide open crumb structure than loaves containing either 2% of the cellulose derivatives or 1% pectin. Therefore, an increase of up to 50% of the amount in the above table of either the cellulose derivatives or pectin is possible if these structural enhancers are used and an open crumb is desired. All of the samples showed excellent moisture retention, that is, 43.5% by weight of the loaf for 5% tapioca after 20 hours at room temperature, good leavening and crumb structure, desirable color both inside and outside the loaf. The pregelatinized-starch-containing samples produced loaves with the best color.

The loaf containing pectin produced a crust color less brown than the loaves containing pregelatinized starch but better than the loaves containing cellulose derivatives, although the crust color on the latter loaves were still deemed acceptable.

The pectin-containing mix also produced a loaf which rose higher in the middle than any of the other loaves thereby producing a loaf more nearly approximating the exact configuration of a conventional gluten-containing wheat bread loaf. Pectin at a level of 0.5% and the methylcellulose present at levels as low as 0.5% produced binding characteristics superior to a 2% level of tapioca or waxy maize. For a relatively dense loaf which has a uniform crumb structure but is still not crumbly, a reduction up to 50% in the amounts of these enhancers is possible.

POLYOSE FOR BAKED GOODS

Solid foods that contain substantial amounts of sugar which can be replaced by an equivalent amount of Polyose include various baked products such as breads and crackers, sweet pastries, cakes, cookies, cake icings, and the like. Also included are solid and semisolid candies of all types, and chewing gum. A Polyose can be used as a bulking agent for diabetic food compositions in place of sugar because it provides the functions, except sweetness, of sugar in these products yet does not have the high calorie and carbohydrate content of sugar. It supplies weight and bulk and serves to build suitable structure in foods.

The description on the use of polyose in the solid dietetic foods named above are given by *P. Jucaitis, I.D. Bliudzius and N.P. Rockwell; U.S. Patent 2,876,106; March 3, 1959; assigned to E.I. du Pont de Nemours and Company.*

Example 1:

Cake

Flour	2½ cups
Shortening	1½ cups
Whole eggs	4
Skim milk	1 cup
Polyose A	1½ cups
Saccharin	*
Baking powder	3½ tsp

*Amount for sweetness.

Sift together flour and baking powder. Mix Polyose A and saccharin and cream this mixture with the shortening. Add the eggs to the shortening mixture, and then carefully mix all the ingredients. Bake in oven at 350°F for 20 to 30 minutes.

Example 2:

Vanilla Wafer

Butter	¼ lb
Polyose C (finely divided)	1 cup
Vanilla	1 tsp
Flour	1¾ cups
Milk	½ cup
Saccharin	*

*Amount for sweetness.

Cream the butter. Sift, then beat in the Polyose until smooth. Add vanilla. Sift and resift flour with saccharin and add same to creamed mixture alternatively with the milk. Beat mix until creamy. Lightly cover a baking sheet with shortening and chill same. Then with a spatula, spread the mixture over the sheet as thin and evenly as possible. Sprinkle with chopped nuts. It is well to press them down a bit so that they will stick. Take a sharp knife and mark off the batter into 1½-inch squares. Bake in 325°F oven until brown. When done, take from oven and, while hot, quickly cut through the marked squares and remove from sheet.

Example 3:

Corn Meal Bran Bread

Butter	½ cup
Polyose D	½ cup
Whole eggs	2
Bran	2 tbsp
All purpose flour	⅔ cup
Baking powder	1⅔ tsp
Table salt	½ tsp
Corn meal	5½ tbsp
Milk	1 cup
Saccharin	*

*Amount for sweetness.

Cream the butter. Add gradually and mix with the Polyose until light and creamy. Then beat in one egg at a time. Stir in the bran. Sift flour; mix baking powder, salt, corn meal, and saccharin, then beat in the sifted ingredients in about three parts alternately with thirds of the milk. Add raisins and nuts. Spread the batter in a greased 9 x 12 inch pan. Bake the bread in a moderate oven at 375°F for about 20 minutes.

POLYOSE THICKENING AGENT

A Polyose can be used as a thickener for low calorie and dietetic food compositions because it provides a viscosity and thickness corresponding to that obtained with usual sugar and carbohydrate-containing thickeners but is nonnutritive compared with usual thickeners. A Polyose is a glucose polymer solid derived from starch by depolymerization followed by heat polymerization (U.S. Patent 2,563,014). The Polyoses have a considerably different susceptibility to amylolytic enzymes than the original starch or its conventional degradation products. The products on test contain groups which act like very small amounts of reducing sugar, not in excess of 5 to 7% in Polyose A. The reducing sugar is not construed as due to the presence of free glucose.

In the process of *P. Jucaitis, I.D. Bliudzius and N.P. Rockwell; U.S. Patent 2,876,107; March 3, 1959; assigned to E.I. du Pont de Nemours and Company* the following foods can be prepared using a Polyose as a thickening agent.

Beverages: Includes all nonalcoholic carbonated and uncarbonated soft drinks, such as drinks containing fruit and vegetable solids and juices, herbs, root extracts, essential oils, and animal solids, extracts, and fluids. Specifically included, for example, are milk and root beer.

Syrups: Includes fountain syrups, table syrups, syrups prepared from natural products and syrups artificially synthesized. Specifically included, for example, are natural and artificial maple syrups, chocolate syrup, fruit-flavored syrups, such as raspberry and cherry.

Toppings and Sauces: Includes all flavors for preparations which can contain fruits, vegetables, meats, and all flavors. Specifically included, for example, are toppings containing strawberry syrup, whipped cream, caramel syrup, raspberry flavors, fresh and canned fruits, for example, peaches and pears.

Canned Fruits and Vegetables: Includes all fruits and vegetables preserved without jelling. Specifically included, for example, are all fresh fruits and candied sweet potatoes, and applesauce.

Puddings and Pie Fillings: Includes non-gelatin-containing mixes of all flavors, as well as those containing materials which are added in addition to sugar to provide greater thickening action than the sugar alone can supply. Specifically included, for example, are custards, tapioca, and lemon chiffon.

Salad Dressings: Includes dressings prepared with oil and water bases of all flavors. Specifically included, for example, are French dressing, mayonnaise, and vinegar-containing dressings.

Soups: Includes chilled, cream, vegetable and fruit stock soups and thick soups such as tomato normally containing sugar.

Thus foodstuffs can be prepared having a normal protein, vitamin and mineral content but having less than normal carbohydrate content. To the diets of persons suffering, for example, from diabetes, obesity, or maybe just too much healthy appetite, can thus be added appetizing, wholesome food compositions containing Polyoses.

The quantity of Polyose used as a thickener in a food composition can be determined by reference to the viscosity of such food composition when prepared with usual sugar, sugar-containing, or, in certain instances, starch or other carbohydrate-containing thickening agents. The normal quantity of thickener can be replaced in whole or in part by appropriate amounts of a Polyose. The result is that dietetic foodstuffs are prepared which have a viscosity simulating that of foodstuffs prepared with customary thickeners but which have a lower calorie content.

Due to the variations in the fluid viscosity of various foodstuffs and in the viscosity of the different types of Polyose, the range of Polyose used in the food products will vary widely. Noncaloric sweetener compositions which can be added to the product to keep it at a normal sweetness level include a noncaloric sweetener such as saccharin.

Polyose C can be described as a glucose polymer derived from starch by depolymerization followed by heat polymerization to such an extent that at 50% solids it gives a viscosity (Brookfield) of 20 to 100 poises at 70°F. More viscous or less viscous Polyoses can be used. Polyose A requires 67%, Polyose B 60%, and Polyose D requires only 40% of solids to give a viscosity equivalent to that of Polyose C.

It may be desirable in certain cases, as when an extremely viscous fluid is needed, to fortify the Polyose thickening action by the use of such substances as gum arabic, cold-water-soluble carrageenan, guar gum, gum tragacanth, gum ghatti, carboxymethylcellulose, methylcellulose, hydrolyzed collagen, degraded gelatin, starch, tapioca and other thickening agents normally used in food. Some of these substances, particularly the last named, have nutritive value and too much should not be used.

Representative examples of foodstuffs using a Polyose as a thickening agent follow. While these examples each use proportions suitable for the particular food composition shown and described, the determination of other suitable proportions for the other food compositions included can readily be made. Standard manufacturing procedures are applicable to all formulations.

Example 1: A dietetic beverage concentrate is prepared having the following ingredients.

Polyose C	6.50 lb
Saccharin	*
Citric acid	0.12 lb
Root beer flavor	0.25 lb
Water	2.91 lb

*Amount for sweetness.

Polyose is dissolved in water and then other ingredients are added. The concentrate is mixed until all solid ingredients go into solution. The resulting beverage concentrate is diluted at the rate of 1 part concentrate to 5 parts water, carbonated or not, to make a finished beverage.

The above composition excluding water can be sold as a powdered beverage concentrate to be dissolved in water and fully prepared by others. Sweeter or less sweet compositions can be prepared using more or less of the sweetening agent proportionately. Also, similar compositions can be prepared using proportionately more or less of a Polyose than shown. Polyose A, B or D can be used in amounts to give viscosity equivalent to that when Polyose C is used.

Example 2: A dietetic carbonated beverage having a thickness comparable to those made with sugar is prepared using a Polyose, as follows. First, a concentrate, having the ingredients shown on the following page, is prepared.

Saccharin	*
Citric acid	0.002 lb
Ginger ale flavoring	0.002 lb
Polyose D	0.08 lb
Water	2.00 oz

*Amount for sweetness.

The Polyose is dissolved in water and then the other ingredients are added. Then, the concentrate is placed in a 12-ounce bottle and about 10 ounces carbonated water are added.

COMPANY INDEX

The company names listed below are given exactly as they appear in the patents, despite name changes, mergers and acquisitions which have, at times, resulted in the revision of a company name.

Abbott Laboratories - 203
Afico SA - 233
Ajinomoto Co., Inc. - 40, 88, 176
Alberto-Culver Co. - 56, 167, 197
Allied Chemical Corp. - 243
Amazon Natural Drug Co. - 128
American Home Products Corp. - 43, 91
American Sugar Refining Co. - 193
American Sweetener Corp. - 158, 161
American Viscose Corp. - 250
Ashe Chemical Limited - 268
Beech-Nut, Inc. - 17
Blaw-Knox Company - 226
Colgate-Palmolive Company - 348
Cumberland Packing Corp. - 45, 47, 150, 151, 155, 156, 159, 160
Dow Chemical Company - 177
E.I. du Pont de Nemours and Co. - 260, 309, 351, 381, 383
Dynapol - 72, 73, 74, 121, 122, 282
Eli Lilly and Co. - 96, 97, 99, 105, 108
Farah Manufacturing Co., Inc. - 280
Foremost Dairies, Inc. - 206
Foremost-McKesson, Inc. - 210
General Foods Corp. - 32, 44, 48, 49, 51, 52, 54, 58, 59, 168, 202, 248, 321, 332, 335, 338, 353
General Mills, Inc. - 163, 378
Guardian Chemical Corp. - 280
Hayashibara Co. - 130, 132, 137
Imperial Chemical Industries Ltd. - 39
International Minerals & Chemical Corp. - 75
Kelco Company - 300
Kellogg Company - 329
Kraftco Corporation - 306
Lever Brothers Co. - 12, 186, 189, 191
Lyckeby Starkelseforadling AB - 272
MacAndrews & Forbes Co. - 23, 25
Maumee Chemical Co. - 94
Meditron, Inc. - 4, 336
Mirlin Corporation - 8, 15, 16
Norse Chemical Corp. - 235
Chas. Pfizer & Co., Inc. - 81, 82, 87, 139 173, 274
Pfizer Incorporated - 359
Pillsbury Company - 180, 185
Procter & Gamble Co. - 65, 68, 69, 71, 119, 123, 361
Ranks Hovis McDougall Limited - 192
Richardson-Merrell Inc. - 262
Roto-Dry Corporation - 238
Sanitas Company Ltd. - 175
G.D. Searle & Co. - 35, 37, 38, 41, 147
Shell Oil Co. - 115
E.R. Squibb & Sons, Inc. - 144, 195, 265
Standard Brands Incorporated - 220
Stanford Research Institute - 112, 125, 126
Takeda Chemical Industries, Ltd. - 79
Unilever Limited - 232
U.S. Secretary of Agriculture - 61, 63, 76, 78, 85, 116, 312, 315
U.S. Secretary of Health, Education and Welfare - 11
Dr. A. Wander, SA - 370
Warner-Lambert Pharmaceutical Co. - 290, 294, 341, 343

INVENTOR INDEX

Acton, E.M. - 112, 125, 126
Adams, J.M. - 168
Allingham, R.P. - 82
Andrews, L.S. - 203
Aoki, H. - 79
Ariyoshi, Y. - 40
Avedikian, A.B. - 356
Avedikian, S.Z. - 356
Baggerly, P.A. - 59
Bahoshy, B.J. - 338
Balmert, C.A. - 290
Barth, J.B. - 348
Battista, O.A. - 250
Berg, J.H. - 51
Beusch, D.W. - 288
Bilotti, A.G. - 341, 343
Bliudzius, I.D. - 260, 309, 351, 381, 383
Bliznak, J.B. - 167
Bouchard, E.F. - 274
Braaten, W.C. - 235
Briggs, J.H. - 305
Brouwer, J.N. - 12
Brown, P.W. - 203
Brunner, G.F. - 361
Capasso, P.J. - 353
Cella, J.A. - 56
Collins, M.C. - 192
Conrad, E. - 272
Cook, M.K. - 25
Crosby, G.A. - 72, 73, 74, 121, 122
D'Alonzo, R.T. - 284
Deadman, L.L.F. - 268
Dobry, R. - 17
Dubois, G.E. - 72, 73, 74
Eguchi, S. - 88
Eisenstadt, M.E. - 45, 47, 151, 155, 156, 159, 160
Endicott, C.J. - 203
Essiet, O.A. - 19
Ewalt, D.J. - 325
Fennell, J.R. - 4, 8, 16
Ferguson, E.A., Jr. - 269
Finley, J.W. - 116
Finucane, T.P. - 168, 353
Francis, B. - 192
Frostell, G. - 272
Fryers, G.R. - 305
Furda, I. - 48, 49
Ganske, W.L. - 180
Garbrecht, W.L. - 105
Gebhardt, H.T. - 233
Gentili, B. - 61, 63, 76, 78
Gidlow, R.G. - 180
Giroux, E.L. - 11
Glicksman, M. - 202
Globus, A.R. - 280
Gregory, H. - 39
Griffin, H.L. - 226
Griffin, J.M. - 87, 173
Grober, N. - 277
Grosvenor, W.M., Jr. - 193
Guadagni, D.G. - 312, 315
Hammond, J.E. - 335
Harvey, R.J. - 4, 8, 15, 16
Hayashibara, K. - 137
Henkin, R.I. - 11
Henneman, H.E. - 91
Henning, G.J. - 12
Herbst, R.M. - 108

Hetzel, C.P. - 274
Hill, J.A. - 144
Hirao, M. - 130, 132
Hodge, J.E. - 85
Horn, L.J. - 306
Horowitz, R.M. - 61, 63, 76, 78
Howard, N.B. - 361
Inglett, G. - 75
Ishii, K. - 79
Ito, M. - 176
Jackson, A.F. - 192
Johnson, T.R., Jr. - 336
Johnston, W.R. - 220
Jucaitis, P. - 260, 309, 351, 381, 383
Kempf, C.A. - 206
Kenyon, R.E. - 321, 325
King, L.D. - 297
Klose, R.E. - 338
Komata, Y. - 88
Koski, D.H. - 24
Koski, J.B. - 24
Kracauer, P. - 150, 156, 158, 161
Krbechek, L.O. - 75
Kronfeld, E.C. - 96, 97, 99
Lapidus, M. - 43
La Via, A.L. - 144, 265
Lawrence, B. - 361
Leaffer, M.A. - 112, 125
Lee, C.-H. - 32
Lemaire, N.A. - 329
Lerom, M.W. - 126
Liggett, J.J. - 153
Long, C.A. - 175
Lontz, J.F. - 284
Lukey, R.A. - 197
Mazur, R.H. - 35
McCarron, F.H. - 180, 185
McGettigan, M.M. - 43
McNaught, J.P. - 189
Menzi, R.F. - 370
Miles, J.J., Jr. - 186
Millard, R. - 290
Miller, R.L. - 81
Mitchell, W.A. - 332
Mitsuhashi, M. - 130, 132
Morris, R.J., Jr. - 23
Muller, R.E. - 23
Murtagh, M.M. - 32
Neely, J.S. - 68, 119, 123
Nelson, E.C. - 85
Newton, K. - 191
Nonomiya, T. - 176
Nordstrom, H.A. - 338
Ojima, T. - 176
O'Laughlin, R.L. - 265
Olsen, R.D. - 274
Osborne, B. - 91
Paterson, R.M.L. - 243
Peder, M. - 186
Peebles, D.D. - 206
Persinos, G.J. - 128
Peters, G.C. - 121
Peterson, R.D. - 329
Pischke, L.D. - 52, 54, 248
Polya, E. - 168
Pryor, J.S. - 305
Radlove, S.B. - 366
Reich, I.M. - 220
Rennhard, H.H. - 139
Rizzi, G.P. - 65, 68, 69, 71
Rockwell, N.P. - 309, 351, 381, 383
Rousseau, P.M. - 58
Rubin, M. - 27
Saffron, P.M. - 122
Saito, T. - 88
Sale, A.J.H. - 191
Scarpellino, R.J. - 32
Schade, H.R. - 44
Schaefer, H.J. - 29
Schapiro, A. - 238
Scharschmidt, R.K. - 321
Schlatter, J.M. - 35, 37, 38, 41
Schmitt, W.H. - 56, 197
Schuppner, H.R., Jr. - 300
Scott, D. - 147
Scott, E.C. - 223
Seiden, P. - 361
Sheneman, J.M. - 97
Shimazaki, H. - 88
Shoaf, M.D. - 52, 54
Singer, R.L. - 372
Skrypa, M.J. - 243
Smith, P.F. - 297
Stahl, H.D. - 332
Stanko, G.L. - 262
Stein, J.A. - 180
Stone, H. - 112, 125, 126
Streckfus, T.K. - 335
Suarez, T. - 97
Sugimoto, K. - 130, 132, 137
Swaine, R.L. - 288
Tate, B.E. - 81, 82
Thompson, J.A. - 123
Tintera, J.W. - 29
Toda, J. - 79

Trumbetas, J.F. - 48, 49, 51
Tsukamoto, S. - 88
Vacek, L.C. - 94
van der Wel, H. - 12
Verlin, M. - 318
Vollink, W.L. - 321
Wakabayashi, H. - 79
Walles, W.E. - 177
Walton, R.W. - 195, 265
Wankier, B.N. - 202
Watson, G.H.R. - 378
Weast, C.A. - 171
Weinshenker, N.M. - 72
White, B. - 294
Windgassen, R.J. - 115
Wookey, N. - 192
Yamaguchi, S. - 176
Yamatani, T. - 40
Yasuda, N. - 40
Yoshida, M. - 165
Yoshikawa, M. - 165
Yueh, M.H. - 163
Zaffaroni, A. - 282

U.S. PATENT NUMBER INDEX

1,091,370 - 277
1,279,392 - 305
2,536,970 - 171
2,554,152 - 91
2,761,783 - 269
2,782,123 - 27
2,788,276 - 220
2,788,281 - 315
2,876,101 - 309
2,876,104 - 351
2,876,105 - 260
2,876,106 - 381
2,876,107 - 383
2,893,871 - 226
2,900,256 - 223
2,971,848 - 168
2,985,562 - 290
3,011,897 - 193
3,014,803 - 206
3,015,654 - 85
3,023,104 - 250
3,025,169 - 312
3,082,091 - 297
3,087,821 - 61
3,097,946 - 370
3,100,909 - 238
3,105,792 - 294
3,130,204 - 81
3,159,652 - 82
3,170,800 - 186
3,170,801 - 189
3,259,506 - 159
3,282,706 - 23
3,285,751 - 150
3,293,045 - 87
3,294,544 - 262
3,294,551 - 108
3,295,993 - 165
3,296,079 - 173
3,320,074 - 233
3,325,296 - 235
3,325,475 - 94
3,340,070 - 115
3,352,689 - 341
3,355,300 - 356
3,356,512 - 329
3,409,441 - 274
3,413,125 - 300
3,433,644 - 180
3,449,339 - 177
3,475,403 - 35
3,476,571 - 168
3,489,572 - 156
3,492,131 - 37
3,497,360 - 29
3,501,319 - 325
3,506,453 - 185
3,510,311 - 288
3,511,668 - 321
3,515,727 - 105
3,535,336 - 96
3,563,769 - 306
3,574,634 - 372
3,583,894 - 63
3,585,044 - 24
3,592,663 - 361
3,592,664 - 318
3,597,234 - 105
3,615,672 - 248
3,615,700 - 99
3,622,349 - 332
3,625,700 - 75
3,625,711 - 160
3,642,491 - 38
3,642,704 - 177
3,647,482 - 163
3,647,483 - 151
3,653,922 - 197
3,653,923 - 79
3,655,866 - 343
3,656,973 - 243
3,658,553 - 366
3,667,969 - 161
3,676,149 - 4
3,681,087 - 336
3,682,880 - 12
3,684,529 - 153
3,687,693 - 19
3,695,898 - 144
3,699,132 - 112
3,702,255 - 116
3,702,768 - 353
3,704,138 - 265
3,705,039 - 132
3,717,477 - 176
3,723,410 - 128
3,730,736 - 280
3,737,436 - 99
3,739,064 - 65
3,741,776 - 130
3,743,518 - 156
3,743,716 - 68
3,746,554 - 203
3,751,270 - 71
3,753,739 - 56

3,761,288 - 202
3,766,165 - 139
3,773,526 - 167
3,778,517 - 119
3,780,189 - 147
3,780,190 - 158
3,780,194 - 125
3,795,746 - 195
3,800,046 - 41
3,814,747 - 43
3,826,795 - 19
3,826,856 - 78
3,833,745 - 284
3,845,225 - 121
3,849,555 - 15
3,851,073 - 25
3,867,557 - 123
3,868,472 - 51
3,875,311 - 45
3,875,312 - 47
3,876,777 - 76
3,876,814 - 122
3,876,816 - 282
3,878,184 - 17
3,898,323 - 16
3,899,592 - 97
3,899,593 - 335
3,916,028 - 32
3,920,626 - 40
3,920,815 - 8
3,922,369 - 202
3,925,547 - 11
3,928,560 - 123
3,928,633 - 52
3,930,048 - 192
3,932,604 - 348
3,932,678 - 69
3,934,047 - 44
3,934,048 - 48
3,943,258 - 338
3,946,121 - 155
3,947,600 - 58
3,950,549 - 191
3,952,058 - 122
3,952,114 - 126
3,955,000 - 59
3,956,507 - 52
3,962,468 - 54
3,971,857 - 49
3,973,050 - 137
3,974,299 - 72
3,976,687 - 73
3,976,790 - 74

BRITISH PATENT NUMBER INDEX

977,482 - 268
999,073 - 232
1,104,251 - 175
1,152,610 - 210
1,169,538 - 272
1,170,590 - 108
1,199,101 - 88
1,233,216 - 39
1,242,350 - 378
1,269,851 - 97
1,275,347 - 359

NOTICE

SPECIALIZED SUGARS FOR THE FOOD INDUSTRY 1976

by Jeanne C. Johnson

Food Technology Review No. 35

There exists a broad spectrum of commercially available nonsucrose sugars which are consumed largely in prepared foods. Such sugar products include invert sugar, corn syrups with varying degrees of sweetness, fructose, dextrose, and maltose. These are the major products, but in addition a number of other sugars and sugar derivatives are produced for commercial, medicinal, and research purposes.

In many cases the nonsucrose sugars are more desirable than ordinary sucrose which does not always have the ideal properties for incorporation into food. Sucrose is not as sweet as some other sugars, and therefore high in calories; it can crystallize out causing a grainy texture in some foods, can cause caries, and is forbidden to diabetics and others.

In recent years interest has concentrated on fructose and fructose-containing syrups as naturally occurring, nonsynthetic sweeteners of the future.

This review describes 241 processes mostly issued since 1969. In addition to the 45 processes on fructose, 74 cover corn syrup and dextrose, 15 are concerned with invert sugar, and the remainder with other sugars and derivatives. A partial and condensed table of contents follows here. Numbers in parentheses are numbers of topics.

1. **INVERT SUGARS (15)**
   Acid Inversion of Molasses
   Chromatographic Separation
   Separation of Isomers
   Invert Sugar Fondants
   Invert Syrups for Bread
2. **LOW DEXTROSE SYRUPS (20)**
   Hydrolysis Processes
   Use in Gum Confections
   In Beverage Mixes
3. **HIGH DEXTROSE SYRUPS (22)**
   Saccharification Methods
   Conversion of Granular Starch
   Soft Sugar Preparations
4. **SOLID DEXTROSE (GLUCOSE) (27)**
   Spraydrying High DE Syrups
   Dextrose Hydrate Crystals
   Solubilizing Food Acids
   Dextrose Icings
   Diacetone Glucose
5. **FRUCTOSE—ENZYMATIC PRODUCTION (23)**
   Isomerase Conversion of Glucose
   Fructose from Starches
   Using Cell-Bonded Enzymes
   Immobilized Isomerase
   Polyelectrolyte-Flocculated Enzymes
6. **FRUCTOSE—OTHER PROCESSES (22)**
   Alkali Isomerization of Glucose
   Agglomerated Anhydrous Fructose
   Nutritional Supplement for Athletes
   Fructose for Avoiding Caries
7. **MALTOSE AND LOWER OLIGOSACCHARIDES (22)**
   Maltose from Wheat Bran
   Icings from Maltose Syrups
   Oligosaccharides with
   Terminal Fructose Units
8. **XYLOSE (8)**
   Xylose from Sulfite Liquors
   From Cottonseed Hulls
   Recrystallization from Acetic Acid
   Purification by Ion Exchange
9. **LACTOSE (8)**
   Amorphous β-Lactose for Tableting
   Lactose-Saccharin Mixtures
   Lactose from Whey
   with Immobilized Enzymes
10. **LACTULOSE (9)**
    From Lactose by Isomerization
    Epimerization with Sulfites
    Use of CaO for Solids
    Use of Konnayaku Powder
    Use of Protein Membranes
11. **OTHER SUGARS (16)**
    Maple Sugar Processing
    Honey Products
    Sorbose & L-Sorbosone
    D-Ribose
    Sugars for Skin Creams
12. **VARIOUS PROCESSES (7)**
    Saccharification of Cellulose
    Fermentation of Hydrocarbons
    Sugar Esters from Hydrocarbons
    Dietetic Saccharide Condensates
    Photochemical Processes
13. **SUGAR ALCOHOLS (17)**
    Mannitol from Hydrocarbons
    Foods Prepared with Maltitol
    Xylitol for Avoiding Caries
    Benzylidene Sorbitols
14. **SUGAR ACIDS (9)**
    Gluconic Acid
    Pangamic Acid Ca-Salt
    2-Keto-L-Gulonic Acid
    D-Gluconolactone
    Aldonic Acids
    Mannuronate Complexes

ISBN 0-8155-0632-5 **360 pages**

CONFECTIONS AND CANDY TECHNOLOGY 1974

by M. E. Schwartz

Food Technology Review No. 12

The scope of the technology covered in this volume is generally taken to include products whose major ingredient is sugar. Thus the range of products includes not only candy per se, but also chewing gum, desserts, sweet snacks, icings and whipped toppings, jellies and syrups. Also included are many processes designed to produce some of the basic ingredients of the industry such as sugars, syrups, fats, milk, and whey products. Technology of additives is similarly included, as far as such additives come within the scope of this survey.

Current concerns of the industry with regard to dietetic demands of the consumer have received special attention. Thus candy and confectionery processes geared to the production of low calorie or nonsugar sweets are covered fully.

The technology of the industry is large as evidenced by the appearance of well over 200 U.S. patents in the last ten years. This vast storehouse of information has been organized so that it may be of maximum use to practitioners in the art of candymaking.

A partial and condensed table of contents follows here. Numbers in parentheses indicate a plurality of processes per topic. Chapter headings are given, followed by examples of important subtitles.

1. **CANDIES (55)**
   - Core Formulations
   - Hard Candies
   - Marshmallows
   - Fudge Mixes
   - Gelled Confections
   - Use of Low Fat Starches
   - Aerated Confections
   - Dietetic Sorbitol Formulations
   - Molding Methods
   - Chocolate Chips
   - Novelty Products
2. **PRODUCTION OF CONFECTIONERY INGREDIENTS (31)**
   - Free-Flowing Sugars
   - Starch Conversion Sugars and Syrups
   - Low DE Starch Hydrolyzates
   - Dehydration of High Fructose Corn Syrup
   - Fats
   - Lauric-Type Hard Butters
   - Partially Hydrogenated Glycerides
3. **CONFECTIONERY ADDITIVES (27)**
   - Low Density Lactose
   - Electrodialyzed Whey Solids
   - Dry Instant Sugar-Fat
   - Butter Flavors
   - Specialty Sugars (Moisture Resistant)
   - Adjuvants
4. **FROZEN DESSERTS (38)**
   - Cream-Type
   - Water-Based Products
   - Dry Mixes
   - Dietary Formulations
5. **MILK PUDDINGS (16)**
6. **GELLED DESSERTS (24)**
   - Gelatin Gels
   - Alginate Gels
   - Pectin Gels
   - Adjuvants
7. **ICINGS (23)**
   - Meringues
   - Bakery Items
8. **WHIPPED TOPPINGS (17)**
   - Dry Mixes
   - Aerosol Products
9. **JELLIES (8)**
10. **NOVELTY PRODUCTS (20)**
    - Sweetened Peanut Butter Compositions
    - Hydrophilic Additives and Emulsifiers
    - Butter-Syrup Emulsions
    - Honey Butter
    - Sweet Snack Foods
    - Candied Produce
11. **CHEWING GUMS (18)**
    - Suitable Polymer Mixtures
    - Plaque-Removing Chewing Gum
    - Sugarless Gums
    - Candy-Coated Gum Balls
    - Miraculin-Containing Formulation

ISBN 0-8155-0524-8 **338 pages**

FRUIT AND VEGETABLE JUICE PROCESSING 1975

by J. K. Paul

Food Technology Review No. 21

The large market for fruit and vegetable juices, because of their relatively low cost and high nutritional value, coupled with modern diet appeal, accounts for the sustained interest in these products and the continued development of improved methods of technology.

It is a primary object of this technology to prepare fruit juices, vegetable juices and related beverages that are able to maintain their natural flavor and aroma characteristics under ordinary, unspecified storage conditions over prolonged periods of time.

Considerable progress has been made in this direction: practically every process in this book is oriented toward the ultimate goal of every juice manufacturer—to retain natural flavors and aromas and to retard indefinitely the formation of unnatural flavors, aromas or colors.

Gone are all those preservatives of the past which, while stabilizing a beverage, also introduced undesirable off-flavors. Approved modern additives not only maximize stability, they also enhance the natural flavors. How to use them in connection with advanced extraction and concentration processes, which were specially designed for juice processing, is the know-how of this book: 164 patent-based processes are described.

A partial and condensed table of contents follows here. Chapter headings are given, followed by examples of important subtitles. Numbers in parentheses indicate the numbers of processes per topic.

1. MANUFACTURING TECHNIQUES (26)
Pome Juice with Pulp
Removal of Bitterness Precursors
Separation of Juice from Pulp
Increasing Viscosity
Clarified Juice Manufacture
Dialysis Methods
When to Change pH
Ultrasonic Treatments

2. CONCENTRATION PROCESSES (42)
Brix to Acid Ratio
Osmotic Processes
Freeze Concentration
Use of Immiscible Coolants
Crystal Purification Columns
Sonic Defoaming

3. DEHYDRATION (16)
Drum Drying
Addition of Amylolytic Enzymes
Foam-Mat Processes
Spray Drying
Vacuum Drying
Other Drying Processes

4. FREEZE DRYING (28)
Fluidized Bed Processes
Continuous Processes
Desiccation of Frozen Particulates
Foam Drying Processes
Slush Freezing
Thermal Shock Process

5. STABILIZATION (19)
Adding O-Methyltransferase
Use of Stannous Ions
Benzhydroxamic Acid
Disubstituted Benzoic Esters and Ketones
Use of Polyphosphates
Ion Exchange Treatments

6. FLAVORS FROM JUICES (14)
Essence Recovery
by Continuous Condensation
by Distillation
Flavor Extraction from Fruit

7. JUICE ENHANCERS (8)
Flavor Enhancers
Glutaminase Hydrolysis of Glutamine to Glutamic Acid
Addition of β-Hydroxybutyrates to Enhance Grape Flavor
2-Ethylpyromeconic Acid
2-Alkylthiazoles
Coloring Agents
Carotenoids in Abietic Acid Melt
Benzopyrilium Compounds
Enrichment of Orange Juice with Chromoplasts

8. VARIOUS PROCESSES (11)
Pear Beverage
Drink from Citrus Juice and Fermented or Acidified Milk
Citrus Juice with Energy Supplement
Licorice-Containing Juices
Juice from Green Leaves of Wheat and Barley
Aeration with Volatile Additive
Aeration with Inert Gas
Dispensing of Semifrozen Comestible

ISBN 0-8155-0565-5 **277 pages**

EDIBLE STARCHES AND STARCH-DERIVED SYRUPS 1975

by Nicholas B. Petersen

Food Technology Review No. 24

The separation of starch from flour and its use as a food product was well known in ancient Greece and Rome. This carbohydrate is widely found in nature, and starches obtained from grains such as corn and sorghum, and from roots and tubers such as tapioca, arrowroot and potato, are of considerable commercial importance.

In the U.S., corn *(Zea mays)* is the major source of starch and enormous amounts of corn starch, corn syrups, and lately, corn sugars are used by the food processing industries (bakeries, breweries, candy makers, soft drink manufacturers, etc.)

Recent developments in enzyme isolation and production have provided improved and cheaper methods of preparing commercially important starch degradation products, such as amylose. The rise in production of fructose, the sweetest of the natural sugars, has been dramatic in recent years, and is now so important that a separate chapter has been devoted to the production of fructose and fructose syrups from corn starch hydrolysates.

This review covers 217 processes disclosed in 233 U.S. patents which have been issued since 1968 on the preparation, purification and use of edible starches and their derivatives and hydrolysates (syrups). The arrangement is by product rather than by process, so that the alert food processor can avail himself of sugar products which are cheaper than sugar from sugarcane or sugar beets. Preparation and uses of edible starches as nutrients have received the same careful attention as a matter of course.

A partial and condensed table of contents follows here. Numbers in parentheses indicate the number of patents per topic.

1. **ISOLATION & USE OF RAW STARCHES (21)**
   From Corn and Sorghum
   From Wheat
   From Potatoes
   Unmodified Starches in Food Processing

2. **PURIFICATION AND MODIFICATION (19)**
   Lipid Removal
   Non-Degradative Enzyme Treatment
   Oxidation
   Starch-Fat Combinations for Food Use

3. **GELATINIZED STARCH (21)**
   Gelatinizing Processes
   Blends with Glycerides, Shortenings and Gums
   Use in Instant Puddings, etc.

4. **STARCH ESTERS (20)**
   Carboxylates
   Phosphates
   Crosslinked Starch Esters
   Use in Puffed Snack Foods
   Use in Simulated Rice

5. **STARCH ETHERS (21)**
   Hydroxypropylation Methods
   Hydroxalkyl Starch + Carrageenan
   Gravy Thickeners
   Baby Food Formulations

6. **AMYLOSE AND AMYLOPECTIN (20)**
   Alpha-1,6-Glucosidase Production
   Amylose and High Amylose-Starch in Food Products
   As Coating for Fried Potatoes
   In Pizza Dough

7. **LOW D.E. STARCH DEGRADATION PRODUCTS (25)**
   Liquefaction with Alpha-Amylase
   Other Enzyme Processes
   Acid Hydrolysis
   Low D.E. Hydrolysates in Processed Foods

8. **STARCH SYRUPS (19)**
   Glucoamylase Production
   High D.E. Hydrolysates
   In Situ Hydrolysis in Processed Foods
   Fructose End Groups by Transfer Reactions

9. **CORN SYRUP SOLIDS (15)**
   Spray Crystallization
   Granular Beta-Dextrose
   Gluconate Products from Starch Hydrolysates

10. **MALTOSE AND LOWER OLIGOSACCHARIDE SYRUPS (21)**
    Production of Maltose Using Amylases
    Maltose Syrups and Their Uses

11. **FRUCTOSE FROM STARCH (30)**
    Production of Glucose Isomerase
    Enzymatic Production of Fructose
    Chromatographic Separation of Fructose
    Drying Fructose Syrups

ISBN 0-8155-0584-1

427 pages

COMMERCIAL PROCESSING OF FRUITS 1976

by L. P. Hanson

Food Technology Review No. 30

Aside from canning, pasteurizing, sterilizing, dehydrating, freezing, and freeze-drying, one prime objective for the modern fruit processor is the retention of the characteristics of freshly picked fruit. A further goal is to minimize the quantities of added sugar, salt or preservatives, when recourse to the use of chemical substances must be taken.

Preventing the growth of undesirable microorganisms in fresh fruit products and thus insuring relatively long "showcase life" is the subject of many of the 170 processes contained in this book.

All processes have been selected with the intention of eliminating the requirement for costly process equipment, notwithstanding the purpose of providing fruits and fruit products that are highly acceptable by organoleptic tests, and which can be considered as "slow spoilers" on the basis of unsightly appearance or off-flavor.

The following partial table of contents indicates the wealth of hard-to-get information collected here. This volume reviews fruit processing technology as depicted in the U.S. patents since 1965. Numbers in parentheses indicate the number of processes in each chapter. Chapter headings are given, and some of the more important subtitles are also included.

1. **HANDLING FRESH FRUIT (21)**
   Removing Volatile Respiratory Products
   Ripening with Controlled Irradiation
   Ethylene Ripening
   Applying Coatings and Films
   Fungicides and Preservatives
   Color Developers & Enhancers

2. **DEHYDRATION PROCESSES (34)**
   Freezedrying
   While Storing
   Dehydrofreezing
   Use of Enzymes While Drying
   Vacuum Drying
   Use of Irradiation Treatments
   to Reduce Rehydration Time
   Amylose Ester-Protein Coatings

3. **PRESERVATION & STABILIZATION (25)**
   Vacuum Peeling
   Canning Processes
   pH Control and Adjustment
   Ultra-Slow Cooling
   Rapid Heat Sterilization
   Preventing Plasmolysis

4. **CITRUS FRUITS (17)**
   Color Additives
   Reduction of Sour Taste
   Canning Citrus Segments
   Extracting Bioflavanoids
   Isolation of Citrus Chromoplasts

5. **POMES (16)**
   Controlling Apple Scald
   Preservatives for Apple Slices
   Chemical & Vacuum Peeling
   Puff Dried Apples & Pears
   Dried Applesauce

6. **DRUPES (17)**
   Fermentation of Green Olives
   Chlorinated Brine for Olives
   Prevention of Cherry Discoloration
   with EDTA
   Cherry Pits for Deodorizing
   Preservation of Peaches
   Prunes from Ente Plums

7. **BERRIES (10)**
   Processing Cranberries
   Immersion Freezing of Strawberries
   Freezedrying Strawberries
   for Incorporation into Cereals
   Processing Perforated Blueberries
   Compression Techniques
   Quick Freezing of Raspberries

8. **GRAPES (4)**
   Destruction of Microorganisms
   Maintaining Softness in Raisins
   Silicone Polymer Washes
   Coloring Raisins

9. **BANANAS (13)**
   Retarding the Ripening
   Drying Banana & Plantain Chips
   Limed Bananas
   Deep Frying

10. **OTHER FRUITS (12)**
    Liquid Nitrogen Treatment of Avocados
    Hydrating Dates
    Cantaloupe Injection with Inert Gas
    Large Scale Pickling of Watermelons
    Fermentation of Coconut Meat
    Delaying Senescence of Pineapples

ISBN 0-8155-0608-2 **302 pages**

FOOD FLAVORING PROCESSES 1976

by Nicholas D. Pintauro

Food Technology Review No. 32

The modern consumer appears to pay more attention than ever to the flavor of food bought and eaten. Because of artifically prolonged shelf life it is important that the flavor and odor of fresh food be maintained. The importance and essentialness of the odor components of most flavors are emphasized throughout this book.

The mechanisms of the chemical excitation of our taste and olfactory receptors are similar. Usually, when a food has lost its original fresh odor, its flavor is gone and our senses classify it as stale. But flavors and odors can be "fixed" by chemical complexing, plating onto solid carriers, frangible or permeable microencapsulation and other subtle techniques.

Because of increasing regulations on the use of synthetics, there is renewed interest in the isolation and application of natural food flavors. It is often possible to extract appetizing flavors from discarded leaves, stems, hulls, skins, bones, etc.

This book is for the food processor who may have to replace aromatic components lost during processing or who may have to reinforce flavors already present. Sometimes off-flavors must be corrected or covered up (e.g. gamy taste). Flavoring skill is very necessary when it is desired to impart meat flavors while texturizing food with fibers made from spun plant proteins. 213 processes in all. An entire chapter is devoted to flavor-active food additives. A partial table of contents follows here. Chapter headings and some important subtitles are given. Numbers in parentheses indicate the number of processes per topic.

1. FRUIT & VEGETABLE FLAVORS (27)
Strawberry Flavors
1-Propenyl-3,4,5-trimethoxybenzene
Hydroxyfuran Derivatives
Pineapple Flavors
6-Ethoxy-2-methylpyrazine
Citrus Flavors
Potato Flavors

2. DAIRY FLAVORS (17)
Cheese Flavor Additives
Flavor Enzymes
Butter Flavors
Citrated Whey Culture
Pyrroline Additive
Lipase Modifier

3. BREAD FLAVORS (11)
Proline Reaction Products
Dry Yeast-Fermented Whey
Sour Dough Flavors
Bakery Flavors
Yeast Autolysate + Maltol

4. MEAT FLAVORS (52)
Thiols
Cyclopentapyrimidines
Cysteine + Thiamine
Ribonucleotide + MSG
Meat Extract Flavors
Corn Gluten Base
Yeast Hydrolysates
Dibutyl Phthalate Extractives
Succinic + Lactic Acids

5. FLAVOR-ACTIVE FOOD ADDITIVES (38)
Enzyme Inactivators
Encapsulated Allylisothiocyanate
Absorption on Silica Gel
Maromi Fermentation Products
Glutamates + Nucleotides
Inosine Phosphates
Maltobionic Acid
Cellulose Carrier
Coatings as Carriers
Herb Extraction Processes

6. ROASTED FLAVORS (15)
Popcorn & Nut Flavors
Fried Flavors
Lard Flavors

7. COFFEE, COCOA, TEA (19)
"Green" Flavor of Pyrazines
Aldimine Compounds
Comprehensive Flavor Listings

8. FRESHNESS & MASKING (13)
Theaspirone
Flavor Modifier-Migrator
Freshness Enhancers
Reducing Bitterness

9. EMULSIONS & FIXATIONS (21)
Larch Gum Emulsions
Aldehyde-Ammonia Fixation
Complexing with Carbohydrates
Hydrophilic Colloid Complexes
Starch Hydrolysate Flavor Complex
Flavoring Powders
Granulating & Coating

ISBN 0-8155-0618-X **205 pages**

TEA AND SOLUBLE TEA PRODUCTS MANUFACTURE 1977

by Nicholas D. Pintauro

Food Technology Review No. 38

In the United States most tea (loose and in bags) is fermented black tea from Eastern Asia. Imports of green or unfermented tea from China and Japan are on the increase to meet the demand for the manufacture of instant teas.

The subject areas in this publication are organized in logical sequence describing all steps in the manufacture of all types of tea products. The opening chapters deal with post-harvest handling and conditioning of tea leaves. Later chapters cover extraction, aroma recovery, and removal of troublesome tea creams. Finally, there is a chapter on tea bag packaging. Nonleaf tea products include:

Instant Tea: Spraydried tea brew, may contain carbohydrates.

Tea Mix: Instant tea with sugar or other sweeteners and flavors.

Canned or Bottled Tea: Ready-to-drink product made from reconstituted soluble tea or liquid concentrate. No refrigeration needed before opening.

Cold-Packed Tea: Ready-to-drink beverage in plastic or paper container. Refrigeration required.

Liquid Concentrate: Concentrated tea product to which water must be added. Shelf-stable in cans or bottles.

Frozen Concentrate: Similar to liquid concentrate. Packaged, stored and reconstituted like frozen fruit juice.

114 processes. A much shortened, partial table of contents follows here. Numbers in parentheses indicate the number of processes per topic.

1. WITHERING & ROLLING (6)
Separating Foreign Matter
Controlled Withering & Drying
Curling, Tearing, Cutting
High Speed Cutter
Rolling to Control Moisture
Rolling Additive

2. SORTING, FERMENTING, "FIRING" (13)
Fiber Separating
Stalk Removal
AFICO Maturing Process
Enzyme Conversion
Suppressed Thearubigens
Oxidative Conversion
Heat and Oxygen
Ozone Process
Hydrogen Peroxide
Potassium Permanganate
Tea Gum Preservative

3. EXTRACTION PROCESSES (17)
Preliminary Leaching
Preliminary Cold Extraction
Countercurrent Extraction
SELTZER and SAPORITO Processes
Extraction under CO_2
Ammonia Extraction
Acetone Extraction
Green Tea Extraction & Conversion
Hard Water Tolerance
Buffers vs. Ion Exchange
Addition of Ascorbic Acid
Irradiation Treatment

4. TANNIN-CAFFEINE PRECIPITATE—TEA CREAM REMOVAL (20)
Stepwise Addition of Gelatin
Suspending Agents
Cream Separation Methods
Oxidation and Bleaching
Tannase Treatment
Calcium Treatments
Use of Polyvinylpyrrolidone
Selective Tannin Removal
Low Temperature Method

5. FILTRATION & CONCENTRATION (9)
Centrifuge Designs
Use of Hydrophilic Membranes
Freeze Concentration Processes

6. DEHYDRATION (14)
Vacuum Drying
Freezedrying Processes
Green Tea Blends

7. AROMA RECOVERY (14)
Recovery During "Firing"
Steam Stripping
Inert Gas Stripping
Aroma Fixation

8. AGGLOMERATION & AROMATIZATION (12)
Various Agglomeration Processes
Black Tea Flavor
Aromatic Extracts
Fruit Extracts
Flavor Protection

9. TEA BAGS (9)
Bag Construction
Single Chamber Bags
Construction from Tubes
Tag Interlock
Slit Opening
Perforation Line
Tea Bag Receptacles
Vending Infusion Apparatus

ISBN 0-8155-0645-7 **266 pages**